Principles of Inorganic and Theoretical Chemistry

Principles of Inorganic and Theoretical Chemistry

Second Edition

C. T. RAWCLIFFE
B.Sc., M.Sc., Grad. Cert. Ed.

and

D. H. RAWSON
B.Sc., F.R.I.C., Ph.D.

Division of Science and Mathematics
Bradford College

HEINEMANN EDUCATIONAL BOOKS
LONDON

Heinemann Educational Books Ltd
LONDON EDINBURGH AUCKLAND MELBOURNE TORONTO
HONG KONG SINGAPORE KUALA LUMPUR NEW DELHI
IBADAN NAIROBI LUSAKA JOHANNESBURG
KINGSTON/JAMAICA

ISBN 0 435 66747 5

First published 1969
Reprinted 1970
Second Edition 1974
Reprinted 1975
Reprinted with corrections 1978

Published by Heinemann Educational Books Ltd
48 Charles Street, London W1X 8AH

Printed and bound in Great Britain at
The Spottiswoode Ballantyne Press by
William Clowes & Sons Limited, London, Colchester and Beccles

Preface to the Second Edition

The second edition of this book conforms to the International System of Units (SI). The chapters concerned with chemical periodicity, physical methods, and general transition metal chemistry have been largely rewritten and expanded. Major additions have been made to the sections on atomic structure, atomic spectra, lattice energy, molecular orbital theory, thermodynamics, acids and bases, noble gases, boron hydrides, graphitic compounds, carbonyls, glasses, cationic iodine compounds, vanadates and organo-metallic compounds. Some recently characterized compounds have been described, e.g. NOF_3, WCl_6, $KBrO_4$, $HMnO_4$, and some of the intriguing properties of polywater receive mention. In view of recent developments, the metallurgical processes for copper and steel have been revised. The nomenclature has been brought up to date throughout, although trivial names have been included where these are still in common usage. Numerous other minor amendments and additions have been made to the text and some questions of a more advanced nature have been appended.

We wish to express our thanks to Dr. P. A. Fryer for reading the amended manuscript and offering some valuable comments and criticisms. It is hoped that the book will continue to serve the needs of students in schools and colleges.

Bradford 1973

C. T. RAWCLIFFE
D. H. RAWSON

Preface to the First Edition

Numbers of excellent texts are now available for University Special Honours students in chemistry but very few cater adequately for the General B.Sc., L.R.I.C., H.N.C., and H.N.D. students. It is for these students that this book has primarily been written, although it is also hoped that it will prove valuable for G.C.E. 'A' and 'S' levels.

An attempt has been made to give a sound, comprehensive, and coherent approach to modern inorganic chemistry. Electronic structure, bonding, the molecular orbital theory, crystal structure, lattice energy, shapes of molecules, electrode potentials, and free energy, are treated in the initial chapters. These concepts are integrated into the remainder of the book which is devoted to the chemistry of the elements and their compounds. The chemistry of the elements has been treated comparatively wherever possible but a considerable amount of detail has been included. In contrast to the modern vogue, the principles and essential details of economically important industrial processes have been included.

The authors will be appreciative of any comments or criticisms from teachers and students who use the book.

C. T. Rawcliffe
D. H. Rawson

Bradford 1969

SI Units

With the advent of metrication, the Système International d'Unités (SI) has been introduced into courses and examinations. The more important units, conversions. and numerical data on the SI system are given below.

Basic units

QUANTITY	UNIT	SYMBOL
Mass	kilogramme	kg
Length	metre	m
Time	second	s
Electric current	ampere	A
Temperature	degree Kelvin	K

Derived units

QUANTITY	UNIT	SYMBOL
Force	newton	$N = kg\ m\ s^{-2}$
Electric charge	coulomb	$C = A\ s$
Quantity of heat or energy	joule	$J = N\ m$
Power	watt	$W = J\ s^{-1}$
Electric potential	volt	$V = W\ A^{-1}$
Electric resistance	ohm	$\Omega = V\ A^{-1}$
Magnetic flux	weber	$wb = kg\ m^2\ s^{-2}\ A^{-1} = V\ s$
Magnetic flux density	tesla	$T = kg\ s^{-2}\ A^{-1} = V\ s\ m^{-2}$
Pressure	pascal	$Pa = N\ m^{-2}$

Other units used

QUANTITY	UNIT	SYMBOL
Entropy (S)	kilojoule per degree Kelvin	$kJ\ K^{-1}$
Enthalpy (H)	joule	J
Gibbs free energy (G)	joule	J
Density (ρ)	kilogramme per cubic metre	$kg\ m^{-3}$
Conductance (G)	siemens	S
Conductivity (σ)	siemens per metre	$S\ m^{-1}$
Frequency (f)	hertz	Hz

QUANTITY	UNIT	SYMBOL
Magnetic field strength (H)	ampere per metre	$\mathrm{A\ m^{-1}}$
Dipole moment (μ)	coulomb metre	$\mathrm{C\ m = A\ s\ m}$
Magnetic moment (μ)	ampere square metre	$\mathrm{A\ m^2}$

Multiples and sub-multiples

UNIT MULTIPLYING FACTOR	PREFIX	SYMBOL
10^9	giga	G
10^6	mega	M
10^3	kilo	k
10^{-3}	milli	m
10^{-6}	micro	μ
10^{-9}	nano	n
10^{-18}	atto	a

Numerical data

Mass of electron	$m = 911 \times 10^{-33}$ kg
Electronic charge	$e = 160 \times 10^{-21}$ C
Velocity of light	$c = 300 \times 10^{6}$ m s^{-1}
Avogadro's number	$N_A = 602 \times 10^{21}$ mol^{-1}
Planck's constant	$h = 663 \times 10^{-36}$ J s
Gas constant	$R = 8{\cdot}31$ JK^{-1} mol^{-1}
Faraday Constant	$F = 9{\cdot}6483 \times 10^{4}$ C mol^{-1}
Permittivity of a vacuum	$\epsilon_0 = 8{\cdot}854 \times 10^{-12}$ kg^{-1} m^{-3} s^4 A^2

Conversion data

1 mm Hg	= 133 Nm^{-2} = 1 torr
1 bar	= 10^5 Nm^{-2}
1 erg	= 10^{-7} J
1 calorie (15°C)	= 4·19 J
1 atmosphere	= 101·325 kNm^{-2} = 101·325 kPa
1 electronvolt	= 0·160 aJ = 96·487 kJ mol^{-1}
1 siemens	= 1 ohm^{-1}
1 Debye	= $3{\cdot}336 \times 10^{-30}$ C m
1 Bohr magneton	= $9{\cdot}273 \times 10^{-24}$ A m^2
1 gauss	= 10^{-4} T
1 Oersted	= $2{\cdot}5 \times 10^{2}$ A m^{-1}

Contents

I

Atomic Structure

The Electron

The first insight into the fundamental nature of the atom resulted from the study of the passage of electricity through gases at low pressure. Metal electrodes were sealed in a glass tube, the tube was evacuated, and the electrodes were connected to a source of high-potential direct current. It was found that at pressures of 0·01 to 0·001 torr the negative electrode (cathode) emitted streaks of purple light called *cathode rays*. Crookes investigated the rays and found that: (*a*) they normally travel in straight lines; (*b*) they possess considerable momentum; (*c*) they are deflected by magnetic and electric fields. In an electric field the rays are deflected towards the positively charged plate, indicating that the rays consist of a stream of negatively charged particles which are called *electrons* (Figure 1.1).

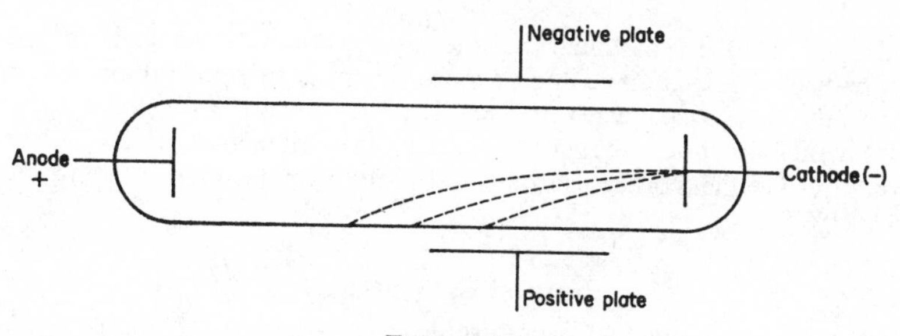

FIGURE 1.1

By observing the extent of deflection in combined electric and magnetic fields, Sir J. J. Thomson (1897) determined the ratio of the charge (e) to the mass (m) of electrons and found that all e/m values were identical irrespective of the gas in the tube. Thus electrons are fundamental particles of all matter. The absolute charge on the electron was evaluated by Millikan

(1911) and from values of e/m and e, the mass of the electron was calculated to be 1/1837 of the mass of a hydrogen atom or 0·00055 units on the relative atomic mass scale ($^{12}_{6}C$) (*see* p. 8).

The Proton

In 1886 Goldstein used a perforated cathode and observed luminous rays penetrating through it. These rays were attracted towards a negatively charged plate and are called *positive rays* (Figure 1.2).

The ratio of charge/mass for these rays was found to be much smaller than for electrons and to vary with the gas in the tube. Particles with the smallest mass were obtained when hydrogen was used in the tube and the magnitude of the charge on these particles was found to be equal but opposite in sign to that of the electron. The charges obtained using gases other than hydrogen were found to be multiples of the charge obtained using

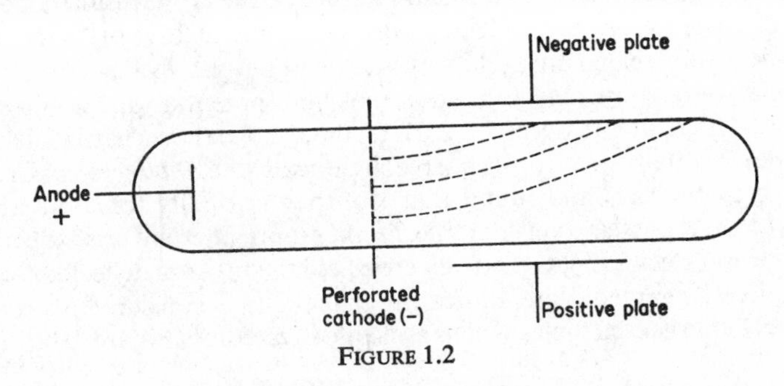

FIGURE 1.2

hydrogen. Thus it is assumed that the neutral atoms or molecules of gases in the discharge tube have one or more electrons stripped off and become positively charged particles or *ions*. The lightest ion is that of hydrogen (H^+) which is called the *proton* and is a fundamental unit of atomic structure. The mass of the proton is 1·0073 units on the relative atomic mass scale ($^{12}_{6}C$).

Natural Radioactivity

Investigations of Pierre and Marie Curie in the early 1900s led to the discovery that the atoms of certain elements such as uranium, radium, and polonium disintegrate spontaneously and continually emit radiation. These elements are said to be *radioactive* and it was found that the rate of disintegration was dependent upon the number and the nature of the atoms present but independent of the external conditions such as temperature and pressure.

Rutherford studied the penetrating power of these radiations and their behaviour in electric and magnetic fields. He deduced that there were three types of radiation:

(*a*) α-rays. These are fast-moving helium nuclei ejected from a radioactive source. They ionize air, produce bright flashes on a fluorescent screen, have a range of only 1–2 cm in air but will not penetrate a thin metal foil.

(*b*) β-rays. These are fast-moving electrons ejected from a radioactive source. They have a range of 1–2 m in air and will pass through 1–2 mm of lead.

(*c*) γ-rays. These are penetrating, short-wave electromagnetic radiations of the X-ray type. They have a long range in air and will pass through 15–20 cm of lead. On passage through matter, γ-rays possess the property of ejecting high-speed electrons.

The effect of a magnetic field on the radiation obtained from radium is shown in Figure 1.3.

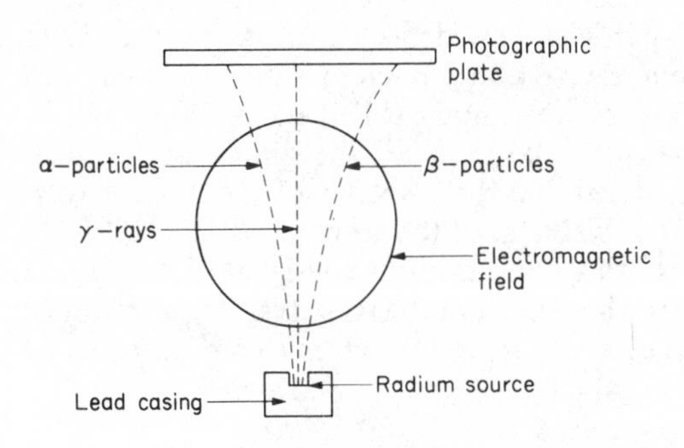

FIGURE 1.3

The Neutron

In 1932 Chadwick bombarded the element beryllium with α-particles and obtained an extremely penetrating radiation which could not be deflected by electric or magnetic fields. These neutral particles are called *neutrons* and have a mass of 1·0087 on the relative atomic mass scale ($^{12}_{6}C$). Since the neutron has approximately the same mass as a hydrogen atom, it is found that, when alkanes are bombarded with neutrons, high-speed protons are ejected due to direct collision of the neutrons with the hydrogen atoms in the alkane.

Rutherford and Marsden Experiment (1911)

A very thin sheet of gold foil was bombarded with α-particles from a radium C source (Figure 1.4).

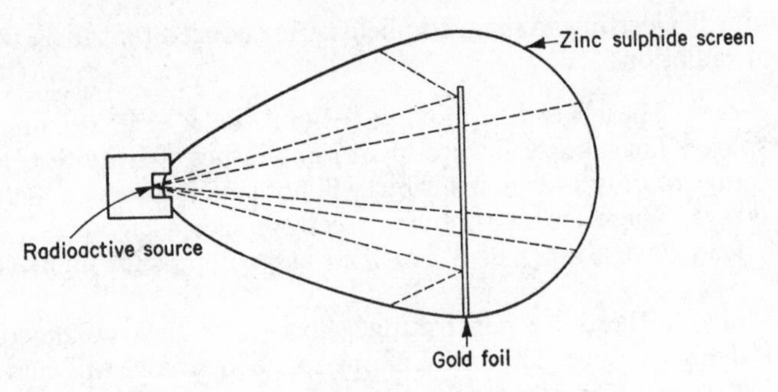

FIGURE 1.4

Rutherford found that:

(*a*) Most α-particles passed straight through the foil and he concluded that the volume occupied by an atom must be largely empty space.

(*b*) A few α-particles were deflected through large angles and this could only result from the close approach of a positively charged α-particle to a minute positively charged body called the *nucleus*. Owing to the rarity of the deflections through large angles, the volume of the nucleus must be very small. He calculated that the diameter of the nucleus was about one ten-thousandth of the diameter of the atom.

Thus Rutherford's nuclear atom consisted of a small, heavy, positively charged nucleus where most of the mass resides, surrounded by the number of electrons necessary to render the atom electrically neutral.

Number of Protons in the Nucleus

In the year prior to the discovery of X-rays, Röntgen found that, when cathode rays strike a solid or metal target, a penetrating radiation is produced. The rays produced were subsequently termed X-rays; they are electromagnetic waves similar in properties to γ-rays but have a slightly longer wavelength.

Moseley (1913) used a number of different metal targets for the cathode rays and passed the resulting X-rays through a crystal of potassium hexacyanoferrate(II) which acted as a diffraction grating. He then analysed the diffracted radiation on a spectrograph and obtained a line spectrum characteristic of the target material. There were two principal lines in the spectra obtained and, with target elements of increasing atomic weight, there was a progressive shift in the position of the lines. Moseley deduced the expression

$$\bar{\nu} = 1/\lambda = c(z - b)$$

where λ is the wavelength of the X-rays, $\bar{\nu}$ is the wave number, c and b are constants, and z is an integer approximately equal to half the relative atomic mass and corresponding to the position of the element in the periodic table (*see* p. 55). This number z was called the *atomic number*. An increase of one unit in the atomic number caused a shift in the position of the line spectrum, as shown in Table 1.1.

TABLE 1.1

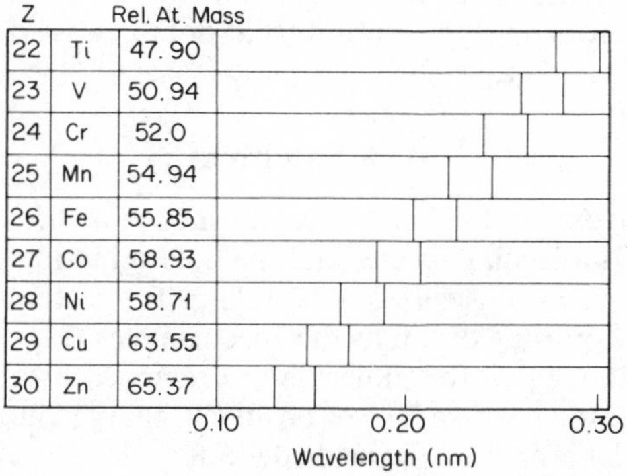

Z		Rel. At. Mass
22	Ti	47.90
23	V	50.94
24	Cr	52.0
25	Mn	54.94
26	Fe	55.85
27	Co	58.93
28	Ni	58.71
29	Cu	63.55
30	Zn	65.37

0.10 0.20 0.30
Wavelength (nm)

Moseley concluded that the atomic number is a fundamental property of the atom and is equal to the number of protons in the nucleus. An increase in atomic number by one unit corresponds to the addition of one proton to the nucleus. Since relative atomic masses are approximately twice atomic numbers, another type of particle must reside in the nucleus with approximately the same mass as the proton. This particle (the neutron) was subsequently discovered by Chadwick (p. 3).

The state of knowledge concerning the atom at this stage is summarized below:

(*a*) The atom consists of a minute, heavy, positively charged nucleus containing protons and neutrons.

(*b*) The nucleus is surrounded by a diffuse pattern of electrons and, since atoms are electrically neutral, the number of electrons is equal to the number of protons.

(*c*) The atomic number of an atom is equal to the number of protons in the nucleus and hence to the number of electrons surrounding the nucleus.

(*d*) The relative atomic mass of an atom = (number of protons × mass of one proton) + (number of neutrons × mass of one neutron) + (number of electrons × mass of one electron). On the relative atomic mass scale the masses of these particles are:

proton	1·0073
neutron	1·0087
electron	0·00055

Thus even for heavy atoms with a large number of protons and neutrons, the relative atomic masses should be fairly close to whole numbers. This is found to be the case for most atoms: the relative atomic masses of carbon and oxygen are respectively 12·01115 and 15·9994, for example.

A number of exceptions were found, however. Chlorine and copper have relative atomic masses of 35·453 and 63·546 respectively. The fact that such relative atomic masses are far from being whole numbers presented a problem to chemists until the development of the mass spectrometer which gave a more detailed knowledge of atomic mass.

Mass Spectrometry

The mass spectrometer has become established as an invaluable instrument in the study of molecular species and the determination of relative atomic and molecular masses. A gaseous sample of the material to be examined leaks through a porous disc into the spectrometer (Figure 1.5(*a*)). One or more outer electrons in the molecule are removed by electron bombardment. The resulting positive ions or positively charged molecular fragments are accelerated by an electric field, deflected along a circular path by a magnetic field, and then fall, in order of increasing mass, on the ion collector. The ion beam current is amplified and recorded.

For the accelerating electric field,

$$Ve = \tfrac{1}{2}mv^2 \qquad (1)$$

For the deflecting magnetic field

$$Bev = \frac{mv^2}{r} \qquad (2)$$

where V = potential difference between plates,
B = magnetic flux density or induction,
m = mass of the ion,
e = charge on the ion,
v = velocity of the ion,
r = radius of the circular path.

From (1) and (2),

$$\frac{m}{e} = \frac{B^2r^2}{2V}$$

Since B, V and r can be controlled experimentally, the m/e ratio obtained can be read directly from the mass spectrum. Although an ion M^{2+} of mass 60 would give a similar peak to an ion N^+ of mass 30 under the usual operating conditions, most of the ions produced are single positively charged species.

Electron bombardment of organic molecules can cause molecular fragmentation and the ionized fragments will form a characteristic pattern known as the mass spectrum of the molecule. This can be used either as a fingerprint for identification or to determine the relative molecular mass (Figure 1.5(*b*)).

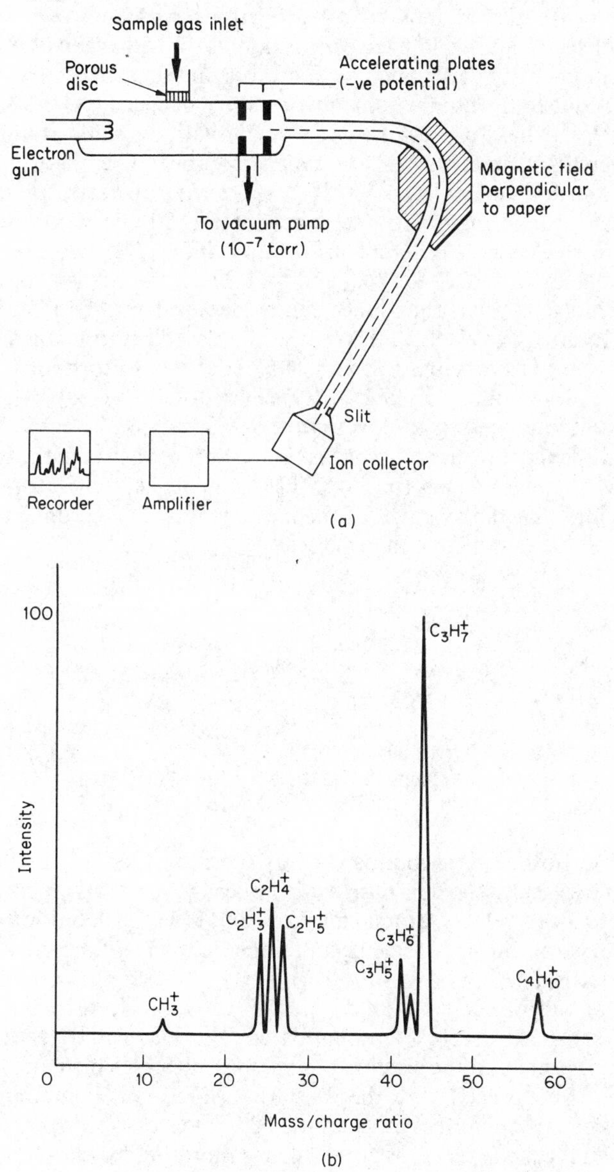

FIGURE 1.5. (a) The mass spectrometer
(b) The mass spectrum of butane

Mass spectrometers are also used for the accurate determination of isotopic atomic masses and relative atomic masses. The mass spectrum of pure oxygen shows three distinct peaks at different *m/e* ratios, indicating that there are three kinds of oxygen atoms with approximate atomic masses of 16, 17 and 18 in fixed proportions. All the oxygen atoms have an atomic number of 8, i.e. they have 8 protons in the nucleus. The difference in mass is due to the number of neutrons in the nucleus. ^{16}O has 8 neutrons, ^{17}O has 9, ^{18}O has 10. Atoms of an element with the same atomic number but different atomic masses are called isotopes. The three isotopes of oxygen are designated $^{16}_{8}O$, $^{17}_{8}O$, $^{18}_{8}O$ where the superscript gives the approximate atomic mass and the subscript the atomic number. Chlorine gives two peaks with different *m/e* ratios, indicating two isotopes with atomic masses 35 and 37, i.e. $^{35}_{17}Cl$ and $^{37}_{17}Cl$. From the intensity of the peaks, the percentage relative abundance can be deduced and the relative atomic mass calculated (Table 1.2). Thus any sample of chlorine gas consists of 75·5 per cent of isotope mass 35 and 24·5 per cent of isotope mass 37, giving a relative atomic mass of 35·453. It will be noticed in Table 1.2 that the observed atomic masses are less than those calculated from the weights of the constituent neutrons and protons. The difference between the two masses is termed the mass defect (p. 13). Most elements exhibit isotopy and it is coincidental that the averages of the individual isotopic atomic masses are close to whole numbers in many cases.

TABLE 1.2

Element	*Protons*	*Neutrons*	*Atomic number*	*Symbol*	*Isotopic mass*	*Relative % abundance*	*Relative atomic mass*
Oxygen 1	8	8	8	$^{16}_{8}O$	15·99491	99·76	
Oxygen 2	8	9	8	$^{17}_{8}O$	16·99914	0·04	15·9994
Oxygen 3	8	10	8	$^{18}_{8}O$	17·99916	0·20	
Chlorine 1	17	18	17	$^{18}_{17}Cl$	34·96885	75·5	35·453
Chlorine 2	17	20	17	$^{20}_{17}Cl$	36·96590	24·5	

The $^{12}_{6}C$ isotope of carbon is used as the new standard for the relative atomic mass scale. Carbon has two isotopes $^{12}_{6}C$ and $^{13}_{6}C$ in such proportions as to give a relative atomic mass of 12·01115. The definition of relative atomic mass on this new scale becomes the number of times an atom of an element is heavier than one-twelfth of the mass of an atom of the $^{12}_{6}C$ isotope of carbon. The choice of carbon was influenced by the ease with which very pure carbon compounds can be made and the very small alterations needed to the existing atomic mass values. The main advantage, however, is concerned with the ease with which ions containing various numbers of carbon atoms can be obtained in the mass spectrometer (Figure 1.5(*b*)). Thus carbon serves as an excellent calibrating standard for the whole mass range.

Since isotopes have the same numbers of protons in the nucleus—and therefore have the same numbers of electrons—and as chemical properties depend upon the number and arrangement of electrons (*see* Chapter 2) isotopes have identical chemical properties. The methods of separation depend upon properties attributable to mass differences. For example, the mass spectrometer can be used to separate isotopes if the recorder is replaced by accurately positioned collecting vessels. Another method of separation (the chemical exchange method) makes use of the fact that isotopic molecules react at different rates. Thus if ammonia gas containing the $^{15}_{7}N$ isotope is bubbled through ammonium nitrate solution containing the $^{14}_{7}N$ isotope, the lighter isotope of nitrogen tends to concentrate in the gaseous phase. The reaction may be represented as follows:

$$^{15}_{7}NH_3(g) + ^{14}_{7}NH_4NO_3(aq) \rightleftharpoons ^{14}_{7}NH_3\,(g) + ^{15}_{7}NH_4NO_3(aq)$$

Radioactive Disintegration Series

All elements with atomic number greater than 83 have one or more naturally occurring radioactive isotopes. These isotopes undergo successive disintegration after emitting radiation until a final stable isotope is produced The duration of time during which half of the atoms originally present disintegrate is called the *half-life* of the isotope and can vary from a millionth of a second to thousands of millions of years.

For any radioactive element the relationship between the number of atoms present initially before decay (N_0) and the number of atoms present after time t in seconds (N_t) is

$$N_t = N_0 e^{-\lambda t}$$

where λ is the radioactive decay constant for a particular nuclear species. The half-life of the species is such that

$$N_t = \tfrac{1}{2} N_0$$

$$\tfrac{1}{2} = e^{-\lambda t}$$

$$t = \frac{\log_e 2}{\lambda}$$

$$= \frac{2{\cdot}303 \log_{10} 2}{\lambda}$$

$$= \frac{0{\cdot}693}{\lambda}$$

where t is the half-life period. The unit of disintegration rate is the Curie (*Ci*). One Curie is that amount of radioactive material in which the number of disintegrations per second is $3{\cdot}7 \times 10^{10}$, i.e.

$$1 \text{ Curie } (Ci) = 3{\cdot}7 \times 10^{10}\ \text{s}^{-1}$$

Part of the uranium disintegration series is shown in Figure 1.6.

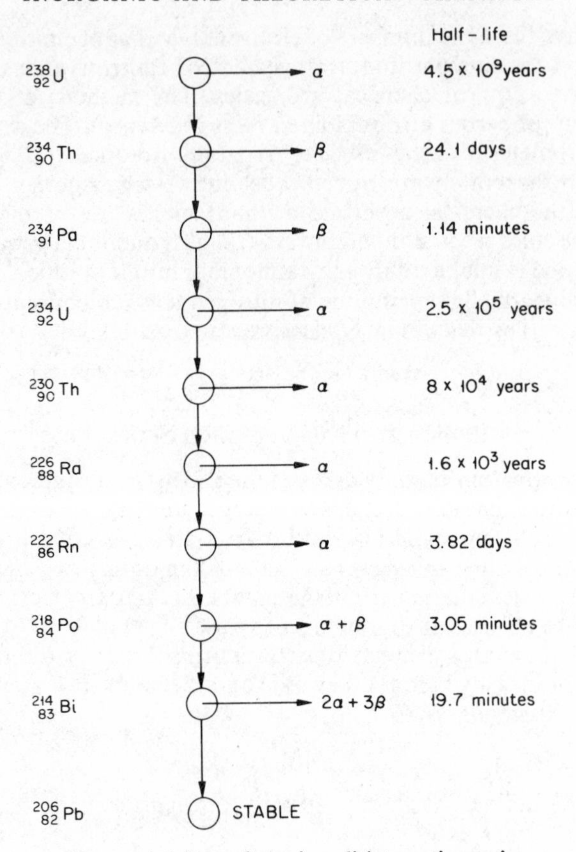

FIGURE 1.6. Part of uranium disintegration series

Note that the emission of an α-particle (^4_2He) from a nucleus causes the isotopic mass to decrease by four units and the atomic number to decrease by two units. The emission of a β-particle causes no change in mass and there is a net gain of a positive charge, possibly by the reaction neutron → proton + electron.

Artificial Nuclear Reactions

The first artificial transmutation of elements was carried out by Rutherford in 1919 by bombarding nitrogen atoms with α-particles from a radioactive source. The following nuclear reaction took place:

$$^{14}_{7}N + ^{4}_{2}He \rightarrow ^{17}_{8}O + ^{1}_{1}H$$

The $^{17}_{8}O$ isotope of oxygen is stable and no further disintegration takes place.

Charged particles such as α-particles, deuterons (nuclei of the heavy hydrogen isotope $^{2}_{1}D$), protons, and electrons can be accelerated to very high speeds by fluctuating electric and magnetic fields in machines such as the cyclotron, bevatron, betatron, and synchrocyclotron. These high-speed particles are more efficient in causing an effective collision, i.e. one which causes a nucleus to disintegrate on impact. Some typical nuclear transmutations are shown below.

$$^{9}_{4}Be + ^{4}_{2}He \rightarrow ^{12}_{6}C + ^{1}_{0}n$$

Since an α-particle is used and a neutron is produced, the reaction may be termed an $\alpha{:}n$ reaction and represented by the notation $^{9}_{4}Be(\alpha{:}n)^{12}_{6}C$.

$$^{27}_{13}Al + ^{4}_{2}He \rightarrow ^{1}_{0}n + ^{30}_{15}P \rightarrow ^{30}_{14}Si + {}_{+1}^{0}e \quad ({}_{+1}^{0}e = \text{positron})$$

This was the first example of radioactivity produced by artificial means and the changes are represented by the notation $^{27}_{13}Al(\alpha{:}n)^{30}_{15}P(\beta^{+})^{30}_{14}Si$.

The stability of a nucleus is largely determined by the ratio of the number of protons (Z) to the number of neutrons (N). For a particular value of Z there are a limited number of N values that permit a nucleus to be stable. Thus for Z values up to 20 the most stable arrangement is when $Z = N$, e.g. $^{4}_{2}He$, $^{16}_{8}O$, $^{12}_{6}C$. If a nucleus contains an excess of neutrons, it often decays by electron emission.

$$^{14}_{6}C \rightarrow ^{14}_{7}N + {}_{-1}^{0}e \qquad ^{19}_{8}O \rightarrow ^{19}_{9}F + {}_{-1}^{0}e$$

If a nucleus is deficient in neutrons, it often decays by positron emission or by electron capture from the inner K shell, e.g.

$$^{14}_{8}O \rightarrow ^{14}_{7}N + {}_{+1}^{0}e \qquad ^{83}_{37}Rb + {}_{-1}^{0}e \rightarrow ^{83}_{36}Kr$$

It is not surprising that the latter process occurs more frequently with isotopes of higher Z values.

Some methods of producing commercially important radioactive isotopes by neutron bombardment are outlined below, together with the main uses of the isotopes.

(*a*) $$^{23}_{11}Na + ^{1}_{0}n \rightarrow ^{24}_{11}Na + \gamma$$

The product has been used to follow the flow of blood in the diagnosis of thrombosis.

(*b*) $$^{31}_{15}P + ^{1}_{0}n \rightarrow ^{32}_{15}P + \gamma$$

$^{32}_{15}P$ is used to relieve leukaemia since it destroys excess of white blood corpuscles. It is also used in agriculture to study the intake of phosphorus by plants.

(*c*) $$^{131}_{52}Te + ^{1}_{0}n \rightarrow ^{132}_{53}I + {}_{-1}^{0}e$$

$^{132}_{53}I$ is used in the treatment of goitre and cancer of the thyroid gland.

(*d*) $$^{59}_{27}Co + ^{1}_{0}n \rightarrow ^{60}_{27}Co + \gamma$$

$^{60}_{27}Co$ is cheaper than radium and is used in the treatment of tumours and cancer. It is necessary to use steel or aluminium shielding during use.

(*e*) $$^{14}_{7}N + ^{1}_{0}n \rightarrow ^{14}_{6}C + ^{1}_{1}H$$

$^{14}_{6}C$ is used to study the intermediate products formed during photosynthesis. The above reaction also occurs when nitrogen in the atmosphere is bombarded by neutrons or cosmic rays which continually enter the atmosphere from outer space. The carbon produced is oxidized to carbon dioxide and absorbed by plants during photosynthesis. The proportion of ^{14}C to ^{12}C in living material is $1:10^{12}$. When carbonaceous material dies the amount of ^{14}C decreases:

$$^{14}_{6}C \rightarrow ^{14}_{7}N + ^{0}_{-1}e \qquad \text{(half-life 5,570 years).}$$

By measuring the proportion of $^{14}C/^{12}C$ it is possible to date the material. This method has been applied to establish the age of items such as the Dead Sea Scrolls, Egyptian mummies, etc.

Synthesis of New Elements

In 1940 the heaviest element known was uranium, but since that time many new radioactive elements have been synthesized with atomic numbers greater than 92. These elements are known as the *transuranic elements* and form part of a radioactive series, the *actinides*, beginning with the element actinium, atomic number 89 (p. 442). Each has a large number of radioactive isotopes. The nuclear reactions used to produce minute amounts of some of these elements are shown below.

$$^{238}_{92}U + ^{1}_{0}n \longrightarrow ^{239}_{92}U \longrightarrow ^{239}_{93}Np + ^{0}_{-1}e$$

$$^{239}_{93}Np \longrightarrow ^{239}_{94}Pu + ^{0}_{-1}e$$

$$^{239}_{94}Pu + ^{4}_{2}He \longrightarrow ^{242}_{96}Cm + ^{1}_{0}n$$

$$^{242}_{96}Cm + ^{4}_{2}He \longrightarrow ^{244}_{98}Cf + 2^{1}_{0}n$$

Isotopes of curium, californium and plutonium have been used to produce the four newest elements, nobelium, lawrencium, rutherfordium and hahnium.

nobelium

$$^{246}_{96}Cm + ^{12}_{6}C \rightarrow ^{254}_{102}No + 4^{1}_{0}n$$

rutherfordium

$$^{242}_{94}Pu + ^{22}_{10}Ne \rightarrow ^{260}_{104}Rf + 4^{1}_{0}n.$$

lawrencium

$$^{250}_{98}Cf + ^{11}_{5}B \rightarrow ^{257}_{103}Lw + 4^{1}_{0}n$$

hahnium

$$^{250}_{98}Cf + ^{15}_{7}N \rightarrow ^{260}_{105}Ha + 5^{1}_{0}n$$

Binding Energy of the Nucleus

The atomic mass of the nucleus of helium which contains two protons and two neutrons may be calculated as follows:

$$\text{Atomic mass of nucleus} = 2 \times 1{\cdot}0073 + 2 \times 1{\cdot}0087$$
$$= 4{\cdot}0342 \text{ atomic mass units (a.m.u.)}$$

The mass of the nucleus is also equal to the atomic mass of the helium atom minus twice the atomic mass of an electron

$$\text{i.e. } 4{\cdot}0320 - 2 \times 0{\cdot}00055 = 4{\cdot}0015 \text{ a.m.u.}$$

The difference between these two values ($4{\cdot}0320 - 4{\cdot}0015 = 0{\cdot}0305$ a.m.u.) is termed the *mass defect.*

Thus when a helium nucleus is formed from two protons and two neutrons there is a loss in mass of 0·0305 atomic mass units. The energy changes accompanying nuclear reactions are extremely large, about a million times more energy being evolved than in a normal chemical reaction where electronic changes occur. The mass of 0·0305 has, therefore, been converted into energy in accordance with Einstein's relationship for the equivalence of mass and energy:

$$\Delta E = \Delta mc^2$$

where ΔE is energy in joules, m is mass in kilogrammes, and c the velocity of light in metres per second.

$$1 \text{ atomic mass unit} = 1/N_A = 1/6{\cdot}023 \times 10^{23} \text{ kg}$$

N_A is the Avogadro number, i.e. the number of atoms in one kilogramme atomic mass.

$$\Delta E = \frac{0{\cdot}0305 \times (3 \times 10^8)^2}{6{\cdot}023 \times 10^{23}}$$
$$= 4{\cdot}55 \times 10^{-12} \text{ J}$$

This amount of energy must be provided to break up the nucleus and the value obtained represents the binding energy of the nucleus. The *specific binding energy* is defined as the binding energy per nuclear particle or *nucleon.* For helium this is

$$\frac{4.55 \times 10^{-12}}{4} = 1.37 \times 10^{-12} \text{ J}$$

Binding energies can be computed for other atoms and, when plotted against the atomic number, give a curve as shown in Figure 1.7.

It is seen that the binding energy per nucleon and hence the stability of the nucleus increases at first with atomic mass to a maximum, and then decreases. There are some subsidiary peaks at $^{4}_{2}He$, $^{12}_{6}C$, and $^{16}_{8}O$ indicating stable nuclear configurations, probably due to the presence of equal numbers of protons and neutrons. Atoms with intermediate atomic masses lie close to the maximum of the curve, indicating considerable nuclear stability.

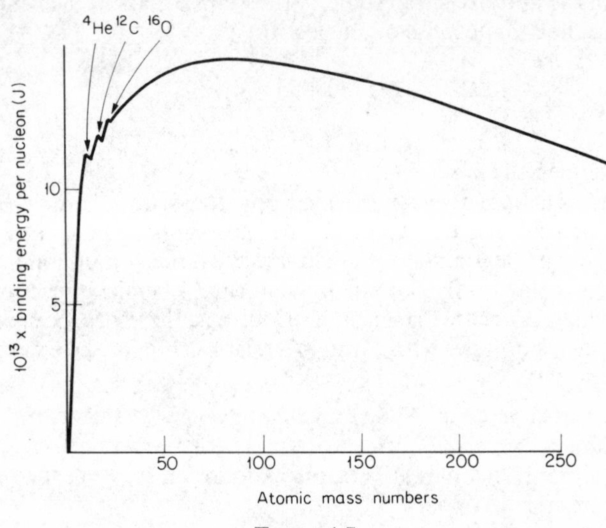

FIGURE 1.7

This suggests the possibility of splitting the heavier, less stable nuclei into fragments which are made up of atoms whose mass numbers are within the maximum region of the curve. This nuclear fission would be accompanied by the release of vast quantities of energy.

The Nature of Nuclear Forces

Gravitational and magnetic forces are too weak to account for the stability of nuclei, and electrostatic forces between charged protons would give rise to repulsive rather than attractive forces. Yukawa in 1935 proposed that nuclear attractive forces which operate over very short distances ($\sim 10^{-14}$ m) arise from a constant rapid exchange of Π mesons between nucleons. Three kinds of meson are involved, Π^{+} and Π^{-} mesons with positive and negative charges equal in magnitude to the charge on the electron but with masses some 273 times greater, and Π^{0} mesons with zero charge and slightly less mass. All nucleons are regarded as consisting of identical positive cores surrounded by meson clouds. The proton–proton and neutron–neutron attractive forces arise from rapid exchange of Π^0 mesons, and the proton–neutron attractive forces from rapid exchange of Π^+ and Π^- mesons. Thus a proton emits a Π^+ meson and becomes a neutron, and the Π^+ meson is absorbed by a neighbouring neutron which becomes a proton.

$$p \rightarrow n + \Pi^+ \qquad n + \Pi^+ \rightarrow p$$

The alternative process also occurs

$$n \rightarrow p + \Pi^- \qquad p + \Pi^- \rightarrow n$$

Π mesons have been detected in certain nuclear reactions, and the application of advanced mathematical techniques has revealed that meson exchange between nucleons can give rise to strong attractive forces.

Nuclear Fission

The three natural isotopes of uranium and the percentage abundance of each is:

$^{238}_{92}U$	$^{235}_{92}U$	$^{234}_{92}U$	
99·274	0·72	0·006	per cent

Of these isotopes, only the $^{235}_{92}U$ isotope undergoes fission with neutrons and its separation on the large scale presented a formidable problem. Electromagnetic separation based on the mass spectrometer principle was used but yielded only small quantities. However, a large-scale fractional diffusion process was adopted using gaseous uranium hexafluoride (UF_6). The principle of the process is that $^{235}_{92}UF_6$ molecules diffuse faster through a porous diaphragm than $^{238}_{92}UF_6$ due to their slightly smaller mass. Enriched $^{235}_{92}UF_6$ can thus be obtained from which uranium is produced by chemical methods. When $^{235}_{92}U$ is bombarded with neutrons the nucleus splits into two main fragments. Thus isotopes of barium and krypton may be produced with an average of $2\frac{1}{2}$ neutrons per fission.

$$^{235}_{92}U + ^{1}_{0}n \rightarrow ^{140}_{56}Ba + ^{93}_{36}Kr + 2{\cdot}5\,^{1}_{0}n$$

A loss in mass occurs releasing a vast quantity of energy ($1{\cdot}93 \times 10^{10}$ kJ mol^{-1} in the above case). With a small lump of ^{235}U most of the neutrons emitted during fission escape but if the mass of the ^{235}U exceeds a few kilogrammes (critical mass), neutrons emitted during fission are absorbed by adjacent nuclei, causing further fission and so producing more neutrons. Hence a self-propagating chain reaction occurs with a sudden release of vast quantities of energy. This is the basis of the atomic bomb, where detonation occurs by rapidly bringing together two subcritical masses of ^{235}U and ^{239}Pu which together make the combined mass greater than the critical mass.

These chain reactions can be controlled in *nuclear reactors* or piles. Here the neutrons from the fissile material are moderated (slowed down) by graphite or heavy water. Also control rods of boron or silicon, which are good neutron absorbers, are inserted into the pile to control what has now become a self-sustaining reaction.

Natural uranium can be used in these reactors, when the following transmutations take place.

$$^{238}_{92}U + ^{1}_{0}n \longrightarrow ^{239}_{92}U \longrightarrow ^{239}_{93}Np + ^{0}_{-1}e$$
$$\qquad\qquad\qquad\qquad \longrightarrow ^{239}_{94}Pu + ^{0}_{-1}e$$

Hence the reactor produces fissile ^{239}Pu from non-fissile uranium.

A new type of *breeder reactor* is now in operation at Dounreay in Scotland. Here natural uranium or thorium (^{238}U, ^{232}Th) surrounds a core of fissile material (^{235}U, ^{239}Pu). Neutrons produced from the fission of the core material are captured by the ^{238}U or ^{232}Th and nuclear transmutation occurs, producing fissile material as in the reactor described above. Thorium, ^{232}Th, produces a fissile isotope of uranium ^{233}U.

$$^{232}_{90}Th + ^{1}_{0}n \longrightarrow ^{233}_{90}Th \longrightarrow ^{233}_{91}Pa + ^{0}_{-1}e$$
$$\qquad\qquad\qquad\qquad \longrightarrow ^{233}_{92}U + ^{0}_{-1}e$$

Thus the breeder reactor produces more fissile material than it consumes. In all reactors, heat from the core is extracted by suitable coolants and is used to convert water into steam; this is then used to drive turbo-alternators for producing electricity. In breeder reactors an alloy of sodium and potassium is used as coolant. The liquid metal gives up its heat to water in a heat-exchanger producing steam.

Nuclear Fusion

Just as the fission of large nuclei is accompanied by large energy changes, the fusion of light nuclei is accompanied by mass losses and the evolution of large quantities of energy. Some examples of nuclear fusions are shown below.

$$4^{1}_{1}H \rightarrow ^{4}_{2}He + 2_{+1}^{0}e \text{ (positrons)} \qquad (1)$$

$$^{2}_{1}D + ^{2}_{1}D \rightarrow ^{4}_{2}He + \gamma \qquad (2)$$

$$^{2}_{1}D + ^{3}_{1}T \rightarrow ^{4}_{2}He + ^{1}_{0}n \qquad (3)$$

Tritium

In reaction (2) above, total mass on left-hand side $= 2 \times 2{\cdot}0141$
$= 4{\cdot}0282$ a.m.u.

Total mass on right-hand side $= 4{\cdot}0026$ a.m.u.

Mass defect $= 4{\cdot}0282 - 4{\cdot}0026 = 0{\cdot}0256$ a.m.u.

$$\text{Gain in energy} = \frac{0{\cdot}0256 \times (3 \times 10^{8})^{2}}{6{\cdot}023 \times 10^{23}}$$
$$= 3{\cdot}82 \times 10^{-12}\ \text{J}$$

When one atom of uranium undergoes fission approximately 32×10^{-12} J are evolved. The atomic mass of uranium is 235. Thus for one unit of atomic mass, the energy produced is

$$\frac{32 \times 10^{-12}}{235} = 1{\cdot}36 \times 10^{-13}\ \text{J}$$

For the fusion of two deuterons, each of atomic mass 2 units, the energy produced per unit of atomic mass is

$$\frac{3{\cdot}82 \times 10^{-12}}{4} = 9{\cdot}55 \times 10^{-11}\ \text{J}$$

Thus, for equal masses of material, much more energy is released in fusion. However, extremely high temperatures are required to initiate fusion reactions (15–30 million degrees Celsius). In the hydrogen bomb, a detonating atomic bomb, which gives the source of high temperature, is surrounded by deuterium. The energies of the sun and stars are supposed to originate from the nuclear fusion reactions just described and a great deal of effort is now being directed towards initiating and controlling fusion reactions, although their use on a commercial scale is very much a thing of the future.

2

Electronic Structure

Atomic Spectra

Chapter 1 was mainly concerned with nuclear structure and no indication was given about the possible distribution and motion of electrons round the nucleus. This is a difficult problem since obviously electrons cannot be stationary, otherwise they would be pulled into the positively charged nucleus. However, if a circular orbit is postulated for electrons to counteract the attractive force of the nucleus, another difficulty occurs. A moving charge continually radiates energy if under the influence of an attractive force. Hence energy would be lost and the electron would eventually slow down and spiral into the nucleus.

The first clue towards a solution of the problem came with the development of spectral analysis. When an electric discharge is passed through a gas at low pressures in a vacuum tube, the gas gives a spectrum which is not continuous, like that of ordinary light, but consists of narrow lines of bright colour. Each line corresponds to light of a definite wavelength (Figures 2.1 and 2.2).

It is found that each element gives a line spectrum, the lines being observed at definite wavelengths and that the spectra of different elements can be correlated. These observations indicate that atoms can only radiate energy with certain definite values and in 1913 Neils Bohr suggested that the total

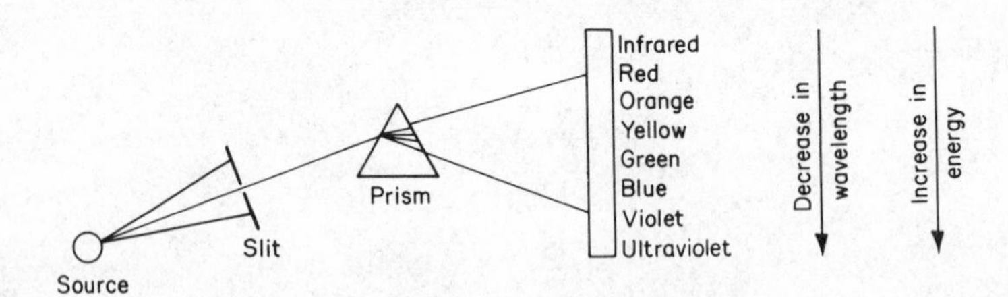

FIGURE 2.1. The continuous colour spectrum from sunlight: colours correspond to light of different wavelengths and different energies

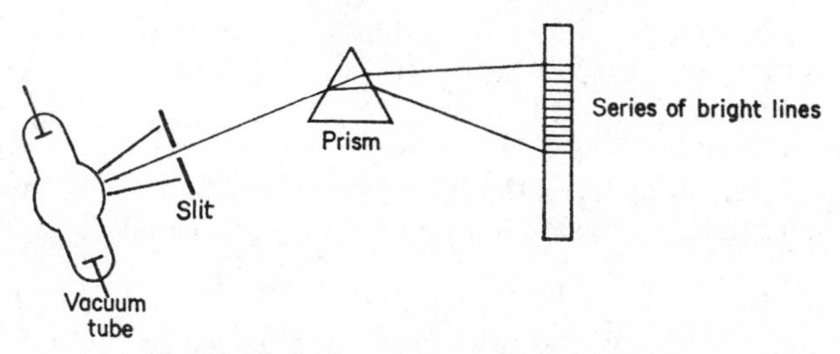

FIGURE 2.2. The bright-line spectrum

energy of an electron is quantized, i.e. can only have certain definite values. According to Bohr, electrons revolved in circular orbits around the nucleus and atoms radiated energy as light only when an electron passed from an orbit of higher energy to one of lower energy. The transition is not gradual but instantaneous and a definite amount of energy is emitted in discrete quantities called *quanta*. The energy of a quantum, E, is given by $h\nu$, where h is a constant known as *Planck's constant* (663×10^{-36} J s) and ν is the frequency of the emitted radiation ($\nu = c/\lambda$).

The energy levels or *shells* within which electrons have definite energy values were designated by a principal quantum number n which had whole-number values 1, 2, 3, 4 . . . n. The shells were also designated by letters:

	K	L	M	N	O	P
$n =$	1	2	3	4	5	6

The K shell ($n = 1$) is closest to the nucleus and electrons in this shell have the least energy. Analysis of the spectral lines revealed that the number of electrons in a particular shell was equal to $2n^2$.

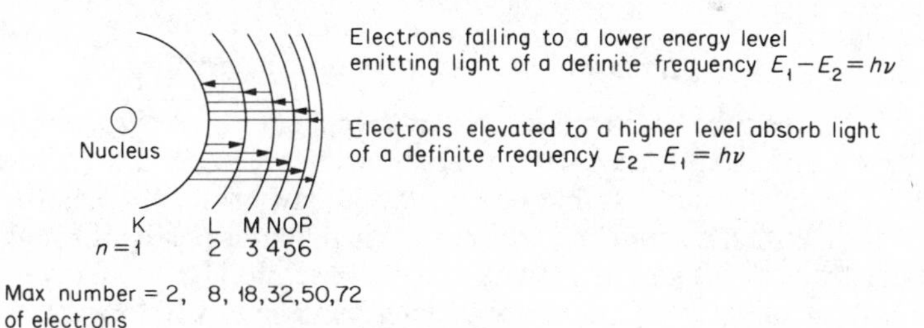

FIGURE 2.3. Electron energy-level diagram. Note that the energy difference between lower levels is greater than the energy difference between higher levels

From the study of spectra it became apparent that each gaseous element emitted its own characteristic spectrum. This led to the technique of spectroscopic analysis of unknown mixtures by comparing the spectrum of a mixture with the spectra of the individual elements. Several series of lines have been observed in different regions of the spectrum, e.g. the Balmer series (visible), the Lyman series (ultraviolet) and the Paschen series (infrared). Elements with small atomic numbers have fairly simple spectra consisting of very few lines, and hydrogen ($Z = 1$) has the simplest spectrum. Balmer and Lyman deduced that the frequency of lines (ν) in the spectrum of hydrogen and hydrogen-like species (He^+, Li^{2+}) fitted a mathematical series

$$\nu = R_H\, c\left(\frac{1}{n_2{}^2} - \frac{1}{n_1{}^2}\right)$$

where R_H is Rydberg's constant ($1{\cdot}097 \times 10^7\ m^{-1}$), c is the velocity of light and n_2 and n_1 are integers. For the Lyman series $n_2 = 1$; $n_1 = 2, 3, 4, 5, \ldots$; for the Balmer series $n_2 = 2$; $n_1 = 3, 4, 5, 6, \ldots$, and for the Paschen series $n_2 = 3$; $n_1 = 4, 5, 6, 7, \ldots$. Thus the Lyman series of lines is associated with the energy changes involved when electrons fall from outer quantum shells ($n > 1$) to the first quantum shell ($n = 1$). Lines in the Balmer and Paschen series involve similar electronic transitions into the second ($n = 2$) and third ($n = 3$) quantum shells respectively.

Consider the frequency (or wavelength) of a line in the Balmer series where $n_2 = 2$ and $n_1 = 3$.

$$\begin{aligned}\nu &= 1{\cdot}097 \times 10^7 \times 2{\cdot}997 \times 10^8(\tfrac{1}{4} - \tfrac{1}{9})\\ &= 1{\cdot}097 \times 2{\cdot}997 \times \frac{5}{36} \times 10^{15}\\ &= 4{\cdot}57 \times 10^{14}\ \text{Hz}\end{aligned}$$

The frequency calculated is that of a red line in the visible Balmer series.

Emission and Absorption Spectra

In a discharge tube, electrons are raised to higher energy levels to give a less stable excited atom. These electrons radiate energy of definite frequency on falling to lower more stable energy levels, giving the characteristic line spectrum termed an *emission spectrum*. When light is passed through a gas, the electrons in the gaseous atoms may be raised to higher energy levels by absorption of light of definite wavelengths. These wavelengths are then missing from the spectrum, giving rise to characteristic black lines on the spectrum which is known as an *absorption spectrum*.

The Sommerfeld Theory

Sommerfeld amended the Bohr theory by postulating elliptical as well as circular electronic orbits. Each elliptical orbit is characterized by a principal quantum number n and a subsidiary quantum number k, the latter having whole-number values up to and including n and where n/k = major axis of ellipse/minor axis of ellipse.

Table 2.1 indicates the relationship between the quantum numbers and the shapes of the orbits to which they refer.

TABLE 2.1

n	k	*Shape of orbit*
1	1	Circular
2	2	Circular
2	1	Elliptical
3	3	Circular
3	2	Elliptical
3	1	Elliptical

Although the subsidiary quantum number can have n values for principal quantum number n, experiment and theory show that one of these values can be zero. Thus k has been replaced by l, the azimuthal quantum number, where $l = k - 1$. Hence l may have values of 0, 1, 2, 3, ..., $n - 1$. Electrons which have l values of 0, 1, 2, 3 are referred to as *s, p, d, f* electrons, the notation being spectroscopic in origin.

The Bohr–Sommerfeld theory was still unable to account for all the spectral lines which could be observed, and it was clear that each energy level must be associated with a number of subsidiary energy levels. When atoms emitting radiation are placed in a strong magnetic field, splitting of the spectral lines occurs and a third quantum number, the magnetic quantum number m, was subsequently used to define the possible orientation of each orbit in a magnetic field. The magnetic quantum number has $2l + 1$ values, $-l$, $(-l + 1)$, $(-l + 2)$, ... 0, 1, 2, 3, ... l.

Under high resolution, spectral lines such as the yellow sodium line were found to consist of a doublet, i.e. two very closely spaced lines. In order to explain this splitting a fourth quantum number, the spin quantum number s, was introduced: s may have the values $\pm\frac{1}{2}$ depending on the direction of spin of the electron which may be either clockwise or anticlockwise.

The allocation of quantum numbers to an electron is restricted by the *Pauli exclusion principle* (1925) which states that in any atom no two electrons can have the same set of quantum numbers. The principle is empirical but agrees fully with experimental observation. With this principle in mind, electrons may be allotted quantum numbers as follows:

When $n = 1$, $l = 0$, $m = 0$, and $s = \pm\frac{1}{2}$

Thus the first quantum shell can hold two electrons (s electrons) with opposed spins.

When $n = 2$, $l = 0$, $m = 0$, and $s = \pm\frac{1}{2}$

The second quantum shell can hold two *s* electrons with opposed spins but may also contain *p* electrons, since *l* may also have the value 1:

$$\text{When } n = 2, l = 1, m = -1, 0, +1$$

and in each of these three cases *s* may be $\pm\frac{1}{2}$. Thus there are six *p* electrons occupying three energy levels, one pair with opposite spins in each. The quantum numbers of the electrons in the first three principal quantum shells are summarized in Table 2.2.

TABLE 2.2

n	*l*	*Type of electron*	*m*	*s*	*Total number in subshell*
1	0	*s*	0	$\pm\frac{1}{2}$	2
2	0	*s*	0	$\pm\frac{1}{2}$	2
2	1	*p*	−1	$\pm\frac{1}{2}$	
2	1	*p*	0	$\pm\frac{1}{2}$	6
2	1	*p*	+1	$\pm\frac{1}{2}$	
3	0	*s*	0	$\pm\frac{1}{2}$	2
3	1	*p*	−1	$\pm\frac{1}{2}$	
3	1	*p*	0	$\pm\frac{1}{2}$	6
3	1	*p*	+1	$\pm\frac{1}{2}$	
3	2	*d*	−2	$\pm\frac{1}{2}$	
3	2	*d*	−1	$\pm\frac{1}{2}$	
3	2	*d*	0	$\pm\frac{1}{2}$	10
3	2	*d*	+1	$\pm\frac{1}{2}$	
3	2	*d*	+2	$\pm\frac{1}{2}$	

The Bohr theory of the atom is essentially pictorial and, as has already been stated, could not explain all the observed spectral lines. In 1926, Heisenberg put forward his *uncertainty principle* which rendered the Bohr theory untenable. This principle states that it is impossible to determine simultaneously the position and the momentum of an electron, since any experiment designed to evaluate both these quantities must, by the very nature of the experimental measurement, alter one of these factors. At best, one can only speak, therefore, of the probability of finding an electron within the atom. Such probability calculations form the basis of *wave mechanics*.

Wave Nature of the Electron

The special study known as wave mechanics originated with the suggestion of de Broglie (1924) that every moving particle is associated with a wave-like motion analogous to the dualism existing for light waves. He described the wave motion by the equation

$$\lambda = \frac{h}{mv}$$

where m is the mass of an electron, v is its velocity, h is Planck's constant, and λ is the associated wavelength.

The wave-like properties were shown by Davisson and Germer (1927) who found that a crystal could act as a diffraction grating for a beam of electrons in the same way that light can be diffracted.

In an atom the path of an electron is closed and hence, if the electron behaves as a wave, the path must contain a whole number of wavelengths (Figure 2.4(d)). If the path does not contain a whole number of wavelengths, the waves will destroy each other by interference. Thus the wave theory leads directly to the concept of definite wavelengths or discrete energy states for the electron.

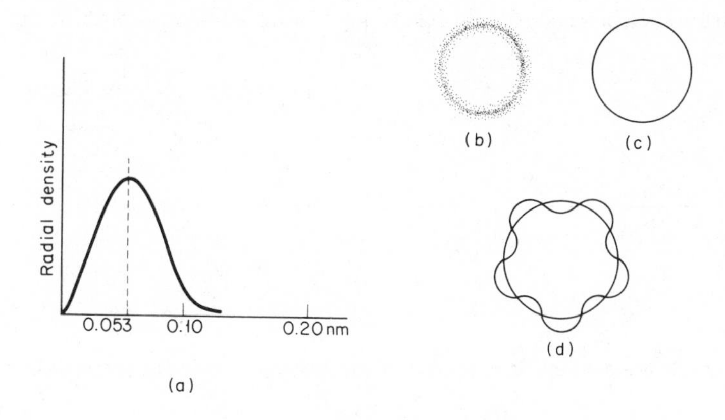

FIGURE 2.4

Schrödinger described the motion of electrons mathematically using classical wave equations. The equations could only be solved for certain definite values of the total energy of the atom (corresponding to the stationary states of the Bohr model). From the wave functions corresponding to these energy values, the probability of finding an electron around the nucleus could be calculated. The region in which an electron has the greatest probability of being located is referred to as an *orbital.* The four quantum numbers which were postulated to explain the spectra of atoms also emerge from the mathematical treatment.

Thus for a hydrogen atom in its lowest energy state (the ground state) the probability of finding the electron is greatest close to the surface of a sphere of definite radius. The probability of finding it either very close to, or at a great distance from, the nucleus is very small. The radial density may be plotted against the distance from the nucleus to give the graph shown in Figure 2.4(a). The maximum density is found by calculation to occur at 0·053 nm from the nucleus, equivalent to the radius of the first Bohr orbit.

The variation in radial density may be represented pictorially as in Figure 2.4(*b*) or more simply as (*c*), which is the boundary surface within which the electron may be found. The electronic configuration for hydrogen is then written $1s^1$. The superscript 1 implies that there is one electron in the spherically symmetrical orbital (*s*) in the first quantum shell.

In the ground state of the helium atom, two electrons occupy the 1*s* orbital and must therefore have opposite spins. It is designated as $1s^2$. The lowest quantum shell is now filled ($2n^2 = 2$, *see* p. 19).

The additional electron which is present in the atom of lithium is in the next quantum shell and the orbital which it occupies in the ground state is again spherically symmetrical. The electronic structure is thus $1s^2 2s^1$. Beryllium similarly has structure $1s^2 2s^2$, where the 2*s* electrons in the second quantum shell have opposed spins. When additional electrons are introduced

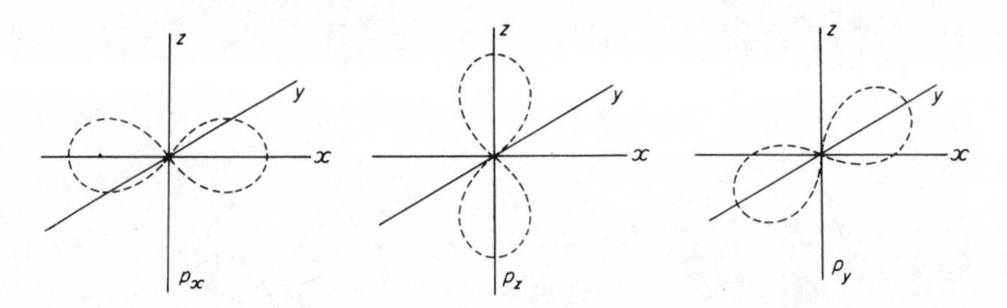

FIGURE 2.5. Three *p* orbitals of equivalent energy lying on mutually perpendicular axes

these go into three *p* orbitals ($n = 2$, $l = 1$, $m = -1, 0, +1$) of equivalent energy. These are not spherically symmetrical but are dumb-bell shaped (Figure 2.5).

Hund's Rules

The entry of electrons into these three equivalent orbitals is governed by empirical rules, Hund's rules, which state:

(*a*) Electrons tend to avoid being in the same orbital and therefore when electrons are added successively as many orbitals as possible are singly occupied before any pairing occurs.

(*b*) In the ground state, two electrons, each occupying singly a pair of equivalent orbitals, tend to have parallel spins.

For more complex atoms, electrons will enter vacant orbitals of lowest energy. For higher principal quantum numbers, more complex orbitals exist termed *d* and *f*, of which there are 5 and 7 respectively. The order of filling energy levels is shown in Figure 2.6. It must be noted that there are some exceptions to the expected energy level sequence, e.g. the 4*s* orbital is filled before the 3*d*, 5*s* before 4*d*, etc.

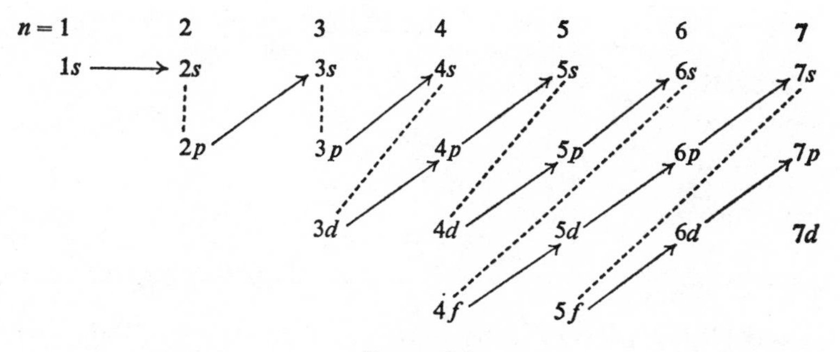

FIGURE 2.6

The electron distribution of the ten lightest elements is shown in the Table 2.3.

TABLE 2.3

Atomic number	*Element*	1*s*	2*s*	$2p_x$	$2p_y$	$2p_z$	*Notation*
1	H	↑					$1s^1$
2	He	↑↓					$1s^2$
3	Li	↑↓	↑				$1s^2.2s^1$
4	Be	↑↓	↑↓				$1s^2.2s^2$
5	B	↑↓	↑↓	↑			$1s^2.2s^2.2p^1$
6	C	↑↓	↑↓	↑	↑		$1s^2.2s^2.2p^2$
7	N	↑↓	↑↓	↑	↑	↑	$1s^2.2s^2.2p^3$
8	O	↑↓	↑↓	↑↓	↑	↑	$1s^2.2s^2.2p^4$
9	F	↑↓	↑↓	↑↓	↑↓	↑	$1s^2.2s^2.2p^5$
10	Ne	↑↓	↑↓	↑↓	↑↓	↑↓	$1s^2.2s^2.2p^6$

↑↓ orbital containing two electrons of opposite spins
↑ orbital containing one electron only

Some Physical Data Concerning Atoms

The probability of finding an electron at a given place is never zero, even at large distances from the nucleus, and thus the idea of atoms having fixed radii is clearly untenable from a wave-mechanical point of view. Moreover, the electron distribution at the surface of an atom may be distorted by neighbouring atoms. The concept of atomic radii is, however, useful and values have been obtained from spectroscopic, electron, and X-ray diffraction measurements of bond lengths. The radii of individual atoms may be determined by taking half the bond length, i.e. the distance between nuclei in homonuclear diatomic molecules such as H—H, F—F. Thus the atomic radius of sulphur is half the bond length between adjacent sulphur atoms (which is found to be 0·206 nm in crystalline sulphur), i.e., 0·206/2 = 0·103 nm (*see* Figure 2.7).

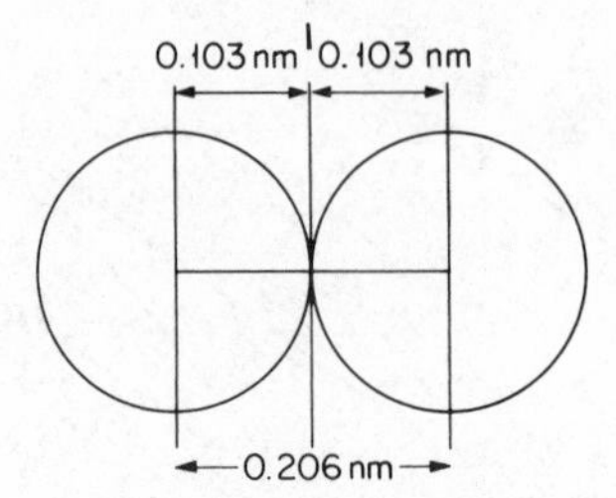

FIGURE 2.7. The atomic radius is half the internuclear distance or bond length

The bond length between carbon atoms in diamond is 0·154 nm, giving a value of 0·077 nm for the atomic radius of carbon.

In heteronuclear diatomic molecules, i.e. those containing two different kinds of atom, it is a simple matter to find the radius of one atom provided the radius of the other is known. For example, the Cl—Cl bond length is 0·198 nm, giving the atomic radius of chlorine as 0·099 nm. Iodine combines with chlorine to give the interhalogen compound ICl where the bond length is 0·232 nm. Hence the atomic radius of iodine = 0·232 – 0·099 = 0·133 nm. In this way a table of atomic radii may be compiled (Table 2.4).

TABLE 2.4

Element	*Atomic number*	*Atomic radius* (*nm*)	*Element*	*Atomic number*	*Atomic radius* (*nm*)
H	1	0·037	Na	11	0·157
He	2	0·093	Mg	12	0·136
Li	3	0·123	Al	13	0·125
Be	4	0·089	Si	14	0·117
B	5	0·080	P	15	0·110
C	6	0·077	S	16	0·103
N	7	0·074	Cl	17	0·099
O	8	0·074	Ar	18	0·193
F	9	0·072			
Ne	10	0·160			

The atomic radius is one of many useful concepts for predicting and explaining chemical behaviour. Thus phosphorus combines with chlorine to form a white solid, phosphorus pentachloride (PCl_5), but no phosphorus pentaiodide (PI_5) has ever been obtained. This is because the iodine atom has a much larger radius than the chlorine atom and it is physically impossible to fit five iodine atoms round the phosphorus atom. The concept of atomic radii is extended in Chapters 3 and 4. When the radius of an atom is determined from the internuclear distance of a covalent molecule, it is often termed the covalent radius.

In this text, however, the term atomic radius is preferred.

Ionic Radii

When one or more electrons are pulled away from an electrically neutral atom, the residual particle acquires a positive charge of one or more units depending upon the number of electrons removed. The positively charged particles are called *cations*, e.g.

$$Na \rightarrow Na^{+} + e$$

$$Mg \rightarrow Mg^{2+} + 2e$$

Similarly, if one or more electrons are pushed on to a neutral atom, the atom acquires a negative charge the magnitude of which depends upon the number of electrons accepted. The negatively charged particles are called *anions*, e.g.

$$Cl + e \rightarrow Cl^{-}$$

$$O + 2e \rightarrow O^{2-}$$

Many crystalline solids consist of a closely packed array of cations and anions held together by electrostatic forces. It is assumed that the outer surfaces of the ions are in close contact. The internuclear distances between the ions are determined by X-ray analysis and since the internuclear distance is equal to the radius of the cation plus the radius of the anion, one radius must be known initially so that others may be determined. In certain halides, e.g. LiCl, it can be assumed that all anions are in contact and the small cation fits into the vacant sites without disturbing the packing. From the diagram (Figure 2.8) it can be seen that:

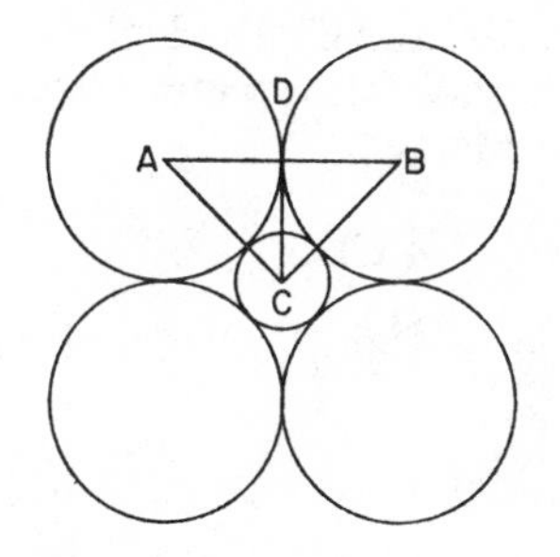

ΔADC is right-angled

$AD = DC =$ radius of anion

$AC =$ internuclear distance between anion and cation (I)

But $DC/AC = \sin 45°$

$DC =$ radius of anion $= I \times 1/\sqrt{2}$

FIGURE 2.8

This, together with other methods, has been used to evaluate ionic radii in crystals; ionic radii are shown in Table 2.5. (A dash indicates that no value is available.)

A comparison of ionic and covalent radii (*see also* p. 55) shows that:

(*a*) Cations have smaller radii than uncharged atoms. The removal of one or more electrons from an atom increases the effective nuclear charge causing a contraction in radius, e.g. Na 0·157 nm but Na^+ 0·095 nm.

TABLE 2.5

Ion	*Atomic number*	*Ionic radius (nm)*	*Ion*	*Atomic number*	*Ionic radius (nm)*
H^-	1	0·208			
He^+	2	—			
Li^+	3	0·060	Na^+	11	0·095
Be^{2+}	4	0·031	Mg^{2+}	12	0·065
B^{3+}	5	0·020	Al^{3+}	13	0·050
C^{4+}	6	0·015	Si^{4+}	14	0·041
N^{3-}	7	0·171	P^{3-}	15	0·212
O^{2-}	8	0·140	S^{2-}	16	0·184
F^-	9	0·136	Cl^-	17	0·181
Ne^+	10	—	Ar^+	18	—

(*b*) Anions have a larger radius than the uncharged atom due to the excess of negative charge resulting in greater electron repulsion, e.g. Cl^- 0·181 nm, but Cl 0·099 nm.

It will be noted that the electronic configurations of Na^+, Mg^{2+}, and Al^{3+} are identical, i.e. $1s^2 2s^2 2p^6$. These ions are said to be isoelectronic and, as expected, the radii decrease with increasing nuclear charge.

	Na^+	Mg^{2+}	Al^{3+}
Atomic number	11	12	13
Ionic radius (nm)	0·095	0·065	0·050

Certain anions are also isoelectronic (e.g. P^{3-}, S^{2-}, and Cl^-) and all have the electronic structure $1s^2 2s^2 2p^6 3s^2 3p^6$. Their radii also contract with increasing nuclear charge but the effect is not as marked.

	P^{3-}	S^{2-}	Cl^-
Atomic number	15	16	17
Ionic radius (nm)	0·212	0·184	0·181

Ionization Energy

The minimum amount of energy required to remove the least strongly bound electron from a gaseous atom, molecule or ion is called the *ionization energy* or potential, measured in kJ mol^{-1}. It is possible to remove more than one electron from an atom, molecule or ion giving the first, second, ... etc., ionization potential. Thus the first ionization potential of a gaseous atom M is the energy required to convert

$$M\,(g) \rightarrow M^{+}\,(g) + e$$

The ionization potential may be determined experimentally either by spectroscopic methods or by measurements of the current passing through a discharge tube with variation in the applied voltage. Marked changes occur in the current passing at certain voltages corresponding to points at which atoms lose one or more electrons.

The value of the ionization potential depends on the distance of the electron from the nucleus: the farther the electron is from the nucleus the more easily it is removed. However, if one electron is removed from a neutral atom, the electrons in the remaining smaller ion are more firmly bound and thus more energy is required to remove other electrons. It must be noted that in a given quantum shell containing *s*, *p*, *d*, and *f* electrons, the *f* electrons are removed more easily than the *d*, the *d* electrons more easily than the *p*, and the *p* more easily than the *s*.

Some values for the lighter elements are given in Table 2.6, the values being given in kJ mol^{-1}.

TABLE 2.6

Element	*Atomic number*	*First ionization potential (kJ mol^{-1})*	*Element*	*Atomic number*	*First ionization potential (kJ mol^{-1})*
H	1	1316			
He	2	2376			
Li	3	525	Na	11	500
Be	4	906	Mg	12	742
B	5	805	Al	13	583
C	6	1096	Si	14	792
N	7	1406	P	15	1066
O	8	1316	S	16	1006
F	9	1686	Cl	17	1266
Ne	10	2086	Ar	18	1526

Electron Affinity

The energy required to remove the valency electrons of a metallic element to form a positive ion is relatively small, e.g. $Na \rightarrow Na^{+} + e$ requires 500 kJ mol^{-1}. The non-metallic elements have little or no tendency to form positive ions (the ionization potential of chlorine is 1266 kJ mol^{-1}) but readily gain an electron to give a negative ion: $X + e \rightarrow X^{-}$.

The energy associated with this process is termed the *electron affinity*. The process may either absorb or evolve energy. Some non-metallic elements evolve large quantities of energy, i.e. their electron affinities are large and negative. For example:

	Electron affinity (*kJ mol*$^{-1}$)
$F(gas) + e \rightarrow F^-(gas)$	−354
$Cl(gas) + e \rightarrow Cl^-(gas)$	−370
$Br(gas) + e \rightarrow Br^-(gas)$	−348
$I(gas) + e \rightarrow I^-(gas)$	−320

Rather unexpectedly chlorine has a higher electron affinity than fluorine, but gaseous fluorine (F_2) is a stronger oxidizing agent than chlorine (Cl_2) because the fluorine molecule is more easily split up into atoms (Chapter 7).

All the electron affinity values quoted above for univalent anions (*see* next chapter) are negative. To push another electron on to a univalent anion usually requires a large amount of energy in order to overcome the electrostatic repulsion between the second electron and the charge on the anion, e.g.

$$O + e \rightarrow O^- \quad \text{Electron affinity} = -148 \text{ kJ mol}^{-1}$$

$$O^- + e \rightarrow O^{2-} \quad \text{Electron affinity} = +850 \text{ kJ mol}^{-1}$$

Values of electron affinities are difficult to determine directly and are usually obtained from the Born–Haber cycle (Chapter 3).

3

The Chemical Bond

The number and arrangement of the electrons in the outermost orbital of an atom determine to a large extent the chemical behaviour of that atom. The electrons in the outermost orbitals which take part in chemical combination are referred to as *valency electrons.*

The noble gases are all relatively stable and chemically unreactive and all have similar electronic configurations with complete outer s and p orbitals:

He $1s^2$

Ne $1s^2 2s^2 2p^6$

Ar $1s^2 2s^2 2p^6 3s^2 3p^6$

Kr $1s^2 2s^2 2p^6 3s^2 3p^6 3d^{10} 4s^2 4p^6$

Xe $1s^2 2s^2 2p^6 3s^2 3p^6 3d^{10} 4s^2 4p^6 4d^{10} 5s^2 5p^6$

The *electronic theory of valency* takes the noble gases as a starting point. It states that when atoms take part in chemical combination, they gain, lose, or share their valency electrons so that their electronic systems become identical with the configuration of a noble gas, either (*a*) at the end of the same period, or (*b*) at the end of the previous period of the periodic table (p. 56).

The Ionic or Electrovalent Link

Here electrons are completely transferred from one atom to another, resulting in the formation of charged ions. Consider the formation of sodium chloride (Na^+Cl^-). When sodium enters into chemical combination with chlorine, the sodium loses its one valency electron and becomes a cation Na^+. The chlorine atom gains this electron to form an anion Cl^-.

$$\underset{1s^2 2s^2 2p^6 3s^1}{Na} \rightarrow \underset{1s^2 2s^2 2p^6}{Na^+} + e$$

$$\underset{1s^2 2s^2 2p^6 3s^2 3p^5}{Cl} + e \rightarrow \underset{1s^2 2s^2 2p^6 3s^2 3p^6}{Cl^-}$$

Hence sodium chloride is written Na^+Cl^-. Other examples of electrovalent compounds are shown below, where only the outer orbital electrons are shown.

(*a*) Magnesium chloride $Mg^{2+}(Cl^-)_2$

$$\underset{2p^6\,3s^2}{Mg} \rightarrow \underset{2p^6}{Mg^{2+}} + 2e$$

$$\underset{3p^5}{2Cl} + 2e \rightarrow \underset{3p^6}{2Cl^-}$$

Hence magnesium chloride is written $Mg^{2+}(Cl^-)_2$.

(*b*) Calcium sulphide $Ca^{2+}S^{2-}$

$$\underset{3p^6\,4s^2}{Ca} \rightarrow \underset{3p^6}{Ca^{2+}} + 2e$$

$$\underset{3p^4}{S} + 2e \rightarrow \underset{3p^6}{S^{2-}}$$

Hence calcium sulphide is written $Ca^{2+}S^{2-}$.

Note that it is incorrect to refer to 'molecules' of an ionic compound. An ionic or electrovalent compound is merely an aggregation of charged ions held together by electrostatic attraction in a close-packed structure: discrete molecules do not exist.

All the alkali metals (Li, Na, K, Rb, Cs) and alkaline earth metals (Mg, Ca, Sr, Ba) form oxides, sulphides, and halides of this type. All the ions produced have a noble-gas configuration and are much less reactive chemically than the original atoms. For example, sodium is vigorously attacked by water but salt solution containing Na^+ and Cl^- ions is quite stable in water.

The energetics of ionic bond formation may be considered thermodynamically. For instance, in the formation of Na^+Cl^- the following reactions occur:

(*a*) $Na(g) \rightarrow Na^+(g) + e$ Ionization energy supplied (I_p) = 500 kJ mol^{-1}

(*b*) $Cl(g) + e \rightarrow Cl^-(g)$ Electron affinity (energy released) (E_A) = −370 kJ mol^{-1}

When two gaseous ions are brought together from a large distance to their positions in a stable crystalline lattice, considerable energy is usually evolved. This energy per mole of crystalline solid is called *lattice energy*, the magnitude of which can be derived theoretically by a consideration of electrostatic repulsion. The lattice energy of sodium chloride is large:

$$Na^+(g) + Cl^-(g) \rightarrow Na^+Cl^-(s) \qquad U = -771 \text{ kJ mol}^{-1}$$

where U is the lattice energy. Since energy is evolved on formation of the crystal lattice and since $-(U + E_A) > I_p$, the reaction between sodium and chlorine should occur readily. However, the above explanation is an over-simplification since the formation of sodium chloride involves a reaction between solid sodium and gaseous chlorine:

$$Na(s) + \frac{1}{2}Cl_2(g) \rightarrow Na^+Cl^-(s)$$

Thus the following stages are involved in the process:

$Na(s) \rightarrow Na(g)$	ΔH (atomization)
$Na(g) \rightarrow Na^+(g) + e$	I_p (ionization potential)
$\frac{1}{2}Cl_2(g) \rightarrow Cl(g)$	$\frac{1}{2}\Delta H$ (dissociation)
$Cl(g) + e \rightarrow Cl^-(g)$	E_A (electron affinity)
$Na^+(g) + Cl^-(g) \rightarrow Na^+Cl^-(s)$	U (lattice energy)

The usual thermodynamic convention is adopted, i.e. energy released is designated negative and energy absorbed is positive. The cycle in Figure 3.1 can be constructed, known as the *Born–Haber cycle.*

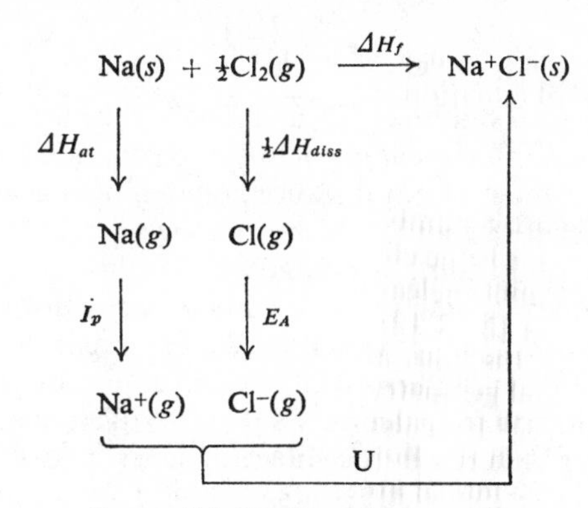

FIGURE 3.1. The Born–Haber cycle

From the first law of thermodynamics:

$$\Delta H_f = \Delta H_{at} + I_p + \frac{1}{2}\Delta H_{diss} + E_A + U$$

where ΔH_f is the heat of formation of sodium chloride. Some numerical values of these quantities are given in Table 3.1. The E_A values are calculated from data for a number of compounds of a given non-metal and the apparent values are averaged.

The Born–Haber cycle can be used to calculate or check any of the quantities enumerated but its great value lies in the determination of E_A which is difficult to determine by other methods.

For the oxides and sulphides of the alkaline earth metals the electron affinities and ionization potentials are large and positive, i.e. energy must be supplied. The lattice energy is, however, numerically large and negative, giving a negative value for ΔH_f. Thus magnesium oxide, etc., form stable crystalline solids.

TABLE 3.1

All values in kJ mol^{-1}

	U	ΔH_f	ΔH_{sub}	ΔH_{at}	I_p	E_A
NaCl	−771	−411	+109	+121	+500	−370
KCl	−701	−436	+90	+121	+424	−370
MgO	−3889	−602	+150	+248	+2198	+691
CaO	−3513	−635	+193	+248	+1752	+685
MgS	−3238	−347	+150	+213	+2198	+330
CaS	−2966	−483	+193	+213	+1752	+325

Lattice energies can be calculated from purely electrostatic considerations. The simplified equation for cubic crystals is

$$U = \frac{N_A M Z_1 Z_2 e^2}{4\Pi \epsilon_0 r_0}\left(1 - \frac{1}{n}\right)$$

where N_A is Avogadro's number, Z_1 and Z_2 are the number of charges on the cation and anion, e is the charge on the electron, ϵ_0 is the permittivity of free space, r_0 is the internuclear distance between oppositely charged ions, n is a repulsive term (8 to 12) which arises from repulsion between the electronic systems of the ions, M is Madelung's constant which is constant for a particular crystal geometry. $M = 1{\cdot}747$ for sodium chloride, 1·763 for caesium chloride, 5·039 for calcium fluoride. Discrepancies between lattice energies evaluated from the Born-Haber cycle and those calculated on the basis of a purely ionic model are very small for alkali metal halides. ($\pm$ 10 kJ mol^{-1}). Larger differences occur with silver and copper(I) halides ($>$ 40 kJ mol^{-1}) where a considerable amount of covalent bonding is indicated.

The Covalent Link

Overlap of Atomic Orbitals

Here atoms are bonded together to form molecules by the sharing of valency electrons, where one or more pairs of electrons are held in common by the combined atoms.

Each shared electron in the molecule is described by a *molecular orbital* similar to an atomic orbital (Chapter 2). Similar contours can be drawn for a molecular orbital as for an atomic orbital but, of course, here the valency electrons move round more than one nucleus. Each molecular orbital is described by quantum numbers which determine its shape, size, and energy. The electrons have the same spin as in a single atom and are fed into molecular orbitals according to the Pauli exclusion principle. Generally, molecular orbitals occupy a larger region of space than atomic orbitals. Hence the electron has a lower energy in a molecular orbital formed by overlapping of atomic orbitals. The resulting molecule is more stable (has lower energy than either of the separate atoms).

The conditions necessary for atomic orbitals to combine effectively to form a molecular orbital are:

(*a*) The separate atomic orbitals must have similar energies.

(*b*) The atomic orbitals should overlap as much as possible (principle of maximum overlapping).

(*c*) To obtain the maximum amount of overlap the atomic orbitals should be in the same plane. The resulting molecular orbital is formed by a *linear combination of atomic orbitals* (L.C.A.O.).

These principles are now applied to several examples of covalent bonding.

The hydrogen molecule

The hydrogen atom has the electronic structure $1s^1$. Two $1s$ orbitals overlap to form a molecular orbital in which the electrons have opposite spins (Figure 3.2).

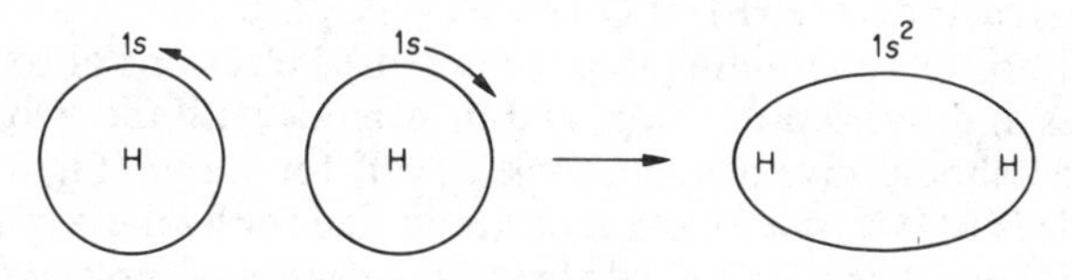

FIGURE 3.2

The probability of finding both electrons near the same nucleus is very small and hence the resulting molecular orbital is doughnut-shaped, with the greatest probability of finding the electron in the region between the two nuclei. The resulting orbital is described as a *localized molecular orbital.* The notation adopted by chemists for a single covalent or sigma (σ) bond is a line connecting both atomic symbols, i.e. for hydrogen it is H—H. The molecule now has the noble-gas structure of the helium atom, He $1s^2$.

The hydrogen chloride molecule

Electronic structures are H $1s^1$; Cl $1s^2 2s^2 2p^6 3s^2 3p_x^2 3p_y^2 3p_z^1$.

In this molecule overlap occurs with the $1s$ orbital of hydrogen and the $3p_z$ orbital of the chlorine atom (Figures 3.3(*a*) and (*b*)).

The region of electron density on the side of the chlorine atom remote from hydrogen indicates that although the electron is most probably located between the two atoms, there is a slight electron drift towards the chlorine atom (Figure 3.3*c*). Thus the molecule is polar ($H^{\delta+}$—$Cl^{\delta-}$) and possesses a dipole moment (p. 84). Chlorine acquires the noble-gas structure of argon $1s^2 2s^2 2p^6 3s^2 3p^6$.

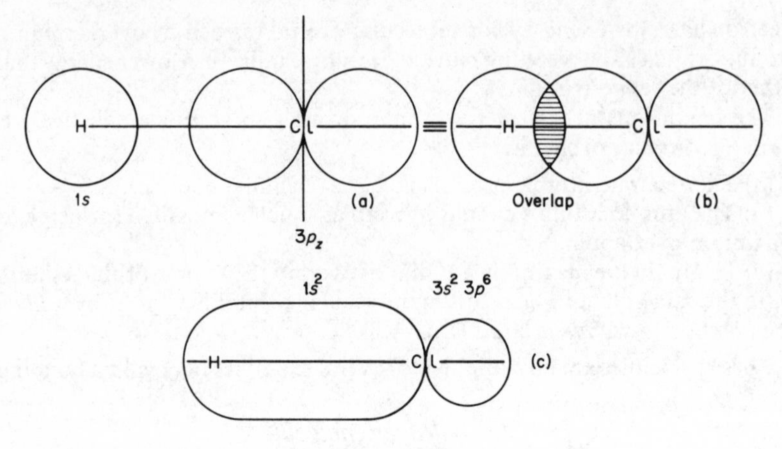

FIGURE 3.3

The water molecule

Electronic structures are H $1s^1$; O $1s^2 2s^2 2p_x^2 2p_y^1 2p_z^1$.

Here the orbitals containing the two unpaired electrons of oxygen are at right-angles and overlap of the p_y and p_z orbitals with the two $1s$ orbitals of hydrogen should give a bond angle of 90° for water (Figure 3.4). The actual angle is $104\frac{1}{2}$°, which cannot be accounted for by the repulsion of the hydrogen atoms. Accurate knowledge of the shapes of molecules obtained from physical data shows that electron pairs, whether unshared (*lone pairs*) or shared (*bonded pairs*) are arranged in space in the same way, the actual shape depending only on the number of electrons (p. 43). Clearly, a pair of

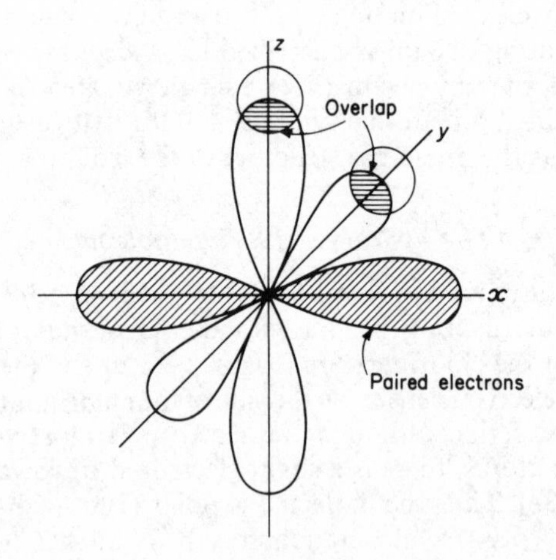

FIGURE 3.4

electrons will tend to repel another pair of electrons whether these are involved in bonding or not. The pairs will therefore be at the maximum distance apart due to this repulsion. Four electron pairs arrange themselves tetrahedrally.

Hence in the above case the four pure atomic orbitals of oxygen are said to be mixed or *hybridized* to form four equivalent tetrahedrally disposed orbitals. The type of mixing or hybridization has involved one *s* and three *p* orbitals and is termed sp^3 hybridization. It must be made clear that hybridization is a descriptive device which can be explained from quantum mechanics. It is used to account for, but is not a reason for, the observed arrangements and bond angles. The resulting picture for the molecule of water is shown in Figure 3.5(*a*).

The deviation from the tetrahedral angle of $109\frac{1}{2}°$ to $104\frac{1}{2}°$ (Figure 3.5(*b*)) can be accounted for by lone-pair repulsion (p. 44). Oxygen has acquired the noble-gas structure of neon $1s^2 2s^2 2p^6$.

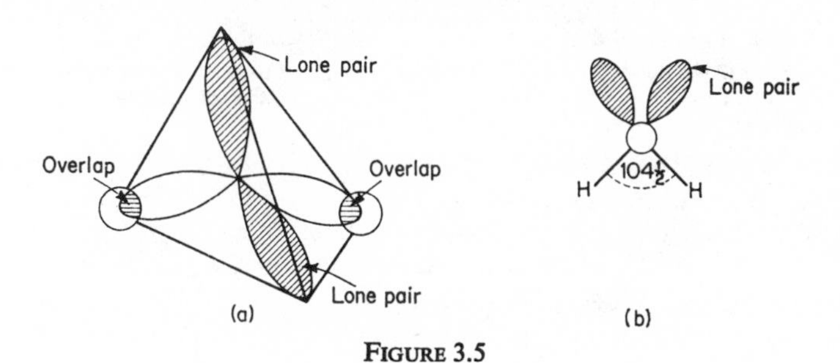

FIGURE 3.5

The ammonia molecule

Electronic structures are H $1s^1$; N $1s^2 2s^2 2p_x^1 2p_y^1 2p_z^1$.

The sp^3 hybridization again occurs giving three bonded pairs and one lone pair of electrons (Figure 3.6). Repulsion between the lone pair and the bond pairs decreases the tetrahedral angle from $109\frac{1}{2}°$ to $107°$.

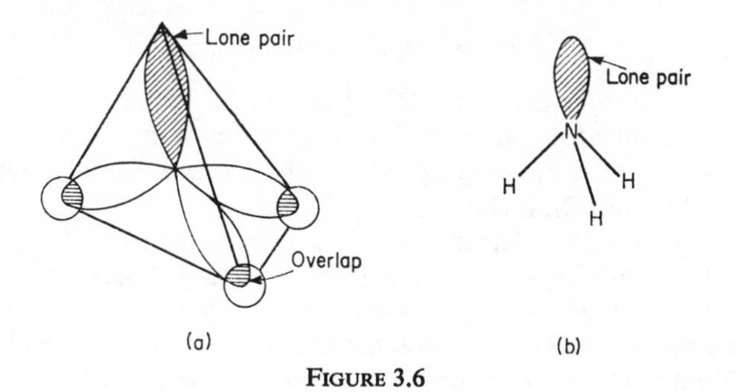

FIGURE 3.6

The boron trichloride molecule

Electronic structures are B $1s^2 2s^2 2p_x^1$; Cl $1s^2 2s^2 2p^6 3s^2 3p_x^2 3p_y^2 3p_z^1$.

Here one of the *s* electrons in the 2*s* orbital of boron is promoted to a $2p_y$ orbital which originally was empty. The *s* and 2*p* orbitals are hybridized to give sp^2 trigonal hybridization, all hybridized orbitals lying in one plane (Figure 3.7(*a*)).

The three p_z orbitals of the chlorine atom now overlap giving three bond pairs directed to the corners of an equilateral triangle with bond angle 120° (Figure 3.7(*b*)).

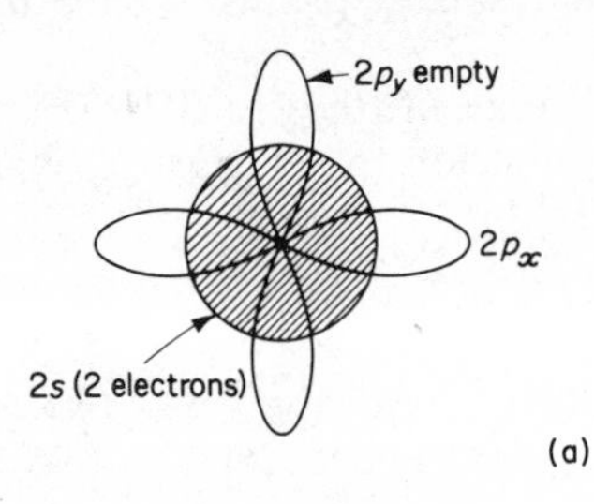

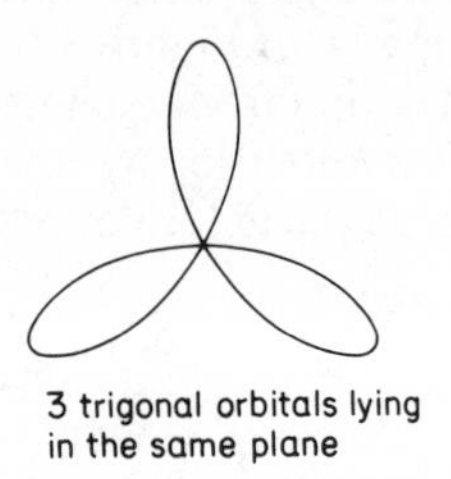

(a)

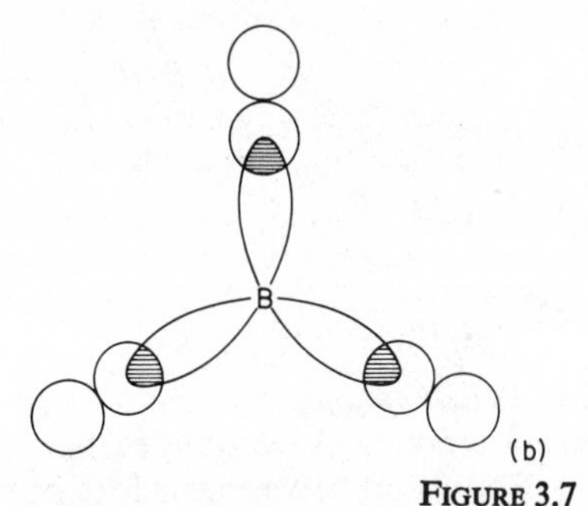

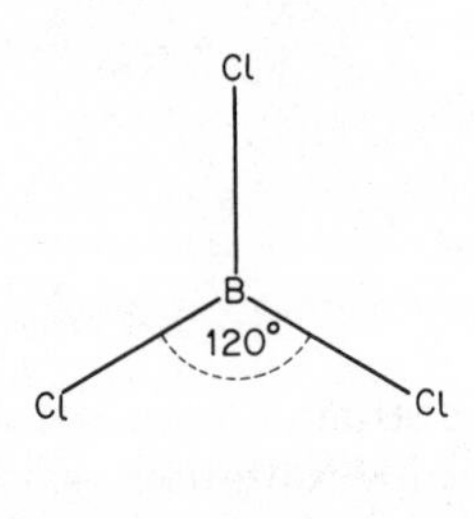

(b)

FIGURE 3.7

The carbon dioxide molecule

Electronic structures are C $1s^2 2s^2 2p_x^1 2p_y^1$; O $1s^2 2s^2 2p_x^2 2p_y^1 2p_z^1$.

In the carbon atom, one of the 2*s* electrons is promoted to the $2p_z$ orbital and then the *s* and p_x orbitals are hybridized giving (Figure 3.8(*a*)) **digonal *sp* hybridization (linear).**

The *sp* hybridized orbitals of carbon overlap with one oxygen p_z and one p_y orbital forming two σ-bonds (Figure 3.8(b)). Apart from this the p_z and p_y orbitals (singly occupied) of carbon overlap with the p_z and p_y orbitals (singly occupied) of oxygen. This is a lateral overlap and the bond formed is termed a π-bond. Thus there are two bonds between each carbon and oxygen

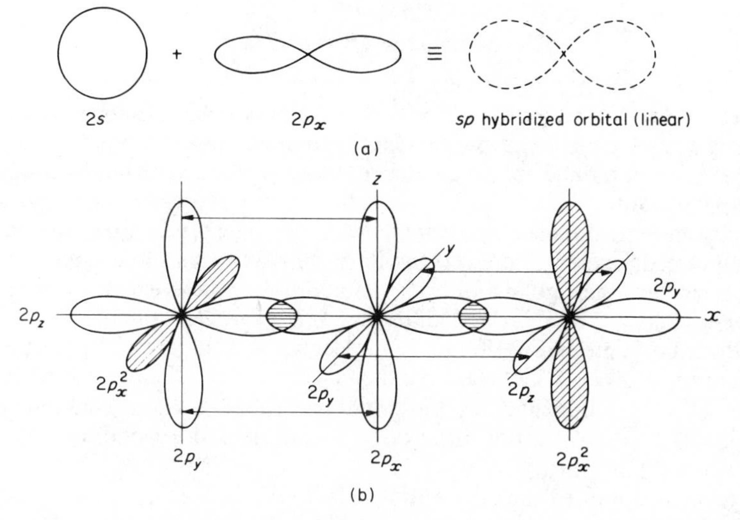

FIGURE 3.8

atom: one σ-bond and one π-bond. Such bonds are commonly called *double bonds* since two electrons are involved per bond. The molecule of carbon dioxide is linear and is written O=C=O.

In certain circumstances *d* orbitals as well as *s* and *p* orbitals can participate in hybridization. Two hybridization schemes involving *d* orbitals are shown below for the SF_4 and SF_6 molecules.

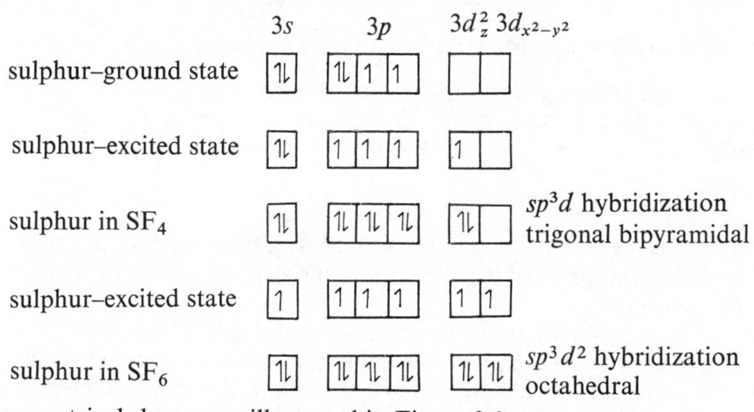

The geometrical shapes are illustrated in Figure 3.9.

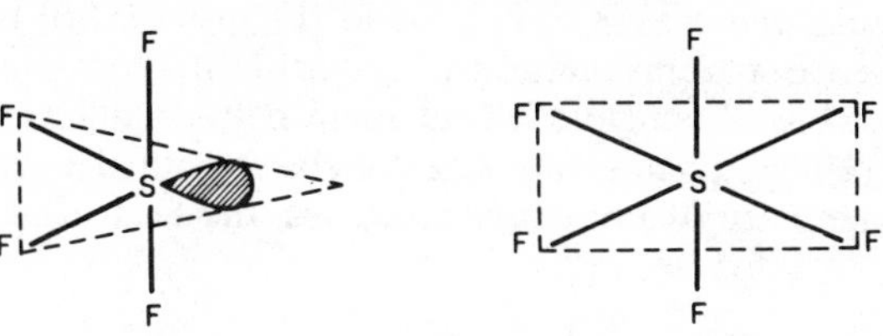

FIGURE 3.9

The Molecular Orbital Theory

It has been seen in the previous section that, for atomic orbitals to combine effectively to form a molecular orbital, the atomic orbitals must have similar energies and must overlap as much as possible. Various types of overlap have been described in which the resulting molecular orbital occupies a larger region of space and is associated with a decrease in energy. Only electrons with opposed spins tend to come together to form a bonding orbital. However, if the electrons have parallel spins the atomic orbitals repel one another, giving an excited state in which the electrons occupy a higher energy orbital termed an *antibonding molecular orbital.*

Both bonding and antibonding molecular orbitals are characterized by quantum numbers which describe their shape and size. The linear combination of a bonding and an antibonding orbital is approximately zero. Figure 3.10(*a*) shows the formation of both σ- and π-bonding and antibonding (*) orbitals.

Electrons are fed into these molecular orbitals according to the *aufbau principle* (p. 58) and the order of increasing energy has been determined mainly from spectroscopic data. For simple homonuclear diatomic molecules, the order of occupancy of the molecular orbitals is in the sequence shown in Figure 3.10(*b*).

Consider some simple molecules:

(*a*) Hydrogen $\sigma 1s^2$.

Here two electrons occupy the $\sigma 1s$ bonding orbital.

(*b*) Nitrogen $\sigma 1s^2 \sigma^* 1s^2 \sigma 2s^2 \sigma^* 2s^2 \sigma 2p_x^2 \pi 2p_y^2 \pi 2p_z^2$.

The net bonding effect of $\sigma 1s$ and $\sigma 2s$ bonding and antibonding orbitals is approximately zero so that the pairs of electrons in the $\sigma 2p_x$, $\pi 2p_y$, and $\pi 2p_z$ orbitals give rise to one σ- and two π-bonds respectively.

(*c*) Fluorine $\sigma 1s^2 \sigma^* 1s^2 \sigma 2s^2 \sigma^* 2s^2 \sigma 2p_x^2 \pi 2p_y^2 \pi 2p_z^2 \pi^* 2p_y^2 \pi^* 2p_z^2$.

The net bonding effect of the $\sigma 1s$, $\sigma 2s$ $\pi 2p_y$, $\pi 2p_z$ bonding and antibonding orbitals is approximately zero and one σ-bond exists between the two fluorine atoms.

This method may, at first inspection, seem a rather unnecessary complication in view of the simple treatment given to some of these molecules in the earlier part of this chapter. However, one of the earlier successes of this theory came with the treatment of bonding in molecular oxygen.

A structure involving a double bond (Figure 3.11(*a*)) is unsatisfactory, since oxygen is paramagnetic, the magnetic moment indicating two unpaired electrons. A single covalent bond between the two oxygen atoms (Figure 3.11(*b*)) still does not adequately explain the structure because, although there are two unpaired electrons, the heat of dissociation of an

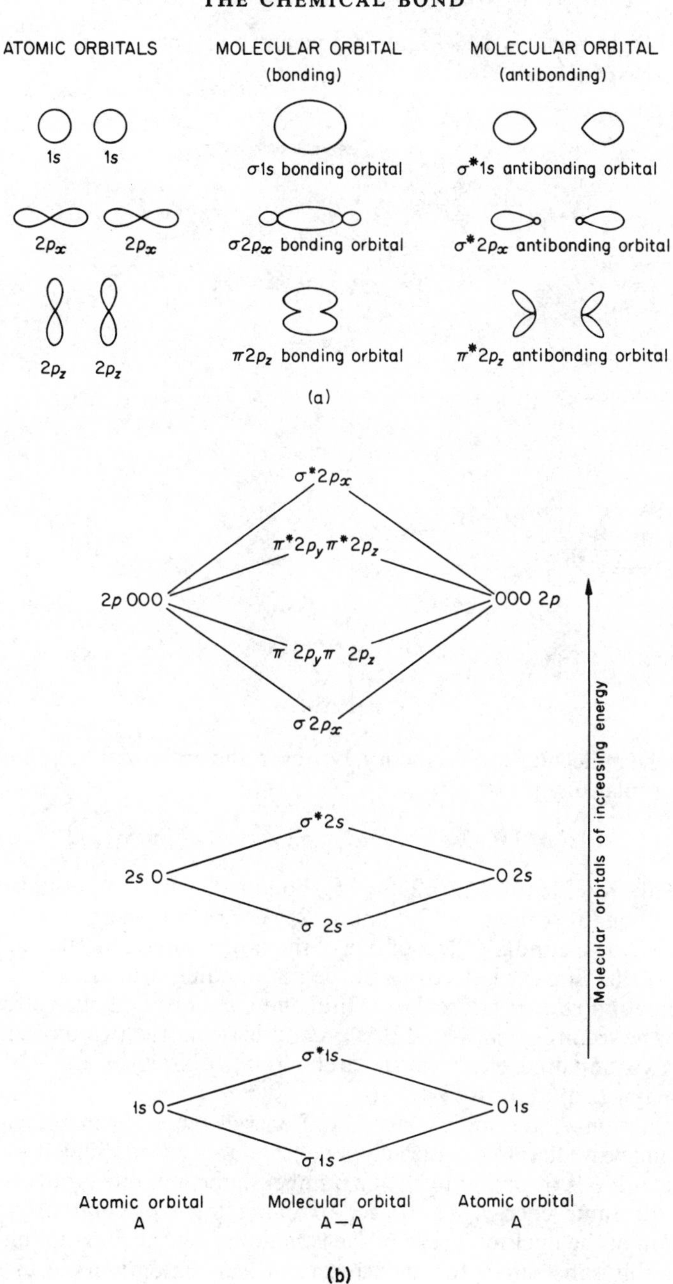

ATOMIC ORBITALS
MOLECULAR ORBITAL (bonding)
MOLECULAR ORBITAL (antibonding)
1s
1s
σ1s bonding orbital
σ*1s antibonding orbital
2px
2px
σ2px bonding orbital
σ*2px antibonding orbital
2pz
2pz
π2pz bonding orbital
π*2pz antibonding orbital
(a)
σ*2px
π*2py π*2pz
2p OOO
OOO 2p
π 2py π 2pz
σ 2px
σ*2s
2s O
O 2s
σ 2s
σ*1s
1s O
O 1s
σ 1s
Molecular orbitals of increasing energy
Atomic orbital A
Molecular orbital A—A
Atomic orbital A
(b)

FIGURE 3.10

O—O single bond is 146 kJ mol^{-1}, whereas the heat of dissociation of oxygen is 496 kJ mol^{-1}.

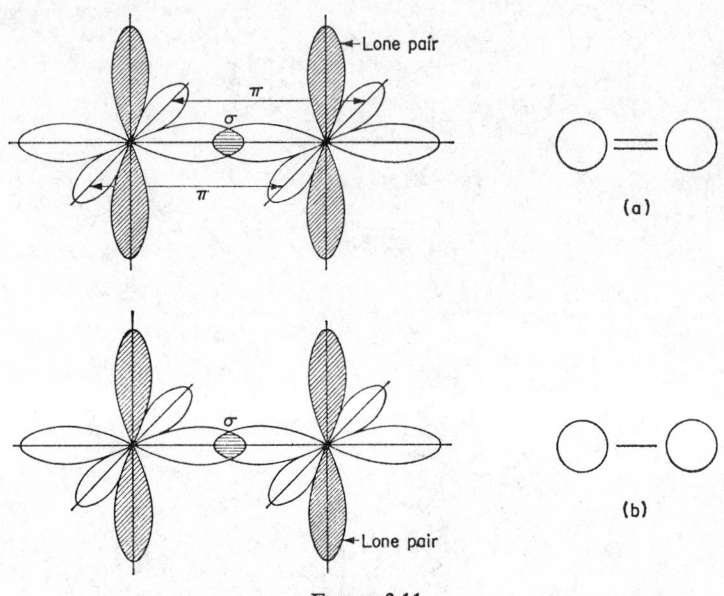

FIGURE 3.11

On the molecular orbital theory, however, the order of filling the orbitals in molecular oxygen is:

$$\sigma 1s^2\, \sigma^* 1s^2\, \sigma 2s^2\, \sigma^* 2s^2\, \sigma 2p_x^2\, \pi 2p_y^2\, \pi 2p_z^2\, \pi^* 2p_y^1\, \pi^* 2p_z^1$$

There are $1\sigma(\sigma 2p_x^2)$ and $2\pi(\pi 2p_y^2 . \pi 2p_z^2)$ bonding orbitals in the molecule and the antibonding orbitals $\pi^* 2p_y^1$ and $\pi^* 2p_z^1$ have equal energy and are singly occupied. The bonding effect of one of the π-orbitals is cancelled out by the effect of the unpaired electrons in the antibonding orbitals, i.e. the linear combination of a π molecular orbital and two antibonding p electrons is zero. The result is one σ- and one π-bond between the two oxygen atoms, with two unpaired electrons in antibonding orbitals to account for the paramagnetism of the molecule.

Evaluation of the molecular orbital energy sequence in the case of a heteronuclear diatomic molecule is considerably more difficult, since the atomic orbitals of similar quantum number in the separate atoms no longer have the same energy. The order of filling the molecular orbitals is a function of the nuclear charge of the two atoms but the diagram on p. 41 is reasonably satisfactory for most heteronuclear diatomic molecules where the combining atoms have low nuclear charges of between 5 and 10, e.g. carbon monoxide (p. 214) and nitrogen(I) oxide (p. 252).

Shapes of Molecules and the Electron-pair Repulsion Theory

Methods used for the elucidation of shapes of molecules have already been discussed. The electron-pair repulsion theory of Sidgwick and Powell explains why molecules assume a particular shape, and supports and clarifies experimental data.

The theory is based on the assumption that the pairs of electrons in the valency shell of a bonded atom in a molecule are arranged in a definite way which depends on the number of electron pairs. The geometrical shapes assumed by different numbers of electron pairs is given in Table 3.2.

TABLE 3.2

Number of electron pairs	*Geometrical arrangement*
2	Linear
3	Triangular
4	Tetrahedral
5	Trigonal bipyramidal
6	Octahedral
7	Pentagonal bipyramidal

On this theory, a multiple bond is regarded as equivalent to a single bond as far as molecular shape is concerned. The above configurations arise due to repulsion between the electron pairs which take up positions as far apart as possible. This is true whether the electron pairs are shared (bonding pairs) or unshared (lone pairs). Deviations from these regular shapes occur because the strength of the repulsion decreases in the order: lone pair—lone pair > lone pair—bond pair > bond pair—bond pair. This decrease in repulsion arises because the bonding electron pairs are slightly drawn away from the nucleus by the other bonding atoms whereas the lone-pair electrons are held fairly close to the nucleus.

The shapes of some simple molecules can be explained using the above assumptions and some examples are given below:

Linear arrangement

The mercury(II) chloride molecule is linear since repulsion between the two bonding pairs causes the chlorine atoms to be as far apart as possible (Figure 3.12). Similarly the carbon dioxide and carbon disulphide molecules are linear.

$$Cl-Hg-Cl \qquad O=C=O$$

FIGURE 3.12

Planar arrangement

In the boron trichloride molecule, the three bonding pairs repel each other giving a symmetrical, triangular, planar arrangement with a bond angle of

120°. The nitrate and carbonate ions also have this symmetrical planar arrangement (Figure 3.13).

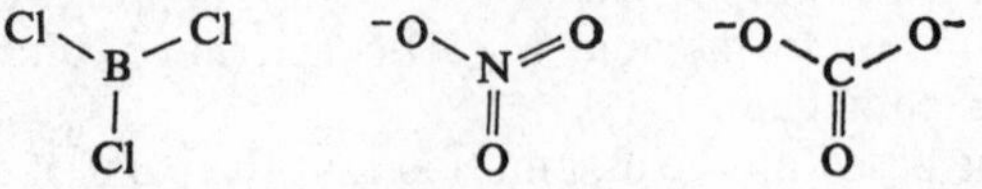

Cl B Cl | Cl $^{-}$O N O ‖ O $^{-}$O C O^{-} ‖ O

FIGURE 3.13

Tetrahedral arrangement

In the methane molecule the carbon atom is surrounded by four bonding pairs of electrons which repel each other giving a regular tetrahedral shape with a bond angle of 109° 28′ (Figure 3.14(*a*)). In the ammonia molecule there are one lone pair and three bond pairs of electrons, the repulsion between the lone pair and the bonding pairs being slightly greater than that between bond pairs. Thus the H—N—H angle is reduced to 106° 7′ (Figure 3.14(*b*)). The water molecule contains two lone pairs of electrons and the above effect is even more marked; the lone pairs repel each other to a greater extent than the bond pairs with a consequent reduction in bond angle to 104° 27′ (Figure 3.14(*c*)).

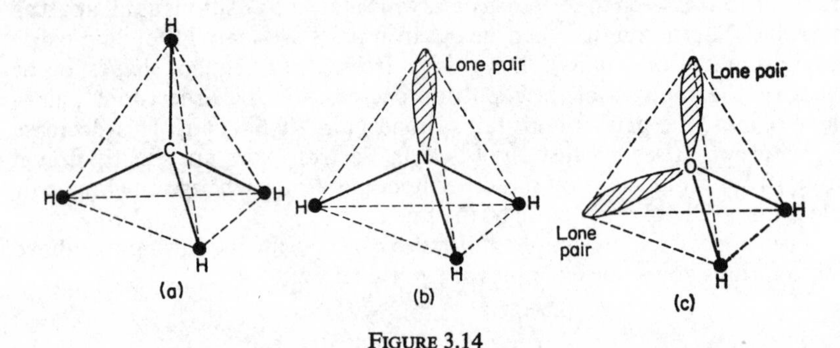

FIGURE 3.14

In the case of a series of compounds of elements in the same periodic group the bond angles differ, e.g. in ammonia the bond angle is 106° 7′ and in phosphine 96° 50′. This reduction in bond angle occurs because the bonds in ammonia are shorter and of greater strength than in phosphine. Thus the repulsion between the lone pair and the bond pairs causes a smaller contraction of bond angle in ammonia than in phosphine.

More complex arrangements

Molecules with five, six or seven pairs of electrons in the valency shell have them directed towards the corners of a trigonal bipyramid, octahedron, and a pentagonal bipyramid respectively. Examples are shown in Figure 3.15 of molecules containing five, six, and seven bonding pairs of electrons.

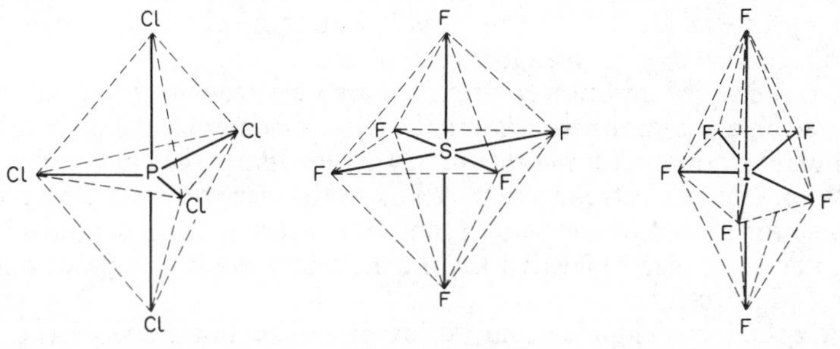

FIGURE 3.15

Where lone-pair electrons exist, in these cases the configuration is more difficult to predict and there may be several alternatives. There are, for example, two possible theoretical configurations for $TeCl_4$ (Figure 3.16).

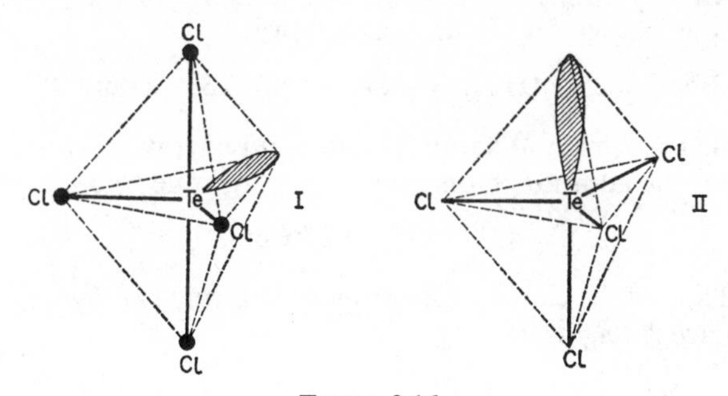

FIGURE 3.16

In I the lone pair is at right-angles to two bonded pairs but in II it is at right-angles to three bond pairs. Hence lone pair—bond pair repulsion energy is lower in I and this is the observed structure.

Bond Polarity and Electronegativity

In the preceding discussion, two types of bonding have been exemplified but very few compounds are either purely ionic or purely covalent. A large number of covalent compounds have a certain percentage of ionic character. This is due to unequal sharing of electrons between different atoms in a molecule. The resulting molecule possesses an electrical dipole and is said to have a dipole moment (p. 83). In the molecule of hydrogen chloride an electron drift occurs towards the chlorine atom. Thus, although the bond is mainly covalent, there is also some ionic character, the hydrogen atom having a slight positive charge and the chlorine atom a slight negative charge. The molecule is written $H^{\delta+}$—$Cl^{\delta-}$, although it must be remembered that the charges are small and not comparable with those obtained by complete electron removal or gain in an ionic compound.

The tendency of an atom to attract shared electrons to itself has been evaluated on a quantitative basis and the relative numerical values obtained for atoms are termed *electronegativities*. This term is not the same as electron affinity (p. 29), as electronegativity refers to the electron-attracting power of atoms for shared electrons in a molecule. Various methods have been used to evaluate electronegativities, the most important being that due to Pauling which utilizes bond energies.

In a polar covalent molecule the experimental bond energy (i.e. the energy required to break a covalent bond) is always greater than the calculated bond energy. This is because calculated bond energies do not take into account the additional bonding energy due to the attraction of the positive and negative ends of a polar molecule. If the electrons are shared equally in a molecule of hydrogen chloride and the bond energies for the H—H and Cl—Cl bonds are 436 and 242 kJ mol^{-1} respectively, then the calculated bond energy for H—Cl is as follows:

$$\text{Bond energy (HCl)} = \tfrac{1}{2}(436 + 242) = 339 \text{ kJ mol}^{-1}$$

However, the experimental value for the bond energy of H—Cl is 431 kJ mol^{-1}, so that the additional ionic binding energy due to bond polarity

$$= 431 - 339 = 92 \text{ kJ mol}^{-1}$$

If x_{Cl} and x_H are the electronegativities of chlorine and hydrogen respectively, then, according to Pauling

$$x_{Cl} - x_H = \sqrt{\frac{92}{96{\cdot}5}}$$

The factor 96·5 converts kJ mol^{-1} into electronvolts which were the units Pauling originally used in deriving his arbitrary scale.

The method clearly gives values for the difference in electronegativities but does not give absolute values. Thus one element must be chosen and an arbitrary value given to it to act as a standard for the derivation of the others. To bring the values within a suitable range, hydrogen is given the value 2·1.

Since $$x_{Cl} - x_H = 0{\cdot}9$$

$$\therefore x_{Cl} = 3{\cdot}0$$

Some examples of electronegativities are given in Table 3.3.

Electronegativity values are very useful in predicting bond type. A large electronegativity difference between two atoms involved in the formation of a chemical bond indicates high polarity and a high degree of ionic character,

TABLE 3.3

Element	*Atomic number*	*Electro-negativity*	*Element*	*Atomic number*	*Electro-negativity*
H	1	2·1			
He	2	—			
Li	3	1·0	Na	11	0·9
Be	4	1·5	Mg	12	1·2
B	5	2·0	Al	13	1·5
C	6	2·5	Si	14	1·8
N	7	3·0	P	15	2·1
O	8	3·5	S	16	2·5
F	9	4·0	Cl	17	3·0
Ne	10	—	Ar	18	—

whereas with atoms of similar electronegativity there is only slight polarity and ionic character. Correlation between electronegativity differences and dipole moment measurement is illustrated in Table 3.4. The greater the

TABLE 3.4

	x_H	x_F etc.	*Difference*	*Dipole moment (Debye)*	
H–F	2·1	4·0	1·9	1·91	↓ Decrease in bond polarity
H–Cl	2·1	3·0	0·9	1·05	
H–Br	2·1	2·8	0·7	0·80	
H–I	2·1	2·5	0·4	0·42	
Cl–Cl	3·0	3·0	0	0	

difference in electronegativity, the greater is the ionic character of the bond formed. This causes the molecule to have a high dipole moment. Deductions of polarity from electronegativity data are only useful qualitatively since dipole moments are sensitive to atomic size and lone pair positions.

The Coordinate (or Coionic) Bond

The third type of bonding is in fact a slight extension of covalent bonding and is best understood by examples. Consider the ionic solid ammonium chloride ($NH_4^+Cl^-$) produced by the action of gaseous ammonia on hydrogen chloride (Figure 3.17).

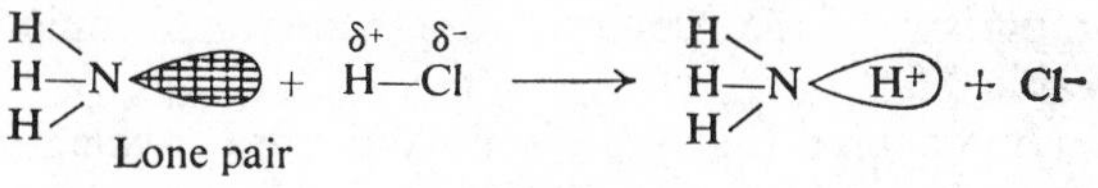

$$H_3N\text{:} + \overset{\delta+}{H}\text{—}\overset{\delta-}{Cl} \longrightarrow H_3N\text{—}H^+ + Cl^-$$

Lone pair

FIGURE 3.17

As the polar hydrogen chloride molecule approaches the lone pair on the nitrogen atom, there is a strong pull on the positive hydrogen end. The covalent bond breaks and the proton becomes embedded in the electron lone-pair cloud. Here the nitrogen atom provides both electrons for bonding and Sidgwick suggested that an arrow pointing from the donor atom should be used to indicate this type of bond (Figure 3.18(*a*)). Note, however, the alternative nomenclature in Figure 3.18(*b*) which is being increasingly used to represent the coordinate link.

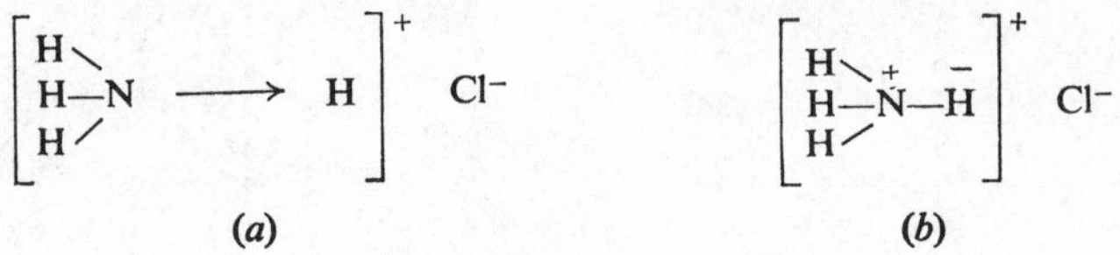

$$\left[\begin{array}{l}H\diagdown \\ H—N \longrightarrow H \\ H\diagup\end{array}\right]^+ Cl^- \qquad \left[\begin{array}{l}H\diagdown \overset{+}{\ } \quad \overset{-}{\ } \\ H—N—H \\ H\diagup\end{array}\right]^+ Cl^-$$

(*a*) (*b*)

FIGURE 3.18

All the hydrogen atoms in the ammonium ion (NH_4^+) are identical and it is impossible to identify the coordinately bonded hydrogen. Thus a coordinate bond, once it is formed, is indistinguishable from a normal covalent bond: only its mode of formation is different.

Another example is the dissolution of mineral acids in water involving the breaking of a polar covalent bond and the hydration of the released proton. This is shown in Figure 3.19.

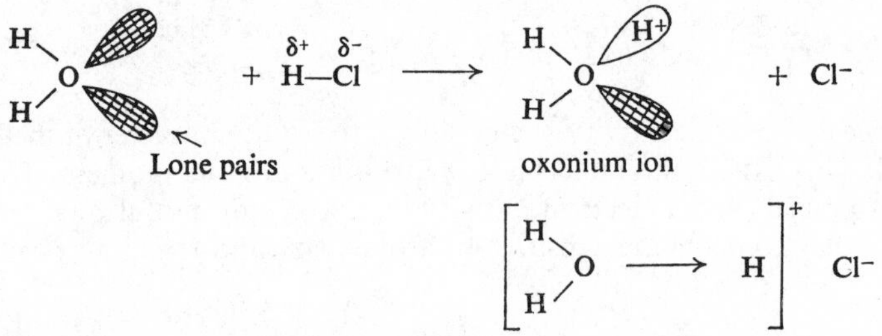

$$H_2O + \overset{\delta+}{H}—\overset{\delta-}{Cl} \longrightarrow H_2O(H^+) + Cl^-$$

Lone pairs

oxonium ion

$$\left[\begin{array}{l}H\diagdown \\ \quad O \longrightarrow H \\ H\diagup\end{array}\right]^+ Cl^-$$

FIGURE 3.19

General Properties associated with Covalent and Ionic Compounds

A general comparison of the properties of covalent and ionic compounds is shown in Table 3.5.

It must be remembered that pure ionic and pure covalent compounds are two extremes which are very rarely found. Most compounds possess bonds of an intermediate type which have both ionic and covalent character and the properties summarized in the Table undergo considerable modification.

TABLE 3.5

Covalent compounds	*Ionic compounds*
1. Bond is rigid and directional: molecules exist and have a definite shape.	1. No directional bond: no molecules, only close packed ions held together by electrostatic attraction.
2. No strong forces holding molecules together. Compounds generally have low melting and boiling points.	2. Strong electrostatic forces hold ions together. Compounds generally have high melting and boiling points.
3. Generally poor electrolytes.	3. Generally good electrolytes in water or when fused.
4. Generally soluble in organic solvents and insoluble in water.	4. Generally insoluble in organic solvents and soluble in water.

The Hydrogen Bond

A study of the physical properties of compounds formed between hydrogen and the most electronegative elements shows certain anomalies which are illustrated by the graphs in Figure 3.20. Boiling points for a number of

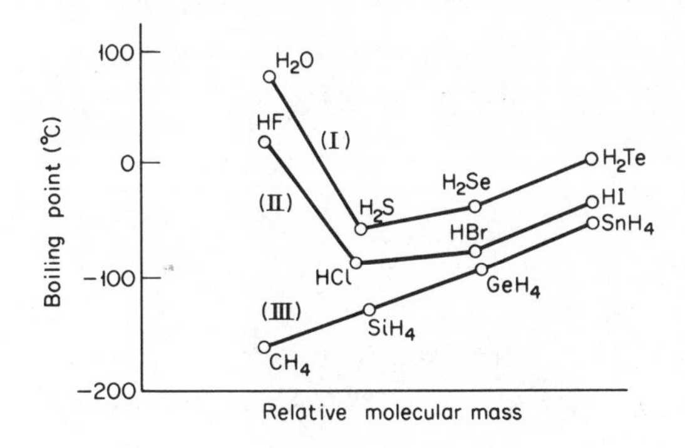

FIGURE 3.20

hydrides are plotted against the relative molecular masses. The boiling points of the hydrides of carbon, silicon, germanium, and tin increase as expected in a linear manner with increasing molecular weight. In graphs I and II exceptionally high boiling points are evident for water and hydrogen fluoride. The electronegativity differences between oxygen and hydrogen and between fluorine and hydrogen are respectively 1·4 and 1·9. These high values indicate strongly polar molecules and, therefore, an additional attractive force binds the molecules together and extra energy is required to volatilize hydrogen fluoride and water (Figure 3.21).

Note that hydrogen chloride is also polar but the electronegativity difference between hydrogen and chlorine is only 0·9 and the effect is much less marked. The two examples quoted above are only two of many examples of this type of interaction. Diagrams showing how hydrogen fluoride and water molecules become grouped together as a result of these electrostatic

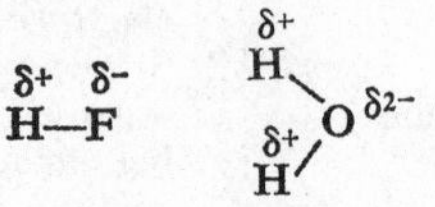

$$\overset{\delta+}{H}-\overset{\delta-}{F} \qquad \overset{\delta+}{H}\diagdown\overset{\delta 2-}{O}\diagup\overset{\delta+}{H}$$

FIGURE 3.21

forces are shown in Figure 3.22, the dotted lines indicating the electrostatic forces operative between the positive and negative ends of adjacent molecules.

The term *hydrogen bond* is used to describe such intermolecular attraction. Thus hydrogen fluoride and water are *polymerized* in the liquid and solid state due to hydrogen bonding. The energies of hydrogen bonds are of the

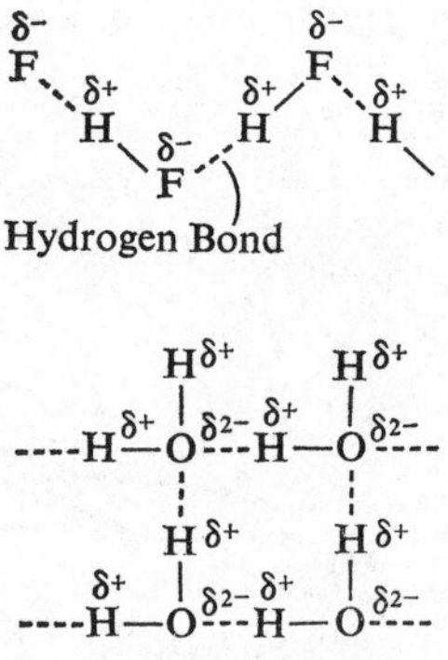

$$\overset{\delta-}{F}\cdots\overset{\delta+}{H}-\overset{\delta-}{F}\cdots\overset{\delta+}{H}-\overset{\delta-}{F}\cdots\overset{\delta+}{H}-$$

Hydrogen Bond

$$\begin{array}{c} H^{\delta+} \qquad\qquad H^{\delta+} \\ \cdots\overset{\delta+}{H}-\overset{\delta 2-}{O}\cdots\overset{\delta+}{H}-\overset{\delta 2-}{O}\cdots \\ \vdots \qquad\qquad \vdots \\ H^{\delta+} \qquad\qquad H^{\delta+} \\ \cdots\overset{\delta+}{H}-\overset{\delta 2-}{O}\cdots\overset{\delta+}{H}-\overset{\delta 2-}{O}\cdots \end{array}$$

FIGURE 3.22

order of 25–35 kJ mol^{-1}, which is approximately one-tenth of the normal covalent bond energy. The length of the hydrogen bond is of the order 0·25–0·275 nm.

Evidence for the existence of hydrogen bonds comes from two sources:

(*a*) Ebullioscopic and cryoscopic studies, which indicate association of a large number of molecules.

(*b*) X-ray crystallography, electron diffraction, infrared, and nuclear magnetic resonance studies have located the position of the hydrogen atoms in a large number of compounds and it has been found that they lie between pairs of highly electronegative atoms, e.g.

$$\diagdown O-H\cdots\diagdown O- \qquad \cdots F-H\cdots F- \qquad >N-H\cdots O$$

The dotted line indicates the hydrogen bond.

Examples of Hydrogen Bonding in Crystals

The crystal may contain:

(*a*) Finite groups or ions as in $K^+HF_2^-$ and carboxylic acids (Figure 3.23(*a*)).

(*b*) Infinite chains of ions as in the bicarbonate ion, the bisulphate ion, and hydrogen fluoride (Figure 3.22).

(*c*) Infinite layers as in crystalline boric acid (p. 194), sulphuric acid, nitric acid, and aluminium hydroxide (Figure 3.23(*b*)).

$$KHF_2 \rightleftharpoons K^+ + HF_2^- \qquad [F{-}H{-}{-}{-}F]^-$$

Potassium hydrogen fluoride

```
       O---H—O
      //       \
CH3C             C—CH3
      \        //
       O—H---O
```

(a)

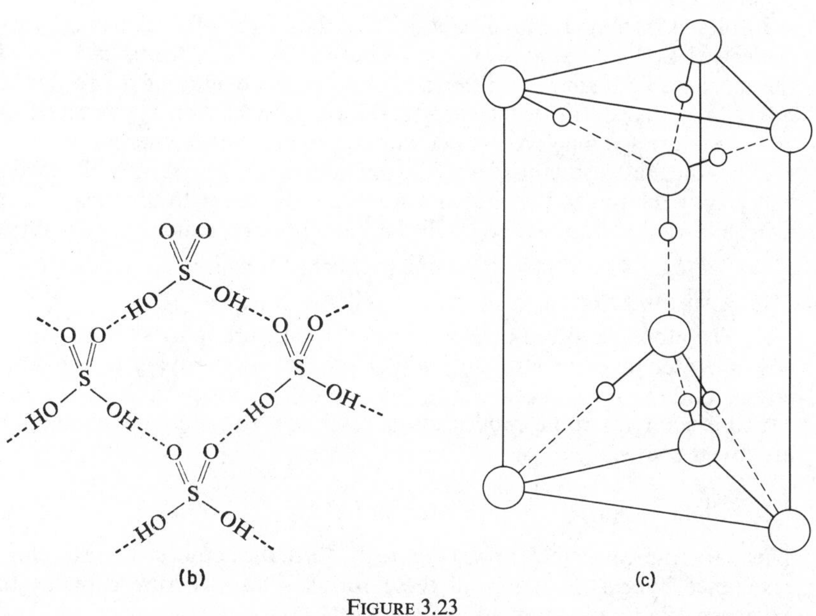

FIGURE 3.23

(*d*) Infinite three-dimensional structures as in ice, hydrogen peroxide, and certain crystalline hydrates. In the continuous ice lattice, each oxygen atom is surrounded by four hydrogen atoms, two attached by covalent bonds and two by hydrogen bonds (Figure 3.23(*c*)).

The Valency Bond Theory and Resonance

Many molecules exhibit properties and possess structures which cannot be explained on any one bond picture. For example, hydrogen chloride cannot be represented by either a purely ionic or a purely covalent structure. In fact the formulation $H^{\delta+}—Cl^{\delta-}$ was used to indicate the bond polarity with a resulting slight negative charge on the chlorine atom and a positive charge on the hydrogen atom. Thus with gaseous hydrogen chloride we have a covalent molecule with some ionic character. The valency bond approach to the problem is to postulate a number of hypothetical and entirely imaginary structures (called *canonical structures*), the actual structure being intermediate between all these canonical forms. The canonical structures for hydrogen chloride are:

I H^+Cl^- (ionic) II H—Cl (covalent)

The molecule is then said to be a *resonance hybrid*. The term resonance is, however, rather misleading and does not mean that structures I and II are rapidly interchanging but that the real structure is intermediate in character between I and II.

It follows from quantum mechanics that the energy of a resonance hybrid is always less than the energy of any one of the canonical forms and hence is the most stable form. The difference between the energy of the resonance hybrid and the energy of the canonical form with least energy is called the *resonance energy*. The greater the amount of resonance energy, the more stable is the final resonance form. Resonance energy is not an absolute value and is only relative to the chosen canonical structure with the least energy. The following rules must be applied when choosing canonical structures:

(*a*) Structures must have the same number of unpaired electrons.

(*b*) Structures must have approximately the same energy.

(*c*) The atoms involved in each canonical structure must all have similar positions. Changes in structure which prevent atoms lying in the same position inhibit resonance.

More examples of the application of resonance to some simple molecules are given below.

Water molecule

There are four canonical forms (Figure 3.24(*a*)), the actual structure being a resonance hybrid involving all these forms. This structure explains the existence of a dipole moment in water (1·84 D).

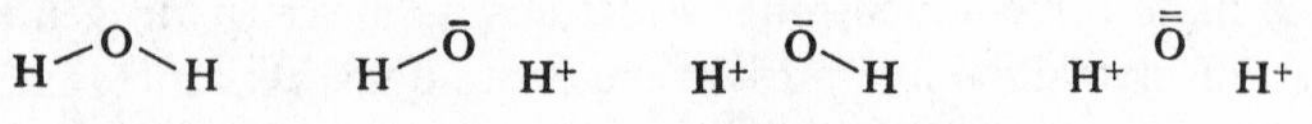

H—O—H (I) H—$\bar{O}$ H^+ (II) H^+ $\bar{O}$—H (III) H^+ $\bar{\bar{O}}$ H^+ (IV)

FIGURE 3.24(a)

Ammonia molecule

Here the actual structure is a resonance hybrid of six canonical forms (dipole moment = 1·48 D). *See* Figure 3.24(*b*)).

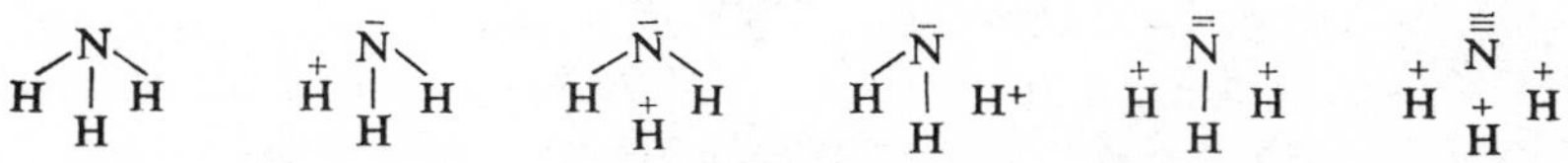

I, II, III, IV, V, VI

FIGURE 3.24(b)

One of the disadvantages of the method is that, for even simple molecules, it is sometimes necessary to postulate several canonical structures and the resulting resonance hybrid does not give a very clear picture of the actual structure.

Carbon dioxide molecule

The molecular orbital bond picture for carbon dioxide was given on p. 39 and the approach indicated two double bonds in the molecule. In fact the bond length between a carbon atom and the oxygen atom in the molecule is 0·116 nm which is intermediate between a double (0·122 nm) and a triple (0·113 nm) bond. The canonical structures in Figure 3.24(*c*) are postulated.

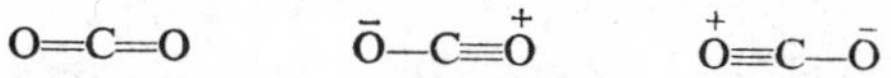

O=C=O (I) $\quad$ $\bar{O}$—C≡$\overset{+}{O}$ (II) $\quad$ $\overset{+}{O}$≡C—$\bar{O}$ (III)

FIGURE 3.24(c)

The dipole moment of carbon dioxide is zero, which would be expected whether the contributing forms were I, II, and III or structure I alone, but as already indicated the bond lengths are shorter than structure I alone would allow. This suggests the existence of resonance and the resonance energy is found to be 134 kJ mol^{-1}.

4

The Periodic Classification

The classification of the chemical elements has a long and tortuous history and only a very short summary is given here. The first classification was made by Lavoisier who divided the elements into two classes, metals and non-metals. Döbereiner and Newlands, using values of recently determined relative atomic masses, noticed a recurrence or periodicity of chemical properties with elements which had:

(*a*) similar relative atomic masses, e.g.

	Fe	Co	Ni
relative atomic mass	55·85	58·93	58·71

(*b*) relative atomic masses which regularly increased, e.g.

	Ca	Sr	Ba
relative atomic mass	40·08	87·62	137·34

The first comprehensive classification of elements was made independently by Mendeléeff in Russia and Lothar Meyer in Germany (1869). They tabulated all the elements known at that time in ascending order of relative atomic masses and a modern version is shown in Table 4.1. A gradation in the properties of elements was noticed in each horizontal row of elements as the relative atomic mass increased. Also many elements in vertical columns had similar chemical properties. Their summarized conclusions were:

(*a*) The relative atomic masses determine the chemical character of an element.

(*b*) Chemically similar elements have either (i) similar relative atomic masses or (ii) relative atomic masses which increase regularly.

(*c*) The most widely diffused elements in nature have small relative atomic masses. For example, aluminium and silicon with relative atomic masses of 26.98 and 28·09 respectively form 82 per cent of the earth's crust.

METALS													NON-METALS					
Period	Gp.IA	Gp.IIA	Gp.IIIA	Gp.IVA	Gp.VA	Gp.VIA	Gp.VIIA	Gp.VIII			Gp.IB	Gp.IIB	Gp.IIIB	Gp.IVB	Gp.VB	Gp.VIB	Gp.VIIB	Gp.O
			Transition metals															
1	1 H 1.0080																1 H 1.0080	2 He 4.003
2	3 Li 6.941	4 Be 9.012											5 B 10.81	6 C 12.011	7 N 14.007	8 O 15.9994	9 F 18.998	10 Ne 20.179
3	11 Na 22.989	12 Mg 24.305											13 Al 26.981	14 Si 28.09	15 P 30.974	16 S 32.064	17 Cl 35.453	18 Ar 39.948
4	19 K 39.102	20 Ca 40.08	21 Sc 44.96	22 Ti 47.90	23 V 50.94	24 Cr 52.00	25 Mn 54.94	26 Fe 55.85	27 Co 58.93	28 Ni 58.71	29 Cu 63.546	30 Zn 65.37	31 Ga 69.72	32 Ge 72.59	33 As 74.92	34 Se 78.96	35 Br 79.904	36 Kr 83.80
5	37 Rb 85.47	38 Sr 87.62	39 Y 88.91	40 Zr 91.22	41 Nb 92.91	42 Mo 95.94	43 Tc 99	44 Ru 101.1	45 Rh 102.91	46 Pd 106.4	47 Ag 107.87	48 Cd 112.40	49 In 114.82	50 Sn 118.69	51 Sb 121.75	52 Te 127.6	53 I 126.90	54 Xe 131.30
6	55 Cs 132.91	56 Ba 137.34	57–71* La 138.91	72 Hf 178.49	73 Ta 180.95	74 W 183.85	75 Re 186.23	76 Os 190.2	77 Ir 192.2	78 Pt 195.09	79 Au 196.97	80 Hg 200.59	81 Tl 204.37	82 Pb 207.19	83 Bi 208.98	84 Po 210	85 At 210	86 Rn 222
7	87 Fr 223	88 Ra 226	89 Ac 227															
Lanthanide series					58 Ce 140.12	59 Pr 140.91	60 Nd 144.24	61 Pm 147	62 Sm 150.35	63 Eu 151.96	64 Gd 157.25	65 Tb 158.92	66 Dy 162.51	67 Ho 164.93	68 Er 167.26	69 Tm 168.93	70 Yb 173.04	71 Lu 174.97
Actinide series					90 Th 232.04	91 Pa 231	92 U 238.03	93 Np 237	94 Pu 242	95 Am 243	96 Cm 247	97 Bk 247	98 Cf 251	99 Es 254	100 Fm 253	101 Md 256	102 No 254	103 Lw 257

TABLE 4.1. The Periodic Table

Core			
Core	1s	1 H 1	2 He 2
$1s^2$	2s	3 Li 1	4 Be 2
$2s^2$ $2p^6$	3s	11 Na 1	12 Mg 2
$3s^2$ $3p^6$	4s	19 K 1	20 Ca 2
$4s^2$ $3d^{10}$ $4p^6$	5s	37 Rb 1	38 Sr 2
$5s^2$ $4d^{10}$ $5p^6$	6s	55 Cs 1	56 Ba 2
$6s^2$ $4f^{14}$ $5d^{10}$ $6p^6$	7s	87 Fr 1	88 Ra 2

3d	21 Sc 1	22 Ti 2	23 V 3	24 Cr 5	25 Mn 5	26 Fe 6	27 Co 7	28 Ni 8	29 Cu 10	30 Zn 10
4d	39 Y 1	40 Zr 2	41 Nb 4	42 Mo 5	43 Tc 6	44 Ru 7	45 Rh 8	46 Pd 10	47 Ag 10	48 Cd 10
5d	57 La 1	72 Hf 2	73 Ta 3	74 W 4	75 Re 5	76 Os 6	77 Ir 7	78 Pt 9	79 Au 10	80 Hg 10
6d	89 Ac 1									

2p	5 B 1	6 C 2	7 N 3	8 O 4	9 F 5	10 Ne 6
3p	13 Al 1	14 Si 2	15 P 3	16 S 4	17 Cl 5	18 Ar 6
4p	31 Ga 1	32 Ge 2	33 As 3	34 Se 4	35 Br 5	36 Kr 6
5p	49 In 1	50 Sn 2	51 Sb 3	52 Te 4	53 I 5	54 Xe 6
6p	81 Tl 1	82 Pb 2	83 Bi 3	84 Po 4	85 At 5	86 Rn 6

4f	58 Ce 2	59 Pr 3	60 Nd 4	61 Pm 5	62 Sm 6	63 Eu 7	64 Gd 7	65 Tb 9	66 Dy 10	67 Ho 11	68 Er 12	69 Tm 13	70 Yb 14	71 Lu 14
5f	90 Th 0	91 Pa 2	92 U 3	93 Np 4	94 Pu 6	95 Am 7	96 Cm 7	97 Bk 8	98 Cf 10	99 Es 11	100 Fm 12	101 Md 13	102 No 14	103 Lw 14

TABLE 4.2. The order of filling atomic orbitals in the periodic table (aufbau principle)

This classification, based upon relative atomic masses, was useful because it stimulated research into new and more accurate methods for determining relative atomic masses and caused the amendment of a large number of hitherto accepted values.

Thus there was no place between carbon and nitrogen for the element beryllium of relative atomic mass 13·5, the only available space in the table which fitted the chemical character of beryllium being between lithium and boron and above the vertical group of elements containing magnesium, calcium, strontium, and barium. Since the relative atomic mass of lithium is 6·94 and that of boron 10·81, the relative atomic mass of beryllium should be approximately $\frac{1}{2}(10{\cdot}81 + 6{\cdot}94) = \frac{1}{2}(17{\cdot}75) = 8{\cdot}87$. Redetermination of the relative atomic mass of beryllium gave a value of 9·01.

The system also predicted the existence and properties of elements not yet discovered. For example in Mendeléeff's original table there were three blank spaces underneath the elements boron, aluminium, and silicon. From the typical properties of the respective vertical groups of elements, he accurately predicted the properties of eka-boron, eka-aluminium, and eka-silicon (eka: underneath). When the elements now called scandium, gallium, and germanium were discovered, Mendeléeff's predictions were found to be correct.

In the periodic table, the horizontal lines of elements are known as *periods*, arranged from left to right in order of increasing relative atomic mass. The vertical columns of chemically similar elements are known as *groups*. Chemical similarity of elements in a group is closest at the two extreme ends of the table with the alkali and alkaline earth metals, the halogens and the noble gases. The elements in the centre of the table are known as the *transition metals* and they have a number of chemical characteristics in common (p. 346). The number above each element in Table 4.1, which was originally thought to be a purely arbitrary number denoting the position of the element in the table, is in fact the *atomic number*.

There are, however, a number of anomalies in the classification based upon relative atomic mass. If elements are arranged in ascending order of relative atomic mass then potassium should precede argon, nickel should precede cobalt, iodine should precede tellurium, and protoactinium should precede thorium. From the point of view of chemical character they are in the correct position: potassium could not be classified as a noble gas, for example. The answer to this problem came from the elucidation of the structure of the atom and the discovery of isotopes (Chapter 1).

Thus it was found that:

(*a*) The arbitrary number (the atomic number) denoting the position of an element in the table was fundamental to the nature of the atom and denoted the number of protons in the nucleus (which is equal to the number of electrons). Hence movement of one place to the right in the table involved the addition of one proton and one electron.

(*b*) Elements in the same vertical group have similar electronic configurations, have the same number of electrons in the outer orbitals, and similar properties. Group 1 (the alkali metals) is an example:

Atomic number	*Symbol*	*Electronic Configuration*
3	Li	$1s^2 2s^1$
11	Na	$1s^2 2s^2 2p^6 3s^1$
19	K	$1s^2 2s^2 2p^6 3s^2 3p^6 4s^1$
37	Rb	$1s^2 2s^2 2p^6 3s^2 3p^6 4s^2 3d^{10} 4p^6 5s^1$
55	Cs	$1s^2 2s^2 2p^6 3s^2 3p^6 4s^2 3d^{10} 4p^6 5s^2 4d^{10} 5p^6 6s^1$

Thus the properties of an element are in periodic dependence upon the atomic number and not on relative atomic mass.

(*c*) The anomalous positions of the pairs of elements K and Ar, Co and Ni, etc., are easily explained from the existence of isotopic forms. Thus potassium consists of three naturally occurring isotopes $^{39}_{19}K$, $^{40}_{19}K$, $^{41}_{19}K$ in fixed proportions so that the atomic weight of potassium is 39·102. Argon, however, has two isotopic forms $^{36}_{18}Ar$ and $^{40}_{18}Ar$ giving a relative atomic mass of 39·948.

In Chapter 3 it was seen that the electrons of an atom are arranged in orbitals. As atoms are built up in order of increasing atomic number electrons are fed into the orbitals in accordance with the aufbau (building-up) principle (*see* Table 4.2):

Period 1 (2 elements): H $1s^1$, He $1s^2$. The $1s$ orbital is full at helium and a new period starts with lithium.

Period 2 (first short period): 8 elements all possessing the helium core $1s^2$. It begins Li $2s^1$, Be $2s^2$. After beryllium, the $2p$ orbital fills from B $2p^1$ to Ne $2p^6$ and the period is complete at neon.

Period 3 (second short period): 8 elements all possessing the neon core $1s^2 2s^2 2p^6$. It begins Na $3s^1$, Mg $3s^2$. After magnesium the $3p$ orbital fills from Al $3p^1$ to Ar $3p^6$ and the period is complete at argon.

Period 4 (first long period): 18 elements all possessing the argon core $1s^2 2s^2 2p^6 3s^2 3p^6$. It begins K $4s^1$, Ca $4s^2$. After calcium, the inner $3d$ orbital fills from Sc $3d^1$ to Zn $3d^{10}$. Note that the filling is not regular as there are anomalies at chromium Cr $3d^5 4s^1$ and copper Cu $3d^{10} 4s^1$. This series of elements is known as the *transition series.* With the $3d$ orbital complete the outer $4p$ orbital fills from Ga $4p^1$ to Kr $4p^6$ and the period is complete at krypton.

Period 5 (second long period): 18 elements all possessing the krypton core $1s^2 2s^2 2p^6 3s^2 3p^6 4s^2 3d^{10} 4p^6$. It begins Rb $5s^1$, Sr $5s^2$. After strontium the inner $4d$ orbital fills from Y $4d^1$ to Cd $4d^{10}$. This is the second series of transition elements. Again a number of anomalies may be noted by consulting the table. With the $4d$ orbital complete the outer $5p$ orbital fills from In $5p^1$ to Xe $5p^6$. The period is complete at xenon.

Period 6 (third long period): 32 elements all possessing the xenon core $1s^2 2s^2 2p^6 3s^2 3p^6 4s^2 3d^{10} 4p^6 5s^2 4d^{10} 5p^6$. This period contains a transition series within a transition series—the *lanthanides*. It begins Cs $6s^1$, Ba $6s^2$. After barium the element La $6s^2 5d^1$ is followed by Ce $6s^2 5d^0 4f^2$ up to Lu $6s^2 5d^1 4f^{14}$ and this latter series is the lanthanide series, involving the development of the inner $4f$ orbital. After lutetium the inner $5d$ orbital fills from Hf $5d^2$ to Hg $5d^{10}$, the normal transition series. With the $5d$ orbital complete the outer $6p$ orbital fills from Tl $6p^1$ to Rn $6p^6$ and the period is complete at radon.

Period 7. This period contains another inner transition series, the *actinides*, and all the elements possess the radon core: $1s^2 2s^2 2p^6 3s^2 3p^6 4s^2 3d^{10} 4p^6 5s^2 4d^{10} 5p^6 6s^2 4f^{14} 5d^{10} 6p^6$. It begins Fr $7s^1$, Ra $7s^2$. Actinium and thorium following radium have electronic structures $7s^2 6d^1$ and $7s^2 6d^2$ respectively. After thorium the $5f$ orbital develops from Pa $5f^2$ to Lw $5f^{14}$.

Group Relationships

The similarity of electronic structures of elements in vertical groups has already been indicated. This structural similarity gives rise to similar chemical properties within a particular group. Progressive differences in properties down a group are due to: (*a*) increase in the magnitude of the positive charge on the nucleus, i.e. atomic number, and (*b*) increase in the radius of the atom.

Group IA

Atomic number	*Symbol*	*Atomic radius (nm)*	*Electronic structure*	*Ionic radius M^+ (nm)*	*First ionization energy ($kJ\ mol^{-1}$)*
3	Li	0·123	(He core) $2s^1$	0·060	525
11	Na	0·157	(Ne core) $3s^1$	0·095	500
19	K	0·203	(Ar core) $4s^1$	0·133	424
37	Rb	0·216	(Kr core) $5s^1$	0·148	408
55	Cs	0·235	(Xe core) $6s^1$	0·169	382

Group VIIB

Atomic number	*Symbol*	*Atomic radius (nm)*	*Electronic structure*	*Ionic radius X^- (nm)*	*First ionization energy ($kJ\ mol^{-1}$)*
9	F	0·072	(He core) $2s^2 2p^5$	0·136	1686
17	Cl	0·099	(Ne core) $3s^2 3p^5$	0·181	1266
35	Br	0·114	(Ar core) $4s^2 3d^{10} 4p^5$	0·195	1146
53	I	0·133	(Kr core) $5s^2 4d^{10} 5p^5$	0·216	1016

The atomic radii increase steadily down a group. Here it is the increase in the number of completed orbitals which results in an increase in radius. As the number of completed inner orbitals increases, more effective screening of the outer electrons from the nucleus occurs. Thus the outer electrons are less strongly bound, i.e. there is an increase in metallic character down a group. This trend becomes much more noticeable for groups in the middle of the periodic table, e.g. Group IVB.

$$\begin{array}{ccccc} \text{C} & \text{Si} & \text{Ge} & \text{Sn} & \text{Pb} \end{array}$$
$$\text{non-metal} \longrightarrow \text{metal}$$

The increase in radius down a group correlates with the decreasing ionization energies. Thus for the alkali metals, Li to Cs, the outer single *s* electron is most readily removed for Cs which is the most metallic element. Fluorine, with the highest ionization potential value for the non-metals, is the most non-metallic element.

As has already been pointed out (Chapter 2): (*a*) the radius of a cation is less than the radius of the parent atom, and (*b*) the radius of an anion is greater than the radius of the parent atom. The ionic radii increase steadily down a group as shown by groups IA and VIIB; the reasons for this are exactly the same as those quoted for atomic radii.

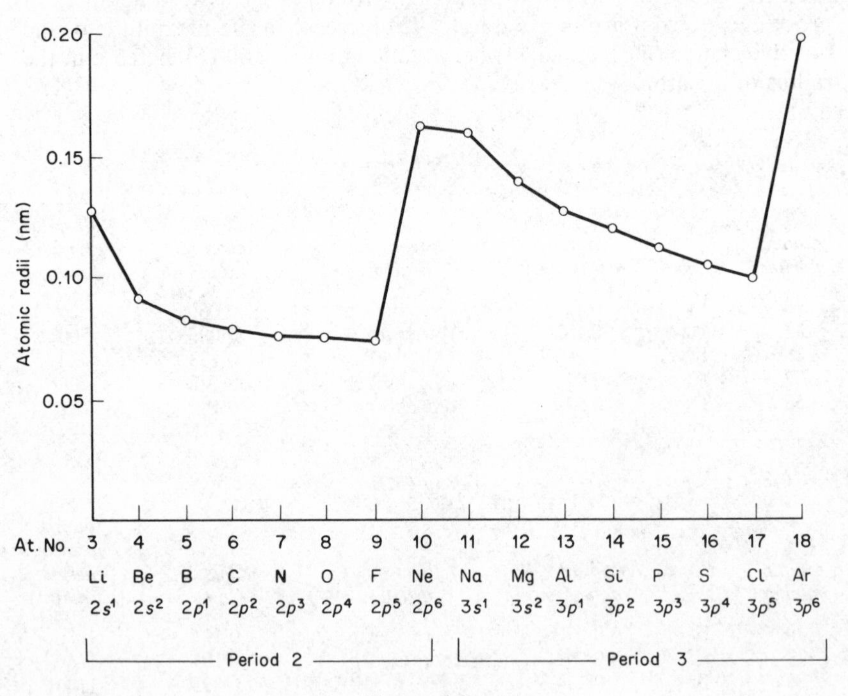

FIGURE 4.1

Periodic Relationships

The atomic radii decrease steadily across each period (Figure 4.1). The addition of one proton and one electron in moving one place to the right causes an increase in the electrostatic attraction between the nucleus and the orbital electrons, hence causing the radii to decrease. The atomic radii are measured from interatomic distances in molecules. The noble gases are

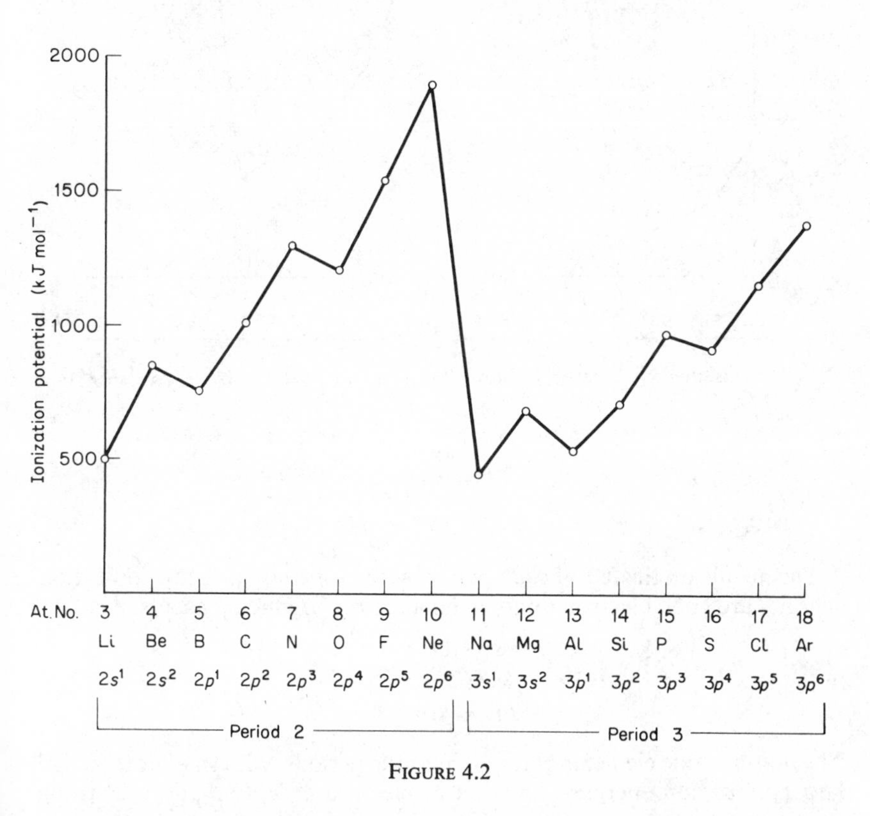

FIGURE 4.2

monatomic and in the solid state are only held by weak van der Waals forces. Thus the measured interatomic distances and calculated van der Waals radii for neon and argon are abnormally high.

Across periods 2 and 3, a contraction in radius correlates with the rise in ionization potential values, but the increase is staggered (Figure 4.2). Abnormally high ionization potential values with beryllium, magnesium, nitrogen and phosphorus are explained on the basis of the extra stability associated with full *s* and half-filled *p* subshells respectively.

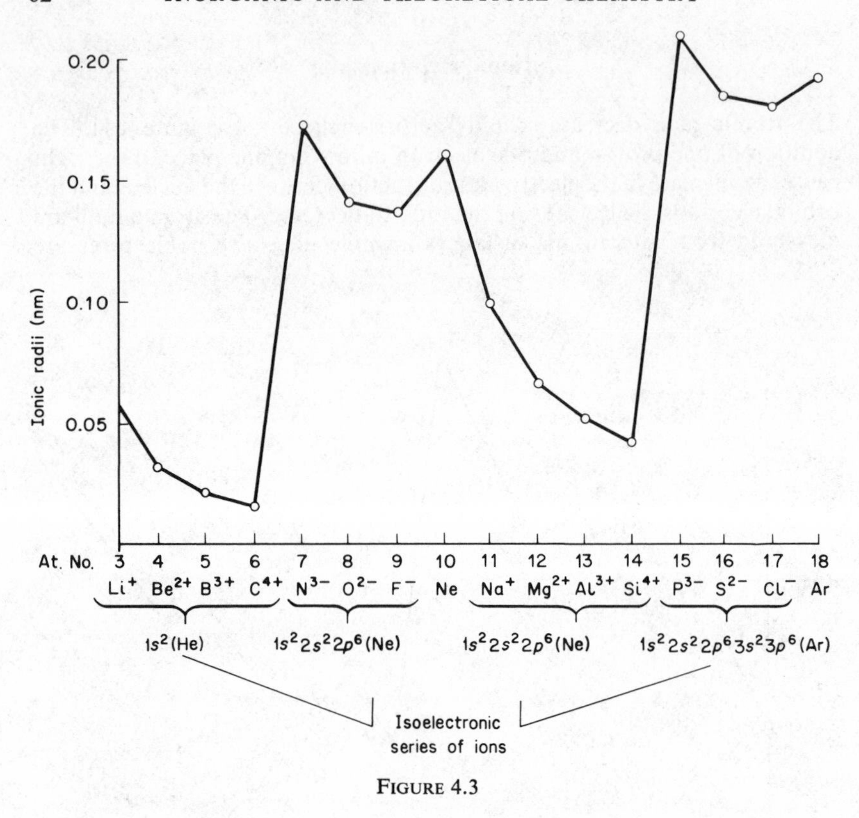

FIGURE 4.3

The metals on the left of each period with large radii and low ionization energies are good electron donors and are strong reducing agents, e.g.

$$\text{Li} \rightarrow \text{Li}^+ + e$$
$$\text{Na} \rightarrow \text{Na}^+ + e$$
$$\text{Al} \rightarrow \text{Al}^{3+} + 3e$$

The non-metallic elements at the right of the periods, with smaller radii and larger ionization energies, function as electron acceptors and are strong oxidizing agents, e.g.

$$\text{F}_2 + 2e \rightarrow 2\text{F}^-$$
$$\text{Cl}_2 + 2e \rightarrow 2\text{Cl}^-.$$

An examination of period 2 shows that the ions Li^+, Be^{2+}, B^{3+} and C^{4+} all have the electronic structure of helium (isoelectronic ions), and an increase in nuclear charge causes a contraction in radius (Figure 4.3). The sudden increase at N^{3-} is due to the addition of three electrons to the neutral atom, giving additional electron repulsion. The ions N^{3-}, O^{2-} and F^- all have the electronic structure of neon (isoelectronic ions), and an increase in nuclear charge causes a contraction in ionic radius. A similar situation can be seen from an examination of the ionic radii of period 3.

The trends in the structures of the elements across each period show some interesting gradations (Figure 4.4). In period 2, the trend is from typical metallic structures (lithium and beryllium), through solid covalent macromolecular structures (boron and carbon) to gaseous covalent diatomic molecules (nitrogen, oxygen and fluorine). In period 3 the trend is more gradual. Sodium, magnesium, and aluminium possess metallic structures,

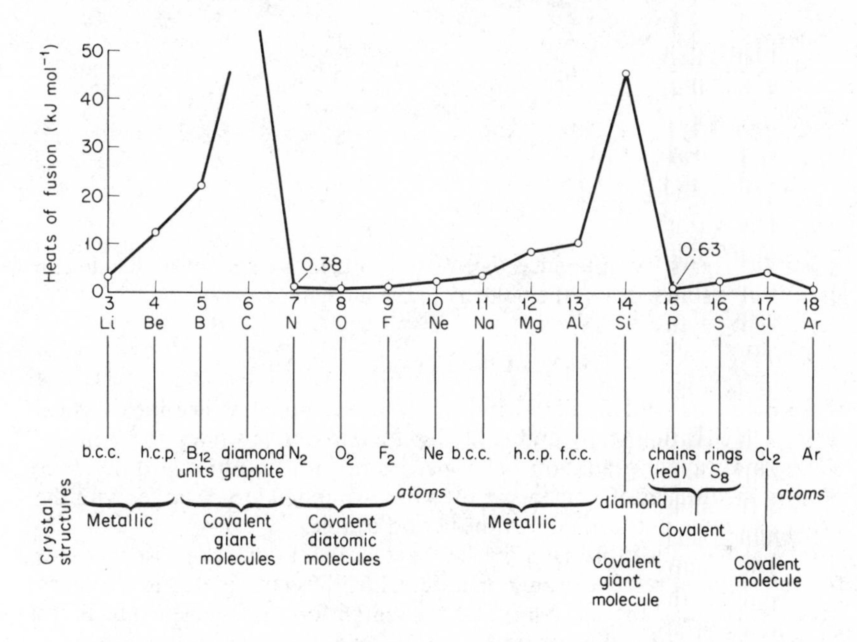

b.c.c. = body centred cubic packing, h.c.p. = hexagonal close packing, f.c.c. = face centred cubic packing

FIGURE 4.4

silicon is macromolecular and covalent with a diamond lattice, red phosphorus consists of long chains of P_4 tetrahedra, sulphur consists of S_8 rings and chlorine has a covalent diatomic molecule. The periodic trends in the heats of fusion can be interpreted in terms of the crystal structures of the elements. Melting points, boiling points, densities, and heats of vaporization also follow a similar pattern.

Electronegativities

The term was defined in Chapter 3. The lowest values are at the left-hand side of the table and the highest values at the right-hand side (Table 4.3).

Large differences in electronegativity values between atoms give rise to ionic compounds, e.g. Na^+Cl^-, $K_2{}^+O^{2-}$, etc. Small differences give rise to covalent compounds, e.g. CO_2, CCl_4.

TABLE 4.3

			H 2.1			
Group IA	IIA	IIIB	IVB	VB	VIB	VIIB
Li 1·0	Be 1·5	B 2·0	C 2·5	N 3·0	O 3·5	F 4·0
Na 0·9	Mg 1·2	Al 1·5	Si 1·9	P 2·1	S 2·5	Cl 3·0
K 0·8	Ca 1·0		Ge 1·7	As 2·0	Se 2·4	Br 2·8
Rb 0·8	Sr 1·0		Sn 1·7	Sb 1·8	Te 2·1	I 2·5
Cs 0·7	Ba 0·9		Pb 1·6	Bi 1·7		

Electronegativity differences between atoms in a covalent molecule can usually give some idea of the polarity in the molecule (p. 47).

Uses of the Periodic Table

(*a*) Elements with similar chemical properties are classified into vertical groups, facilitating study and enabling a comparative survey to be made. There is always a gradation in properties down a group and differences between properties of elements in the same group tend to be larger with the groups situated in the middle of the periodic table.

(*b*) The periodic table has stimulated research into more precise methods for determining relative atomic masses, which has culminated in the mass spectrometer and the ${}^{12}_{6}C$ scale, and the search for missing elements. Since Mendeléeff's original classification forty-one new elements have been discovered.

There are, however, some inherent weaknesses in the system:

(*a*) There is no suitable place for hydrogen, which resembles the alkali metals in forming a positive ion H^+ and the halogens in forming an anion H^-. It also resembles many other non-metallic elements in forming covalent compounds such as H—Cl.

(*b*) The group position usually emphasizes one valency or oxidation state. For example, the alkali metals all have an oxidation state of +1. However, this does not always hold for other groups. Nitrogen in group V shows oxidation states of +5 to −2, i.e.

N_2O_5	NO_2	N_2O_3	NO	N_2O	NH_2OH	N_2H_4
+5	+4	+3	+2	+1	−1	−2

(*c*) The lanthanide and the actinide elements have properties so similar that they cannot be included without considerable expansion of the table. These elements are usually included separately, as shown in Tables 4.1 and 4.2.

5

Some Structural Considerations

The structure of chemical compounds has been elucidated by the application of advanced physical techniques. A summary of some of the more important experimental methods is given below.

X-ray Diffraction

When electrons from a hot filament strike a copper metal target, X-rays are produced. The rays emitted are not all of the same wavelength and a nickel filter is necessary to cut out extraneous radiation. X-rays have wavelengths of the same order as interatomic distances in crystals (0·1 nm) and hence the atoms or ions in a crystal can function as a diffraction grating and produce an X-ray spectrum. The close-packed atoms or ions in a crystal can be regarded as being arranged in a series of parallel and equidistant planes. The X-ray reflections obtained from these planes exhibit maximum or minimum intensity depending upon whether the reflecting waves are in

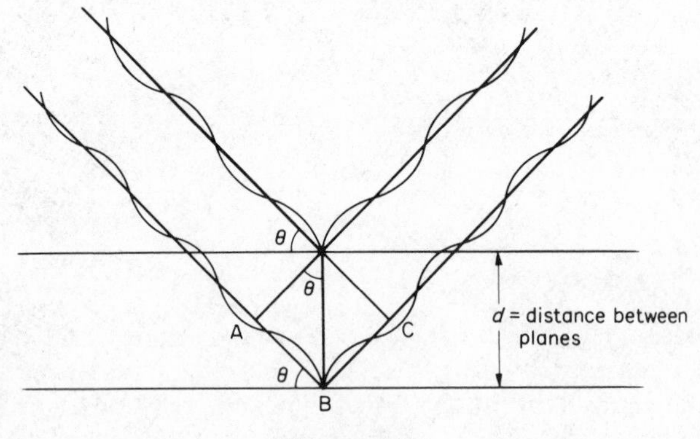

FIGURE 5.1

phase (constructive interference) or out of phase (destructive interference). Consider a homogenous beam of X-rays (wavelength λ) incident to a set of parallel crystal planes at an angle θ. (Figure 5.1). The path difference of the rays between successive layers is AB + BC, i.e. $2d\sin\theta$. For constructive interference $2d\sin\theta$ must be equal to a whole number of wavelengths.

$$n\lambda = 2d\sin\theta \qquad (n = 1, 2, 3, 4, \ldots)$$

This equation is known as the Bragg equation and is the fundamental basis for all structural determinations using X-rays.

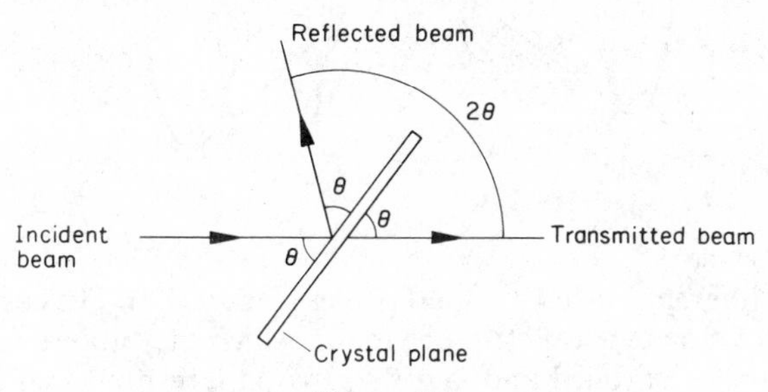

FIGURE 5.2

The powder (Debye Scherrer) method is the technique most commonly used in inorganic chemistry for the determination of crystal systems and unit cell dimensions. The crystals to be examined are reduced to a fine powder and placed in a beam of X-rays. The reflection of an X-ray beam from a single crystal plane is shown in Figure 5.2. If the reflecting plane is rotated about the incident beam so that θ remains constant, the reflected beam will travel over the surface of a cone (Figure 5.3).

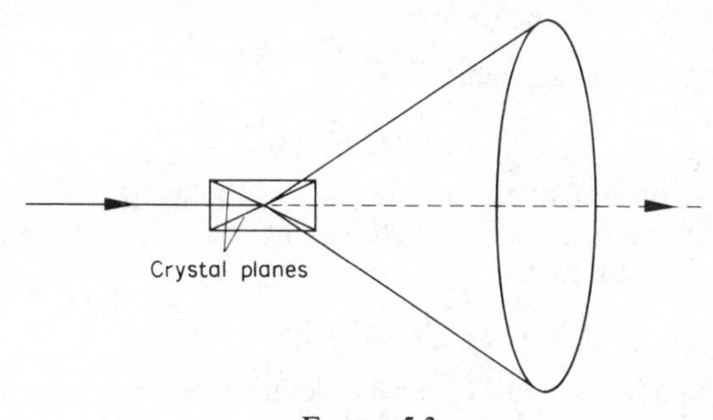

FIGURE 5.3

This rotation is not necessary for a powder since it contains millions of randomly orientated crystals which produce the same effect as a single crystal rotated about all possible axes. Thus the reflections obtained from the powder consist of cones of radiation and a separate cone is formed from

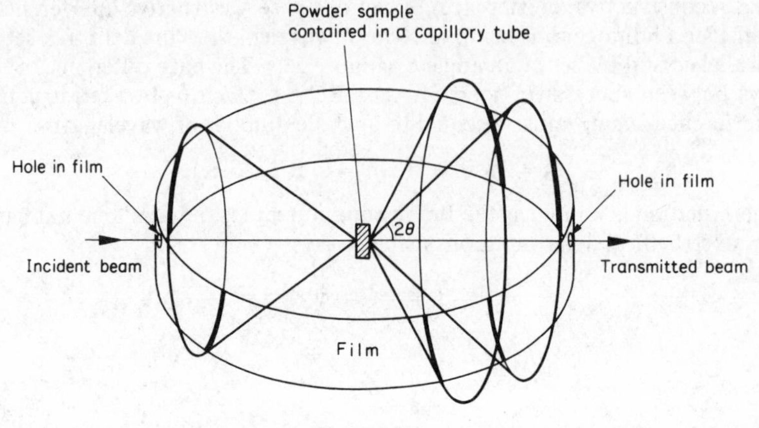

FIGURE 5.4

each set of differently spaced crystal planes (Figure 5.4). The cones of radiation intersect a cylindrical strip of film on which they appear as lines and, when the film is unrolled and laid flat, the pattern produced is shown in Figure 5.5.

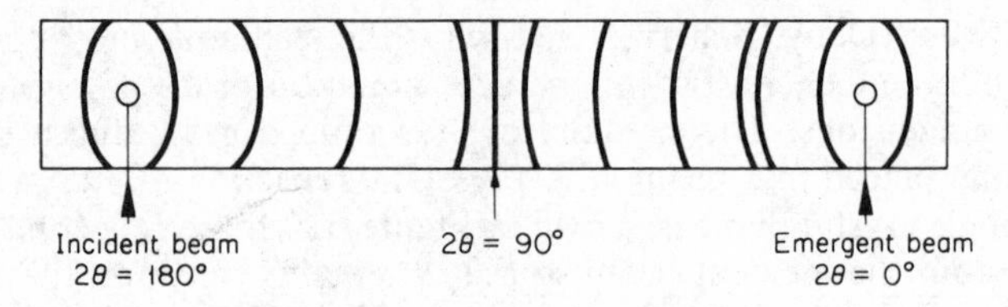

FIGURE 5.5

According to the Bragg equation,

$$n\lambda = 2d \sin \theta \tag{1}$$

If the diameter of the camera containing the film is $2R$ and S is the distance from a particular diffraction line to the point at which $2\theta = 0$ measured from the film (Figure 5.6), then

$$S = R \times 2\theta \tag{2}$$

where θ is in radians. Thus θ can be determined from equation (2) and, knowing the wavelength λ, the d spacings can be calculated from equation (1).

An examination of the $\sin \theta$ and d values reveals the type of unit cell to which the crystal belongs: cubic, hexagonal, tetragonal etc. Mathematical relationships have been derived for each type of unit cell relating the d or $\sin^2 \theta$ values to the unit cell dimensions which can subsequently be evaluated.

Once the unit cell dimensions are known, the volume of the unit cell can be deduced. The density of the solid (ρ) can be determined by standard pyknometric methods and since

$$\rho = \frac{\text{weight of atoms (molecules) in the unit cell}}{\text{volume of unit cell}}$$

$$= \frac{\text{no. of atoms (molecules)} \times \text{mass of one atom (molecule)}}{\text{volume of unit cell}}$$

$$= \frac{\text{no. of atoms (molecules) in unit cell}}{\text{volume of unit cell}} \times \frac{\text{relative atomic (molecular) mass}}{\text{Avogadro number}}$$

Thus the number of atoms present in the unit cell can be calculated.

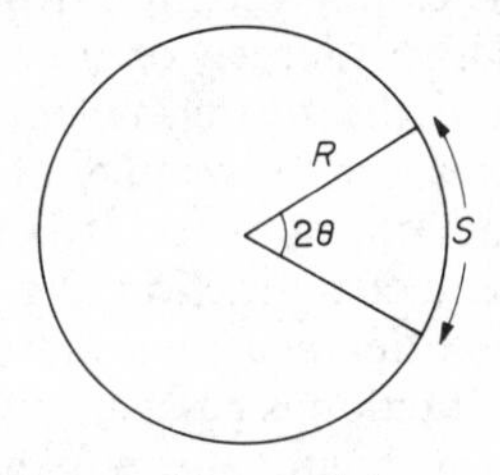

FIGURE 5.6

The unit cell for sodium chloride is cubic with a unit cell edge of 0·5641 nm. The relative molecular mass is 58·442 and the density is 2·165 g cm^{-3}. The Avogadro number (N_A) = 6·023 × 10^{23} mol^{-1}.

$$\text{Volume of unit cell} = (0{\cdot}5641 \times 10^{-7})^3 \text{ cm}^3$$

$$2{\cdot}165 = \frac{x}{(0{\cdot}5641 \times 10^{-7})^3} \times \frac{58{\cdot}44}{6{\cdot}023 \times 10^{23}}$$

$$x = 4{\cdot}004$$

The sodium chloride unit cell has four molecules, i.e. eight ions per unit. Although the unit cell for sodium chloride shown on p. 94 has 27 ions, it must be noted that, apart from the ion in the centre of the cell:

(*a*) six ions in the centre of each face are common to one other unit cell.

(*b*) eight ions at the corners of the cell are common to eight other unit cells.

(*c*) twelve ions in the middle of each edge are common to four other unit cells.

$$\text{The unit cell content} = 1 + 6 \times \tfrac{1}{2} + 8 \times \tfrac{1}{8} + 12 \times \tfrac{1}{4}$$
$$= 8$$

It will be appreciated that, if all the crystalline data is available, the method can be used to evaluate the Avogadro constant (N_A).

For a complete structural investigation, or for crystals of very low symmetry where the powder pattern is very complex, the rotating crystal method is used. A small crystal is exposed to a beam of X-rays and slowly rotated, bringing various planes of atoms into position for diffraction to occur and producing a corresponding series of spots on the photographic film. Rotations are made about the three principal axes of the crystal and, from the pattern and intensity of the spots obtained, the three-dimensional structure of the crystal may be computed.

Neutron Diffraction

De Broglie's equation, $\lambda = h/mv$, can be applied to neutrons as well as to electrons, and the associated wavelengths are found to be of the order 0·1–0·2 nm. Thus neutrons can be used for structural investigations. The scattering of X-rays depends very largely on the number of orbital electrons of the scattering atom, and the scattering power of an atom increases proportionally with an increase in the number of electrons, i.e. with atomic number. In neutron diffraction, however, it is scattering by the nucleus which determines the extent of diffraction and some light elements, e.g. hydrogen and beryllium scatter neutrons more effectively than heavier elements. Thus neutron diffraction is particularly useful for the location of protons in a compound, a technique not possible with X-rays due to the weak scattering. It is also possible, using neutrons, to discriminate between atoms of approximately the same atomic number (e.g. iron, cobalt and nickel) which is again not possible with X-rays. Although the work is limited because a strong beam of neutrons requires the use of an atomic pile, neutron diffraction has proved to be a valuable complement to X-ray work.

Electron Diffraction

This method depends upon the wave property of the electron and has been used to study the structure of gases, metallic films, and colloids. Although

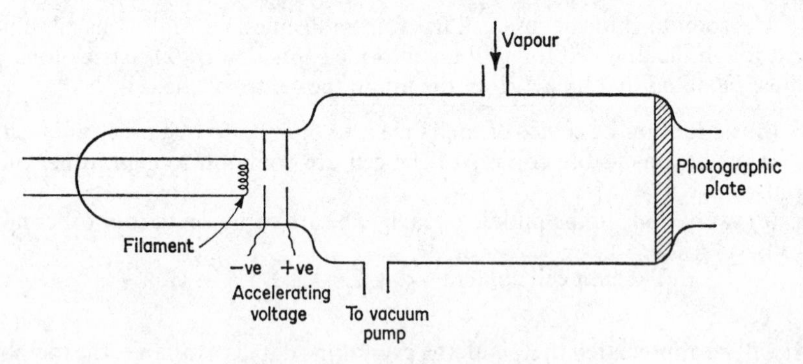

FIGURE 5.7. Electron diffraction

the penetrating power of electrons is low the intensity of the diffracted beam is high and this gives a much more detailed picture of surface structure. A beam of electrons from a filament is accelerated by a potential of 50 kV and meets a stream of vapour at low pressure (Figure 5.7). The diffracted beam falls on to a photographic plate which is removed and developed.

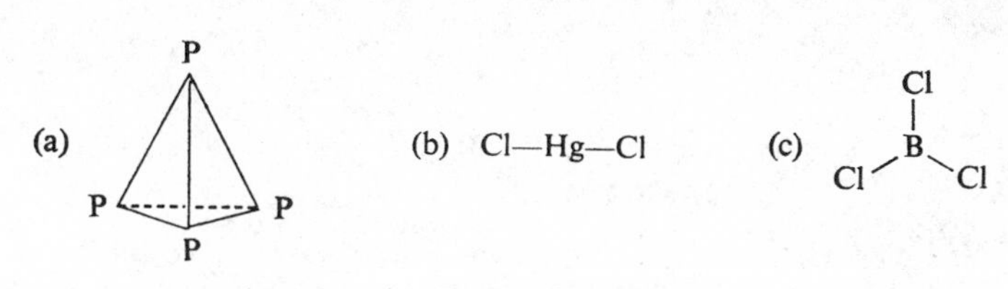

FIGURE 5.8

A series of light and dark concentric rings round a central spot is produced. Analysis of the intensity and position of these rings gives information regarding molecular configuration and bond distance. Thus it has been shown (Figure 5.8) that (*a*) a molecule of yellow phosphorus is tetrahedral; (*b*) a molecule of mercury(II) chloride is linear; and (*c*) a molecule of boron trichloride is triangular.

The Electromagnetic Spectrum

The various regions of the electromagnetic spectrum (Figure 5.9) are not sharply divided. The various types of radiation may be identified in terms of:

(*a*) wave numbers ($\bar{\nu}$), which are the reciprocal of the wavelength and which are expressed in metres^{-1} (m^{-1}).

$$\bar{\nu} = \frac{1}{\lambda}$$

(*b*) frequency (ν) which is the number of waves per second passing a given point expressed as Hertz (Hz).

$$\nu = \frac{c}{\lambda}$$

where c is the velocity of light. Thus a wavelength of 5×10^{-6} m is equivalent to a wave number ($\bar{\nu}$) of $2{\cdot}0 \times 10^5$ m^{-1} and a frequency (ν) of $6{\cdot}0 \times 10^{13}$ Hz.

The absorption frequencies in infrared spectroscopy are still quoted as wave numbers expressed in cm^{-1}. Thus wave numbers quoted in the text for infrared spectra conform with this procedure, e.g. $2{\cdot}0 \times 10^5$ m^{-1} = 2000 cm^{-1}.

The absorption of radiant energy by molecules can cause (*a*) electronic transitions, (*b*) vibrational energy changes, (*c*) rotational energy changes.

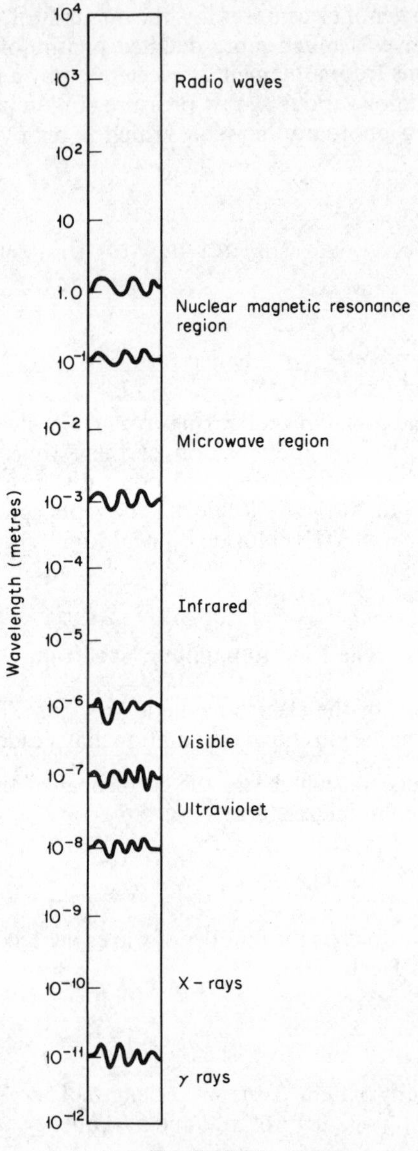

FIGURE 5.9

The energy absorbed during vibration and rotation is less than that absorbed during electronic transitions. Thus absorption in the microwave region gives rise to rotational energy changes, in the infrared region vibrational energy changes, and in the ultraviolet and visible all three types of energy change.

Infrared Spectroscopy

Molecular vibrations correspond in frequency to the infrared region of the spectrum and, when radiation of resonant frequency is incident on a molecule, vibration is stimulated and energy is lost from the incident radiation. There must, however, be a change in the dipole moment of the molecule during vibration before absorption of this type can occur. The oscillating dipole can then interact with the electric field of the radiation, enabling energy to be transferred from the radiation to the vibrating

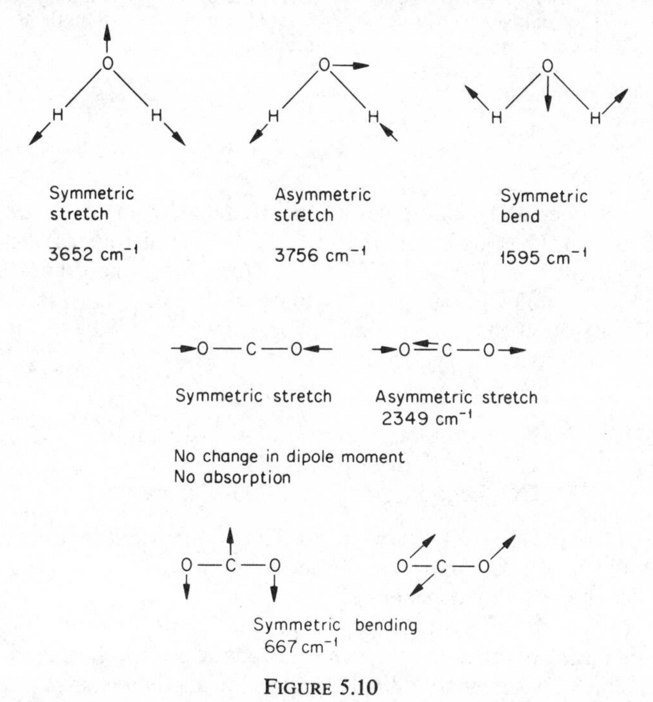

FIGURE 5.10

molecules. Thus hydrogen, nitrogen, and oxygen with zero dipole moment will not be infrared active.

It can be shown that, for a molecule containing n atoms, there are $3n-6$ ($3n-5$ for a linear molecule) stretching and bending vibrations. The vibrations for the carbon dioxide and water molecules and the observed absorptions in the infrared are shown in Figure 5.10.

One of the possible vibrations in carbon dioxide does not involve a change in dipole moment and therefore infrared absorption does not occur. The two bending vibrations only differ in being at right angles and give rise to single absorption at 667 cm^{-1} and are said to be doubly degenerate.

The frequency of absorption in the infrared depends upon the mass of the atoms linked by the chemical bond and the strength of that bond. The effect of these two factors on the infrared absorption frequency can be seen in Table 5.1.

TABLE 5.1

C–halogen vibration		*Wave no. (cm^{-1})*	*C–oxygen vibration*		*Wave no. (cm^{-1})*
C–Cl	Increase	800–600	—C—O (ethers)	Increase	1150–1060
C–Br	in size	700–500	—C=O (ketones)	in bond	1850–1700
C–I	of halogen	600–400	C≡O (carbon monoxide)	strength ↓	2143
	decrease in bond strength ↓				

Applications and examples

Gaseous nitrogen(IV) oxide gives an infrared spectrum with three vibrational bands at 750, 1323, and 1616 cm^{-1} whereas nitrogen(I) oxide only shows two bands at 1725 and 2250 cm^{-1}. Some rotational bands are also observed since molecules are free to rotate in the gaseous state. From the above discussion, it can be seen that nitrogen(IV) oxide must be an angular and nitrogen(I) oxide a linear molecule.

The infrared spectrum of the carbonates shows three bands:

CO asymmetric stretch 1450–1410 cm^{-1}
CO_2 bend out of plane 880–850 cm^{-1}
CO_2 bend in plane 720–680 cm^{-1}

For the carbonate ion (CO_3^{2-}) $3n-6=6$. The symmetric stretching vibration is not infrared active. The asymmetric stretching and one of the bending modes are each doubly degenerate.

It is possible to differentiate between the different forms of calcium carbonate by means of infrared spectra e.g. aragonite gives a characteristic band with very fine structure in the 880–850 region. Bicarbonates show frequencies due to bonded —OH groups above 3000 cm^{-1} and also give two bands in the regions 665–655 and 710–690 cm^{-1} but do not absorb in the 1430 region. Basic carbonates give complex spectra. The 1430 cm^{-1} band is a doublet and there are a number of bands in the region 1110–700 cm^{-1}. They also show frequencies due to bonded —OH groups. Figure 5.11 shows the recorded infrared spectrum of calcium carbonate (calcite form).

Crystalline hydrates absorb strongly in the region 1670–1600 cm^{-1}. This band can be used as a means of identifying water of crystallization. Thus there are two possible formulations for sodium stannate: $Na_2SnO_3.3H_2O$ or $Na_2Sn(OH)_6$. Chemical studies show that water cannot be driven off without decomposition. Sodium stannate shows bands at 3597 cm^{-1}

consistent with —OH stretching and bands below 576 cm^{-1} due to Sn–OH stretching. No band is observed in the region 1670–1600 cm^{-1}. Hence the correct formulation for sodium stannate is $Na_2Sn(OH)_6$. For very complex

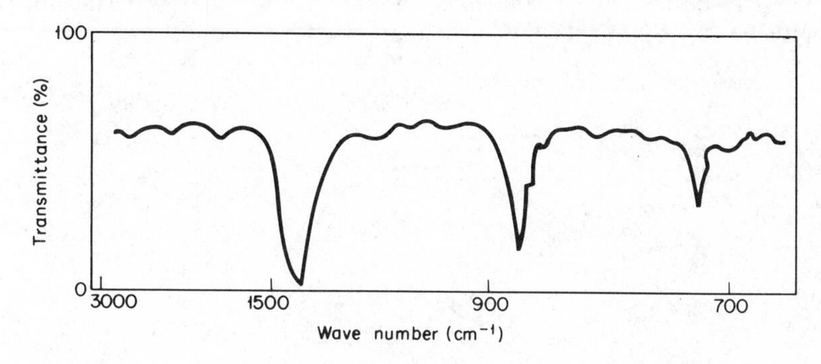

FIGURE 5.11

molecules, it is rarely possible to determine the structure wholly from a consideration of infrared spectra, but because a very large collection of spectra is now available these can act as 'fingerprints' for the identification of unknown substances.

Instrumentation

The source of infrared radiation is an electrically heated filament usually made from rare earth metal oxides. The system usually employs a combination of a prism and grating to resolve the incident light into its spectrum. Glass prisms do not transmit infrared radiation and those employed are usually made of NaCl, KBr, CsBr, As_2Se_3, or As_2Te_3. Reflectance optics are used to limit radiation losses and to avoid the use of fragile NaCl, KBr, and CsBr lens systems. Most instruments employ a double beam system (Figure 5.12) to cancel out interfering bands, e.g. CO_2 and H_2O in air, etc. The source beam is split into two, a sample and a reference beam, and passes pulses of each alternately to the spectrometer. Two other features are also important:

(*a*) A system of automatically adjustable slits is required to ensure uniformity in the intensity of the radiation over the various wavelengths employed.

(*b*) A servo system of recording is used. On absorption of radiation in the sample beam, a comb moves automatically into the reference beam under the driving force of the servomechanism and reduces its energy until the two beams are again equal. The movement of the comb is a measure of the absorption by the sample and can be linked mechanically to the pen of the recorder.

Liquids can be examined as thin films between NaCl or KBr plates. Solids can be melted, ground to a fine mull with Nujol or hexachlorobutadiene or ground with KBr and compressed into a disc.

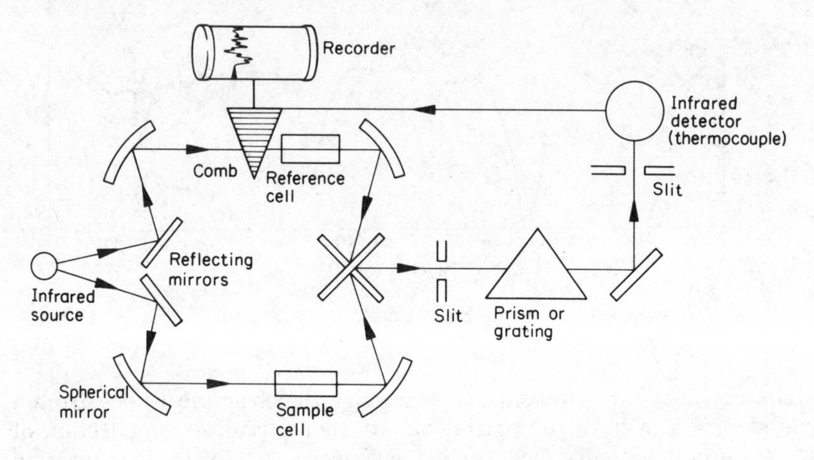

FIGURE 5.12

Magnetic measurements

A spinning electron generates a magnetic field and if two electrons are paired (i.e. have opposite spins), the resultant magnetic field produced by them is zero. If an atom, ion, or molecule possesses unpaired electrons, there will be a resultant magnetic field and the species will possess a magnetic moment measured in Bohr magnetons (1 Bohr magneton $= 9{\cdot}27 \times 10^{-24}$ A m^2). In a substance with unpaired electrons, the resultant spin moment aligns itself with an externally applied magnetic field resulting in a slight attraction and the substance is said to be paramagnetic. Materials which are

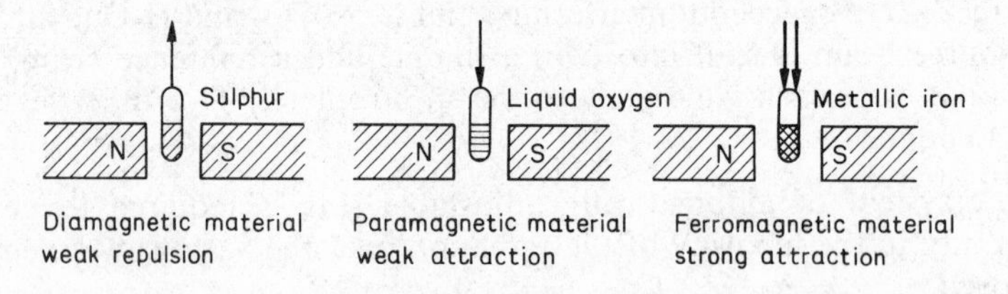

FIGURE 5.13

attracted strongly into a magnetic field are said to be ferromagnetic. This is an extreme form of paramagnetism which is due to the aggregation of atoms or ions and is not the property of an individual species, e.g. metallic iron is ferromagnetic whereas the iron(III) cation (Fe^{3+}) is paramagnetic. Compounds which are weakly repelled from a magnetic field are classed as diamagnetic and contain no unpaired electrons. In the presence of an applied field, the planes of the full orbitals tilt slightly so that a small orbital moment is set up in opposition to the field causing repulsion (Figure 5.13). It is important to note that paramagnetism due to a single unpaired electron largely outweighs the diamagnetism due to all the paired electrons.

Magnetic measurements are made by attaching the sample in a sealed tube to one arm of a microbalance and suspending it between the poles of an electromagnet. The change in weight of the sample is then determined by counterbalancing with weights before and after the application of a magnetic field (Figure 5.14). An identical sequence of operations is also carried out using a calibrating substance, e.g. mercury(II) cobalt(II) thiocyanate, $HgCo(NCS)_4$. From the measured changes in weight, and knowing the force exerted by the magnetic field per unit mass, a quantity termed the paramagnetic susceptibility (χ_M) for $HgCo(NCS)_4$, the paramagnetic susceptibility of the sample can be evaluated and finally corrected for the diamagnetism due to all the paired electrons. The paramagnetic susceptibility of the sample is related to the magnetic moment (μ_{eff}) by the equation

$$\mu_{eff} = 7{\cdot}98 \times 10^2 \left[\frac{\chi_M MT}{\rho}\right]^{1/2}$$

where M is the relative molecular mass (kg mol^{-1}), ρ is the density (kg m^{-3}), T is the temperature (K).

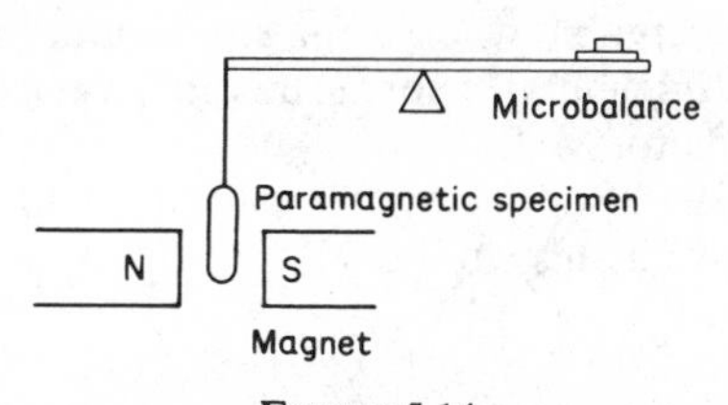

FIGURE 5.14

Studies of this kind indicate that the molecules of NO, NO_2, ClO_2 and O_2 are paramagnetic and possess one or more unpaired electrons. This gives an insight into the nature of the chemical bond between the atoms. The method may also be applied to distinguish between two valency states of an element, e.g. salts of the copper(I) ion (electronic configuration $3d^{10}$) are diamagnetic but those of the copper(II) ion ($3d^9$) are paramagnetic.

It can also be shown theoretically that the spin only magnetic moment ($\mu_{s.o.}$) is related to the number of unpaired electrons by the expression

$$\mu_{s.o.} = [n(n+2)]^{1/2}$$

Experimental (μ_{eff}) and calculated ($\mu_{s.o.}$) magnetic moments for transition metal cations show good agreement before any electron pairing takes place, e.g. $V^{2+}(3d^3)$, $Cr^{2+}(3d^4)$, $Fe^{3+}(3d^5)$. However, when electron pairing occurs ($Fe^{2+}(3d^6)$, $Co^{2+}(3d^7)$, $Ni^{2+}(3d^8)$), μ_{eff} is greater than $\mu_{s.o.}$ due to the orbital motion of the electrons contributing to the magnetic moment. Thus the spin only equation ($\mu_{s.o.}$) is not valid. Before electron pairing occurs, it appears that the orbital moments are almost completely quenched by the fields set up by neighbouring ions.

Nuclear Magnetic Resonance

In a nucleus containing an odd number of protons, the odd unpaired proton spins and hence causes the nucleus to behave like a miniature spinning magnet. Thus the nucleus has an intrinsic magnetic moment which is a vector quantity possessing magnitude and direction. In presence of an external magnetic field, the magnetic field due to the nucleus can align itself to the field (low energy state) or against it (high energy state). Thus the presence of an external magnetic field establishes the existence of two energy levels E_1 and E_2 where

$$E_1 - E_2 = \frac{h\gamma B}{2\pi}$$

where γ is the constant for the nucleus, B is the magnetic flux density, h is Planck's constant.

In the past, nuclear magnetic resonance (NMR) spectroscopists have incorrectly used H (magnetic field strength) instead of B (magnetic flux density). This was acceptable when using e.m.u. units since H and B both have the same magnitude, but it is not permissible when using SI units where H and B differ by a factor of $4\pi \times 10^{-7}$.

Since $E_1 - E_2 = h\nu$, the frequency of radiation which will induce transitions between the two states will be

$$\nu = \frac{\gamma B}{2\pi}$$

Thus for an applied flux density of 1 tesla, the frequency (ν) will be of the order of 10^7 Hz (10 MHz).

The radiations required for determining the nuclear magnetic resonance frequency are in the radio frequency range. Thus when certain nuclei are subjected to radiation in this range, energy is absorbed as a result of transitions from lower to higher nuclear energy levels. The flux density B is altered and the radio frequency is kept constant until absorption takes place. The absorption signal is picked up on an oscilloscope and transmitted to an automatic recorder (Figure 5.15). The conditions required to produce absorption will depend to some extent on the environment of the nucleus and, from the spectrum obtained, conclusions can be drawn about this

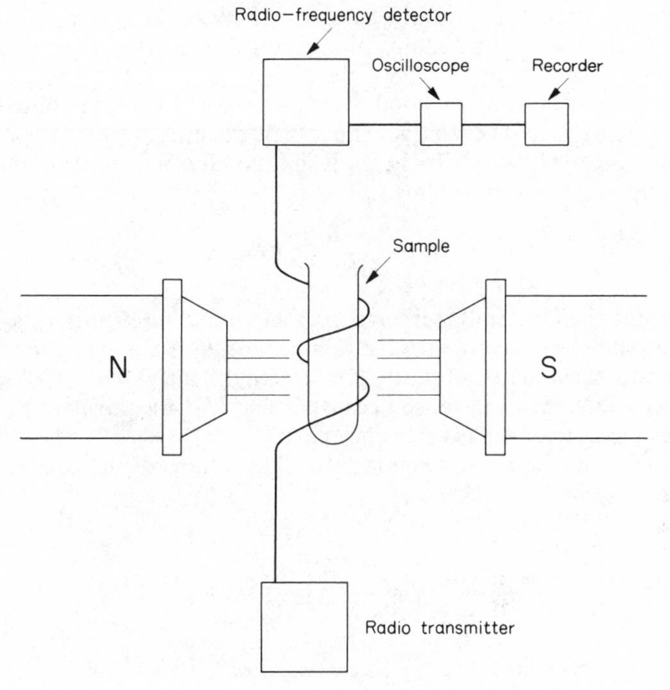

FIGURE 5.15

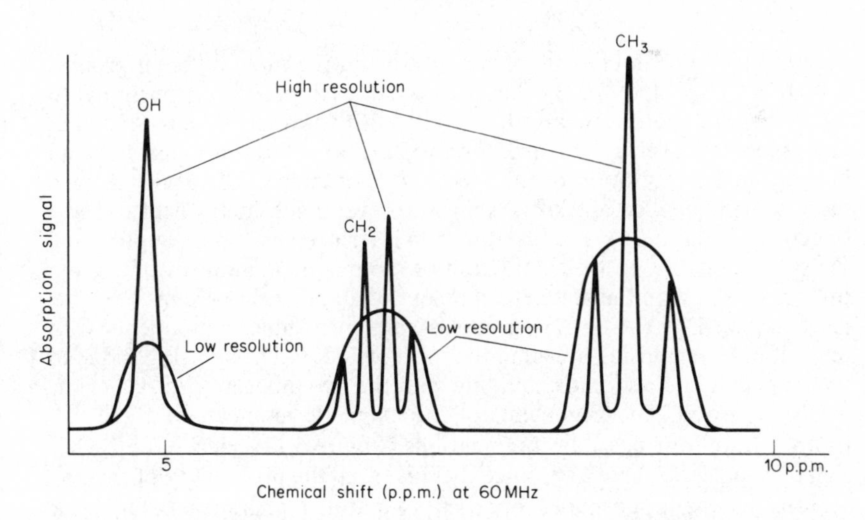

FIGURE 5.16

environment. The technique has been successfully applied to the case of the proton. Thus the hydrogen atoms of the methyl, methylene, and hydroxyl groups of ethanol (CH_3CH_2OH) absorb energy of radio frequency at different values of flux density and the low resolution spectrum obtained is shown in Figure 5.16. The relative change in flux density required to produce absorption is called the NMR chemical shift and the magnitude of the shift depends upon proton environment.

$$\delta = \frac{B_r - B_s}{B_r} \times 10^6$$

where δ is the chemical shift measured in parts per million, B_s is the magnetic flux at resonance absorption for the sample, and B_r is the magnetic flux at resonance for a standard substance. Tetramethylsilane (TMS), $(CH_3)_4Si$, is chosen as a reference substance because it has twelve equivalent protons and gives a single very sharp reference peak.

The spectrum of ethanol (Figure 5.16) shows three distinct peaks with

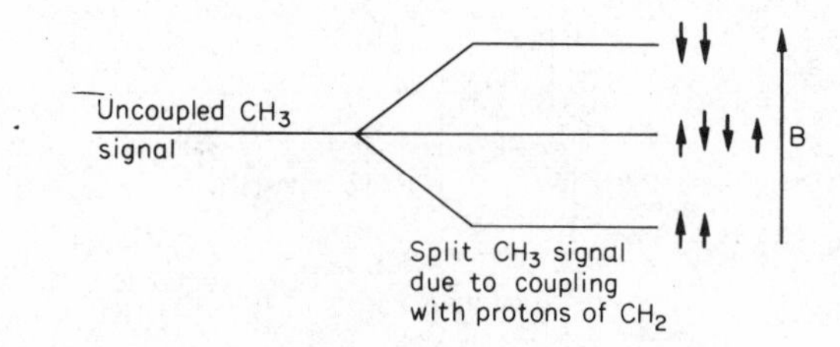

FIGURE 5.17

chemical shifts characteristic of the protons in the three different groups, —OH, —CH_2— and —CH_3. The area under each peak is proportional to the number of protons involved, i.e. 1:2:3. If the sample tube is rotated at high speed to average out any non-uniformity of the applied flux, an increase in the resolution of the spectrum is obtained. The peaks become more sharply defined and often split into several subsidiary peaks of fine structure (Figure 5.16). The phenomenon is termed spin–spin coupling and the splitting is due to the interaction of the magnetic moment of a given proton system with the magnetic moment of an adjacent system. Thus the peak assigned to the —CH_3 group is split into three components with intensities (or enclosing areas) in the ratio of 1:2:1. The splitting is due to the magnetic interaction or coupling with the two adjacent protons of the —CH_2— group. The alignments of the magnetic moments of these two protons have four possible combinations (Figure 5.17). Hence the —CH_3 signal is split into three and, since there is twice the probability of the two protons having opposite as opposed to parallel alignments, the peak intensities are in the ratio 1:2:1.

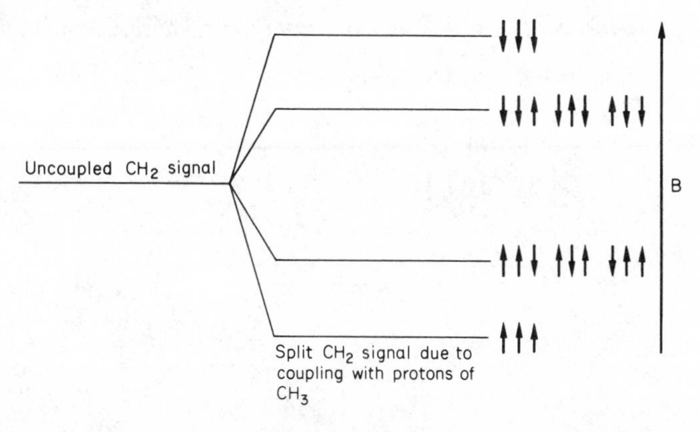

FIGURE 5.18

The peak assigned to the —CH_2— group is split into four with intensities in the ratio of 1:3:3:1. Here the splitting is due to the magnetic coupling of the three protons in the adjacent —CH_3 group. There are eight possible combinations for the alignment of the magnetic moments of three protons (Figure 5.18). Hence the —CH_2— signal is split into four components. The —OH signal should be split into a triplet by the —CH_3 group and the —CH_2— signal into a doublet by the proton on the —OH group (Figure 5.19). In fact this does occur with very pure ethanol with no basic or acidic impurities. A triplet is obtained for the —OH signal and the combined coupling of —CH_3 and —OH on the —CH_2— signal splits it into an octet. However, the NMR spectrum for ethanol is usually carried out in weakly acidic or basic solution. In this medium, very rapid proton exchange occurs with the proton in the —OH group and the exchange is so fast that alignment of the magnetic moment in the applied field does not occur. Thus splitting of the —CH_3 signal by the —OH group and the —OH signal by the —CH_2— group is not obtained, and the —OH and —CH_2— signals remain a singlet and quadruplet respectively.

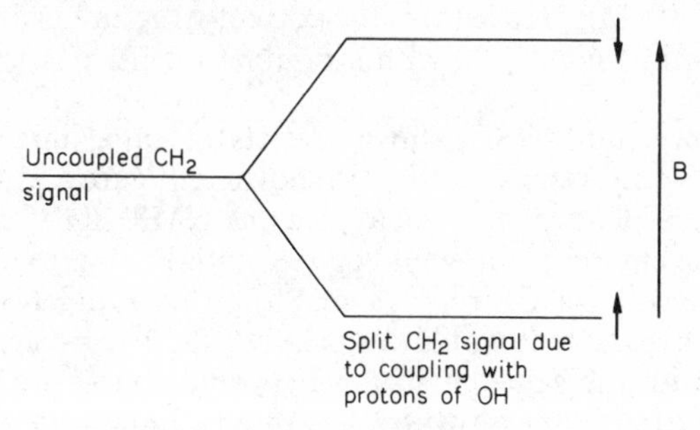

FIGURE 5.19

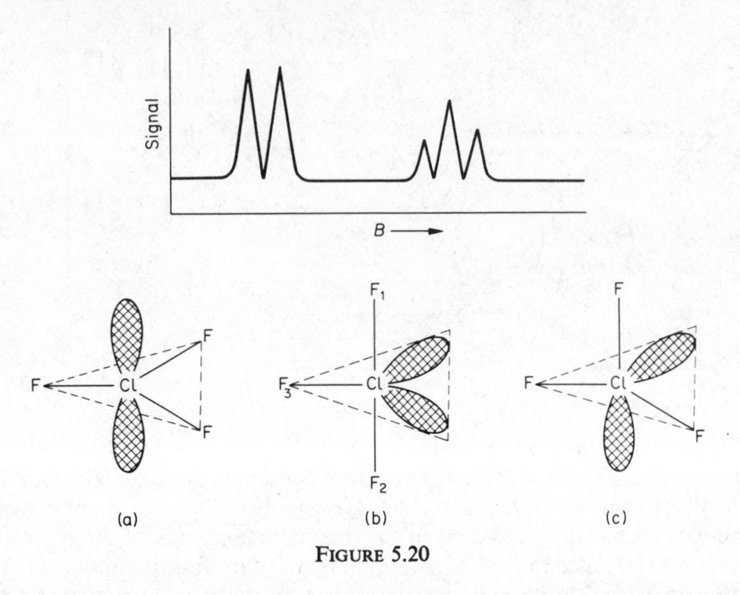

FIGURE 5.20

The NMR spectrum for fluorine in ClF_3 is shown in Figure 5.20. The spectrum consists of a doublet and a triplet, the triplet intensity being half that of the doublet. Although hydrogen atoms are not involved here, fluorine (atomic number 9) possesses an odd number of protons and gives an NMR spectrum. The signals due to chlorine occur at a different flux density and are not shown. Since the chlorine atom in ClF_3 possesses two lone pairs, three structures are possible for ClF_3 (Figure 5.20). In (*a*), all the fluorine atoms have the same environment and only one NMR signal would be observed. In (*b*), F_1 and F_2 with the same environment would couple with F_3 to give a triplet and F_3 would couple with F_1 and F_2 to give a doublet. If lone pair–lone pair repulsion is considered, structure (*c*) can be regarded as being identical with (*b*). Hence (*b*) is the correct structure giving the NMR spectrum shown in Figure 5.20 with a doublet intensity twice that of the triplet.

The NMR spectrum of SF_4 shows triplets of equal intensity (Figure 5.21). The possible structures for SF_4 are shown in Figure 5.21. In (*a*) all the fluorine atoms are identical and would give one NMR signal. In (*b*) F_1, F_2, F_3 with the same environment would couple with F_4 to give a quadruplet. F_4 would couple with the F_1, F_2, F_3 signal to give a doublet. In (*c*), F_1, F_4 with the same environment would couple with the F_2, F_3 signal to give a triplet. F_2, F_3 with the same environment would couple with the F_1, F_4 signal to give a triplet. Since two triplets of equal intensity are obtained, (*c*) is the correct structure.

The bridged configuration of the boron hydrides and the structures of certain polyphosphates have also been elucidated by the application of NMR. The method has also shown that certain monohydrates originally formulated as $HClO_4.H_2O$, $HNO_3.H_2O$, are in fact oxonium salts $H_3O^+ClO_4^-$, $H_3O^+NO_3^-$.

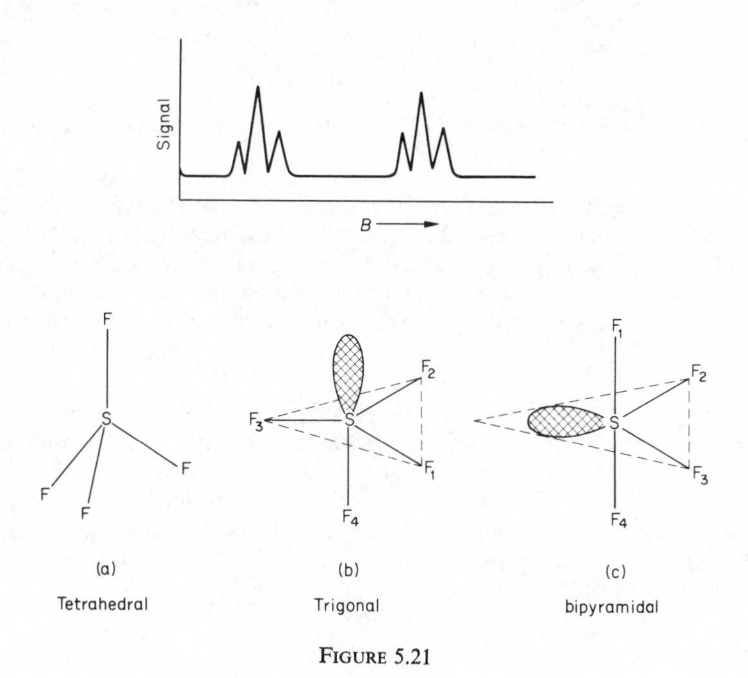

FIGURE 5.21

Dipole Moments

The existence of a permanent electrical dipole in a molecule results from the unequal sharing of the electrons forming the bond between the atoms. This system is said to have an electrical *dipole moment*, the magnitude of which is determined by the product of the charge and the distance between the positive and negative centres.

$$q+ \overset{d}{\longleftrightarrow} q-$$

The unit is the debye (D) and is $3{\cdot}336 \times 10^{-30}$ Cm. When placed between the plates of a condenser, polar molecules line up as shown in Figure 5.22, and

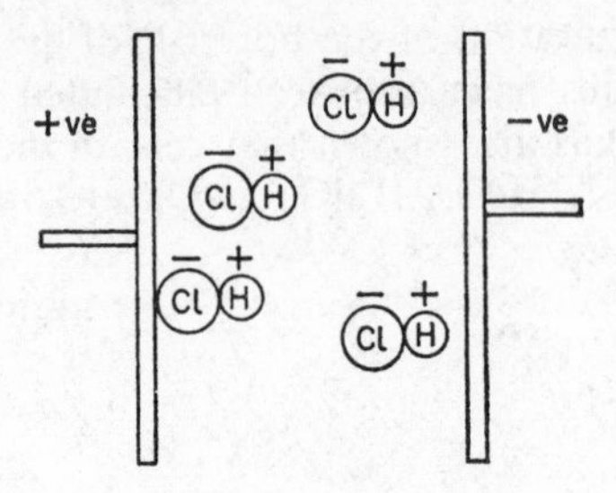

FIGURE 5.22. Polar molecules of HCl lined up between the plates of a condenser

increase the capacitance of the condenser. From this increase, the dielectric constant and hence the *polarization* can be calculated. This total polarization (P_m) originates from two sources: the orientation polarization (P_u) due to alignment of permanent dipoles in the electrostatic field, and the induced polarization (P_d) which is due to distortion of molecules by the field. Hence

$$P_m = P_u + P_d$$

P_u varies with temperature, due to thermal agitation of the molecules, but P_d is not temperature-dependent. Thus a plot (Figure 5.23) of P_m versus $1/T$, where T is the absolute temperature, gives a straight line, the slope of which measures the magnitude of P_u. Molecules with no permanent dipole give a horizontal straight line. Some typical values for dipole moments are given in Table 5.2.

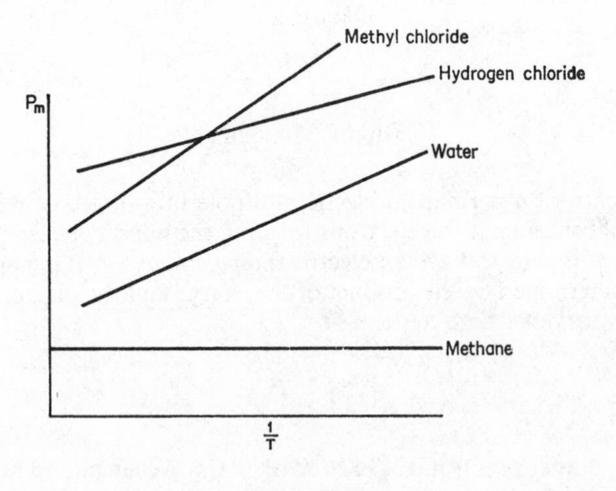

FIGURE 5.23

These values give information regarding the nature of the bond and the shape of the molecule. By evaluating the ratio of the observed dipole moment to the calculated theoretical value for an ionic structure, Pauling determined the percentage ionic character of the bonds in the halogen hydrides, as follows: HF 43 per cent, HCl 17 per cent, HBr 12 per cent, and HI 5 per cent. The covalent character of the bond increases markedly from HF to HI. Triatomic molecules with zero dipole moment must be linear, e.g. CO_2.

TABLE 5.2

Molecule	*Dipole moment, D*	*Molecule*	*Dipole moment, D*	*Molecule*	*Dipole moment, D*
HF	1·91	H_2O	1·84	NH_3	1·48
HCl	1·05	H_2S	0·92	BCl_3	0·00
HBr	0·80	CO_2	0·00		
HI	0·42				

Although there is an electron drift towards each oxygen atom, the resultant effect is zero. On the other hand the molecules of water and hydrogen sulphide must be angular. Boron trichloride, which has zero moment, must be planar and ammonia must be pyramidal (Figure 5.24).

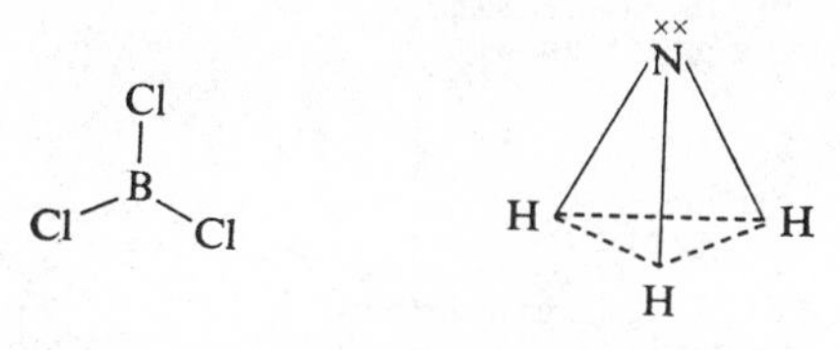

FIGURE 5.24

Crystal Structure

Results of X-ray analysis confirmed the existing theories on crystal structure. All crystals can be classified into the six fundamental systems shown in Table 5.3.

These crystal shapes arise from the internal arrangement of atoms, molecules or ions which make up the crystal structure. These atoms, molecules or ions are arranged in a series of repeating units termed *unit cells*, these cells forming a continuous *lattice* through the entire crystal.

TABLE 5.3

Type	Characteristics	Example
Cubic	Three axes equal and mutually perpendicular	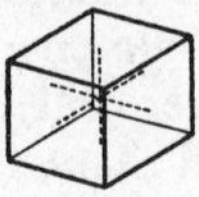 Galena (PbS) Halite (NaCl)
Tetragonal	Two equal axes and one unequal mutually perpendicular	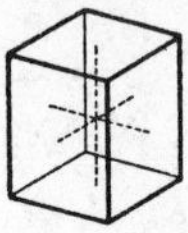 Zircon (ZrO_2) Rutile (TiO_2)
Orthorhombic	Three unequal axes mutually perpendicular	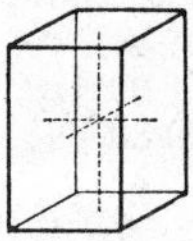 Sulphur (S) Olivine $(MgFe)_2SiO_4$
Hexagonal	Three equal axes inclined at 120° with a fourth unequal and perpendicular to the other three	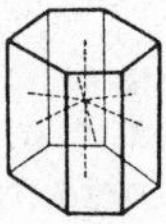 Quartz (SiO_2) Calcite ($CaCO_3$)
Monoclinic	Two axes at an oblique angle with a third perpendicular to the other two	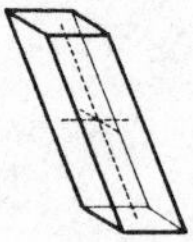 Gypsum $CaSO_42H_2O$ Micas $KAl_2(OH)_2Si_3AlO_{10}$
Triclinic	Three unequal axes intersecting obliquely	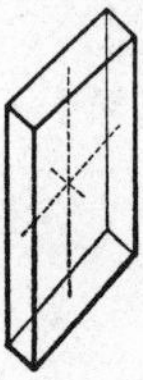 Feldspars e.g. :– $KAlSi_3O_8$

The Metallic State

Metallic elements crystallize in one of three forms corresponding to one of the following types of packing:

(*a*) *Hexagonal close packing*, e.g. Mg, Zn, Ca, Ti.
(*b*) *Face-centred cubic packing*, e.g. Cu, Ag, Au, Ca, Sr, Al, Ni, Pb.
(*c*) *Body-centred cubic packing*, e.g. Li, Na, Ba, Fe, V.

Types (*a*) and (*b*) represent the most efficient packing of spherical units and, together with the unit cell, are shown diagrammatically in Figure 5.25. Each atom in these structures is surrounded by twelve others and is said to have *coordination number* 12. In the body-centred cubic structure packing is less efficient and the coordination number is 8.

It may be seen from the diagrams that the holes or interstices between the packing of the metallic atoms are of two kinds:

(*a*) *octahedral*, made by six atoms at the corners of a regular octahedron (O),
(*b*) *tetrahedral*, made by four atoms at the corners of a regular tetrahedron (T).

These are not indicated in Figure 5.25(*b*) to avoid confusion.

Thus, under suitable conditions many transition metals admit small atoms (hydrogen, carbon, boron, and nitrogen) into these interstices forming interstitial compounds (p. 348). *See also* the spinels (p. 372).

The metallic bond

In the metallic structures discussed above each atom is surrounded by eight or twelve other atoms depending upon the type of packing. Each metallic atom can only furnish one, two or at the most three valency electrons for binding purposes. Hence normal covalent bonds cannot be formed as insufficient electrons are available. The metallic bonding theory postulates that all the metal atoms lose some or all of their valency electrons which form a mobile loose network binding the cations together. Lorentz described this as 'an array of cations in a sea of free electrons'. This type of binding explains the high thermal and electrical conductivity associated with metals. Normally electrons in a metal do not escape because they are held electrostatically by the cations but, when heat energy is supplied, electrons are emitted (*thermionic emission*) and when a metal is irradiated with ultraviolet light emission again occurs (*photoelectric effect*). Although ordinary light does not cause such emission, radiations of certain wavelengths are absorbed by the loose electrons, giving rise to the characteristic lustrous appearance of metals. Other characteristic properties of metals can be explained by the mode of packing. The face-centred cubic structure possesses the largest number of planes of atoms which can glide over other planes. Hence metals which crystallize in this form are usually malleable and ductile.

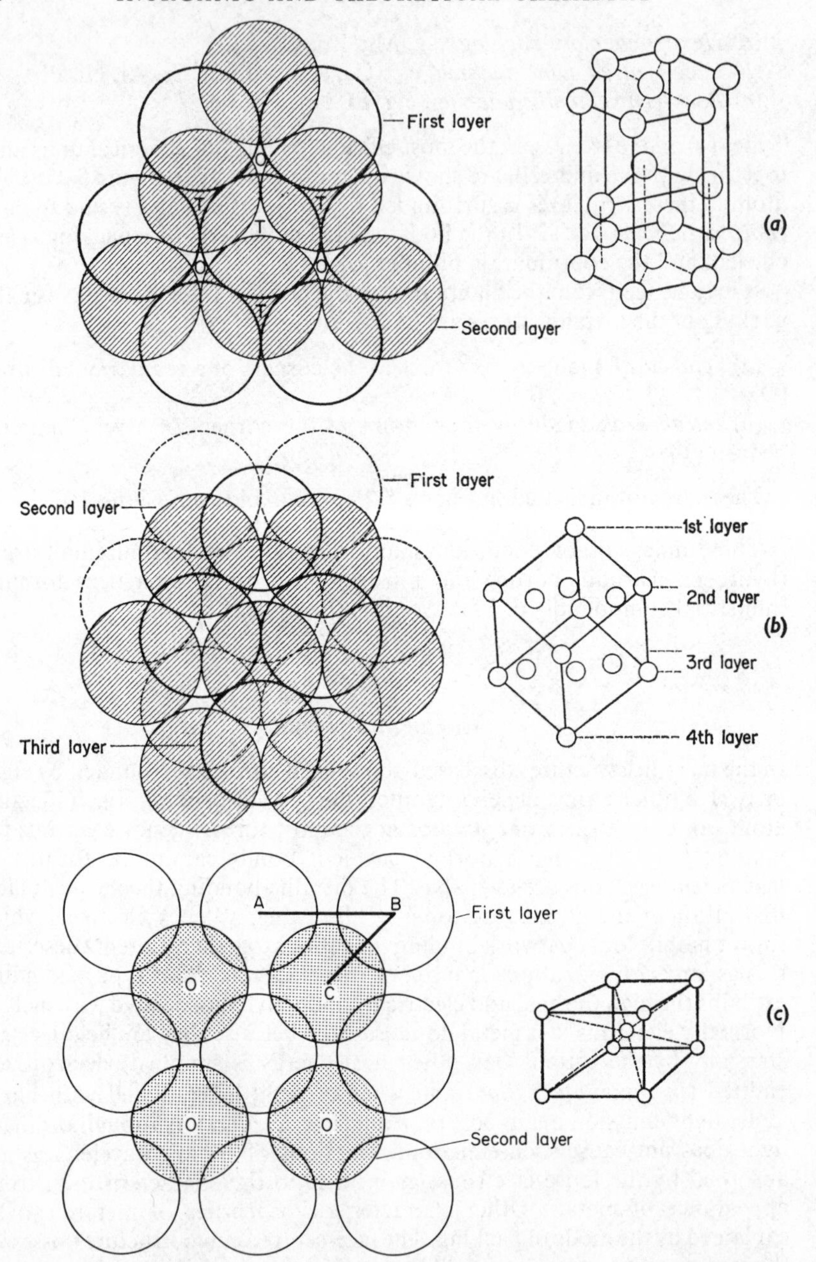

FIGURE 5.25. (*a*) hexagonal close packing—alternate layers identical; (*b*) face-centred cubic packing—1st and 4th, 2nd and 5th, 3rd and 6th layers identical; (*c*) body-centred cubic packing—alternate layers identical. Distance AB > BC

The band theory of metals

As indicated above, many metallic properties can be explained using the idea that metals are composed of an aggregate of cations floating in a mobile sea of electrons. The band theory of solids is a further development used to explain the nature of insulators, semi-conductors, and conductors.

In an atom the electrons occupy definite energy levels (Figure 5.26(*a*)). A solid, however, consists of a three-dimensional system of close-packed

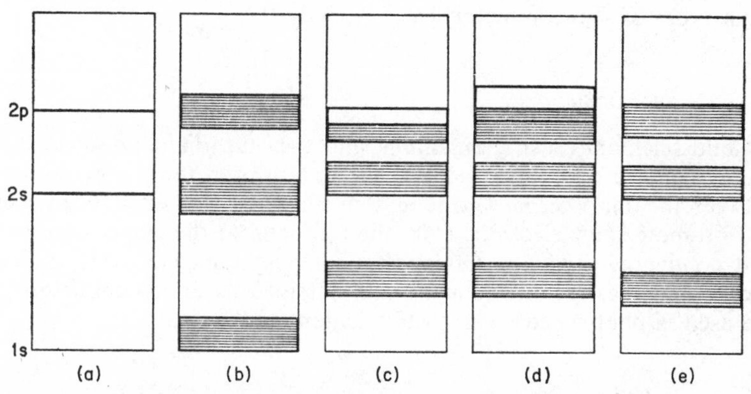

FIGURE 5.26. (*a*) the definite energy levels of an individual atom; (*b*) the energy levels of a solid—all the bands are full, separated by large energy gaps—thus an *insulator*, e.g. S_8, P_4, C (diamond); (*c*) solid with closely spaced bands and an incompletely filled outer energy band—thus a *conductor*, e.g. Cu, Na; (*d*) solid with closely spaced bands and an empty outer band overlapping with a full inner band—thus a *conductor*, e.g. Mg; (*e*) solid with small energy gaps between full bands—thus a *semiconductor*, e.g. Ge, Se, Si

atoms or ions and these interact with one another and the definite energy levels of single atoms are replaced by bands of energy levels (*Brillouin zones*). For *n* atoms in a solid lattice, there will be *n* possible energy levels within each band. Each band is separated from another band by a region of forbidden energy known as an *energy gap*.

In an insulator all the energy bands are full and are separated by large energy gaps. Thus the promotion of electrons to a higher empty band to enable electron transfer between atoms to occur is almost impossible (Figure 5.26(*b*)). Metallic conductors are characterized by closely spaced energy bands and, although the inner bands are full, the outermost bands are only partially occupied. The application of an external e.m.f. causes the electrons in these bands to move to similar bands in other atoms (Figure 5.26(*c*)). In some metallic conductors all the bands are full but an empty outer band overlaps with an inner full band and free electron movement can again occur to similar bands in other atoms (Figure 5.26(*d*)). The conductivity of metals generally decreases with increase in temperature. Absorption of thermal energy causes vibration of the atoms or ions within the crystal lattice which scatters and thus hinders the free movement of electrons through the bands. Some metalloids, particularly those containing impurities, however, possess a low conductivity at ordinary temperatures which rises as the temperature increases. In these materials, classified as *semiconductors,* the energy bands are separated by small energy gaps (Figure 5.26(*e*)). Increase of temperature supplies some electrons in a full band with sufficient energy to jump the energy gaps into the next empty band, causing a considerable increase in the conductivity. There are three main types of semiconductor:

Intrinsic semiconductors

In solid selenium (Se_8), germanium, and grey tin (diamond lattices), the energy bands are full and do not overlap. However, the energy difference between the full outermost band and an empty band is small. Heat energy will promote some electrons from the full band to the empty outer band and conductivity results. Often ultraviolet light may provide sufficient energy and these semiconductors are classified as photoconductors and are used in photoelectric cells, e.g. selenium and germanium.

N-type semiconductors

When germanium is crystallized from a melt containing a small quantity of arsenic, some arsenic atoms are incorporated into the diamond-type lattice of germanium. Only four of the five valency electrons of arsenic ($4s^2 4p^3$) are required to bond with germanium ($4s^2 4p^2$). On heating, the fifth electron enters an upper energy band and can readily pass into an outer empty band of germanium, the atoms of the latter thus acquiring a negative charge. This type of semiconductor is known as an N-type semiconductor since the germanium metal acquires a negative (N) charge.

P-type semiconductor

When gallium ($4s^2 4p^1$) or indium ($5s^2 5p^1$) is incorporated into the germanium, absorption of heat energy causes the extra germanium electron to enter an upper energy band which can pass readily into an empty gallium (or indium) band, the germanium atoms thus acquiring a positive charge. This is known as a P-type semiconductor.

All three types of semiconductor are used in the electronics industry in the manufacture of transistors. The solar battery makes use of N- and P-type semiconductors in converting light energy into electricity. Although still in the experimental stage the solar battery has been used as a source of power for telephones and for instruments in rocket vehicles.

Alloys

An alloy is composed of two or more metals, or of one metal plus a non-metallic element of small radius, e.g.

Atom	H	B	C	N
Radius (nm)	0·037	0·080	0·077	0·074

The usual method of preparation is to fuse the metals together and allow to cool. Electrolytic methods can also be used, for example, iron articles are coated with brass by electrodeposition from a solution containing the complex anions $Cu(CN)_2^-$ and $Zn(CN)_4^{2-}$. Almost all alloys are solids, exceptions being some liquid alloys of mercury. X-ray analysis has shown alloys to be of three types, as considered below.

Simple mixtures

Here the two metals are completely insoluble, the resulting solid alloy being an intimate mixture of crystals of both metals. A common example is solder, an alloy of tin and lead.

Solid solutions

If two metals have the same structure and have atoms of approximately the same size a continuous series of solid solutions is formed without any change in structure. For example, silver and gold (atomic radii ~0·13 nm) both have face-centred cubic structures. The silver dissolves the gold, when the gold atoms replace silver atoms at the cube corners in the face-centred lattice. When the conditions mentioned above are not fulfilled, more complicated structures may result. Thus copper has a face-centred cubic lattice and will dissolve up to 38 per cent zinc, when its structure changes to body-centred cubic. Above 67 per cent dissolved zinc this structure changes to hexagonal close packing.

Small atoms (C, H, B, N) can occupy the holes in the lattice of a metal giving interstitial solid solutions. These compounds are extremely hard and have a pronounced metallic character. Usually no simple formula can be given and these compounds are described as *non-stoichiometric* or *Berthollide* compounds (named after the chemist who was opposed to the Daltonian idea of fixed composition). For example, γ-iron which crystallizes as a face-centred cube dissolves carbon to form a solid solution. When this solution is quenched a solid solution of carbon in α-iron (body-centred cubic) is obtained and is called martensite (Figure 5.27).

Intermetallic compounds

In many alloys the constituents appear to be present in certain stoichiometric proportions. These ratios cannot be related to ordinary valencies and quite often involve large numbers, e.g. Cu_5Zn_8, $Cd_{11}K$. Some attempts have been

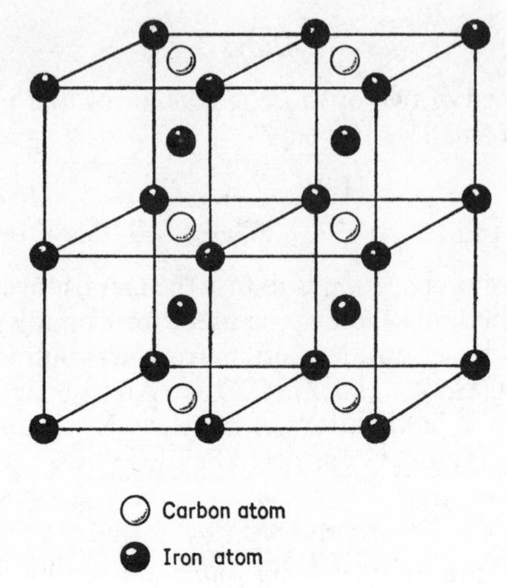

FIGURE 5.27. Structure of martensite

made to explain the ratios in terms of valency binding but so far no satisfactory quantitative results have been obtained.

Properties of alloys

Alloys are often harder and tougher than the constituent elements, e.g. bronze, 90 per cent Cu, 10 per cent Sn; brass, 80 per cent Cu, 20 per cent Zn. However, the thermal and electrical conductivity of a metal is usually considerably reduced when alloyed, hence the necessity for copper of the highest purity for electrical transmission wires. The melting point of the alloy is usually much lower than the melting points of the constituent elements and use is made of this property in the manufacture of easily moulded alloys such as solder and type metal (alloys of lead and tin). Certain alloys of iron, cobalt, and nickel are very strongly magnetic (ferromagnetic). Soft magnets are usually made from iron, cobalt, and nickel alloys, whereas hard permanent magnets are constructed from iron, cobalt, nickel, and aluminium.

Inorganic Compounds

Ionic crystals

The type of crystal structure shown by inorganic compounds is determined by two main factors: (i) the ratio of the number of cations to anions, e.g. in calcium fluoride there are two fluoride anions per cation; (ii) the ratio of the radius of cation to anion.

Consider the ionic compound A^+B^- in which the total numbers of anions and cations must be equal so that the structure as a whole is neutral. If the radius of the ion A^+ is small compared to B^- the coordination number will be low as only a small number of anions can be accommodated round a cation. Large coordination numbers result when the radius of A^+ is large in comparison with that of B^-. Stable efficient packing of ions in crystals corresponding to one particular coordination number is only possible when the radius ratio (r_{cation}/r_{anion}) lies within certain limits. The values of r_c/r_a corresponding to various coordination numbers are given below:

Coordination number	8	6	4
r_c/r_a	>0·73	0·41–0·73	0·22–0·41
Arrangement	body-centred cubic	octahedral	tetrahedral

Coordination numbers of 5 and 7 are not possible geometrically if the charges are to balance.

Crystal structures of A^+X^- type

(i) Caesium chloride: $r_c/r_a = 0{\cdot}169/0{\cdot}181 = 0{\cdot}93$, corresponding to a coordination number of 8, giving rise to a body-centred cubic lattice (Figure 5.28(*a*)). Coordination of this type also occurs with caesium bromide and iodide and with ammonium chloride and bromide.

(ii) Sodium chloride: $r_c/r_a = 0{\cdot}095/0{\cdot}181 = 0{\cdot}64$, corresponding to a coordination number of 6, giving rise to a cubic lattice (Figure 5.28(*b*)). This type of coordination is characteristic of lithium, sodium, potassium, and rubidium halides, ammonium iodide, and the alkaline earth metal oxides.

(iii) Zinc blende, ZnS: $r_c/r_a = 0{\cdot}074/0{\cdot}184 = 0{\cdot}40$, corresponding to a coordination number of 4, giving rise to a structure in which each sulphide ion is surrounded tetrahedrally by four zinc ions and each zinc ion by four sulphide ions (Figure 5.28(*c*)). Other compounds with this structure are beryllium oxide and sulphide.

(iv) Wurtzite, ZnS: in this second crystal form of zinc sulphide the coordination number is again 4, the sulphur atoms being hexagonal close packed instead of cubic close packed, as in zinc blende (Figure 5.28(*d*)). Other compounds with this structure are cadmium sulphide and zinc oxide.

In view of the fact that the ions are regarded as hard impenetrable spheres, radius ratio calculations can be regarded as rough approximations valid only for simple ionic species. Thus a simple ionic model for copper(I) chloride would give

$$\frac{r_{Cu^+}}{r_{Cl^-}} = 0{\cdot}53$$

and indicate octahedral coordination. However, copper(I) chloride has considerable covalent character and possesses a zinc blende structure with tetrahedral coordination.

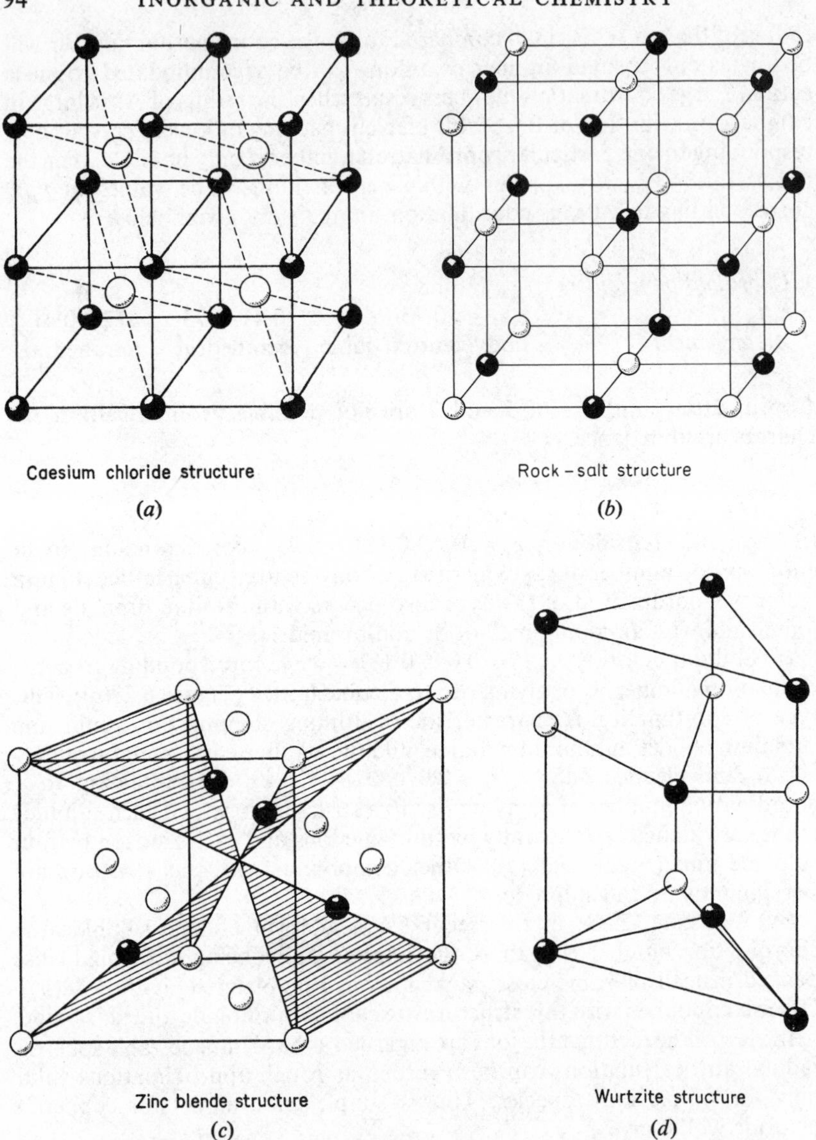

FIGURE 5.28. Crystal structures of the $A^+ X^-$ type

Crystal structures of $A^{2+}(X^-)_2$ type

(i) Fluorite (calcium fluoride): $r_c/r_a = 0{\cdot}099/0{\cdot}136 = 0{\cdot}73$, corresponding to 8 coordination. Each calcium ion is surrounded by eight fluoride ions at the corners of a cube and to maintain electrical neutrality, each fluoride ion is surrounded by four calcium ions situated at the corners of a tetrahedron (Figure 5.29(*a*)). Essentially the structure is body-centred cubic, the calcium ions occupying half the body-centre positions. Other compounds exhibiting this structure are barium and cadmium fluorides and thorium oxide.

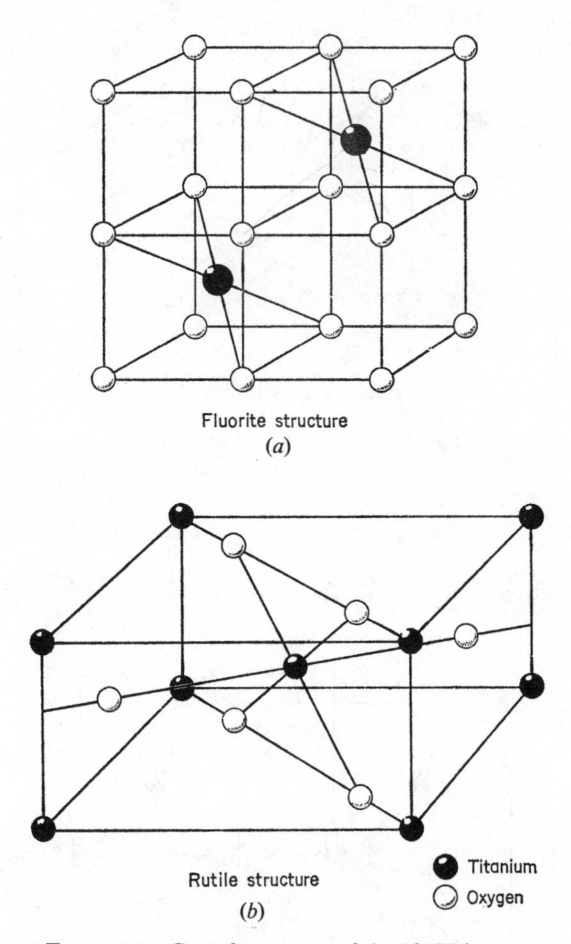

FIGURE 5.29. Crystal structures of the $A^{2+}(X^{-})_2$ type

(ii) Rutile (titanium dioxide): $r_c/r_a = 0{\cdot}068/0{\cdot}140 = 0{\cdot}48$, corresponding to 6 coordination. Here each titanium ion is surrounded by six oxide ions in octahedral positions and each oxide ion is surrounded by three titanium ions in a triangular arrangement (Figure 5.29(*b*)). The unit cell is not cubic. Other compounds exhibiting this structure are stannic oxide, manganese and lead dioxides.

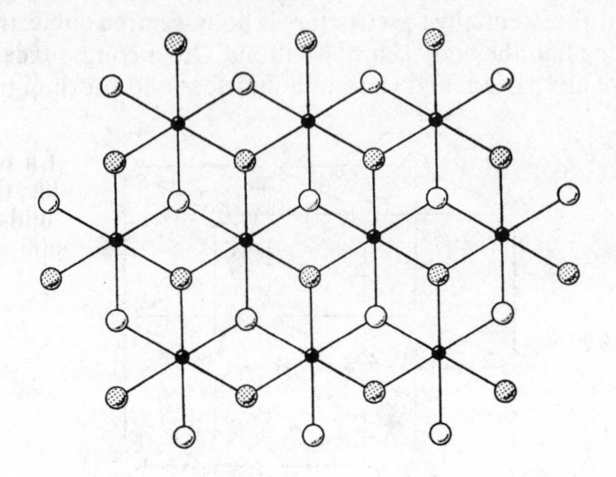

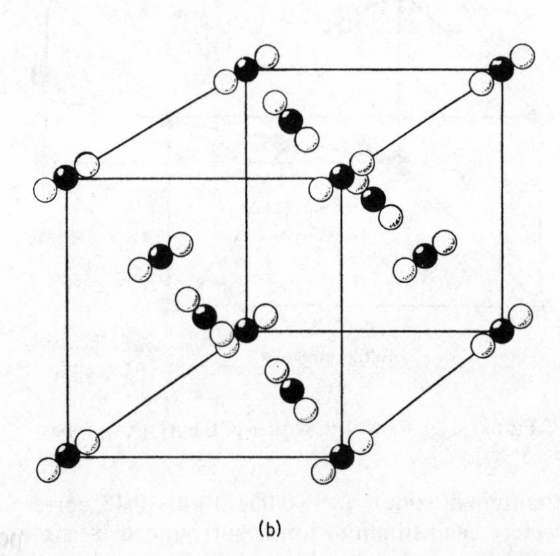

FIGURE 5.30. Covalent crystals: (*a*) cadmium chloride; (*b*) solid carbon dioxide

Covalent crystals

Layer structures

In this type of crystal, the cadmium chloride and cadmium iodide type, binding between atoms in a layer is essentially covalent and the layers are weakly held by van der Waals forces (Figure 5.30(*a*)). Each cadmium atom is bonded to six halogen atoms and each halogen atom to three cadmium atoms in each composite layer. In the cadmium chloride lattice the layers of halogen atoms are cubic close packed but in the cadmium iodide lattice they are hexagonal close packed.

Molecular crystals

These consist of discrete molecules, e.g. solid carbon dioxide which has a face-centred cubic structure (Figure 5.30(*b*)). This type of crystal is characterized by its low heat of sublimation as the intermolecular forces are weak.

Lattice defects

In the crystallization of a compound, the ions or molecules in solution come together and take up fixed positions in the solid crystal. Inevitably the build-up of the crystal is not perfect and some irregularities occur in the positions taken up by the ions or molecules. The resulting defect in the crystalline lattice may be a Schottky defect (Figure 5.31) involving anion and cation

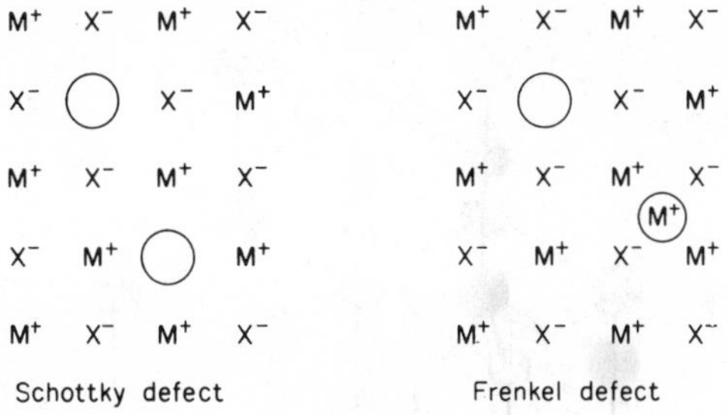

M^+ X^- M^+ X^- ⠀⠀⠀ M^+ X^- M^+ X^-

X^- ○ X^- M^+ ⠀⠀⠀ X^- ○ X^- M^+

M^+ X^- M^+ X^- ⠀⠀⠀ M^+ X^- M^+ X^- (M^+)

X^- M^+ ○ M^+ ⠀⠀⠀ X^- M^+ X^- M^+

M^+ X^- M^+ X^- ⠀⠀⠀ M^+ X^- M^+ X^-

Schottky defect ⠀⠀⠀ Frenkel defect

FIGURE 5.31

vacancies in the lattice which may or may not give deviations from stoichiometry. Alternatively a Frenkel defect (Figure 5.31) may occur where ions occupy interstitial positions leaving the correct sites unoccupied. When an electron is trapped in a lattice site, it is termed an F (F = farbre, meaning colour) centre and these electrons can often be excited to higher energy levels by visible light, thus imparting colour to the compound. When chlorine is treated with an excess of sodium, the sodium chloride produced is blue and contains an excess of sodium ions in interstitial positions with an equal number of electrons trapped in lattice sites to maintain electrical neutrality. When white zinc oxide is heated, there is a slight oxygen loss and the colour changes to yellow. Surplus zinc ions migrate to interstitial sites and an equal number of electrons become trapped in lattice sites.

Non-metallic Elements

X-ray analysis of the non-metallic elements has revealed a variety of structures. In these elements the atoms are linked by covalent bonds, each atom exerting valency n or $8-n$, where n is the periodic group number. This is illustrated by the following examples.

Group IV: Carbon

Carbon exists in two allotropic forms, diamond and graphite.

Diamond

This is a giant molecule (macromolecule) in which each carbon atom is surrounded tetrahedrally by four others. The C—C distance is 0·154 nm,

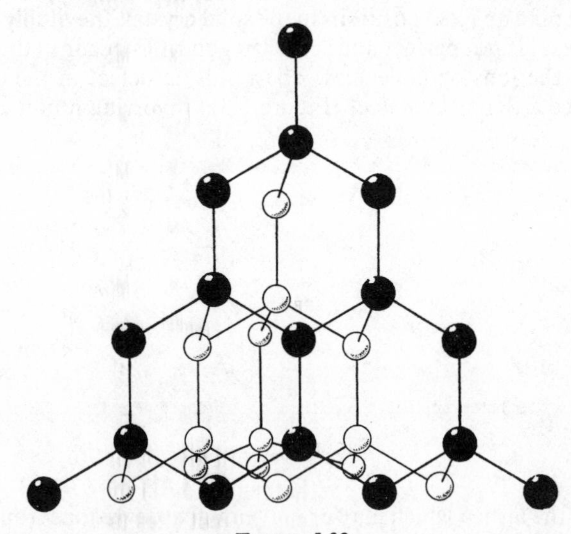

FIGURE 5.32

indicating a single covalent bond between adjacent carbon atoms (Figure 5.32). The structure explains the extreme hardness and lack of chemical reactivity. Germanium and silicon have diamond-like structures but these are fragmentary and not continuous, due to weaker bonds.

Graphite

The atoms are arranged in flat planar hexagons in which the C—C distance is 0·142 nm, thus showing considerable double-bond character as in benzene (C—C distance = 0·139 nm). Sheets of these atoms form layers 0·341 nm apart, held together by weak van der Waals forces (Figure 5.33). Thus

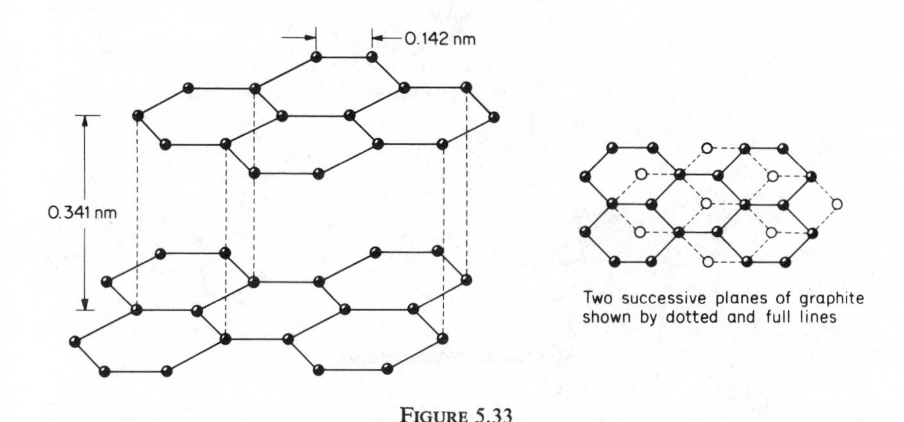

FIGURE 5.33

graphite is not as hard as diamond and the openings between the planes are modes of entry for attack by strong chemical reagents. The ability of the planes to glide over each other gives graphite its lubricating power.

Group V: Phosphorus

Phosphorus exists in a number of allotropic forms, the most important of which are white and red phosphorus.

White phosphorus

The molecule here is tetratomic (P_4) with each phosphorus atom at the corner of a regular tetrahedron. The tetrahedra are held together by weak forces of the van der Waals type (Figure 5.34(*a*)).

Red phosphorus

Here the P—P bonds in the tetrahedral unit of white phosphorus are broken. The free valencies are used to link units together giving rise to long chains cross-linked through atoms (1) to other chains (Figure 5.34(*b*)).

Arsenic, antimony, and bismuth also form layer lattices of puckered hexagonal rings, the distance between the atoms increasing from arsenic to bismuth (Figure 5.35).

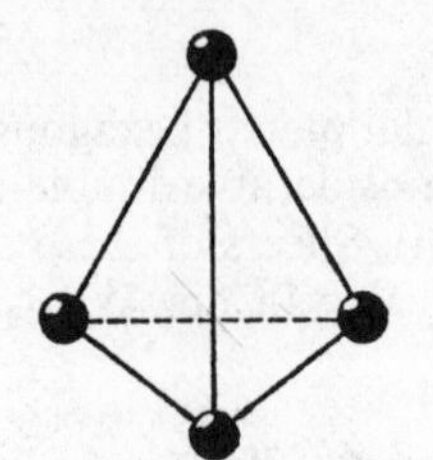

Tetrahedral molecule of white phosphorus

(a)

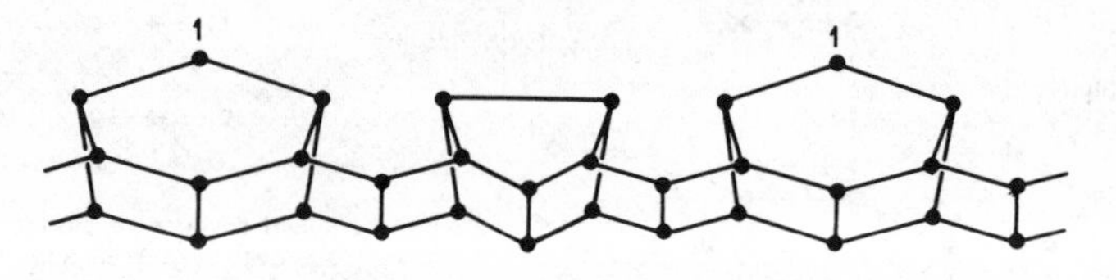

Structure of red phosphorus

(b)

FIGURE 5.34

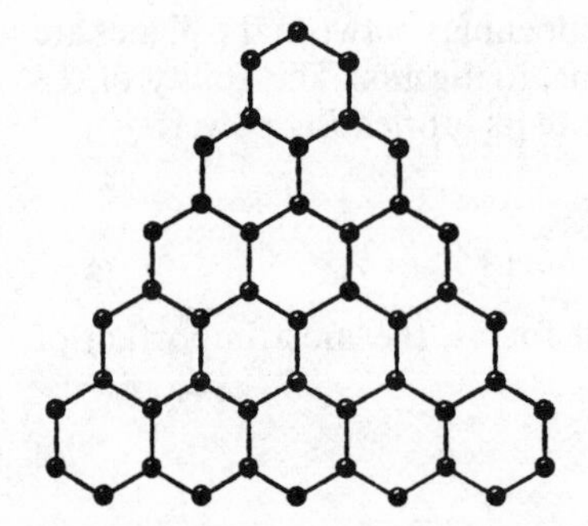

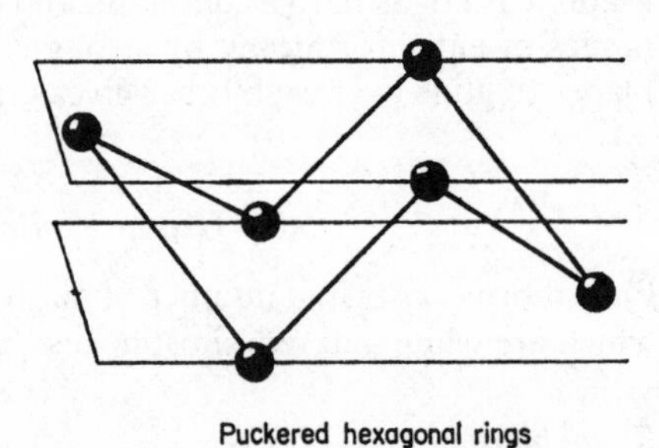

Puckered hexagonal rings

FIGURE 5.35

Group VI: Sulphur

Sulphur exists in three main allotropic forms.

Rhombic and monoclinic sulphur

These consist of puckered eight-membered rings, each atom being covalently bound to two others (Figure 5.36(*a*)). The different crystalline forms are entirely due to the different methods of stacking S_8 rings.

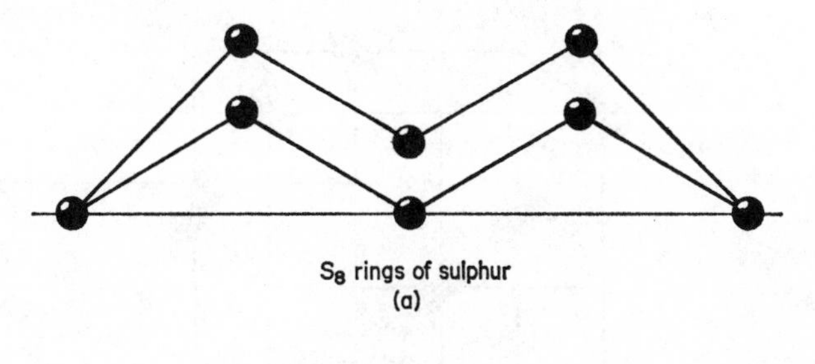

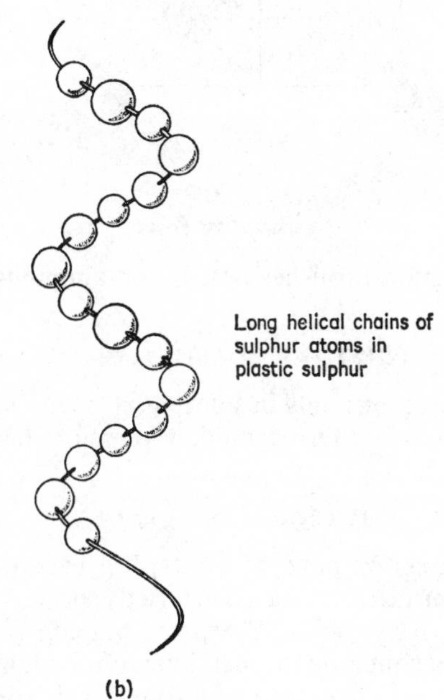

FIGURE 5.36

Plastic sulphur

Here the sulphur atoms are arranged in spiral chains giving the characteristic plastic properties (Figure 5.36(*b*)). Selenium and tellurium have similar structures to plastic sulphur, the chains being held parallel to each other by van der Waals forces.

Group VII: Iodine

This is the only halogen to be examined in the solid state and is an example of a molecular crystal. The structure is shown in Figure 5.37. The solid is easily vaporized, indicating weak intermolecular forces.

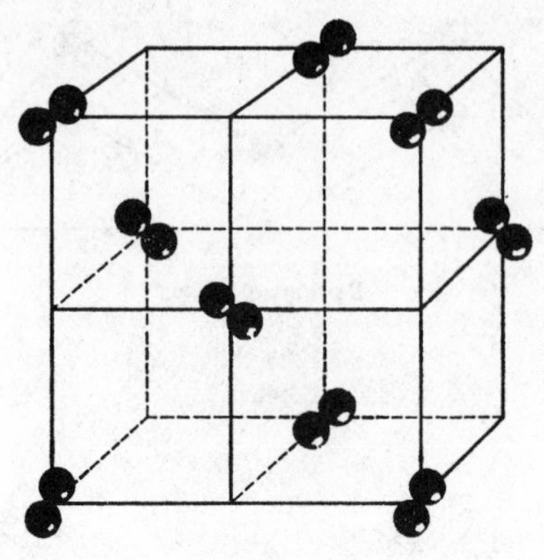

FIGURE 5.37

Complex Ions

Potash alum in solution furnishes potassium, aluminium (hydrated), and sulphate ions:

$$KAl(SO_4)_2.12H_2O \rightleftharpoons K^+ + Al(H_2O)_6^{3+} + 2SO_4^{2-} + 6H_2O$$

Because it gives simple ions only in solution, it is termed a *double salt*.

A solution of potassium ferrocyanide (potassium hexacyanoferrate(II)) furnishes potassium and ferrocyanide ions:

$$K_4Fe(CN)_6 \rightleftharpoons 4K^+ + Fe(CN)_6^{4-}$$

The $Fe(CN)_6^{4-}$ ion shows none of the typical reactions of iron(II) and cyanide ions and is termed a *complex ion*. Strictly speaking this term refers to such ions as SO_4^{2-}, NO_3^-, and CO_3^{2-}, etc., but convention has restricted its use to complexes containing a metallic atom or ion, e.g. $Cu(NH_3)_4^{2+}$, $Co(NH_3)_6^{2+}$. Complexes may also be formed which are uncharged, e.g. nickel dimethylglyoxime.

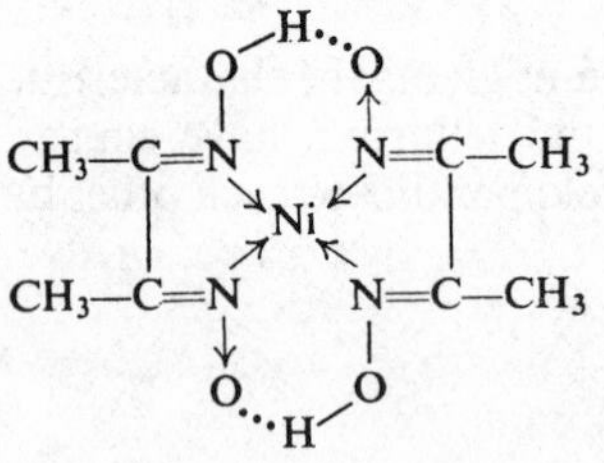

nickel dimethylglyoxime

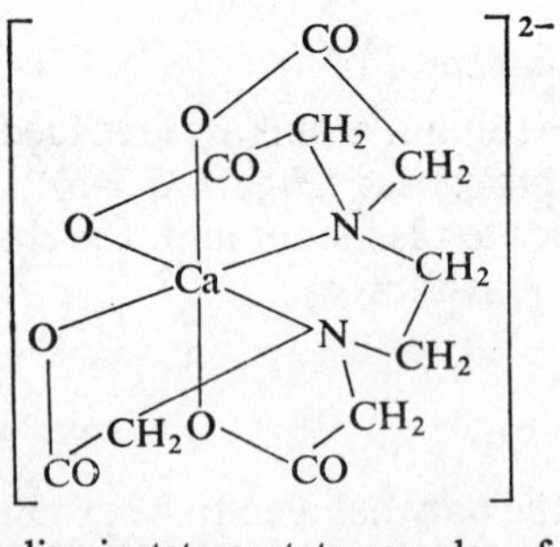

ethylenediaminetetraacetate complex of calcium

FIGURE 5.38

In the hexacyanoferrate(II) ion the iron atom is called the *central atom* and the cyanide anions are called *ligands*. These ligands are linked to the central cation by coordinate links. Complexes may also be formed when one ligand is bound to the central cation by two, three or more coordinate links, the ligand being termed bi-, tri-, or poly-dentate and the resulting complex is called a *chelate complex*. Hence dimethylglyoxime and ethylenediamine-tetraacetic acid (EDTA) are respectively bi- and hexa-dentate ligands. The EDTA complex of calcium is shown in Figure 5.38.

The following series of complex compounds can be obtained by air oxidation of solutions of cobaltous and ammonium chlorides:

(*a*) $CoCl_3.6NH_3$—yellow—hexamminecobalt(III) chloride.
(*b*) $CoCl_3.5NH_3$—purple—chloropentamminecobalt(III).
(*c*) $CoCl_3.4NH_3$—green—dichlorotetramminecobalt(III) chloride.

These compounds are unreactive to hydrochloric acid even when boiled and cold caustic soda solution does not precipitate the hydroxide. This indicates that free cobalt ions are no longer present. Treatment of (*a*) with silver nitrate causes all the chlorine to be precipitated as AgCl. In the case of (*b*) and (*c*) only two-thirds and one-third respectively of the total chlorine is precipitated by silver nitrate solution.

The significance of these observations was explained by Alfred Werner, a Swiss Nobel Prizewinner, who developed the concept of the coordination of ligands round the central metallic atom or ion. He represented the structure of the three complexes mentioned above as:

(*a*)
$$\left[\begin{array}{ccc} & NH_3 \quad NH_3 & \\ & \searrow \quad \swarrow & \\ NH_3 \rightarrow & Co & \leftarrow NH_3 \\ & \nearrow \quad \nwarrow & \\ & NH_3 \quad NH_3 & \end{array}\right]^{3+} 3Cl^-$$

(*a*) Six neutral ligands (NH_3) are coordinated to the central cation Co^{3+} giving an overall charge of +3 on the complex.

(*b*)
$$\left[\begin{array}{ccc} & NH_3 \quad NH_3 & \\ & \searrow \quad \swarrow & \\ NH_3 \rightarrow & Co & \leftarrow NH_3 \\ & \nearrow \quad \nwarrow & \\ & Cl \quad NH_3 & \end{array}\right]^{2+} 2Cl^-$$

(*b*) Five neutral ligands and one chloride ion are coordinated to the central cation Co^{3+} giving an overall charge of +2 on the complex.

(*c*)
$$\left[\begin{array}{ccc} & NH_3 \quad NH_3 & \\ & \searrow \quad \swarrow & \\ NH_3 \rightarrow & Co & \leftarrow NH_3 \\ & \nearrow \quad \nwarrow & \\ & Cl \quad Cl & \end{array}\right]^{+} Cl^-$$

(*c*) Four neutral ligands and two chloride ions are coordinated to the central cation giving an overall charge of +1 on the complex.

The number of coordinate bonds to the central atom/ion is called the coordination number of the central atom or ion which, in the case of cobalt, is six. The linkage to the central atom is not always coordinate. Thus in the chloroplatinic acid complex, the complex anion is $PtCl_6^{2-}$:

$$PtCl_4 + 2Cl^- \rightleftharpoons [PtCl_6]^{2-}$$

Here the two chloride ions coordinate while the other four are covalently bound to the central atom, although the two types of bond are indistinguishable once they are formed.

In many complex ions the number of electrons associated with the central atom corresponds to the number associated with a noble gas. For example, in $Co(NH_3)_6^{3+}$ the cobalt (III) ion has 24 electrons plus 12 electrons (two from each ammonia molecule), giving a total of 36, the same number as that possessed by the noble gas krypton. Similarly in the hexacyanoferrate(II) ion the total number of electrons associated with the central cation is $24 + 12 = 36$. Although the central cation often attains the effective atomic number of the next noble gas, this is not a necessary requirement of complex formation. Thus in $Cu(NH_3)_4^{2+}$ the copper atom has 35 electrons and in $Cr(H_2O)_4Cl_2^+$ the chromium atom has 33 electrons.

Coordination number and structure

The structure of inorganic complexes has been elucidated from X-ray, dipole moment, and magnetic susceptibility measurements. Results show that the complexes have a particular structural arrangement which is closely dependent on the coordination number of the central atom. Only two coordination numbers will be discussed: four and six.

Coordination number 4

(i) Tetrahedral structure (Figure 5.39(*a*)). This has been found for a large number of complexes of metals, e.g. $[CoCl_4]^{2-}$, $[Zn(CN)_4]^{2-}$, $[Cd(CN)_4]^{2-}$, $[Cu(CN)_4]^{3-}$, $[NiCl_4]^{2-}$, $Ni(CO_4)$.

$$Cu^+ + 4CN^- \rightleftharpoons [Cu(CN)_4]^{3-}$$

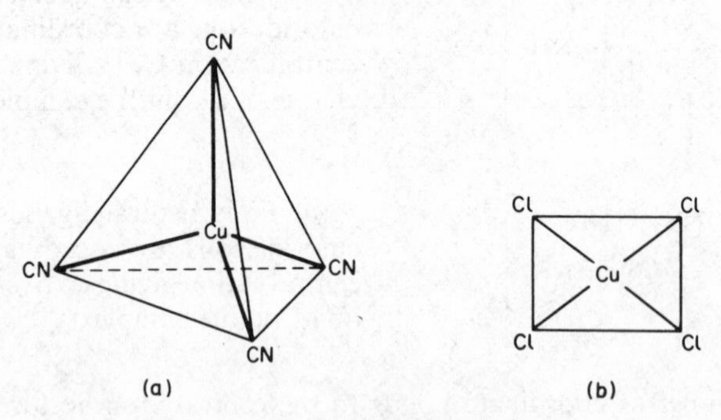

FIGURE 5.39. Coordination number 4

(ii) Planar structure (Figure 5.39(*b*)). This has been established for $[CuCl_4]^{2-}$, $[PtCl_4]^{2-}$, $[Ni(CN)_4]^{2-}$, $[Cu(NH_3)_4]^{2+}$, $[PdCl_4]^{2-}$.

Coordination number 6

This octahedral structure has been established for $[SnCl_6]^{2-}$, $[PtCl_6]^{2-}$, $[Cr(NH_3)_6]^{3+}$, $[Fe(CN)_6]^{3-}$, $[Fe(CN)_6]^{4-}$, $[Co(CN)_6]^{3-}$. (*See* Figure 5.40.) The octahedral structure accounts for the existence of isomers in many complexes. Two isomers exist for the complex $[Co(NH_3)_4Cl_2]^+$, one form

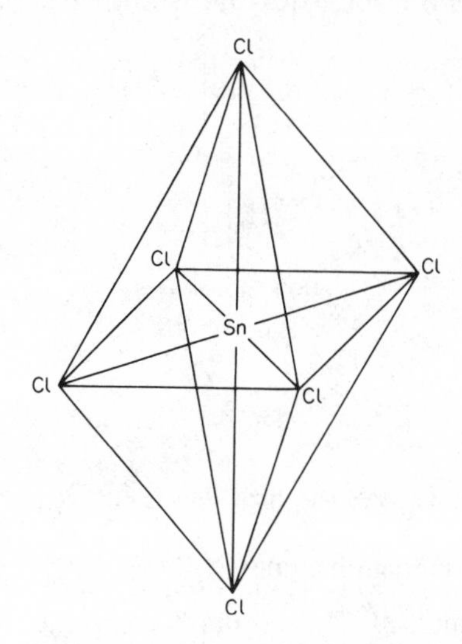

FIGURE 5.40. Coordination number 6

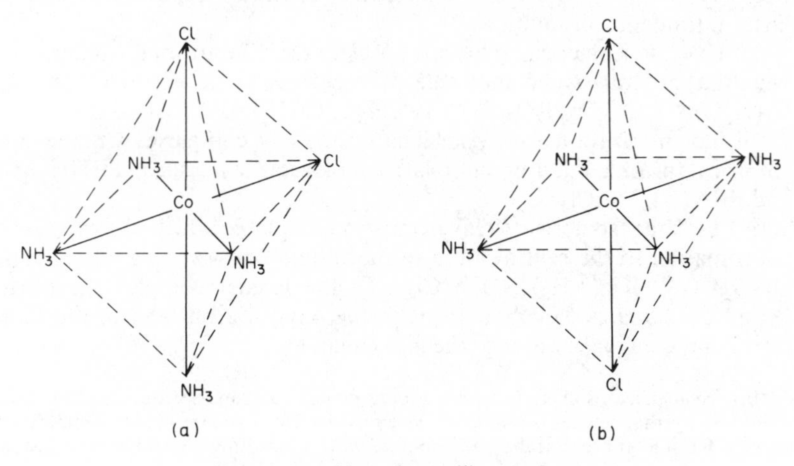

FIGURE 5.41 (*a*) *cis*-form; (*b*) *trans*-form

of which is green and the other violet (Figure 5.41). This is an example of *geometrical isomerism.*

The complex $Co(en)_2Cl_2^+$ (where en = ethylene diamine) is another example of geometrical isomerism. Here, however, there are two different arrangements of the *cis* form, one being a non-superimposable mirror image of the other. Such isomerism is known as *optical isomerism* and the isomers are called optical isomers. Optical isomers rotate the plane of plane polarized light either to the right (dextro-rotatory) or to the left (laevo-rotatory). The isomers of $Co(en)_2Cl_2^+$ are shown in Figure 5.42.*

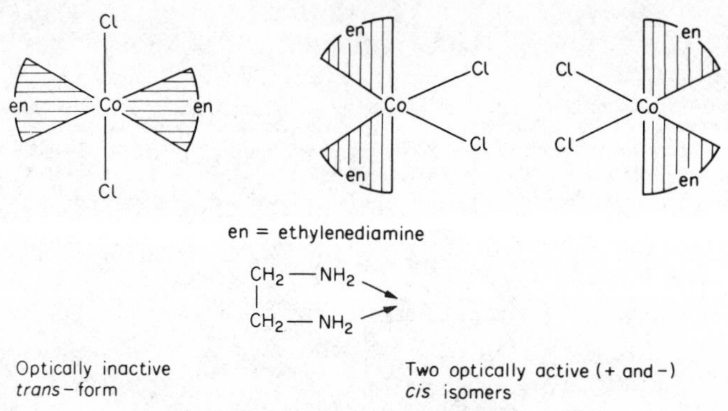

FIGURE 5.42. The three isomers of $Co(en)_2Cl_2^+$

Other types of isomerism include:

(i) Ionization isomerism. This occurs in complexes which have the same formula but yield different ions in solution, e.g. $[Co(NH_3)_5NO_3]^{2+}SO_4^{2-}$ $[Co(NH_3)_5SO_4]^+NO_3^-$. The former isomer furnishes sulphate ions and the latter nitrate ions in solution.

(ii) Hydrate isomerism. This is a special case of ionization isomerism in which a neutral water molecule is replaced by a negative ion, e.g. $Cr(H_2O)_6^{3+}Cl_3^-$, $Cr(H_2O)_5Cl^{2+}Cl_2^-.H_2O$, $Cr(H_2O)_4Cl_2^+Cl^-.2H_2O$.

(iii) Polymerization isomerism. This occurs with complexes with the same empirical formula but different relative molecular masses, e.g. $Pt(NH_3)_2Cl_2$ and $Pt(NH_3)_4^{2+}PtCl_4^{2-}$.

(iv) Linkage isomerism. This occurs in complexes which are capable of coordinating to the central atom in more than one way, e.g. $[Co(NH_3)_5\text{-}ONO]^{2+}Cl_2^-$ $[Co(NH_3)_5NO_2]^{2+}Cl_2^-$. In the former complex the nitrito ligand coordinates to cobalt through the oxygen atom and in the latter through the nitrogen atom of the nitro ligand.

*For geometrical isomerism with a square planar complex, see p. 357. The reader should verify that optical isomerism is not possible for a square planar complex and is only feasible for a tetrahedral complex with four different unidentate ligands. Geometrical isomerism is not possible for a tetrahedral complex.

(v) Coordination isomerism. This occurs in complexes in which the ligands in a combined anion–cation complex can be interchanged between different central atoms, e.g. $Co(NH_3)_6^{3+}Cr(CN)_6^{3-}$, $Cr(NH_3)_6^{3+}Co(CN)_6^{3-}$.

The arrangement in space of the coordinating groups can be explained on the basis of hybridization of orbitals (*see* Figure 5.43). The Ni^{2+} ion has an observed magnetic moment of 3·2 Bohr magnetons, indicative of two unpaired electrons. The $[Ni(CN)_4]^{2-}$ complex has zero moment, dsp^2 hybridization occurring to give a planar structure. The $[NiCl_4]^{2-}$ ion has a moment, indicating two unpaired electrons, showing that sp^3 hybridization of orbitals occurs giving a tetrahedral arrangement. The magnetic moments of the Fe^{2+} and Fe^{3+} ions indicate the presence of four and five unpaired electrons respectively. The $[Fe(CN)_6]^{4-}$ ion has zero moment and hence has no unpaired electron: d^2sp^3 hybridization occurs giving an octahedral structure. This type of hybridization also occurs with the $[Fe(CN)_6]^{3-}$ complex which contains one unpaired electron and hence has a magnetic moment of 1·7 Bohr magnetons. Another theoretical approach to bonding in complexes is the crystal field theory which is outlined in Chapter 18.

The stability of complexes

A measure of the stability of a complex is given by its equilibrium or stability constant. Consider the addition of cyanide ions to a solution containing hydrated iron(II) cations:

$$Fe(H_2O)_6^{2+} + 6CN^- \rightleftharpoons Fe(CN)_6^{4-} + 6H_2O$$

$$K_{stab} = \frac{[Fe(CN)_6^{4-}]}{[Fe(H_2O)_6^{2+}][CN^-]^6} = 10^{37}$$

where K_{stab} is the stability constant for the complex at 298 K. The greater the numerical value of K_{stab}, the greater is the stability of the complex. Some examples of complex equilibria, together with the K_{stab} and $\log_{10} K_{stab}$ (pK_{stab}) values at 298 K, are given in Table 5.4.

One ligand will usually substitute another in a complex if the ligand replacement results in the formation of a complex with a much greater stability constant than that of the original complex. During the addition of EDTA solution to a yellow solution containing $CuCl_4.2H_2O^{2-}$, ligand exchange occurs and the colour changes to blue:

$$CuCl_4.2H_2O^{2-} + EDTA^{4-} \rightleftharpoons CuEDTA^{2-} + 4Cl^- + 2H_2O$$

TABLE 5.4

Equilibria	K_{stab}	pK_{stab}
$Ni(H_2O)_6^{2+} + 6NH_3 \rightleftharpoons Ni(NH_3)_6^{2+} + 6H_2O$	$4{\cdot}8 \times 10^7$	7·7
$Ni(H_2O)_6^{2+} + 3en \rightleftharpoons Ni(en)_3^{2+} + 6H_2O$	$4{\cdot}0 \times 10^{18}$	18·6
$Cu(H_2O)_6^{2+} + 4Cl^- \rightleftharpoons CuCl_4.2H_2O^{2-} + 4H_2O$	$4{\cdot}2 \times 10^5$	5·6
$Cu(H_2O)_6^{2+} + EDTA^{4-} \rightleftharpoons CuEDTA^{2-} + 6H_2O$	$6{\cdot}3 \times 10^{18}$	18·8

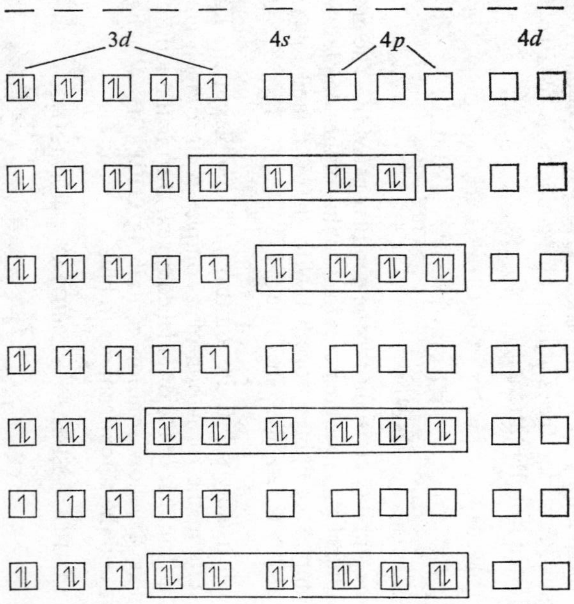

	3d					4s	4p			4d		*Hybridization*	*Structure*	*Magnetic moment (Bohr magneton)*
Ni^{2+}	↑↓	↑↓	↑↓	↑	↑									3·2
$Ni(CN)_4^{2-}$	↑↓	↑↓	↑↓	↑↓	[↑↓	↑↓	↑↓	↑↓]				*d sp²*	Square planar	0
$NiCl_4^{2-}$	↑↓	↑↓	↑↓	↑	↑	[↑↓	↑↓	↑↓	↑↓]			*sp³*	Tetrahedral	3·2
Fe^{2+}	↑↓	↑	↑	↑	↑									5·3
$Fe^{II}(CN)_6^{4-}$	↑↓	↑↓	↑↓	[↑↓	↑↓	↑↓	↑↓	↑↓	↑↓]			*d²sp³*	Octahedral	0
Fe^{3+}	↑	↑	↑	↑	↑									5·9
$Fe^{III}(CN)_6^{3-}$	↑↓	↑↓	↑	[↑↓	↑↓	↑↓	↑↓	↑↓	↑↓]			*d²sp³*	Octahedral	1·7

FIGURE 5.43

Generally chelation results in the formation of very stable complexes, the most stable being those in which a strainless ring is formed consisting of five or six atoms. In the chelation of the hydrated calcium ion, two reactant molecules form seven molecules of product on ligand exchange resulting in a large increase in entropy (ΔS is positive) (p. 115).

$$Ca(H_2O)_6^{2+} + EDTA^{4-} \rightleftharpoons CaEDTA^{2-} + 6H_2O$$

Since $\Delta G = \Delta H - T\Delta S$, a positive ΔS value renders ΔG more negative and thus increases the possibility of reaction.

The nomenclature of complexes

The International Union of Pure and Applied Chemistry (I.U.P.A.C.) has recommended the following system of nomenclature for complexes:

(*a*) In naming the complex the ligand is named first and then the central atom or ion.

(*b*) Ligands of a mixed complex are named in the following order: negative ligands, neutral ligands, and then positive ligands. Where there is more than one ligand of the same kind, the prefixes di (bi), tri, tetra, etc., are used. Where the ligand also possesses a number (as in ethylene *di*amine) the prefixes bis, tris, and tetrakis are used. Negative anionic ligands end in -o, e.g. X^- is halo (chloro, bromo, iodo), CN^- cyano, NO_3^- nitrato, NO_2^- nitro, ONO^- nitrito, OH^- hydroxo, CO_3^{2-} carbonato, $\begin{matrix} COO^- \\ | \\ COO^- \end{matrix}$ oxalato, etc. Neutral or cationic ligands have the same name apart from aquo for water and ammine for ammonia.

(*c*) When the coordination sphere (i.e. central atom or ion plus ligand) is positive or neutral the name of the central atom or ion remains unaltered, followed by a roman numeral to indicate oxidation state of the atom or ion. When negative, the suffix -ate is added to the central atom or ion followed again by the oxidation number in roman numerals. The rules enumerated above are illustrated by the following examples:

$[Co(NH_3)_6]Cl_3$	Hexamminecobalt(III) chloride
$[Cr(H_2O)_4Cl_2]Cl$	Dichlorotetraaquochromium(III) chloride
$K_4Fe(CN)_6$	Potassium hexacyanoferrate(II).
$Fe(CN)_6^{3-}$	Hexacyanoferrate(III) anion.
$[Co(NO_3)_3(NH_3)_3]$	Trinitratotriamminecobalt(III).
$[Co(NO_2)_6]^{3-}$	Hexanitrocobaltate(III) anion.
$[Co(en)_3^{3+}]Cl_3^-$	Tris(ethylenediamine) cobalt(III) chloride.

(*d*) Bridging ligands are indicated by prefixing the names by μ:

$K_2^+CrO_3.O.CrO_3^{2-}$ Potassium μ-oxobistrioxochromate(VI)

6

The Motive Force behind Chemical Reactions

Before proceeding to discuss the chemistry of the elements, their compounds, and the reactions which they undergo, it is necessary to consider briefly the motive or driving force behind chemical reactions. Such considerations form a part of chemistry called *thermodynamics* and it is the task of thermodynamics to show why chemical reactions occur and to measure the tendency for reaction under stated conditions.

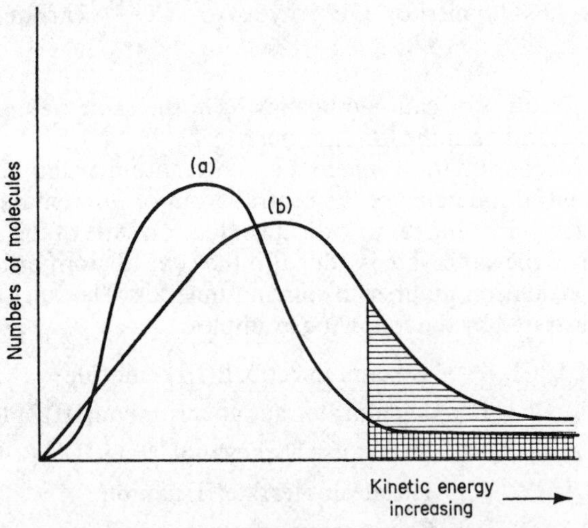

FIGURE 6.1

Consider any molecular system. At a particular temperature, the distribution of kinetic energy among these molecules is described by a distribution curve (Figure 6.1(*a*)) where the majority of the molecules have kinetic energies close to some average value but few have very much lower or higher than the average kinetic energy. However, as the temperature is raised, the number of molecules with high kinetic energy increases (Figure 6.1(*b*)). All collisions between low-energy molecules will simply result in rebound, but collisions between high-energy molecules may cause bond breakage with subsequent chemical reaction. The extra energy above the average which molecules must have in order to cause the breakage of chemical bonds is called the *activation energy* for the reaction.

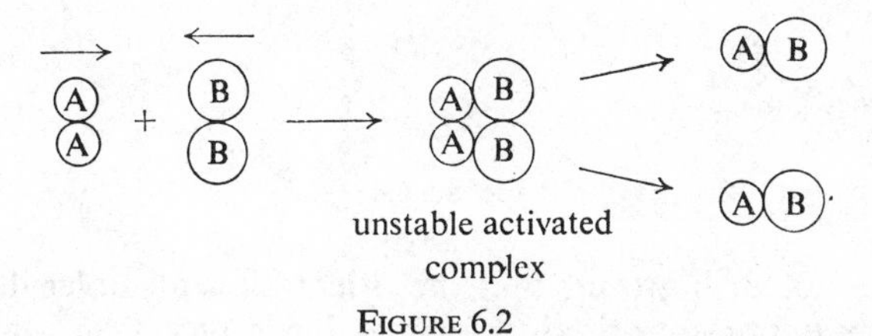

FIGURE 6.2

In a reaction between molecules A—A and B—B, collisions must occur between the molecules of A—A and those of B—B and energy is required in the first instance to stretch and break the bonds in the reactants before the reaction can proceed to give AB. One possible reaction mechanism is shown in Figure 6.2. The energy necessary to do this is the activation energy

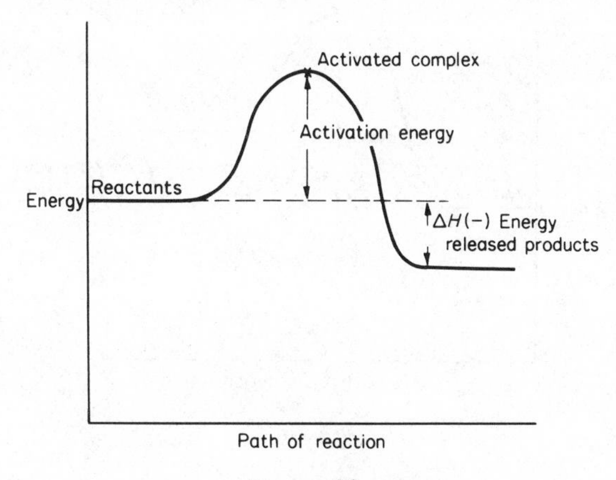

FIGURE 6.3

for the reaction and only those collisions which have this quantity of energy available will be effective. The collisions often result in the formation of an intermediate activated complex which is unstable and breaks down to give the reaction products. The reaction sequence may be represented pictorially (Figure 6.3).

In the reaction shown, the products have a lower energy than the reactants and the excess energy (ΔH) appears as the heat of reaction, i.e. the reaction is *exothermic*. The quantity ΔH is the change in heat content (or *enthalpy*) of the system at constant pressure and since, in an exothermic reaction, the system loses heat, it is designated *negative*. Exothermic reactions give products with a lower energy content (and, therefore, of greater stability) than the reactants and, therefore, generally proceed quite readily.

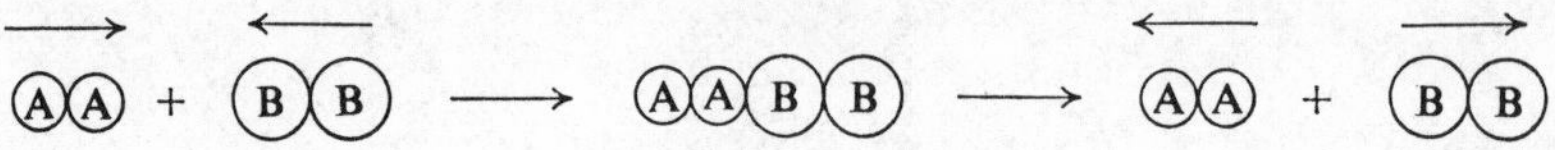

$$\overrightarrow{\text{AA}} + \overleftarrow{\text{BB}} \longrightarrow \text{AABB} \longrightarrow \overleftarrow{\text{AA}} + \overrightarrow{\text{BB}}$$

FIGURE 6.4

It must be noted, however, that the orientation of molecules prior to collision is an important criterion for reaction. Thus, even if high energy molecules collide, reaction will not occur unless the molecules are correctly orientated (*see* Figure 6.4).

Although increase of temperature increases the kinetic energy of the molecules and the likelihood of reaction, it has no effect whatsoever on the possible orientation of the molecules. The rate of a reaction can be altered

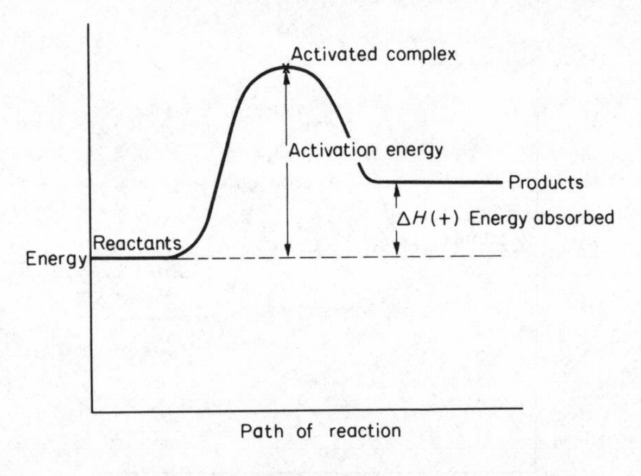

FIGURE 6.5

by the introduction of certain substances called *catalysts* which remain chemically unchanged at the end of the reaction. The catalyst may provide a new reaction path for the reacting molecules with lower activation energy, or orientate the reacting molecules in the correct position for reaction. Both factors allow for a more rapid reaction at a particular temperature.

In a reaction where the products have a greater energy than the reactants, heat is absorbed and the reaction is *endothermic* (Figure 6.5). Since the system gains heat, the change in heat content, ΔH (enthalpy), is designated *positive*. All endothermic reactions give products of higher energy than the reactants and generally do not proceed very readily.

Thermochemistry is the study of the relationship between heat changes in chemical reactions. In many cases, the direct determination of the heat of a reaction is not possible. Thus the *heat of formation* (ΔH_f) (the heat of reaction when 1 mole of a compound is formed from its elements) of a hydrocarbon cannot be measured directly since carbon will not combine directly with hydrogen. The quantity ΔH_f can, however, be found if the *heats of combustion* (the heat of reaction when 1 mole of a compound is burnt completely in oxygen) of carbon, hydrogen, and the hydrocarbon are known, since it has been shown experimentally that the heat change in any chemical reaction is the same whether it is carried out in one or several stages. This is in fact a statement of *Hess's law of constant heat summation*, which implies that heats of reaction can be added and subtracted algebraically. Thus the heat of formation of benzene may be found from the heats of combustion of carbon, hydrogen, and benzene.

$$C(s) + O_2(g) \rightarrow CO_2(g) \qquad \Delta H = -394 \text{ kJ mol}^{-1} \qquad (1)$$

$$H_2(g) + \tfrac{1}{2}O_2(g) \rightarrow H_2O(l) \qquad \Delta H = -286 \text{ kJ mol}^{-1} \qquad (2)$$

$$C_6H_6(l) + 7\tfrac{1}{2}O_2(g) \rightarrow 6CO_2(g) + 3H_2O(l) \qquad \Delta H = -3273 \text{ kJ mol}^{-1} \qquad (3)$$

The heat of formation (ΔH_f) of benzene

$$6C(s) + 3H_2(g) \rightarrow C_6H_6(l)$$

is obtained by multiplying equation (1) by 6, and equation (2) by 3, adding, and then subtracting equation (3).

$$\Delta H_f \text{ (benzene)} = 6 \times (-394) + 3 \times (-286) - (-3273)$$
$$= +51 \text{ kJ mol}^{-1}$$

Hess's law is a manifestation of the *first law of thermodynamics* which states that energy can neither be created nor destroyed. Thus the energy change in a chemical process is independent of the nature of the intermediate steps and depends only upon the initial and final states.

Hess's law can also be used to construct thermodynamic cycles which are used for the evaluation of bond enthalpies. For a diatomic molecule X—Y, the bond enthalpy (ΔH) is defined as the enthalpy change for the process

$$X\text{—}Y(g) \rightarrow X(g) + Y(g)$$

Consider the following cycle in which ΔH_f is the molar enthalpy of formation of liquid water from its gaseous elements, ΔH_a is the enthalpy of atomization i.e. the enthalpy of formation of one mole of gaseous atoms from the gaseous element, ΔH_v is the molar enthalpy of vaporization for the transition liquid to vapour.

$$\begin{array}{ccc} H_2(g) + \tfrac{1}{2}O_2(g) & \xrightarrow{\Delta H_f} & H_2O(l) \\ \downarrow 2\times\Delta H_{a1} \quad \downarrow \Delta H_{a2} & & \downarrow \Delta H_v \\ 2H(g) + O(g) & \xleftarrow{\Delta H} & H_2O(g) \end{array}$$

Using Hess's law:

$$\Delta H_f + \Delta H_v + \Delta H = 2 \times \Delta H_{a1} + \Delta H_{a2}$$

Substituting the appropriate numerical thermochemical quantities in kJ mol^{-1},

$$-286 + 41{\cdot}1 + \Delta H = +2 \times 218 + 248$$
$$\Delta H = 928{\cdot}9 \text{ kJ mol}^{-1}$$

Since the water molecule contains two —OH bonds the bond enthalpy of a single —OH bond is

$$\frac{928{\cdot}9}{2} = 464{\cdot}4 \text{ kJ mol}^{-1}$$

Thermochemical cycles can also be used to evaluate hydration enthalpies for a pair of ions, e.g. Na^+ and Cl^-. The hydration enthalpy for an ion (ΔH_{hyd}) refers to the process

$$M^{z+}(g) + aq \rightarrow M^{z+}(aq)$$

Consider the following cycle in which U is the lattice energy and ΔH_{sol} the enthalpy of solution when one mole of substance is dissolved in sufficient water so that dilution produces no further heat change.

$$\begin{array}{ccc} Na^+Cl^-(s) & \xrightarrow{-U} & Na^+(g) + Cl^-(g) \\ \downarrow \Delta H_{sol} & \swarrow \Delta H_{hyd}^{Na+} \quad \swarrow \Delta H_{hyd}^{Cl-} & \\ Na^+(aq) + Cl^-(aq) & & \end{array}$$

Using Hess's law,

$$\Delta H_{sol} = -U + \Delta H_{hyd}^{Na+} + \Delta H_{hyd}^{Cl-}$$

Unfortunately, it is much more difficult to evaluate single hydration energies. Consider the cycle below involving the hydration enthalpy of sodium ions:

$$\begin{array}{ccccc} Na(s) & \xrightarrow{\Delta H_a} & Na(g) & \xrightarrow[\text{Ionization potential}]{I_p} & Na^+(g) + e(g) \\ & \searrow \Delta H_{sol} & & & \downarrow \Delta H_{hyd}^{Na+} \quad \downarrow \Delta H_{hyd}^{e} \\ & & & & Na^+(aq) + e(aq) \end{array}$$

The difficulties involved in the evaluation of accurate absolute values for the hydration energy of the electron and the heat of solution (ΔH_{sol}) of aqueous sodium ions from solid sodium can be appreciated and it is only comparatively recently that acceptable values have become available.

As stated above, exothermic reactions are more likely to occur spontaneously (i.e. without the help of an external influence) than endothermic ones. Nevertheless there are many examples of spontaneous processes which are endothermic even though energy is required for the reaction, e.g. the dissolution of ammonium chloride in water. Thus energy is not the only criterion for chemical reactivity.

If we consider a process in which there is no change in energy of the system, such as the mixing of two gases which do not react, there is an increase in the degree of disorder (or randomness) of the system. The mixing will occur spontaneously when the gases are placed in contact. Since the reverse process of separating the gases would be difficult, it is clear that the more random state is the more probable one. A measure of the degree of randomness of a system is given by a thermodynamic function, the *entropy* factor (S).

For the change: crystalline solid $\rightarrow$ liquid $\rightarrow$ gas the randomness and, therefore, the entropy is increasing. The reverse process, however, involves an increase in orderliness and hence a decrease in entropy. A crystalline solid at absolute zero is said to possess zero entropy.

In all natural spontaneous processes, such as the mixing of gases, the dissolution of a solute in a solvent, the burning of coal or wood, etc., there is an increase in the entropy of the system. These generalizations are embodied in the *second law of thermodynamics* which states that the total energy of the universe remains constant but the entropy approaches a maximum.

Chemical reactions tend to take place in such a direction as to give a system with the minimum of energy but the maximum randomness or entropy. The tendency of a chemical reaction to occur at constant temperature and pressure can be expressed in a mathematical form by the equation:

$$\Delta G = \Delta H - T\Delta S$$

where ΔH is the change in heat content (J mol^{-1}), T is the temperature in degrees Kelvin, and ΔS is the change in entropy (JK^{-1} mol^{-1}). The quantity ΔG (J mol^{-1}) is the change in *free energy* of the system, i.e. the free energy available for work, and is a measure of the tendency of a reaction to occur.

The change in free energy ΔG in a chemical reaction is the sum of the free energies of the products (G_p) less the sum of the free energies of the reactants (G_r). Thus if:

(1) $\Delta G = 0$, $G_p = G_r$, no work is available and the system is in equilibrium.

(2) ΔG is positive, $G_p > G_r$ and the reaction will not occur.

(3) ΔG is negative, $G_p < G_r$ and the reaction is thermodynamically feasible.

Thus reactions in which ΔH is positive may still be feasible if $T\Delta S$ is positive and larger than ΔH, giving a negative value for ΔG.

At 0 K or where there is no change in entropy, $T\Delta S$ is zero and $\Delta G = \Delta H$. In this case, the reaction may still be feasible if ΔH is negative. Cases in which ΔS is zero are mainly of interest to the physicist.

It must be noted, however, that, although ΔG may be negative, the rate of reaction may be very slow since the thermodynamic approach above has taken no account of reaction rates.

In the production of water gas (p. 209):

$$C + H_2O \rightarrow CO + H_2 \qquad \Delta H = +131{\cdot}4 \text{ kJ mol}^{-1}$$

the entropy change (ΔS) for the reaction, calculated from thermodynamic data, is +134 JK^{-1} mol^{-1}. At 25 °C

$$\Delta G = 131\ 400 - (298 \times 134)$$
$$= +91\ 400 \text{ J mol}^{-1}$$

Thus the reaction is not feasible at 25 °C. At 1000 °C

$$\Delta G = 131\ 400 - (1273 \times 134)$$
$$= -39\ 700 \text{ J mol}^{-1}$$

The net free energy change is large and negative and the reaction proceeds.

Fuel Cells

Fuel cells are voltaic cells in which gases are consumed to produce electricity. Although still in the pilot state, they may be very important commercially in the near future as storage devices and small power generators.

In one type of cell, gaseous hydrogen and oxygen under pressure are bubbled through porous nickel electrodes immersed in an electrolyte of potassium hydroxide solution at about 200°C. The electrode reactions which occur are:

Anode	$H_2 + 2OH^- \rightarrow 2H_2O + 2e$
Cathode	$\frac{1}{2}O_2 + H_2O + 2e \rightarrow 2OH^-$
Overall reaction	$H_2 + \frac{1}{2}O_2 \rightarrow H_2O$

For the overall reaction: the free energy change $\Delta G = -237$ kJ mol^{-1}.

It can be shown that the change in free energy is related to the e.m.f. of a cell (E) by the expression:

$$\Delta G = -nFE$$

where n is the number of electrons involved in the reaction (=2), F is the Faraday (96 483 coulombs), and E is the e.m.f. of the cell in volts.

$$237\,000 = 2 \times 96\,483 \times E$$

$$\text{and } E = 1{\cdot}23 \text{ V}$$

The thermal efficiency of a fuel cell will depend upon ΔH where

$$\Delta G = \Delta H - T\Delta S$$

Whether the thermal efficiency of a fuel cell is greater or less than its free energy will depend upon the magnitude and sign of $T\Delta S$.

Other fuels such as hydrocarbons, alcohols, etc., have also been employed at high temperatures using chromium and manganese oxide catalysts.

$$CH_3OH + 1\tfrac{1}{2}O_2 \rightarrow CO_2 + 2H_2O$$
$$\Delta G = -702 \text{ kJ mol}^{-1}$$

$$C_3H_8 + 5O_2 \rightarrow 3CO_2 + 4H_2O$$
$$\Delta G = -2105 \text{ kJ mol}^{-1}$$

A recent use of the hydrogen/oxygen fuel cell has been to supply power (and drinking water) for astronauts in the Apollo manned space programme to the moon.

Free Energy and Chemical Equilibrium

The following relationship between the change in free energy in a reaction (ΔG) and the equilibrium constant for that reaction can be derived from thermodynamic data:

$$\Delta G = \Delta G^{\ominus} + RT\log_e K$$

where ΔG is the change in free energy and $\Delta G^{\ominus}$ is the standard free energy change for substances in their standard states at 298 K and 101·325 kN m^{-2} (1 atmosphere). For substances in solution all concentrations are at unit activity (1 mol l^{-1}). The equilibrium constant K can be expressed in terms of partial pressures (K_p) or in terms of molar concentrations (K_c). For a system in equilibrium $\Delta G = 0$; therefore,

$$\Delta G^{\ominus} = -RT \log_e K = -2{\cdot}303\, RT \log_{10} K$$

This relationship is important in that it may be used to calculate equilibrium constants from thermodynamic data or vice versa. Also, since $\Delta G^{\ominus} = -nFE^{\ominus}$, electrode potentials can be used to evaluate equilibrium constants for ionic reactions.

Consider the evaluation of the equilibrium constant, K_c for the reaction between aqueous solutions of iron(II) sulphate and silver(I) nitrate:

$$Ag^+ + Fe^{2+} \rightleftharpoons Ag + Fe^{3+}$$

The half reactions are:

$$Fe^{3+} + e \rightarrow Fe^{2+} \qquad E^{\ominus} = +0{\cdot}76\text{ V},\ \Delta G_1^{\ominus} = -0{\cdot}76F$$
$$Ag^+ + e \rightarrow Ag \qquad E^{\ominus} = +0{\cdot}80\text{ V},\ \Delta G_2^{\ominus} = -0{\cdot}80F$$

The change in standard free energy for the reaction ($\Delta G^{\ominus}$) is given by

$$\Delta G^{\ominus} = \Delta G_2^{\ominus} - \Delta G_1^{\ominus}$$

$$= -0{\cdot}80F - (-0{\cdot}76F)$$

$$= -0{\cdot}04F$$

$$= -0{\cdot}04 \times 96\,480$$

$$= -3859\text{ kJ}$$

Since $\Delta G^{\ominus} = -2{\cdot}303\ RT \log_{10} K_c$,

$$-3859 = -2{\cdot}303 \times 8{\cdot}31 \times 298 \log_{10} K_c$$

The equilibrium constant K_c is 4·75 at 298 K.

Suppose it is required to calculate the standard free energy change for the reaction

$$N_2(g) + 3H_2(g) \rightleftharpoons 2NH_3(g)$$

(*a*) at 298 K where $K_p = 6{\cdot}76 \times 10^5$, and
(*b*) at 773 K where $K_p = 3{\cdot}55 \times 10^{-2}$.

$$\Delta G_{298}^{\ominus} = -2{\cdot}303\ RT \log_{10} 6{\cdot}76 \times 10^5$$
$$= -2{\cdot}303 \times 5{\cdot}83RT$$
$$= -33{\cdot}2\text{ kJ}$$

$$\Delta G_{773}^{\ominus} = -2{\cdot}303RT \log_{10} 3{\cdot}55 \times 10^{-2}$$
$$= +2{\cdot}303 \times 1{\cdot}45RT$$
$$= +21{\cdot}45\text{ kJ}$$

Generally, if $\Delta G^{\ominus}$ is numerically larger and more negative than –60 kJ, the reaction will go to completion, but if $\Delta G^{\ominus}$ is numerically larger and more positive than +60 kJ, no reaction takes place. The $\Delta G_{298}{}^{\ominus}$ value (–33·2 kJ) indicates that the product, ammonia, predominates in the equilibrium mixture. Unfortunately, the reaction rate is too slow and a long time is taken to reach equilibrium. The $\Delta G_{773}{}^{\ominus}$ value (+21·45 kJ) indicates that the reactants predominate in the equilibrium mixture. However, at this temperature a greater reaction rate is achieved and hence, even though the equilibrium position is unfavourable, this is the temperature actually employed in the process (p. 245).

7

Extractive Metallurgy

The Structure of the Earth

Information concerning the composition and structure of the earth's crust comes mainly from two sources, meteorites and seismology.

Meteorites

These give information about the internal structure of other planets from which deductions can be made regarding our own. Meteorites consist mainly either of a nickel–iron alloy or of silicates (olivine) and glassy minerals, although grains of metallic iron often occur. Other constituents include cobalt, nickel, graphite, and the sulphides and phosphides of iron, cobalt, and nickel.

Seismology

Shock waves from earthquakes are refracted and reflected as they travel through the earth and on passing through different sections of the earth the wavelength and frequency of the waves alters. Studies of these waves using an instrument known as the *seismograph* indicate three main zones in the earth's interior. These zones and their probable composition are shown in Figure 7.1.

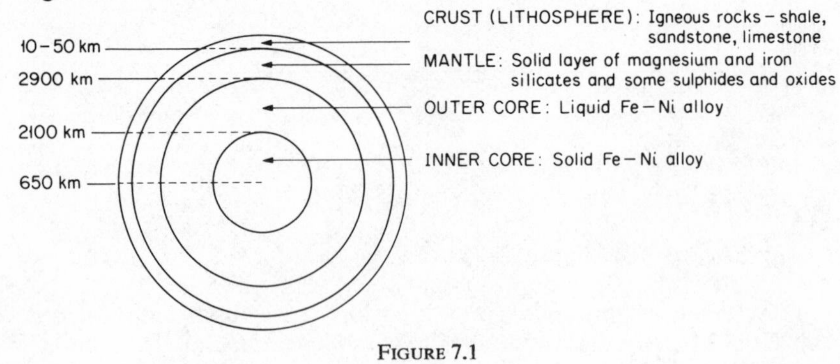

FIGURE 7.1

Temperatures and pressures increase markedly as the core is reached, values of 4 000 000 times atmospheric pressure and 4000 °C being estimated for the core itself.

The Origin and Chemical Nature of the Zones

In the early stages of its formation the earth was in a molten state and the most abundant elements were probably oxygen, silicon, aluminium, iron, calcium, sodium, magnesium, nickel, and sulphur. As the earth cooled the bulk of the iron and nickel sank towards the centre of the earth and a liquid layer of magnesium and iron silicates and sulphides formed round the core. Further cooling caused the crystallization of the iron and magnesium silicates (the olivine minerals) at the inner zone between the mantle and the outer core, which formed a crust round the core. This crust reduced the rate of cooling of the core and maintained the outer core in the liquid state. As crystallization in this zone continued, the remaining mantle liquid became richer in oxygen, aluminium, calcium, potassium, and sodium. Crystallization then began to take place in the outer zone of the mantle, forming minerals such as quartz (SiO_2), olivine $(MgFe)_2SiO_4$, feldspar $KAlSi_3O_8$, and sodium, potassium, and calcium aluminosilicates.

Aggregation of these minerals then took place forming igneous rocks of two main types:

(*a*) *light rocks*, i.e. granite and related types (specific gravity 2·7, containing 70 per cent SiO_2), together with Al_2O_3—aluminium is the second most abundant element in these rocks.

(*b*) *heavy rocks*, i.e. basalt (specific gravity 2·9–3·0, containing 40–50 per cent SiO_2) together with MgO—magnesium is the second most abundant element in these rocks.

Convection currents both in the air above and in the molten melt below the crust probably caused a continual break up and rebuilding of the crust giving rise to high (continent) and low (ocean) regions. Support for this theory comes from the nature of the continents where the upper crust is granite, whereas ocean beds and volcanoes are basaltic. Hence it is assumed that underneath the granite layer there is a basaltic layer. The earth's crust, however, also suffered erosion by water, carbon dioxide, and air, producing sedimentary material (clay, sand, and mud) which deposited in rivers, lakes, and oceans. Continual deposition of sedimentary material on the earth's surface caused pressure to be exerted on the lower layers, squeezing out the water and cementing the particles together, forming sedimentary rocks. These rocks comprise about 5 per cent of the earth's crust and consist mainly of shale, limestone, and sandstone. It is these sedimentary rocks which provide the most readily available source of ores.

The primitive atmosphere consisted of carbon dioxide, methane, ammonia, hydrogen, and water vapour. As the temperature fell the water vapour condensed, falling as heavy rain for thousands of years and forming the oceans which once again subjected the crust to erosion. The methane, hydrogen, ammonia, and water vapour, under the influence of heavy electrical storms formed some free nitrogen and simple organic molecules—the forerunners of life. The oxygen in the atmosphere may have been produced by plants containing chlorophyll, during photosynthesis:

$$6CO_2 + 6H_2O \xrightarrow{\text{sunlight}} C_6H_{12}O_6 + 6O_2$$

The Abundance of the Elements

Some 103 elements are known at the present time: eighty-eight have been found in nature and fifteen have been synthesized. Oxygen constitutes approximately one-half and silicon one-quarter of the total weight of elements in the earth's crust. The sixteen most abundant elements in the earth's crust and their percentages by weight are shown in Table 7.1.

TABLE 7.1

Atomic number	*Element*	*Percentage*	*Atomic number*	*Element*	*Percentage*
8	Oxygen	49·5	1	Hydrogen	0·9
14	Silicon	25·7	22	Titanium	0·6
13	Aluminium	7·5	17	Chlorine	0·2
26	Iron	4·7	15	Phosphorus	0·1
20	Calcium	3·4	25	Manganese	0·09
11	Sodium	2·6	6	Carbon	0·08
19	Potassium	2·4	16	Sulphur	0·06
12	Magnesium	1·9	56	Barium	0·04

Total percentage weight = 99·77

Two points may be noted regarding the relative abundance of elements: (*a*) elements with low atomic numbers constitute a major portion of the earth's crust; and (*b*) elements with even atomic numbers are generally more abundant than elements with odd atomic numbers.

There are, however, some notable exceptions to (*b*). The noble gases (even atomic numbers) occur only to a minute extent and probably escaped into the atmosphere during the earth's molten stage. Also the elements chromium, sulphur, selenium, and tellurium only occur in small quantities in the crust due to absorption into the earth's interior during the cooling process. It is important to realize, however, that in extracting elements from the earth's crust, availability is much more important than abundance. Uranium, for example, is more abundant than silver or mercury but is uniformly distributed and there are no large concentrated deposits.

The Occurrence of Metals

Metals with electrode potentials greater than hydrogen are rarely found in the free state, whereas those with values below hydrogen often occur in the native condition, e.g. silver, gold, copper, and mercury. Another useful generalization is that metallic compounds found in the crust usually have a low solubility in water. The more soluble, naturally occurring compounds are found either in sea water or in the large salt-bed deposits formed by the evaporation of inland seas. The naturally occurring compounds in the earth's crust are known as *minerals* and those which can be used as a source of commercial materials are termed *ores*. The ores usually contain a high percentage of earthy material called *gangue*.

Some important classes of ores are listed in Table 7.2.

TABLE 7.2

Ore	*Metals present*
Native ores	Cu, Ag, Au, Hg, As, Sb, Bi
Oxide ores	Fe, Al, Mn, Sn, Ti, Zr
Sulphide ores	Zn, Cd, Hg, Cu, Pb, Ni, Mo, As, Sb, Bi, Ag
Carbonate ores	Fe, Pb, Zn, Cu (usually less important than oxide and sulphide ores except for the alkaline earth metals Mg, Ca, Sr, Ba)
Halide ores	Na, K, Rb, Cs, Mg, Ca, Ag
Sulphate ores	Ca, Sr, Ba, Mg, Pb
Silicate ores	Li, K, Na, Mg, Ca, Be, Al, Fe, Zn, Ni

The silicate ores are not generally worked at the present time because of difficulties attending the extraction of the element but, as conventional mineral resources become exhausted, these may become more important once more and economical extraction techniques become available (p. 416).

Vast quantities of metallic salts are also held in solution in the oceans. The principal ions in solution are Na^+, Mg^{2+}, Ca^{2+}, K^+, Sr^{2+}, Cl^-, SO_4^{2-}, HCO_3^-, F^-, Cl^-, Br^-. One hundred kilogrammes of sea water yield 2·46 kg of sodium chloride, 0·11 kg of magnesium, 0·033 kg of calcium, and 0·045 kg of bromine. Already sea water is an important commercial source of 'solar salt' on the west coast of the U.S.A., where sea water is allowed to evaporate. Bromine and magnesium are now also extracted on a large scale from sea water (p. 178). In the future sea water may become an important source of many materials such as uranium, gold, etc.

Extractive metallurgy

There are three main stages in the extraction of metals from ores in the earth's crust: (*a*) concentration of the ore, (*b*) smelting, and (*c*) refining.

Concentration of the ore

These processes consist of removing most of the gangue, thus rendering the concentrate suitable for smelting. The ore is graded, pulverized, and ground to a powder in crushers and grinders and is then concentrated to remove the gangue. The methods employed utilize:

(*i*) *Vibrators*: inclined platforms rock rapidly to and fro separating the lighter gangue from the heavier ore.

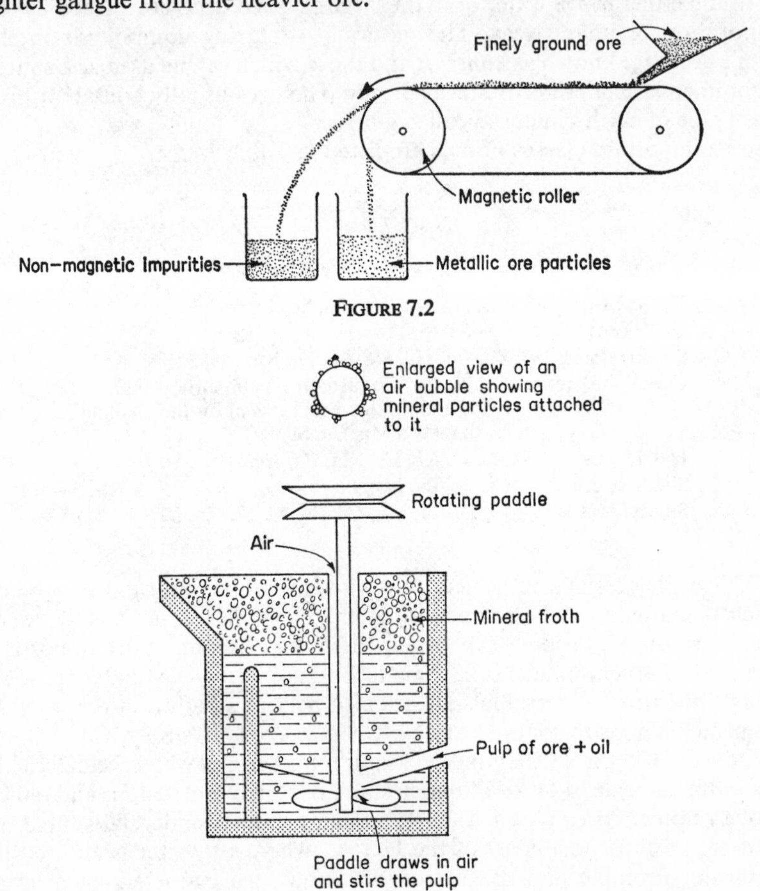

FIGURE 7.2

FIGURE 7.3

(*ii*) *Magnetic separators*: metallic ore is often separated from non-metallic gangue by passing the ground ore through a magnetic or electrostatic field (Figure 7.2).

(*iii*) *Oil flotation*: this is a very useful technique for separating zinc, copper, and lead sulphide ores from the gangue. A suspension of the finely powdered ore is made with water. To the suspension are added: *collectors* (e.g. xanthates, fatty acids, and pine oils), which enhance the non-wettability of the mineral particles, and *froth stabilizers* (e.g. cresols and aniline) which stabilize the resulting froth. The principle of the process is that gangue is wetted by water and the mineral particles by the oils. A rotating paddle agitates the pulped ore and draws in air, which causes a froth of tiny oil bubbles to which the mineral particles adhere. This light froth floats to the surface and is skimmed off. By suitably adjusting the proportion of oil to water, it is often possible to separate one ore from another, e.g. zinc and lead sulphides. A diagram of the process is shown in Figure 7.3.

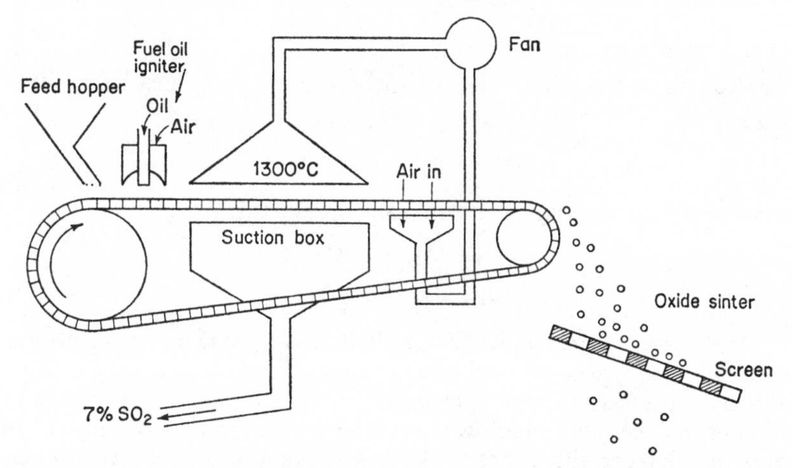

FIGURE 7.4. The Dwight-Lloyd sintering machine. Fine concentrated ore is roasted in air on a moving conveyor belt to give an oxide sinter and sulphur dioxide

After concentration, the ore is usually ready for smelting but if the smelting is to be carried out in a blast furnace, the ore is too fine and would either be blown out or choke up the furnace. The ore is, therefore, made into coarse lumps by a *sintering* process which converts fine sulphide powders into semi-fused coarse lumps of oxide. The process is used extensively in the metallurgy of zinc, copper, lead, nickel, and iron. A diagram of a sintering machine is shown in Figure 7.4.

Smelting

This process involves the reduction of the fused metallic compound to the metal. Most ores still contain a little gangue at this stage and this is removed by the addition of a flux, i.e. a compound which reacts with the gangue to give a low melting point, easily fusible slag. Thus if the gangue is mainly sand, a flux of limestone is added to form a fusible slag of calcium silicate which is molten at the high temperature of the furnace and can be run off.

A summary of methods employed in smelting is shown below:

(*i*) Heat the carbonate or hydroxide to convert it into the more readily reducible oxide, e.g.

$$CuCO_3 \rightarrow CuO + CO_2$$

(*ii*) Roast the sulphide in air to the oxide, e.g.

$$2ZnS + 3O_2 \rightarrow 2ZnO + 2SO_2$$

(*iii*) Reduce the sulphide by heating with iron, e.g.

$$Sb_2S_3 + 3Fe \rightarrow 2Sb + 3FeS$$

$$HgS + Fe \rightarrow Hg + FeS$$

(*iv*) Reduce oxides by heating with carbon or carbon monoxide, e.g.

$$ZnO + C \rightarrow Zn + CO$$

$$Fe_2O_3 + 3CO \rightarrow 2Fe + 3CO_2$$

However, other reducing agents are sometimes used:

$$Cr_2O_3 + 2Al \rightarrow Al_2O_3 + 2Cr$$

$$WO_3 + 3H_2 \rightarrow 3H_2O + W$$

(*v*) Reduction of the halides of the alkali and alkaline earth metals and aluminium chloride is impossible with carbon (p. 132) and very difficult with other metals, although calcium has been used for the reduction of rubidium and caesium chlorides. Because of the very large amount of energy needed for the process $Na^+ + e \rightarrow Na$ (*metal*), electrochemical methods utilizing the fused salts are usually employed (p. 166).

The smelting processes mentioned above all make use at some stage of the *reverberatory furnace* and a diagram of a modern furnace is shown in Figure 7.5. Oil burners shoot a long intensely hot flame down the furnace, melting the charge. On the end walls there are suitable tap holes for the crude metals and the slag. Such a furnace may melt up to 10^6 kg of charge per day and this is an essential process where it is necessary for the heating fuels to be kept away from the material to be reduced.

Smelting operations can also be carried out in *converters*. Air is blown through the molten material, when impurities are largely removed as volatile oxidation products. Also magnesite linings are used in converters to slag off any remaining metallic impurities. Converters may be of two types

(*i*) *The Bessemer converter*, similar to that originally used in the metallurgy of iron.

(*ii*) *The Pierce-Smith converter*, which consists of a large horizontal steel drum resting upon rollers. Rows of steel tubes (tuyères) pass into the converter and are connected to an air duct. Air is forced into the molten bath of

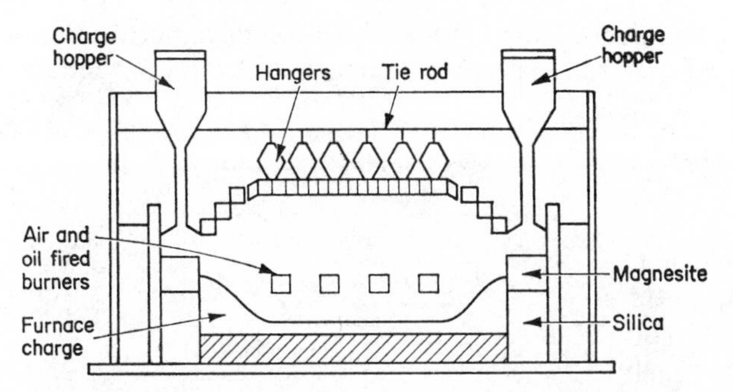

FIGURE 7.5. Section of a modern reverberatory furnace

crude metal. The process provides its own heat due to the oxidation of impurities and the temperature rises to 2400 °C.

Some metals can be extracted from their ores *hydrometallurgically*, i.e. by bringing the metal into aqueous solution as a complex ion. The metal is then precipitated by adding a suitable reagent, e.g. in the extraction of silver:

$$AgCl + 2CN^- \rightarrow Ag(CN)_2^- + Cl^-$$

$$2Ag(CN)_2^- + Zn \rightarrow 2Ag + Zn(CN)_4^{2-}$$

Refining

Several methods can be used to prepare the pure metal:

(*i*) *Distillation*. This is useful for removing low boiling point impurities as in the purification of zinc and mercury.

(*ii*) *Liquation*. A low melting point metal can be made to flow away on a sloping hearth from higher melting point impurities as in the purification of tin.

(*iii*) *Electrolysis*. This is widely used in refining copper, gold, silver, zinc, aluminium, lead, etc. The impure metal is made the anode and a pure strip of metal the cathode in a suitable electrolytic bath. The electrode processes which take place are:

$$\text{Anode} \qquad M \rightarrow M^{n+} + ne$$

$$\text{Cathode} \quad M^{n+} + ne \rightarrow M$$

The net result is the transfer of the metal from the anode to the cathode, leaving the impurities behind as an anode sludge.

(*iv*) *Zone refining*. This is used to obtain metals in an extremely high state of purity and is utilized in the purification of germanium, silicon, and boron for the manufacture of transistors and semiconductors. A rod of the metal is heated at one end: the impurities are more soluble in the melt than in the solid and concentrate in the melt. The melted zone is moved very slowly along the length of the rod, carrying the impurities with it. The process is repeated a number of times until the desired state of purification is reached. The end of the rod where the impurities have concentrated is then removed (Figure 7.6).

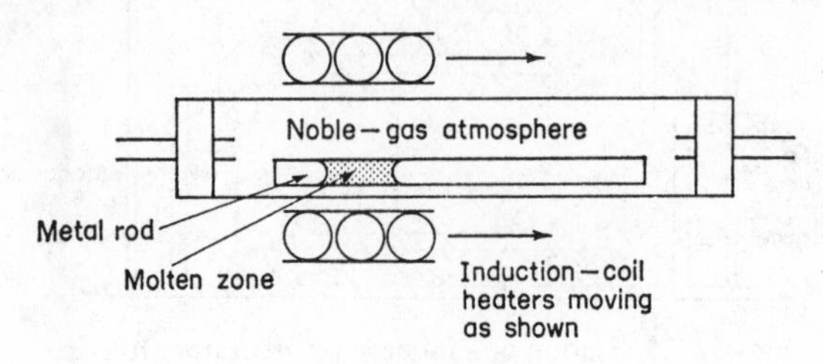

FIGURE 7.6

Thermodynamics of Metallurgical Processes

From the equation:

$$\Delta G = \Delta H - T\Delta S$$

ΔG is shown to be temperature-dependent. Although ΔH varies with temperature, the variation is sufficiently small to render the relationship between ΔG and T almost linear. In order to compare the changes in free energy for reactions, these are often calculated so that the reactants and products are in their standard states (at 298 K and 101·325 kN m^{-2}). The change in free energy taking place under these conditions is known as the *standard free energy* ($\Delta G^{\ominus}$). The graphical method of plotting $\Delta G^{\ominus}$ against T is shown for oxides, sulphides, and chlorides.

Diagrams for Oxide Formation

Figure 7.7 shows a number of oxide plots with slopes defined by $\Delta G^{\ominus}/T = -\Delta S$. The $\Delta G^{\ominus}$ values for reactions involving one mole of oxygen at atmospheric pressure are plotted on the graph. It will be noted that:

(*a*) Most lines slope upwards from left to right indicating entropy losses which follow the disappearance of one mole of gaseous oxygen for the reaction in question. At the melting points and the boiling points of the metals (points marked ○), the slope of the curves steepens. At the boiling point there are large entropy losses due to the disappearance of both gaseous oxygen and metal vapour. A smaller effect is seen at the melting point. If, however, the oxide undergoes a phase change, there will be an increase in the entropy of the oxide and, at such a point, the curve becomes less steep. For example, in the case of lead, the oxide (PbO) boils while lead is liquid. In these instances the entropy change becomes positive for the reaction and hence the slope of the $\Delta G^{\ominus}/T$ line changes sign, the situation reverting to normal once the boiling point of lead is reached. The carbon–carbon monoxide line slopes downwards from left to right since, for every mole of oxygen lost two moles of carbon monoxide appear, i.e. the entropy increases.

(*b*) At a particular temperature, high negative $\Delta G^{\ominus}$ values indicate oxides of great stability. It is possible to obtain $\Delta G^{\ominus}$ values for a reduction reaction by subtracting $\Delta G^{\ominus}$ values for one oxidation reaction from another. Thus $\Delta G^{\ominus}$ for the reaction:

$$2MgO + 2C \rightarrow 2Mg + 2CO$$

can be obtained by subtracting the $\Delta G^{\ominus}$ value for the reaction:

$$2Mg + O_2 \rightarrow 2MgO$$

from that for the reaction:

$$2C + O_2 \rightarrow 2CO$$

An examination of Figure 7.7 reveals that

(*i*) Below 1600 °C the free energy difference is positive and no reaction occurs.

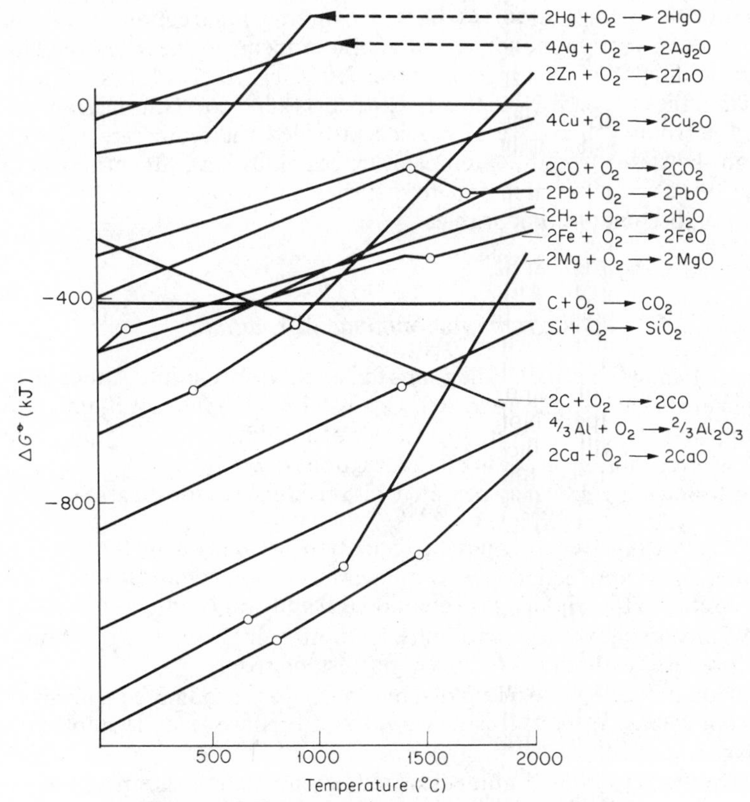

FIGURE 7.7. Free-energy/temperature plots for oxides

(*ii*) At 1600 °C the lines intersect and the free energy difference is zero.

(*iii*) Above 1600 °C the free energy difference becomes negative and the reaction becomes thermodynamically feasible and has been carried out commercially at 1900 °C. It can also be seen that reduction with silicon (Pidgeon process) is also feasible at temperatures greater than 1700 °C where the free energy change becomes negative.

$$2MgO + Si \rightarrow 2Mg + SiO_2$$

Actually, however, the temperature used is 1200 °C and the reaction is operated at reduced pressure which makes $\Delta G^{\ominus}$ for the reduction reaction more negative.

(*c*) Oxides of silver and mercury have very small negative $\Delta G^{\ominus}$ values and can be reduced by heating to temperatures where the free energy becomes zero.

(*d*) The reducing power of hydrogen gas is very limited and the slope does not increase markedly with increase in temperature. Oxides reducible by hydrogen are found above this line.

(*e*) Carbon oxidation curves have a common intersection at 710 °C. Below 710 °C carbon monoxide is the best reducing agent (as in the blast furnace reduction of iron oxide). Above 710 °C carbon is the best reducing agent and the cheapest. Note that the line cuts right across the diagram and, in fact, carbon will reduce any oxide provided the temperature is high enough. However, very high temperature reactions have limitations in the design of suitable furnaces and refractories and reduction of Al_2O_3 and CaO, etc., becomes impracticable.

Diagrams for Sulphide Formation

Figure 7.8 shows the $\Delta G^{\ominus}/T$ diagrams for sulphide formation. Although the S_8 puckered ring is stable up to 800 °C, it will be noted that the S_2 unit is the common reactant. However, this does not invalidate the diagrams since sulphur is a common factor in all the reactions quoted.

The following points may be noted with reference to the diagram:

(*a*) Decreasing free energies of formation are shown in the sequence calcium, zinc, iron, lead, copper, and mercury at low temperature.

(*b*) Mercury(II) sulphide is decomposed readily on heating.

(*c*) Calcium shows a great affinity for both sulphur and oxygen and is, therefore, used as both a deoxidizer and desulphurizer.

(*d*) Compared to the oxide plots, hydrogen and carbon are ineffective as reducing agents. In fact, it is just possible to reduce lead(II) sulphide by hydrogen.

(*e*) Only a very limited number of reduction reactions are possible, e.g. the reduction of Sb_2S_3 by iron.

However, many important ores are sulphides and roasting to the oxide is the usual preliminary. From the diagram, it is possible to obtain the free energy change for the reaction: $S_2 + 2O_2 \rightarrow 2SO_2$ at a particular temperature.

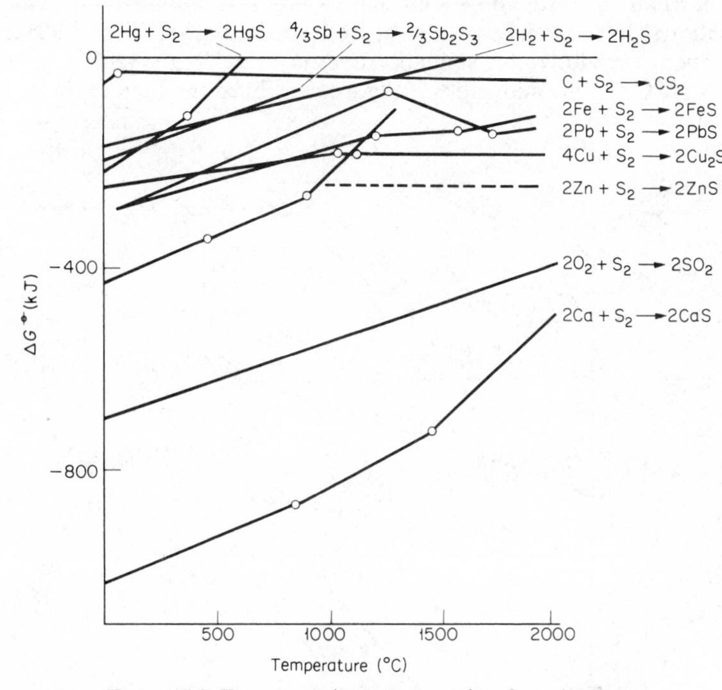

FIGURE 7.8. Free-energy/temperature plots for sulphides

This gives a link with the oxide diagram. For example at 800°C, $\Delta G^{\ominus}$ for the above reaction is −544 kJ. From the oxide curve:

$$2Zn + O_2 \rightarrow 2ZnO \qquad \Delta G^{\ominus} \text{ at } 800\ °C = -480 \text{ kJ}$$

From the sulphide curve:

$$2Zn + S_2 \rightarrow 2ZnS \qquad \Delta G^{\ominus} \text{ at } 800\ °C = -293 \text{ kJ}$$

Hence $\Delta G^{\ominus}$ at 800 °C for the reaction:

$$2ZnS + 3O_2 \rightarrow 2ZnO + 2SO_2 \text{ is } \Delta G^{\ominus} = -544 - 480 - (-293) = -731 \text{ kJ}$$

Hence the above reaction is thermodynamically feasible. It is also possible to combine the two diagrams for free energy data concerning reductions such as:

$$PbS + 2PbO \rightarrow 3Pb + SO_2$$

$$Cu_2S + 2Cu_2O \rightarrow 6Cu + SO_2$$

These reactions are used in the production of lead and copper.

Chlorine Metallurgy

Some $\Delta G^{\ominus}$ calculations for the free energies of formation of chlorides from one mole of chlorine at one atmosphere pressure are plotted against temperature in Figure 7.9. It can be seen that carbon is quite useless for chloride reduction. The reducing power of hydrogen increases steadily with increase in temperature (due to the small positive ΔS value for the reaction). Hence $SiCl_4$ and $CrCl_3$ are reducible by hydrogen. Since the halides, in general,

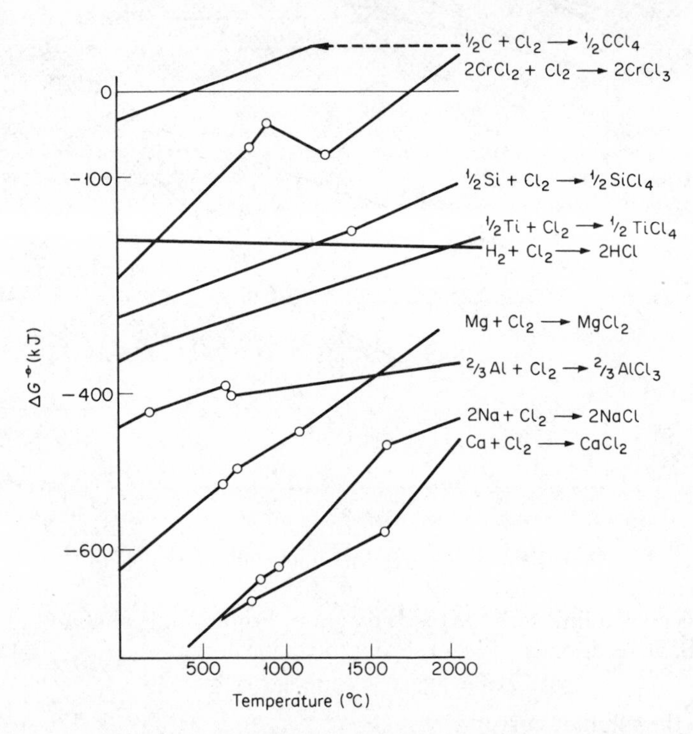

FIGURE 7.9. Free-energy/temperature plots for chlorides

have low melting points and boiling points and are good electrolytes, electrolysis is the principal method for producing the active metals, e.g. Na, Mg, Ca, etc. From $\Delta G^{\ominus} = -nFE^{\ominus}$ (p. 117), the diagram can be used to determine the minimum e.m.f. required to carry out an electrolysis, although this is of little practical use, since polarization and internal resistance of the electrolyte cause large alterations in $E^{\ominus}$ values.

The electrolytic production of sodium, magnesium, and calcium is dealt with in the appropriate chapter. From the curve, it can be seen that there is considerable scope for replacement reactions, e.g. the reduction of $TiCl_4$ with magnesium:

$$TiCl_4 + 2Mg \rightarrow Ti + 2MgCl_2 \text{ (Kroll process)}$$

In addition, a great deal of information can be gained by combination of the free energy diagrams. For example, by combining the oxide and chloride diagrams, information can be obtained on the chlorination of oxides. Thus for the reaction:

$$MO + Cl_2 \rightarrow MCl_2 + \tfrac{1}{2}O_2$$

$\Delta G^{\ominus}$ is negative for silver, mercury, lead, copper, manganese, zinc, tin, iron, cadmium, and nickel but is positive for magnesium, chromium, titanium, aluminium, and silicon between 500 and 1000 °C.

8

Oxidation and Reduction

Oxidation was originally defined as a chemical process involving gain of oxygen whereas a loss of oxygen was termed *reduction*. For example,

$$S + O_2 \rightarrow SO_2 \text{ (oxidation)}$$

$$2KClO_3 \rightarrow 2KCl + 3O_2 \text{ (reduction)}$$

However, the two processes can occur together:

$$PbO + H_2 \rightarrow Pb + H_2O$$

$$2Mg + CO_2 \rightarrow 2MgO + C$$

where lead oxide is reduced to lead by the reducing agent hydrogen, which is oxidized to water, and where carbon dioxide is reduced to carbon by the reducing agent magnesium, which is oxidized to magnesium oxide.

Since hydrogen burns in oxygen to produce water and, during electrolysis, hydrogen is often the cathode product and oxygen the anode product, hydrogen was regarded as the chemical opposite of oxygen. Hence removal of hydrogen was regarded as an oxidation and addition of hydrogen as a reduction, e.g.

$$H_2S + Cl_2 \rightarrow 2HCl + S$$

where hydrogen sulphide is oxidized by the oxidizing agent chlorine to sulphur *or* chlorine is reduced by the reducing agent hydrogen sulphide to hydrogen chloride.

Consider the reactions outlined below.

$$4FeO + O_2 \longrightarrow 2Fe_2O_3$$

$$\downarrow HCl \qquad\qquad \downarrow HCl$$

$$2FeCl_2 + Cl_2 \longrightarrow 2FeCl_3$$

iron(II) iron(III)

$\uparrow$ —— $SnCl_2$ —— (iron(III) → iron(II))

When iron(II) oxide is heated in air, iron(III) oxide is produced, the process being one of oxidation. Dissolution of these oxides in hydrochloric acid produces iron(II) and iron(III) chloride respectively. Chlorination of the iron(II) chloride produces iron(III) chloride and this is, therefore, regarded as an oxidation. Treatment of iron(III) chloride with tin(II) chloride produces iron(II) chloride by a reduction reaction. Hence oxidation and reduction are now defined more generally in terms of an increase or decrease in the valency of the metallic atom or ion.

If the above reactions are examined from the point of view of ionic changes we have, for the oxidation process:

$$2Fe^{2+} + Cl_2 \rightarrow 2Fe^{3+} + 2Cl^-$$

Electrons are transferred from the iron(II) ions to the chlorine molecules which become anions. Thus chlorine, the oxidizing agent, is an electron acceptor and the iron(II) ion which is being oxidized suffers electron loss.

In the reduction process:

$$2Fe^{3+} + Sn^{2+} \rightarrow 2Fe^{2+} + [Sn^{4+}]$$

the iron(III) ions accept electrons from the tin(II) ions which are the reducing agents. There are always two ways of looking at an electron transfer, and in the above example the iron(III) ion has accepted an electron and is the oxidizing agent which has oxidized the tin(II) ion to tin(IV). The tin(IV) ion is bracketed as in fact it does not really exist, a covalent Sn^{IV} compound being formed, but this does not affect the basic principles involved.

Thus oxidation is a process involving electron loss and an oxidizing agent is an electron acceptor. Reduction is a process involving electron gain and a reducing agent is an electron donor. Some examples of oxidizing agents and reducing agents are shown below, the half reactions being written electronically.

Oxidizing agents	*Reducing agents*
$Cl_2 + 2e \rightarrow 2Cl^-$	$2I^- \rightarrow I_2 + 2e$
$Fe^{3+} + e \rightarrow Fe^{2+}$	$S^{2-} \rightarrow S + 2e$
$H_2O_2 + 2H^+ + 2e \rightarrow 2H_2O$	$Fe^{2+} \rightarrow Fe^{3+} + e$
$MnO_4^- + 8H^+ + 5e \rightarrow Mn^{2+} + 4H_2O$	$SO_3^{2-} + H_2O \rightarrow SO_4^{2-} + 2H^+ + 2e$
$Cr_2O_7^{2-} + 14H^+ + 6e \rightarrow 2Cr^{3+} + 7H_2O$	$Na \rightarrow Na^+ + e$
$NO_3^- + 2H^+ + e \rightarrow NO_2 + H_2O$	$Sn^{2+} \rightarrow Sn^{4+} + 2e$

The full ionic equations for some reactions are outlined below:

(*a*) Chlorine gas oxidizes iron(II) salts to iron(III).

$$2Fe^{2+} \rightarrow 2Fe^{3+} + 2e$$

$$Cl_2 + 2e \rightarrow 2Cl^-$$

$$\overline{2Fe^{2+} + Cl_2 \rightarrow 2Fe^{3+} + 2Cl^-}$$

(*b*) Potassium permanganate liberates iodine from potassium iodide.

$$MnO_4^- + 8H^+ + 5e \rightarrow Mn^{2+} + 4H_2O \quad (1)$$

$$2I^- \rightarrow I_2 + 2e \quad (2)$$

In order to balance the number of electrons multiply (1) by two and (2) by five and add.

$$2MnO_4^- + 16H^+ + 10I^- \rightarrow 2Mn^{2+} + 8H_2O + 5I_2$$

(*c*) Sulphite solution reduces iron(III) ions to iron(II).

$$SO_3^{2-} + H_2O \rightarrow SO_4^{2-} + 2H^+ + 2e$$

$$2Fe^{3+} + 2e \rightarrow 2Fe^{2+}$$

$$SO_3^{2-} + 2Fe^{3+} + H_2O \rightarrow SO_4^{2-} + 2H^+ + 2Fe^{2+}$$

Oxidation State and Oxidation Numbers

Another method of balancing oxidation-reduction (*redox*) reactions is the oxidation number method. Each element or ion is given an oxidation number and oxidation occurs when the oxidation number is increased; reduction occurs when the oxidation number is decreased. The following points should be noted in giving an element its oxidation number:

(*a*) The oxidation number of ions in electrovalent compounds is equal to the charge on the ions, e.g. in $Ca^{2+}Cl_2^-$, the oxidation number of calcium (Ca^{2+}) is +2 and that of chlorine (Cl^-) is −1. For covalent compounds such as NH_3, CH_4, and H_2O, the oxidation numbers are assigned by assuming that the bonds are ionic and giving shared electron pairs to the atom with the greater electronegativity. Thus in water, ammonia, and methane, the oxidation numbers are shown above the symbols

$$\overset{-2}{O}(\overset{+1}{H})_2 \qquad \overset{-3}{N}(\overset{+1}{H})_3 \qquad \overset{-4}{C}(\overset{+1}{H})_4$$

(*b*) The algebraic sum of oxidation numbers of elements in a compound is zero, e.g.

$$\underset{+1+7+(-2\times 4)}{KMnO_4} \qquad \underset{(+1\times 2)+6+(-2\times 4)}{H_2SO_4}$$

(*c*) The algebraic sum of oxidation numbers in an ion is equal to the charge on the ion, e.g.

SO_4^{2-} has an oxidation number of $+6 - 8 = -2$

MnO_4^- has an oxidation number of $+7 - 8 = -1$

(*d*) All elements in the uncombined state have oxidation numbers of zero: Sn, S_8, F_2, all have oxidation numbers of zero since we assume that the electron pairs are shared equally.

Note also that (*a*) hydrogen normally has an oxidation number of +1 except in hydrides of metals, e.g. $\underset{+1}{Na^+}\underset{-1}{H^-}$, where the oxidation number is −1; (*b*) oxygen normally has an oxidation number of −2 except in peroxides, e.g. $\underset{+1}{Na_2^+}(\underset{-1}{O}-\underset{-1}{O})^{2-}$, where it is −1.

Application of the Oxidation Number Concept in Balancing Redox Reactions

Consider the oxidation of iron(II) ions by acidified dichromate. The unbalanced ionic equation may be written:

$$\underset{+6}{Cr_2O_7^{2-}} + H^+ + \underset{+2}{Fe^{2+}} \rightarrow \underset{+3}{Fe^{3+}} + H_2O + \underset{+3}{Cr^{3+}}$$

The dichromate ion is reduced to the chromium(III) ion giving a change of three units in the oxidation number of chromium. The iron(II) ion is oxidized to the iron(III) ion giving a change of one unit in the oxidation number of iron. Since the dichromate ion contains two atoms of chromium, the total change in oxidation number is six and hence six iron(II) ions will be needed for each dichromate ion.

$$Cr_2O_7^{2-} + H^+ + 6Fe^{2+} \rightarrow 6Fe^{3+} + H_2O + 2Cr^{3+}$$

The charges are then balanced:

$$Cr_2O_7^{2-} + 14H^+ + 6Fe^{2+} \rightarrow 6Fe^{3+} + 7H_2O + 2Cr^{3+}$$

Consider the oxidation of sodium nitrite by acidified potassium permanganate solution.

$$\underset{+7}{MnO_4^-} + H^+ + \underset{+3}{NO_2^-} \longrightarrow \underset{+2}{Mn^{2+}} + H_2O + \underset{+5}{NO_3^-}$$

gain of 2 units ($NO_2^- \rightarrow NO_3^-$); loss of 5 units ($MnO_4^- \rightarrow Mn^{2+}$)

Hence two permanganate ions are required for five nitrite ions.

$$2MnO_4^- + H^+ + 5NO_2^- \rightarrow 2Mn^{2+} + H_2O + 5NO_3^-$$

Balancing the charges:

$$2MnO_4^- + 6H^+ + 5NO_2^- \rightarrow 2Mn^{2+} + 3H_2O + 5NO_3^-$$

The oxidation number method is usually the simplest to apply in deriving relationships used for analytical purposes. The ion–electron method, however, is more accurate as far as reaction mechanisms are concerned and is more suitable for discussions based on fundamental principles.

Quantitative Aspects—Standard Electrode Potentials ($E^{\ominus}$)

If a metal M is placed in a solution containing M^{n+} ions, a potential difference is set up between the metal and the solution. In the case of a zinc rod in contact with a molar solution of zinc ions, the following equilibrium

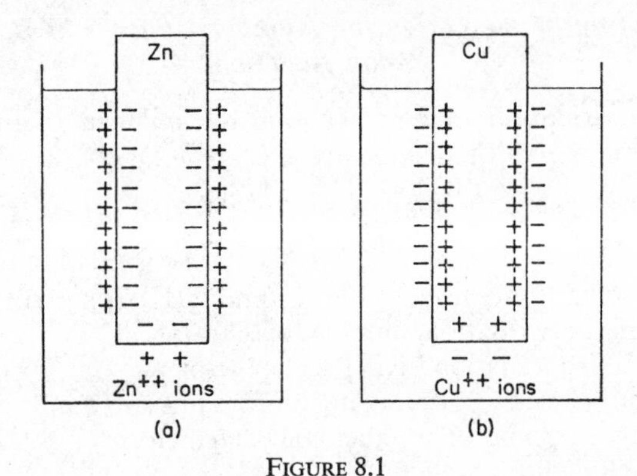

FIGURE 8.1

exists: $Zn \rightleftharpoons Zn^{2+} + 2e$ (Figure 8.1*a*). The electrons remain on the metal whilst zinc ions pass into solution and stay close to the metal surface giving an *electrical double layer*. The magnitude of the potential difference between the metal and the solution is called the *electrode potential* of zinc and is designated negative.

With a copper rod (Figure 8.1*b*), copper ions are deposited on the metal giving it a positive charge with respect to the solution. At equilibrium copper is said to have an electrode potential which is designated positive.

The magnitude of electrode potentials depends upon:

(*a*) The physical state of the metal.
(*b*) The concentration of the metal ions in solution.
(*c*) The temperature.

It is impossible to measure the electrode potential of a single metal: a reference electrode is needed under standard conditions. The first reference electrode chosen was the hydrogen electrode which was arbitrarily given zero potential at all temperatures (Figure 8.2). Hydrogen gas at atmospheric pressure is bubbled over an electrode of platinum foil coated with platinum black. The electrode is immersed in a solution containing hydrogen ions of unit activity. The activity (a) of an ion in a solution at high dilution is equal to the concentration in gramme ions per litre. Hence a solution containing hydrogen ions of unit activity is equivalent to a concentration of one gramme ion per litre of hydrogen ions.

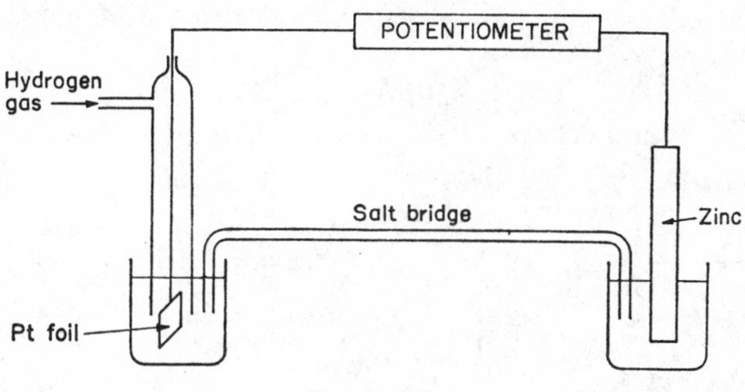

FIGURE 8.2

The following equilibrium is set up: $\frac{1}{2}H_2 \rightleftharpoons H^+ + e$. The other element in its normal state (say, zinc metal) is in contact with a solution of metal ions at unit activity (approximately one gramme ion per litre) at 25 °C.

The two electrodes are connected by a salt bridge containing potassium chloride in a gel which prevents diffusion of the two solutions but allows passage of ions establishing electrical contact. Non-metal electrodes can be constructed in the same way as the hydrogen electrode, for example, oxygen gas at one atmosphere in OH^- ions of unit activity at 25 °C. The e.m.f. generated by the cell is measured by potentiometric methods, details of which are not shown in the diagram. Other reference electrodes are often used instead of hydrogen since they are easier to handle and are more permanent. A standard calomel electrode may be constructed which has an electrode potential of +0·334 V relative to the hydrogen electrode. Therefore 0·334 V must be added to any electrode potential value obtained using the calomel electrode.

The sign convention is that adopted by the International Union of Pure and Applied Chemistry. The electrode which is positively charged with respect to the contact solution is given a positive electrode potential. The electrode which is negatively charged with respect to the contact solution is given a negative electrode potential.

On this basis the $Zn^{2+}|Zn$ electrode has a negative potential and the $Cu^{2+}|Cu$ electrode has a positive potential. Hence the cell

$$\xrightarrow{\text{path of electrons}}$$

$$\underset{-\text{ve}}{Zn}|Zn^{2+}\ (a=1) \qquad Cu^{2+}\ (a=1)|\underset{+\text{ve}}{Cu}$$

has an e.m.f. of 1·1 V since:

$$Zn^{2+} + 2e \rightarrow Zn \quad E^{\ominus} = -0{\cdot}76\ V$$

$$Cu^{2+} + 2e \rightarrow Cu \quad E^{\ominus} = +0{\cdot}34\ V$$

and $E^{\ominus}_{cell} = E^{\ominus}$ (electrode accepting electrons) $- E^{\ominus}$ (electrode donating electrons)

Values of some standard redox potentials are given in Table 8.1.

TABLE 8.1

Standard Electrode Potentials $E^{\ominus}$ (volts) at 298 K

Electrode reaction	$E^{\ominus}$ (*volts*)	*Electrode reaction*	$E^{\ominus}$ (*volts*)
$Li^+ + e = Li$	−3·04	$H^+ + e = \frac{1}{2}H_2$	−0·00
$K^+ + e = K$	−2·92	$Sn^{4+} + 2e = Sn^{2+}$	+0·15
$Ba^{2+} + 2e = Ba$	−2·90	$Cu^{2+} + e = Cu^+$	+0·15
$Ca^{2+} + 2e = Ca$	−2·87	$Cu^{2+} + 2e = Cu$	+0·34
$Na^+ + e = Na$	−2·71	$Cu^+ + e = Cu$	+0·52
$Mg^{2+} + 2e = Mg$	−2·37	$Fe^{3+} + e = Fe^{2+}$	+0·76
$Al^{3+} + 3e = Al$	−1·67	$Ag^+ + e = Ag$	+0·80
$Mn^{2+} + 2e = Mn$	−1·18	$Hg^{2+} + 2e = Hg$	+0·80
$Zn^{2+} + 2e = Zn$	−0·76	$Hg^{2+} + e = \frac{1}{2}Hg_2^{2+}$	+0·92
$Cr^{3+} + 3e = Cr$	−0·74	$Pb^{4+} + 2e = Pb^{2+}$	+1·70
$Fe^{2+} + 2e = Fe$	−0·44	$Ag^{2+} + e = Ag^+$	+1·98
$Cr^{3+} + e = Cr^{2+}$	−0·41		
$Co^{2+} + 2e = Co$	−0·28		
$Ni^{2+} + 2e = Ni$	−0·25		
$Sn^{2+} + 2e = Sn$	−0·14		
$Pb^{2+} + 2e = Pb$	−0·13		

Electrode Potentials for some Important Half Reactions

Half reaction	$E^{\ominus}$ (*volts*)
$2H_2O + 2e = 2OH^- + H_2$	−0·83
$S + 2H^+ + 2e = H_2S$	+0·14
$SO_4^{2-} + 4H^+ + 2e = H_2SO_3 + H_2O$	+0·20
$Fe(CN)_6^{3-} + e = Fe(CN)_6^{4-}$	+0·36
$\frac{1}{2}I_2 + e = I^-$	+0·54
$MnO_4^- + e = MnO_4^{2-}$	+0·56
$O_2 + 2H^+ + 2e = H_2O_2$	+0·68
$\frac{1}{2}Br_2 + e = Br^-$	+1·06
$\frac{1}{2}O_2 + 2H^+ + 2e = H_2O$	+1·23
$\frac{1}{2}Cr_2O_7^{2-} + 7H^+ + 3e = Cr^{3+} + \frac{7}{2}H_2O$	+1·33
$\frac{1}{2}Cl_2 + e = Cl^-$	+1·36
$MnO_4^- + 8H^+ + 5e = Mn^{2+} + 4H_2O$	+1·52
$MnO_4^- + 4H^+ + 3e = MnO_2 + 2H_2O$	+1·69

The Application of Electrode Potential Values

The arrangement of elements in order of their standard electrode potentials ($E^{\ominus}$) is known as the *electrochemical series*. The series gives information on displacement reactions, as any metal will displace one below it, e.g.

$$Fe + Cu^{2+} \rightarrow Fe^{2+} + Cu$$

and any non-metal will displace one above it, e.g.

$$Cl_2 + 2Br^- \rightarrow 2Cl^- + Br_2$$

Generally a reaction goes to completion when the difference in $E^\ominus$ values is greater than 0·4 V.

Elements above hydrogen in the electrochemical series displace hydrogen from dilute acids:

$$Zn + 2H^+ \rightarrow Zn^{2+} + H_2$$

Metals high in the series may even displace hydrogen from water:

$$Na^+ + e \rightarrow Na \quad E^\ominus = -2{\cdot}71\ V$$

$$2H_2O + 2e = 2OH^- + H_2 \quad E^\ominus = -0{\cdot}83\ V$$

Electrode potentials do not, however, give information concerning reaction rates. Thus although magnesium might be expected to liberate hydrogen from water, there is no reaction because of the formation of an insoluble protective film of magnesium hydroxide. The reaction will occur if the magnesium is first amalgamated, thus inhibiting the formation of a coherent film.

Values of $E^\ominus$ can also be obtained for ion systems, e.g. Fe^{3+}/Fe^{2+}. This is determined by immersing a platinum electrode into a solution containing one gramme ion per litre (activity = unity) of both iron(II) and iron(III) ions. Thus the cell may be represented as:

$$Pt \left.\begin{matrix} Fe^{3+} \\ Fe^{2+} \end{matrix}\right\} (a = 1) \underset{\text{salt bridge}}{||} H^+ (a = 1) | Pt$$

The $E^\ominus$ value for the reaction $Fe^{3+} + e \rightarrow Fe^{2+}$ is +0·76 V (1)

The $E^\ominus$ value for the reaction $Fe^{2+} + 2e \rightarrow Fe$ is –0·44 V (2)

Therefore, Fe^{2+} is a poorer reducing agent than Fe, or Fe^{3+} is a better oxidizing agent than Fe^{2+}. By a similar method to that adopted for the system Fe^{3+}/Fe^{2+} the $E^\ominus$ values for the systems Sn^{4+}/Sn^{2+} and $Hg^{2+}/\frac{1}{2}Hg_2{}^{2+}$ may be determined.

$$Sn^{4+} + 2e \rightarrow Sn^{2+} \qquad E^\circ = +0{\cdot}15\ V \qquad (3)$$

$$Hg^{2+} + e \rightarrow \tfrac{1}{2}Hg_2{}^{2+} \qquad E^\circ = +0{\cdot}92\ V \qquad (4)$$

Thus, from the four values quoted in (1), (2), (3), (4),

(*a*) Sn^{2+} will reduce Fe^{3+} to Fe^{2+} but not Fe^{2+} to Fe, whereas Fe will reduce Sn^{4+} to Sn^{2+}.

(*b*) Fe and Sn^{2+} will reduce Hg^{2+} to $Hg_2{}^{2+}$.

Oxidizing agents in acid solution

Potassium permanganate and potassium dichromate are widely used in acid solution for volumetric work, e.g.

$$Cr_2O_7^{2-} + 14H^+ + 6e \rightarrow 2Cr^{3+} + 7H_2O$$

Here the $E^\ominus$ value applies to the solution where the hydrogen ion activity is unity

$$E^\ominus \text{ for } Cr_2O_7^{2-}|Cr^{3+} \text{ is } +1{\cdot}33 \text{ V}$$

$$E^\ominus \text{ for } MnO_4^-|Mn^{2+} \text{ is } +1{\cdot}52 \text{ V}$$

Hence MnO_4^- is a better oxidizing agent than $Cr_2O_7^{2-}$.

The $E^\ominus$ value Fe^{3+}/Fe^{2+} is +0·76 V, that for $\frac{1}{2}Cl_2/Cl^-$ is +1·36 V, and that for $\frac{1}{2}I_2|I^-$ is + 0·54 V. Hence both permanganate and dichromate will oxidize iron(II) and iodide ions but only permanganate will oxidize chloride. Thus dichromate may be used in volumetric titrations with hydrochloric acid but permanganate may not.

Electrode Potential and Concentration

Since $\Delta G = \Delta G^\ominus + RT\log_e K_c$, $\Delta G = -nFE$, and $\Delta G^\ominus = -nFE^\ominus$ (p. 117), the electrode potential (E) is related to the absolute temperature (T) and the equilibrium constant (K_c) by the expression

$$E = E^\ominus - \frac{RT}{nF}\log_e K_c$$

where $E^\ominus$ is the standard electrode potential for an ion system $M^{n+} + ne \rightleftharpoons M$, $K_c = [M]/[M^{n+}]$, F is the Faraday constant ($9{\cdot}648 \times 10^4$ C mol^{-1}), and R is the gas constant (8·314 JK^{-1} mol^{-1}).

Substituting values of F, R, and T at 25 °C, the expression reduces to:

$$E = E^\ominus + \frac{8{\cdot}314 \times 298}{n \times 96{,}483}\log_e\frac{[M^{n+}]}{[M]}$$

but

$$\log_e\frac{[M^{n+}]}{[M]} = 2{\cdot}303\log_{10}\frac{[M^{n+}]}{[M]}$$

$$\therefore E = E^\ominus + \frac{0{\cdot}059}{n}\log_{10}\frac{[M^{n+}]}{[M]}$$

Since the activity of a solid is unity*, as the metal goes into solution as its ions, $[M^{n+}]$ increases and hence E becomes less negative (more positive) and the reducing power of the metal decreases.

* By convention. all substances in standard states, i.e. pure crystalline solids, liquids, and gases at 101·325 kN m^{-2} and 298 K have unit activity.

Consider the displacement reaction mentioned above:

$$Zn + Cu^{2+} \rightarrow Zn^{2+} + Cu$$

At equilibrium,

$$E_1\ Zn^{2+}|Zn = E_2\ Cu^{2+}|Cu$$

therefore

$$E^{\ominus}_1 + \frac{0{\cdot}059}{2} \log_{10} [Zn^{2+}] = E^{\ominus}_2 + \frac{0{\cdot}059}{2} \log_{10} [Cu^{2+}]$$

$$\log_{10} \frac{[Zn^{2+}]}{[Cu^{2+}]} = (E^{\ominus}_2 - E^{\ominus}_1) \frac{2}{0{\cdot}059}$$

But $E^{\ominus}_1 = -0{\cdot}76$ V and $E^{\ominus}_2 = +0{\cdot}34$ V.
Therefore

$$\log_{10} \frac{[Zn^{2+}]}{[Cu^{2+}]} = + \frac{2{\cdot}2}{0{\cdot}059}$$

and

$$\frac{[Zn^{2+}]}{[Cu^{2+}]} = 2 \times 10^{37}$$

$\therefore$ $[Cu^{2+}]$ is negligible and the precipitation of copper is complete.
Consider now the oxidation-reduction reaction:

$$MnO_4^- + 8H^+ + 5e = Mn^{2+} + 4H_2O \qquad E^{\ominus}_1 = +1{\cdot}52 \text{ V}$$

$$5Fe^{3+} + 5e = 5Fe^{2+} \qquad E^{\ominus}_2 = +0{\cdot}76 \text{ V}$$

At 25 °C,

$$E_1 = 1{\cdot}52 + \frac{0{\cdot}059}{5} \log_{10} \frac{[MnO_4^-][H^+]^8}{[Mn^{2+}]}$$

Since the solution is dilute, water can be considered to be in its standard state and the activity is unity

$$E_2 = 0{\cdot}76 + \frac{0{\cdot}059}{5} \log_{10} \frac{[Fe^{3+}]^5}{[Fe^{2+}]^5}$$

At equilibrium $E_1 = E_2$

$$\therefore 0{\cdot}76 + \frac{0{\cdot}059}{5} \left(\log_{10} \frac{[MnO_4^-][H^+]^8}{[Mn^{2+}]} - \log_{10} \frac{[Fe^{3+}]^5}{[Fe^{2+}]^5} \right) = 0$$

$$\therefore -64 = \log_{10} \frac{[MnO_4^-][H^+]^8[Fe^{2+}]^5}{[Mn^{2+}][Fe^{3+}]^5}$$

$$\therefore 10^{64} = \frac{[Mn^{2+}][Fe^{3+}]^5}{[MnO_4^-][H^+]^8[Fe^{2+}]^5}$$

$$= \text{Equilibrium constant } (K_c) \text{ for the reaction at 298 K}$$

and therefore once again oxidation is complete.

From the above equations for $MnO_4^-|Mn^{2+}$ it is obvious that the electrode potential of the system depends upon the hydrogen ion concentration of the solution, i.e. the pH. Thus $[H^+] = 10^{-5}$ moles per litre at a pH of 5.

If the $[MnO_4^-]$ and $[Mn^{2+}]$ are kept constant at unit activity,

$$E_1 = 1{\cdot}52 + \frac{0{\cdot}059}{5} \log_{10} [10^{-5}]^8$$

$$= 1{\cdot}52 + \frac{0{\cdot}059}{5} (-40) = 1{\cdot}52 - 0{\cdot}47 = +1{\cdot}05 \text{ V}$$

whereas when $[H^+] = 1$, i.e. pH $= 0$

$$E_1 = +1{\cdot}52 \text{ V}$$

Disproportionation

This term describes a reaction in which an atom or ion undergoes self oxidation and reduction simultaneously, i.e. its oxidation number is both increased and decreased in the same reaction. An example is the decomposition of nitrous acid:

$$\underset{+3}{3HNO_2} \rightarrow \underset{+5}{HNO_3} + \underset{+2}{2NO} + H_2O$$

The conversion of green potassium manganate into purple permanganate with strong acid is another example of disproportionation:

$$\underset{+6}{3MnO_4^{2-}} + 4H^+ \rightarrow \underset{+7}{2MnO_4^-} + \underset{+4}{MnO_2} + 2H_2O$$

Copper(I) ions in aqueous solution undergo disproportionation:

$$\underset{+1}{2Cu^+} \rightleftharpoons \underset{+2}{Cu^{2+}} + \underset{0}{Cu}$$

The half reactions in the above case are:

$$Cu^+ + e \rightarrow Cu \qquad E^\ominus = +0{\cdot}52 \text{ V}$$

$$Cu^{2+} + e \rightarrow Cu^+ \qquad E^\ominus = +0{\cdot}15 \text{ V}$$

The copper(I) ion is, therefore, a better oxidizing agent than the copper(II) ion or copper is a poorer reducing agent than the copper(I) ion. From thermodynamics it can be shown that $\Delta G^\ominus = -nFE^\ominus$ (p. 117). $E^\ominus$ in the above reaction is $0{\cdot}52 - 0{\cdot}15 = +0{\cdot}37$ V. Therefore ΔG is negative and disproportionation will occur.

Acids and Bases

With the development of the ionic theory, an acid became defined as a molecule or ion which furnished an hydroxonium ion in water ($H^+ + H_2O \rightarrow H_3O^+$) and a base as a molecule or ion which furnished a hydroxyl ion

(OH^-). The process of neutralization involved the union of a hydroxyl and an hydroxonium ion to form water ($H_3O^+ + OH^- \rightarrow 2H_2O$). However, many acid–base reactions involve neither H_3O^+ nor OH^-, and yet undergo neutralization reactions in other solvents such as liquid ammonia (p. 247). It was in an attempt to broaden the concept of acids and bases that Lowry and Brönsted extended the acid–base theory.

According to the Lowry and Brönsted theory an acid is a molecule or ion which donates a proton; a base is a molecule or ion which accepts a proton.

Thus

$$\underset{\text{acid 1}}{HA} + \underset{\text{base 1}}{H_2O} \rightleftharpoons \underset{\text{acid 2}}{H_3O^+} + \underset{\text{base 2}}{A^-}$$

For every acid there exists a base which is produced when the acid loses a proton. The acid and base so related are called *conjugates*.

A strong acid (one which readily loses a proton) must have a weak conjugate base and a weak acid (one which loses a proton with difficulty) must have a strong conjugate base. Thus in the reaction:

$$\underset{\text{acid 1}}{HCl} + \underset{\text{base 1}}{H_2O} \rightleftharpoons \underset{\text{acid 2}}{H_3O^+} + \underset{\text{base 2}}{Cl^-}$$

the equilibrium lies well over to the right-hand side; hence HCl (strong acid) has a weak conjugate base Cl^-, while H_2O (weak base) has a strong conjugate acid H_3O^+.

In a weak (little-ionized) acid such as acetic acid:

$$\underset{\text{acid 1}}{CH_3COOH} + \underset{\text{base 1}}{H_2O} \rightleftharpoons \underset{\text{acid 2}}{H_3O^+} + \underset{\text{base 2}}{CH_3COO^-}$$

acetic acid is a weak acid to the strong conjugate base CH_3COO^- and H_2O is a weak base to the strong conjugate acid H_3O^+.

On this definition the hydroxyl ion becomes one of many bases, the distinguishing feature of the base OH^- being its great strength since the reaction:

$$OH^- + H_3O^+ \rightleftharpoons 2H_2O \qquad \Delta H \approx -57{\cdot}1 \text{ kJ mol}^{-1}$$

can, for practical purposes, be regarded as irreversible. The OH^- is a strong base to the weak conjugate acid H_2O.

Water is unusual in showing both acidic and basic behaviour and is said to be *amphiprotic*:

(*a*) with acids ($HCl + H_2O \rightleftharpoons H_3O^+ + Cl^-$) it behaves as a base or proton acceptor.

(*b*) with ammonia ($NH_3 + H_2O \rightleftharpoons NH_4^+ + OH^-$) it behaves as an acid or proton donor.

Thus water has both proton-accepting (*protophilic*) and proton-donating (*protogenic*) properties.

On this definition the basicity of an acid becomes the number of protons which it can lose per molecule of acid: e.g.

$$H_2SO_4 + 2H_2O \rightleftharpoons 2H_3O^+ + SO_4^{2-} \quad \text{diprotic acid}$$

$$H_3PO_4 + 3H_2O \rightleftharpoons 3H_3O^+ + PO_4^{3-} \quad \text{triprotic acid}$$

Note that H_3PO_3 is diprotic:

$$H_3PO_3 + 2H_2O \rightleftharpoons 2H_3O^+ + HPO_3^{2-}$$

Hydrolysis of Salts

The hydrolysis of salts may be explained in terms of the Lowry and Brönsted theory.

(*a*) The salt of a strong acid and a strong base, e.g. Na^+Cl^-. The hydrated Na^+ ion has little tendency to donate a proton to water. The Cl^- ion has little tendency to receive a proton from water. Hence the solution remains neutral.

(*b*) The salt of a strong acid and a weak base, e.g. $NH_4^+Cl^-$. NH_4^+ is a fairly strong cationic acid whereas Cl^- is a very weak anionic base:

$$NH_4^+ + H_2O \rightleftharpoons NH_3 + H_3O^+$$

Hence the resulting solution is acidic.

(*c*) The salt of a strong base and a weak acid, e.g. $Na_2^+CO_3^{2-}$. Hydrated Na^+ is a weak cationic acid but CO_3^{2-} is a strong anionic base:

$$CO_3^{2-} + H_2O \rightleftharpoons HCO_3^- + OH^-$$

Hence the resulting solution is alkaline.

(*d*) The salt of a weak acid and a weak base, e.g. $(NH_4^+)_2CO_3^{2-}$. NH_4^+ is a strong cationic acid, as mentioned in (*b*), and CO_3^{2-} is a strong anionic base.

If $K_{acid} = K_{base}$, where K is the dissociation constant, the solution is neutral: $(OH^- + H_3O^+ \rightarrow 2H_2O)$. If $K_{acid} > K_{base}$, the solution is slightly acidic. If $K_{acid} < K_{base}$, the solution is slightly alkaline.

The Lowry and Brönsted theory also explains why a number of metallic salts are acidic in solution, for example, $Al_2(SO_4)_3$, $FeCl_3$, etc. In the hydrated salts and in solution, the small Al^{3+} and Fe^{3+} ions are hydrated with six molecules of water. These hydrated cations then function as acids by donating a proton to water:

$$Al(H_2O)_6^{3+} + H_2O \rightleftharpoons Al(H_2O)_5OH^{2+} + H_3O^+$$

Successive donations eventually deposit the hydrated oxides.

$$Al(H_2O)_5OH^{2+} + H_2O \rightleftharpoons Al(H_2O)_4(OH)_2^+ + H_3O^+$$

$$Al(H_2O)_4(OH)_2^+ + H_2O \rightleftharpoons Al(OH)_3.3H_2O + H_3O^+$$

The cation withdraws electronic charge from the oxygen atoms in co-ordinated water, rendering the hydrogen atoms more positive than in free water molecules. Oxygen lone pairs in uncoordinated water molecules remove these positive hydrogen atoms (as protons) to form hydroxonium ions.

Quantitative equilibrium calculations and expressions using the Lowry and Brönsted theory are, of course, identical with the classical methods.

Lewis Theory of Acids and Bases

This is a much more generalized concept of acids and bases. Here an acid is identified as a molecule or ion which can accept a pair of electrons and a base as a molecule or ion which can donate a pair of electrons.

Thus in the reaction

$$H-Cl+H-\underset{\displaystyle H}{\underset{|}{O}} \rightarrow \left[H-\underset{\displaystyle H}{\underset{|}{O}} \rightarrow H\right]^{+} + Cl^{-}$$

water donates an electron pair to the proton and is a Lewis base; HCl is a Lewis acid. Note that the process involves breaking a covalent bond between H and Cl and forming a new coordinate bond. Other examples include:

$$\underset{\text{base}}{NH_3} + \underset{\text{acid}}{H-Cl} \longrightarrow \left[H_3N\rightarrow H\right]^{+} + Cl^{-}$$

$$\underset{\text{acid}}{BF_3} + \underset{\text{base}}{F^{-}} \longrightarrow \left[F_3B\leftarrow F\right]^{-}$$

$$\underset{\text{base}}{Ca^{2+}O^{2-}} + \underset{\text{acid}}{SO_3} \longrightarrow Ca^{2+}\left[O-SO_3\right]^{2-}$$

One advantage of the Lewis theory is that it classifies as acids many non-hydrogen-containing molecules because they have the same function. One serious disadvantage is that the approach cannot be applied on a quantitative basis.

Oxo Acid Strengths

For any series of oxo acids of general formula $(HO)_xMO_y$, the acid strength increases as y increases. When x and y remain constant, the acid strength increases with increasing electronegativity of M. The validity of these rules can be seen from an examination of the pK values for a selection of representative acids (Table 8.2).

TABLE 8.2

Name of acid	*Formula*	*Electronegativity of M*	pK_1 *value*
Nitrous	$HNO_2(HONO)$	3·0	+3·37
Nitric	$HNO_3(HONO_2)$	3·0	−1·14
Tellurous	$H_2TeO_3((HO)_2TeO)$	2·1	+5·3
Selenious	$H_2SeO_3((HO)_2SeO)$	2·4	+2·4
Sulphurous	$H_2SO_3((HO)_2SO)$	2·5	+1·7
Sulphuric	$H_2SO_4((HO)_2SO_2)$	2·5	−3·0
Hypoiodous	HOI	2·5	+10·6
Hypobromous	HOBr	2·8	+8·7
Hypochlorous	HOCl	3·0	+7·5
Chloric	$HClO_3(HOClO_2)$	3·0	−1·0
Perchloric	$HClO_4(HOClO_3)$	3·0	−10·0

In the dissociation of an acid, the greater the magnitude of the δ^+ charge on the hydrogen atom of the hydroxy group, the greater is the degree of ionization and the acid strength

$$M\text{—}\overset{\delta-}{O}\text{—}\overset{\delta+}{H} \rightleftharpoons M\text{—}O^- + H^+$$

Thus in the hypohalous acids, the δ^+ charge on the hydrogen atom is largest in HOCl ($x_{Cl} = 3{\cdot}0$) and least in HOI ($x_I = 2{\cdot}5$) and acid strengths are in the order HOCl > HOBr > HOI (*see* Table 8.2). A similar argument applies to sulphurous, selenious, and tellurous acids where the decreasing electronegativities of M gives a decrease in acid strength. The presence of a multiply bonded oxygen atom in the acid increases the magnitude of the δ^+ charge on the hydrogen atom as a result of electron withdrawal:

$$^{\delta-}O = M \leftarrow O\text{—}\overset{\delta+}{H}$$

Hence the acid strength increases with an increase in the number of multiply bonded oxygen atoms. In fact, where there are no multiply bonded oxygen atoms the acid is weak, e.g. hypohalous acids. Although other factors, such as hydrogen bonding and relative bond strengths in free acids and oxo anions, are often influential, the above generalizations are useful and correlate many differences in acid strengths.

9

The Noble Gases

Physical Properties

	Helium He	Neon Ne	Argon Ar	Krypton Kr	Xenon Xe	Radon Rn
Atomic mass	4·0026	20·179	39·948	83·80	131·30	222
Atomic number	2	10	18	36	54	86
Isotopes (in order of abundance)	4, 3	20, 21, 22	36, 40, 38	84, 86, 82, 83, 80, 78	129, 132, 131, 134, 136, 128, 130, 126, 124	222,220
Electronic structure	$1s^2$	$2s^2 2p^6$	$3s^2 3p^6$	$3d^{10}$ $4s^2 4p^6$	$4d^{10}$ $5s^2 5p^6$	$5d^{10}$ $6s^2 6p^6$
Van der Waals radius (nm)	0·093	0·160	0·192	0·197	0·217	—
Ionization potential (kJ mol^{-1})	2376	2086	1526	1356	1176	1046
Density (gl^{-1})	0·178	0·900	1·781	3·736	5·887	9·73
Melting point (°C)	−271·9 (26 atm)	−248·5	−189·3	−156·6	−111·5	−71
Boiling point (°C)	−268·9	−245·9	−185·9	−152·9	−108	−61·8
Heat of vaporization (kJ mol^{-1})	0·084	1·80	6·53	9·04	12·60	16·40

General Characteristics

All noble gases except helium have the stable outer electronic configuration ns^2np^6; the small radii and high values for the ionization potentials indicate strongly bound electrons in these full orbitals, giving rise to atoms which show extreme chemical inertness. Thus there are no molecules and the noble gases are all monatomic. Low values for the heats of vaporization and boiling points indicate that all forces of attraction between atoms are absent apart from the weak forces of the van der Waals type. These forces arise as a result of the induction of a weak dipole in one electronic system by another and increase in magnitude with increase in atomic radius.

Van der Waals forces are extremely weak ($2-20$ kJ mol^{-1}) and are inversely proportional to the sixth power of the internuclear distance ($F \propto 1/d^6$). Thus these forces only become significant for internuclear distances of the order 0·2 to 0·5 nm.

Helium possesses the lowest boiling point of any known substance and forms a true solid (under pressure) at 4·2 K. Liquid helium cooled to 2·2 K undergoes remarkable changes in properties: the thermal conductivity increases (800 times that of copper) and the viscosity decreases (0·001 per cent of that of hydrogen). The liquid can in fact flow in thin films without friction. It is interesting to note that only the isotope ^4_2He shows this peculiar property and in this 'superfluid state' it is probable that the motion of the atoms has ceased altogether but the forces between the atoms (van der Waals forces) are insufficient to cause solidification.

Occurrence

The atmosphere is the only source of neon, argon, krypton, and xenon. Helium is found in natural gases in the U.S.A., probably formed during the radioactive decay processes with subsequent solution of the gaseous products. The amount of the noble gases present in the earth's atmosphere in parts per million according to Paneth (1939) is given in Table 9.1.

TABLE 9.1

Helium	5·24 parts per million
Neon	18
Argon	9·3
Krypton	1·0
Xenon	0·08

Extraction

The United States operates plants for the extraction of helium from natural gas which contains up to 8 per cent of helium. The gas does not contain either hydrogen or neon so that rapid cooling by adiabatic expansion condenses out all the gases except helium. Recovery from the atmosphere involves complex liquefaction and rectification processes. Air is liquefied (p. 281) and in the subsequent fractional distillation the liquid oxygen obtained also contains argon, krypton, and xenon. Crude argon containing krypton and xenon is obtained by refractionation, further separation being carried out by selective adsorption on charcoal cooled in liquid air. Helium, neon, and some nitrogen remain gaseous where the liquid nitrogen collects. The nitrogen is removed by cooling in liquid nitrogen and passing over heated magnesium. The remaining gases are cooled in liquid hydrogen when the neon solidifies.

Properties of the Noble Gases

The chemical inactivity of the noble gases may be expected from the absence of any bonding orbitals and no compounds are formed with the most vigorous oxidizing and reducing agents. However, certain compounds have been made, some very recently, and these are briefly summarized below.

The Fluorides of Xenon and Radon

It has been known for some time that platinum hexafluoride vapour and oxygen produced a red compound by the reaction

$$PtF_6 + O_2 \rightarrow O_2^+PtF_6^-$$

Since the ionization potentials of oxygen and xenon are approximately the same, N. Bartlett in 1962 carried out a similar reaction with xenon and produced the first stable yellow compound of a noble gas.

$$Xe + PtF_6 \rightarrow Xe^+PtF_6^-$$

Further work showed that xenon also formed a fluoride XeF_4, prepared as shown in Figure 9.1. XeF_6 has also been produced under pressure and XeF_2 by a photochemical reaction. All three fluorides are colourless volatile solids.

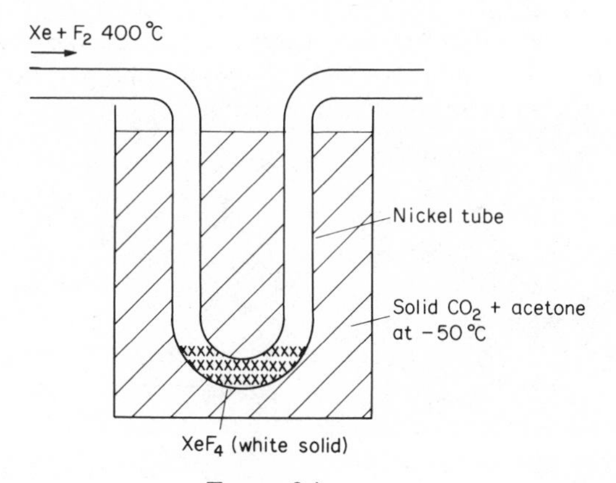

FIGURE 9.1

Xenon hexafluoride, XeF_6, reacts separately with both silica and water to produce a liquid oxo salt, $XeOF_4$.

$$2XeF_6 + SiO_2 \rightarrow 2XeOF_4 + SiF_4$$
$$XeF_6 + H_2O \rightarrow XeOF_4 + 2HF$$

The action of water on $XeOF_4$ or XeF_4 produces an acidic aqueous solution thought to contain H^+ and $HXeO_4^-$. Careful evaporation of this solution produces a solid explosive trioxide XeO_3.

$$XeOF_4 + 2H_2O \rightarrow XeO_3 + 4HF$$
$$3XeF_4 + 6H_2O \rightarrow 2Xe + XeO_3 + 12HF + \tfrac{3}{2}O_2$$

If the aqueous solution containing $HXeO_4^-$ is treated with alkali, an oxo anion of xenon(VIII) is produced. Solid salts containing this anion have been isolated, e.g. $Na_4XeO_6.8H_2O$, $Ba_2XeO_6.1\tfrac{1}{2}H_2O$, which produce unstable XeO_4 gas with concentrated H_2SO_4.

$$2HXeO_4^- + 2OH^- \rightarrow Xe^{VIII}O_6^{4-} + Xe + O_2 + 2H_2O$$

Krypton and radon fluorides have also been prepared, e.g. —KrF_2, RnF_2, but no compounds of helium, neon, or argon have yet been reported.

Structural investigations have shown that XeF_2 is linear, XeO_3 is trigonal pyramidal, XeF_4 is square planar, $XeOF_4$ is square pyramidal, and XeO_6^{4-} is octahedral. The stable form of XeF_6 consists of cyclic tetramers $(XeF_6)_4$ containing distorted XeF_6 octahedra with fluorine bridges.

The explanation of the bonding on an orbital basis is difficult, but one theory suggests that the strong attraction of fluorine atoms for electrons causes the promotion of one or more electrons from the full $5p$ orbital of xenon into the empty $5d$ orbital. Hybridization can then take place involving $5s$, $5p$ and $5d$ orbitals. Thus in XeF_2, sp^3d hybridization occurs, giving a linear molecule possessing three lone pairs of electrons. In XeF_4, the hybridization is sp^3d^2, giving a square planar molecule with two lone pairs of electrons. (Figure 9.2) An sp^3d^3 hybridization scheme in XeF_6 involves six bond pairs and one lone pair which, due to lone-pair bond-pair repulsion gives a distorted octahedral arrangement.

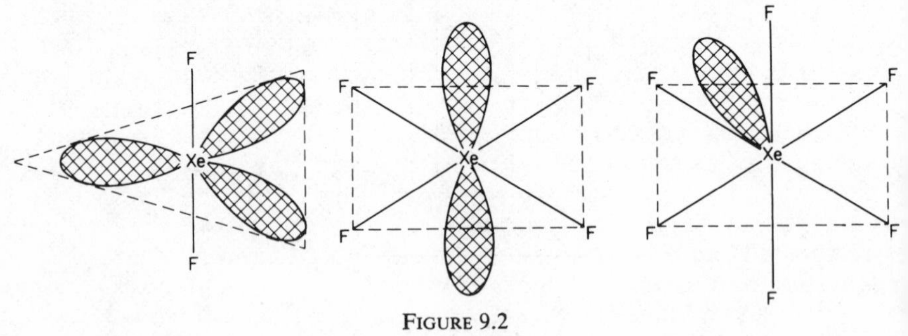

FIGURE 9.2

Compounds have not yet been prepared from helium, neon, and argon by reaction with fluorine.

Clathrate (Cage) Compounds

If β-quinol is crystallized from water in a noble gas atmosphere of xenon, krypton or argon under pressure, a three-dimensional hydrogen bonded structure of quinol is produced which contains cavities or holes, three-quarters of which are filled by noble gas atoms. On heating or dissolving in alcohol, the hydrogen-bonded cages break down and the noble gas is liberated. Size is the important factor in these compounds and helium and neon do not form clathrates since their atoms are too small and can escape from the cavities. The structure of a clathrate compound is shown in Figure 9.3.

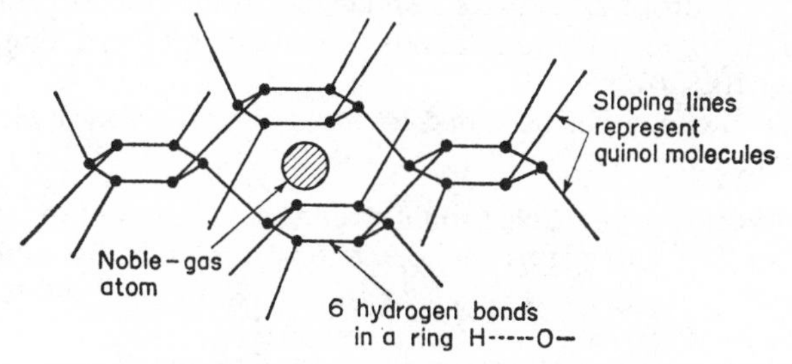

FIGURE 9.3. A noble-gas clathrate with quinol

The noble gas hydrates are in fact clathrate compounds. When solidified with noble gases under pressure, water forms an open structure containing cavities which can trap the noble gas atoms of argon, krypton, and xenon. The formulae of these hydrates vary depending upon the number of cavities occupied: $X(5{\cdot}76 - 7{\cdot}67)H_2O$.

Uses

Helium is less soluble than nitrogen in the blood and mixtures of helium and oxygen are used by divers and tunnel workers working under high pressures. Since helium diffuses more rapidly than nitrogen, helium/oxygen mixtures permit easier breathing. They are used to relieve attacks of asthma. Liquid helium is used in some gas-cooled reactor systems. Fluorescent tubes have a coating of fluorescent material on the inside surface and are filled with a noble gas. Argon, krypton, and xenon are also used in gas-filled incandescent lamps. Radon is collected from the radioactive decay of radium salts, sealed in small tubes, and used in the treatment of malignant growths and tumours.

10

Hydrogen

The atom of hydrogen possesses a single electron in the 1*s* orbital and, in the majority of hydrogen compounds, it is used to form a single covalent link (Figure 10.1(*a*)).

An electron can either be added to or removed from the hydrogen atom to give respectively:

(*a*) The hydride ion H^-, electronic structure $1s^2$, present in the dry ionic hydrides Na^+H^-, $Ca^{2+}(H^-)_2$ but decomposed rapidly by water:

$$H^- + H_2O \rightarrow OH^- + H_2$$

(*b*) The hydrogen ion H^+ or free proton. This small ion, with its high charge density, rapidly hydrates in water producing the hydroxonium ion (Figure 10.1(*b*)).

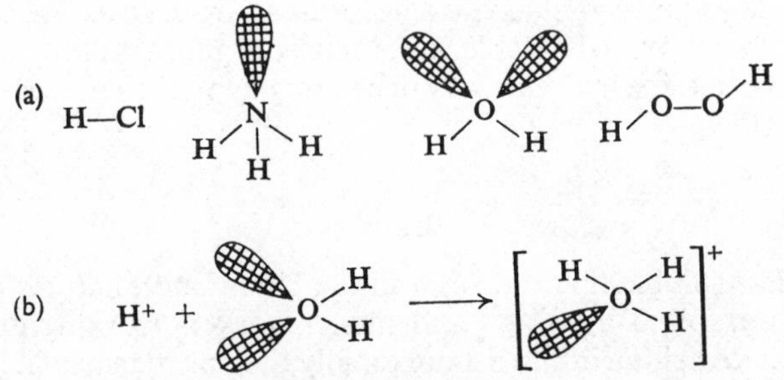

(a) H—Cl

(b) $H^+ +$

FIGURE 10.1

Certain crystalline salts also contain the hydroxonium ion. Thus $HClO_4.H_2O$ is better written as $H_3O^+ClO_4^-$. The hydrates originally formulated as $HCl.2H_2O$ and $HClO_4.2H_2O$ contain the ion $H_5O_2^+$, e.g. $H_5O_2^+Cl^-$, $H_5O_2^+ClO_4^-$.

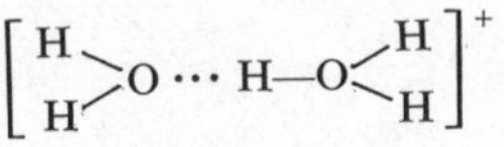

$$\left[\begin{matrix} H \\ H \end{matrix} > O \cdots H—O < \begin{matrix} H \\ H \end{matrix} \right]^+$$

Occurrence

The hydrogen molecule possesses a velocity which enables it to escape from the earth's gravitational field and hence only occurs to the extent of 1 part per million in the earth's atmosphere. Combined hydrogen occurs as water in hydrated salts and in almost all materials of organic origin.

Laboratory Preparation

Many methods are available and a summary of some of the reactions producing hydrogen is given below:

I. The addition of electrons to protons

(*a*) The action of dilute acids on electropositive metals (Fe, Zn, or Mg).

$$Zn + 2H^+ \rightarrow Zn^{2+} + H_2 \qquad (E^{\ominus}_{Zn} = -0{\cdot}76\ V)$$

(*b*) Electrolytic hydrogen may be made by the electrolysis of water containing a small amount of sulphuric acid or sodium hydroxide.

path of electrons

negative cathode		positive anode
	$\leftarrow Na^+ \quad OH^- \rightarrow$	
$2H^+ + 2e \rightarrow H_2$	$\leftarrow H^+ + OH^- \rightarrow$	$4OH^- \rightarrow 2H_2O + O_2 + 4e$
	$\rightleftharpoons H_2O$	

II. Other methods involve the addition of electrons to water

$$2H_2O + 2e \rightarrow 2OH^- + H_2$$

(*a*) Ionic hydrides on water

$$Ca^{2+}(H^-)_2 + 2H_2O \rightarrow Ca^{2+}(OH^-)_2 + 2H_2$$

(*b*) Alkali and alkaline earth metals on water

$$2Na + 2H_2O \rightarrow 2Na^+OH^- + H_2$$

(*c*) Aluminium, tin, and zinc on hot strong caustic alkali solution

$$Zn + 2OH^- + 2H_2O \rightarrow Zn(OH)_4^{2-} + H_2$$

zincate ion in sodium zincate $Na_2Zn(OH)_4$

The gas produced by any method may be dried over calcium chloride or phosphorus pentoxide.

Industrial Manufacture

(*a*) Hydrogen is an important by-product formed during the electrolysis of brine for the manufacture of caustic soda (p. 168).

(*b*) In the Bosch process steam is passed over white hot coke, producing water gas.

$$C + H_2O \rightleftharpoons CO + H_2 \qquad \Delta H = +131{\cdot}4 \text{ kJ}$$

The water gas is mixed with steam and passed over an iron(III) oxide/chromium(III) oxide catalyst at 300 °C to produce more hydrogen.

$$CO + H_2 + H_2O \rightarrow CO_2 + 2H_2 \qquad \Delta H = -42{\cdot}7 \text{ kJ}$$

Carbon dioxide is removed by bubbling the gas into water under pressure and carbon monoxide by bubbling it through ammoniacal copper(I) chloride.

(*c*) Modern methods for producing hydrogen include the catalytic oxidation of methane by oxygen or steam.

$$CH_4 + H_2O \rightleftharpoons CO + 3H_2 \qquad \Delta H = +214 \text{ kJ}$$

$$2CH_4 + O_2 \rightleftharpoons 2CO + 4H_2 \qquad \Delta H = -67 \text{ kJ}$$

A temperature of 900 °C with a nickel catalyst gives about 98 per cent conversion.

The Uses of Hydrogen

These include: (i) the oxyhydrogen and atomic hydrogen torch; (ii) the conversion of vegetable and fish oils to solid cooking fats (hydrogenation); (iii) the synthesis of ammonia (Haber process); (iv) the conversion of coal dust to petroleum hydrocarbons for use as fuel; and (v) the synthetic production of methanol.

Properties of Hydrogen

Hydrogen is a colourless, odourless, sparingly water-soluble gas of melting point −259 °C and boiling point −253 °C. Its reactions with some non-metallic elements are outlined below.

$2H_2 + O_2 \rightarrow 2H_2O$ (burn hydrogen in oxygen)

$H_2 + X_2 \rightarrow 2HX$ (X = F, Cl, Br, I) (hydrogen explodes with fluorine, burns in chlorine but reacts reversibly with bromine and iodine)

$H_2 + S \rightleftharpoons H_2S$ (burn sulphur in hydrogen)

$3H_2 + N_2 \rightleftharpoons 2NH_3$ (at 500 °C, 200 times atm, iron catalyst)

With compounds, hydrogen may either add on or remove oxygen, e.g.

$$\begin{array}{ccc} CH_2 + H & \xrightarrow{Ni} & CH_3 \\ \| \quad\;\; | & & | \\ CH_2 \quad H & & CH_3 \\ \text{ethylene} & & \text{ethane} \end{array}$$

$$CO + 2H_2 \xrightarrow[\text{under pressure}]{Al_2O_3} CH_3OH \text{ (methanol)}$$

$$CuO + H_2 \longrightarrow Cu + H_2O$$

Hydrogen combines with a large number of elements on heating to form hydrides and these can be conveniently classified as follows:

Salt-like Hydrides

These are formed by the Group IA and IIA metals and contain the hydride ion H^-, e.g. Na^+H^-. The hydrides of zinc, and cadmium, are also probably ionic but have to be prepared indirectly by reduction of the halide with an ethereal solution of lithium aluminium hydride ($LiAlH_4$). Group IA and IIA hydrides possess high melting points, characteristic of an ionic lattice. The Group I hydrides have a sodium chloride type lattice. All ionic hydrides liberate hydrogen from water.

$$H^- + H_2O \rightarrow OH^- + H_2$$

Covalent Hydrides

These are generally hydrides of non-metals or weakly electropositive metals, e.g. PH_3, CH_4, SbH_3, BiH_3, SnH_4, H_2Te. The methods of preparation are usually very specific for the particular compound. In these hydrides, molecules are held together by weak van der Waals forces and hence compounds are usually gases, liquids, or solids with low melting points and boiling points. The stability of covalent hydrides usually decreases progressively down a group, e.g. Group V

$$\underrightarrow{NH_3 \quad PH_3 \quad AsH_3 \quad SbH_3 \quad BiH_3}$$

Stability decreasing

Interstitial Hydrides of the Transition Metals

The small hydrogen atoms (atomic radius = 0·037 nm) enter the holes or interstices between the atoms comprising the metallic lattice without destroying the original crystal structure of the metal. These hydrides are often both variable in composition and non-stoichiometric, e.g. $VH_{0.6}$, $TiH_{1.5}$–$TiH_{1.9}$. They have a metallic appearance and their properties are closely related to those of the parent metal. Recent work has shown that the arrangement of the metal atoms in these hydrides is often different from that of the parent metal, i.e. the hydrides are not interstitial but have a definite structure. Thus Ti and Cr are respectively hexagonal close packed and body-centred cubic in the metal but face-centred cubic in the hydrides, MH_2.

Atomic Hydrogen

Molecular hydrogen can be decomposed into its atoms by a platinum wire at 1500 °C. Although recombination of hydrogen atoms does not readily occur, since the energy liberated is sufficient to cause immediate separation, it can occur at the surface of a metal such as iron, copper or platinum with the subsequent liberation of heat. This is the basis of the atomic hydrogen torch. Hydrogen is atomized by an electric arc struck between tungsten electrodes and the atoms recombine on a metal surface giving out sufficient heat to raise the temperature of the metal to 3500 °C. A layer of molecular hydrogen forms just above the metal surface and this excludes all air; thus the torch is particularly useful for welding readily oxidized metals such as aluminium (Figure 10.2).

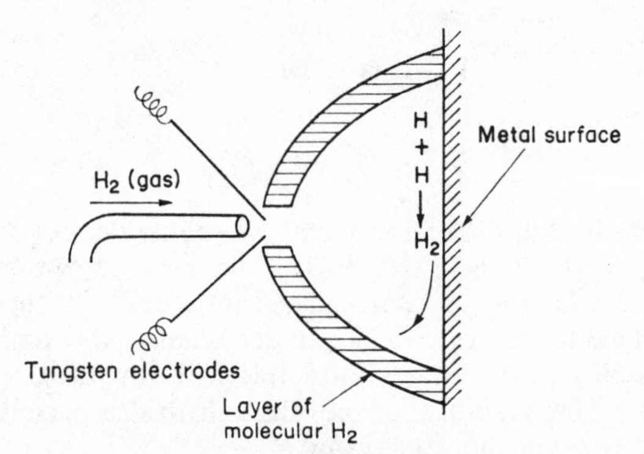

FIGURE 10.2. The atomic hydrogen torch

Atomic hydrogen is a very powerful reducing agent and reduces many metal oxides, chlorides, and sulphides to the free metal.

Nascent Hydrogen

Molecular hydrogen has little reducing action in aqueous solution but hydrogen generated in aqueous solution by a chemical reaction does possess considerable reducing power and was formerly called nascent (new-born) hydrogen. Thus gaseous hydrogen bubbled through acidified iron(III) chloride solution does not reduce it, whereas hydrogen generated in the solution from zinc and hydrochloric acid reduces the iron(III) chloride to iron(II) chloride:

$$2Fe^{3+} + H_2\,(\textit{nascent}) \rightarrow 2Fe^{2+} + 2H^+$$

At one time it was thought that the reducing power was due to atomic hydrogen but it is now clear that the chemical reaction used to produce the nascent hydrogen is important. Before any reduction can take place with molecular hydrogen the H—H bond must be broken and the failure of molecular hydrogen to reduce iron(III) chloride is due to the large amount of energy needed to break the H—H bond (436 kJ mol^{-1}). In the zinc and hydrochloric acid reduction of iron(III) chloride, zinc metal alone effects the reduction by supplying electrons.

$$Zn^{2+} + 2e \rightarrow Zn\ (E^{\ominus} = -0{\cdot}76\ V)$$

$$2Fe^{3+} + 2e \rightarrow 2Fe^{2+}\ (E^{\ominus} = +0{\cdot}76\ V)$$

Similar considerations apply to other reducing agents, e.g. sodium amalgam.

$$Na^+ + Hg + e \rightarrow Na/Hg\ (E^{\ominus} = +1{\cdot}84\ V)$$

Ortho and Para Hydrogen

In the hydrogen molecule two 1*s* electrons are in a molecular orbital bonding together two protons (p. 35). The protons can either have their spins in the same or opposite directions giving two different kinds of hydrogen molecules in equilibrium:

(*a*) *Ortho hydrogen* (75 per cent of ordinary hydrogen) with the protons spinning in the same direction.

(*b*) *Para hydrogen* (25 per cent of ordinary hydrogen) with the protons spinning in opposite directions.

There is no difference in chemical properties between these two forms, only slight differences in specific heats, melting points, and boiling points, etc. These are listed in Table 10.1.

TABLE 10.1

	Ordinary Hydrogen	*Para*	*Ortho*
Specific heat (100 K)	0·431	1·504	0·073
Melting point	13·92 K	13·88 K	13·93 K
Boiling point	20·38 K	20·29 K	20·41 K

Almost pure (99·8 per cent) para form can be obtained by adsorbing hydrogen at its boiling point on to charcoal. On warming, the para form is liberated.

Isotopy of Hydrogen

The three isotopic forms of hydrogen are shown below:

	$^{1}_{1}H$ (protium)	$^{2}_{1}D$ (deuterium)	$^{3}_{1}T$ (tritium)	
Nuclear structure	1	1	1	proton
	0	1	2	neutrons

Deuterium and tritium are respectively twice and three times as heavy as protium and only occur to a very minute extent in ordinary hydrogen ($H:D \approx 6000:1; H:T \approx 10^{17}:1$). All the isotopic forms have almost identical chemical properties but deuterium and tritium have higher melting and boiling points and densities. Oxides are known and deuterium oxide (*heavy water*) is important as a moderator in nuclear reactors (p. 16). The melting point of heavy water is 3·8 °C, its boiling point is 101·4 °C, and its dissociation constant is 3×10^{-15} at 25 °C.

Deuterium oxide is prepared by the prolonged electrolysis of water, when 'light' hydrogen is evolved leaving a liquid richer in deuterium oxide. Later fractions of gaseous hydrogen become progressively richer in deuterium which is burned to the oxide and returned to the cell. On an average, 20 litres of water, on electrolysis, yield 1·5 cm^3 of 60 per cent D_2O. Deuterium is used as an isotopic indicator in the study of reactions, particularly in organic chemistry.

For every hydrogen compound, there is a corresponding deuterium compound and these are generally obtained from deuterium oxide:

$$2D_2O + 2Na \rightarrow 2NaOD + D_2$$

$$D_2O + SO_3 \rightarrow D_2SO_4$$

$$D_2O + N_2O_5 \rightarrow 2DNO_3$$

$$CuSO_4 + 5D_2O \rightarrow \underset{\text{(green)}}{CuSO_4.5D_2O}$$

Deuterium is used to elucidate structures and determine the mechanism of chemical reactions by exchange with hydrogen atoms.

Tritium is radioactive and is a weak β emitter with a half-life of 12·5 years. It is prepared by the action of slow neutrons on the $^{6}_{3}Li$ isotope:

$$^{6}_{3}Li + ^{1}_{0}n \rightarrow ^{3}_{1}T + ^{4}_{2}He$$

$$^{3}_{1}T \rightarrow ^{3}_{2}He + ^{0}_{-1}\beta$$

11

The Alkali Metals
Group IA

Physical Properties

	Lithium Li	Sodium Na	Potassium K	Rubidium Rb	Caesium Cs
Atomic mass	6·941	22·9898	39·102	85·47	132·905
Atomic number	3	11	19	37	55
Isotopes (in order of abundance)	7, 6	23	39, 40, 41	85, 87	133
Electronic structure	$1s^2 2s^1$	$2p^6 3s^1$	$3p^6 4s^1$	$4p^6 5s^1$	$5p^6 6s^1$
Atomic radius (nm)	0·123	0·157	0·203	0·216	0·235
Ionic radius (nm) (M^+)	0·060	0·095	0·133	0·148	0·169
Ionization potential (kJ mol^{-1})	525	500	424	408	382
Electronegativity	1·0	0·9	0·8	0·8	0·7
Density (g cm^{-3})	0·53	0·97	0·86	1·53	1·90
Melting point (°C)	179	98	63·5	39	28·5
Boiling point (°C)	1330	882·9	774	688	690
Electrode potential (V)	−3·04	−2·71	−2·92	−2·99	−3·02
Heat of atomization at 298 K (kJ mol^{-1})	161	109	90	86	79
Heat of hydration (kJ mol^{-1})	−519	−406	−322	−301	−276

(Francium, atomic number 87, is not included above as little data is available.)

General Characteristics

All the atoms of the alkali metals except lithium have the configuration $np^6(n+1)s^1$ in their outermost orbitals. The low values of the ionization potentials indicate easy removal of the single *s* electron. In the elementary state these electrons can readily pass from one atom to another giving rise to the conductivity and lustre characteristic of metals. All the metals have a body-centred cubic lattice and, since the metallic atoms have a large radius with only one electron per atom available for bonding, the metals are soft with low melting points. Increase in atomic radius down the group correlates with the decrease in the values of the ionization potentials. Thus caesium is the most reactive of the alkali metals. Francium is, in fact, radioactive but little is known concerning its chemical properties.

The chemistry of the alkali metals is the chemistry of ions and all compounds are ionic. A study of alkali metal vapours indicates the presence of 1 per cent double molecules which is almost the only example in this group of covalent bonding involving overlap of *s* orbitals. The electrode potential of lithium appears abnormally high in contrast with the high ionization potential. However, it must be remembered that values of ionization potential refer to the reaction $Li(gas) \rightarrow Li^{+}(gas) + e$, whereas electrode potential values refer to a reaction in aqueous solution which involves the hydration of a unipositive cation (p. 138):

$$Li^{+}(g) + xH_2O \rightarrow Li(H_2O)_x^{+} \qquad (\Delta H = -519 \text{ kJ mol}^{-1})$$

Lithium is the smallest ion and attracts polar water molecules most strongly releasing a large quantity of energy as heat of hydration. Thus a considerable quantity of energy is recovered giving rise to a high electrode potential.

Solubility

For an ionic compound to be soluble in water, the crystal lattice must be broken down into its constituent ions. This process requires energy which can only be recovered as the heat of hydration of the ions. Thus, if the energy of hydration is greater than the lattice energy of the crystal, the substance should be soluble. Water, having a high relative permittivity, diminishes forces between ions and is particularly effective as a solvent. Unfortunately, predictions involving solubility are extremely involved since:

(*a*) Increase in ionic size decreases both the lattice energy and the energy of solvation by different amounts, the former being affected to a greater extent.

(*b*) Increase in the charge on an ion increases the lattice energy to a greater extent than the energy of solvation. A graphic examination of the solubilities of the alkali metal chlorides shows the difficulty of relating solubility to ionic size (Figure 11.1).

With an increase in charge on the cation it is usually safe to predict a decrease in solubility (Figure 11.2).

The process of dissolution involves the breakdown of the crystalline solid to give solvated ions, i.e. there is an increase in the degree of disorder of the system and ΔS is positive. For spontaneous dissolution to occur, ΔG must be negative and, since $\Delta G = \Delta H - T\Delta S$, this can only occur if ΔH is

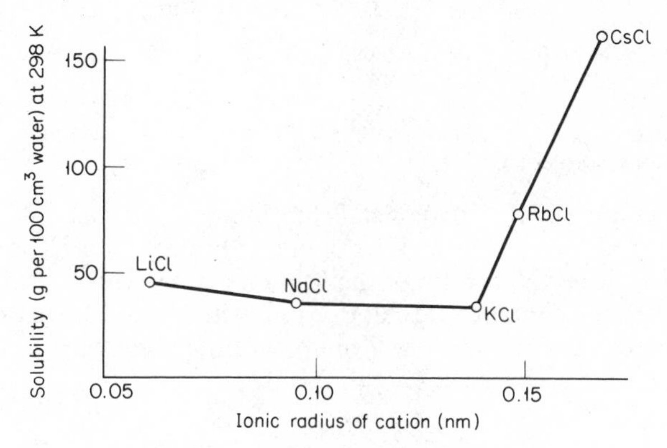

FIGURE 11.1

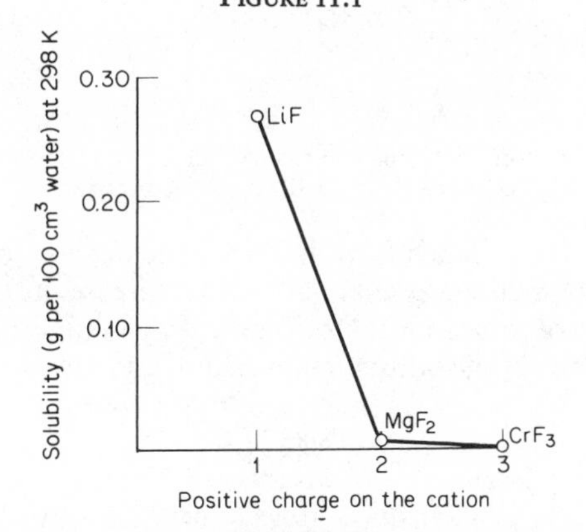

FIGURE 11.2

negative (as in exothermic processes) or if ΔH is positive but less than $T\Delta S$. Thus it is possible that in some cases in which the lattice energy exceeds the solvation energy (ΔH is positive), spontaneous dissolution can occur provided that the entropy increase is sufficiently large.

Melting Points

The large ionic radii (apart from lithium) make the alkali metal ions poor electron-cloud distorters or polarizers. The external field at the surface of an ion depends both upon the charge and the radius of an ion and a good quantitative measure of the external polarizing effect of an ion is the ratio of charge to radius (Table 11.1).

TABLE 11·1

	Li^+	Na^+	K^+	Rb^+	Cs^+
Ionic radius (nm)	0·060	0·095	0·133	0·148	0·169
Charge	+1	+1	+1	+1	+1
Charge/radius	16·7	10·5	7·5	6·8	5·9
Melting point of chlorides (°C)	613	801	776	715	645

The melting points of the alkali metal chlorides fall with the decrease in the charge/radius ratio. The low melting point* of lithium chloride is explained by the large charge/radius ratio of the lithium cation. This causes polarization of the electron clouds of the chloride ions with an increase in the electron density between the nucleus of the lithium ion and the chloride ion, giving an increase in covalent character (Figure 11.3).

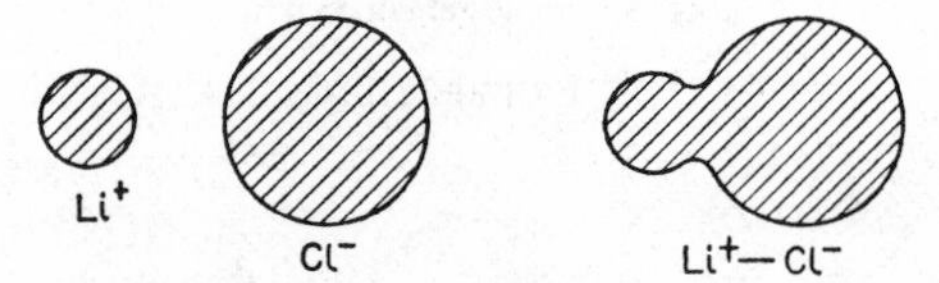

FIGURE 11.3. Small cation distorting the electron cloud of a large anion, giving rise to some covalent bonding between the ions

All the other alkali metals form halides which can be regarded as being mainly ionic and melting points fall with decrease in the charge/radius ratio. This effect becomes more noticeable on examination of the melting points of the chlorides of sodium, magnesium, and aluminium (Table 11.2).

TABLE 11.2

	Na^+	Mg^{2+}	Al^{3+}
Ionic radius (nm)	0·095	0·065	0·050
Charge	+1	+2	+3
Charge/radius	10·5	30·8	60·0
Melting point of chlorides (°C)	801	712	180 (sublimes)

The increase in charge/radius ratio is accompanied by a corresponding decrease in the melting points of the chlorides. The distortion and increase in covalent character becomes very marked with aluminium chloride, which is usually regarded as covalent.

* Lithium chloride has a lower melting point than expected from the charge/radius ratio figures (cf. Mg Cl_2) because in solid lithium chloride all the Cl^- ions surrounding the small Li^+ ions are in contact. The resulting interanionic repulsive forces reduce the lattice energy of the crystal and hence reduce the melting point.

Lithium is also anomalous in the group by showing chemical similarities with its diagonal neighbour magnesium. From Table 11.3 it is interesting to

TABLE 11.3

	Li^+	Mg^{2+}
Ionic radius (nm)	0·060	0·065
Charge	+1	+2
Charge/radius	16·7	30·8

note the similarity in radii but the large difference in the charge/radius ratio. Resemblances between lithium and magnesium are usually due to a similarity in ionic size although the charge/radius ratio is also effective. The following properties may be noted for lithium and magnesium which are in contrast to the properties of the other alkali metals:

(*a*) Both metals are harder and have a higher melting point than the other metals in their respective groups.

(*b*) Lithium and magnesium only slowly react with cold water, whereas the other alkali metals react vigorously (potassium, rubidium, and caesium ignite).

(*c*) Their hydroxides are less soluble and decompose at red heat.

(*d*) A nitride is formed by direct combination with elementary nitrogen (sodium, potassium, rubidium, and caesium do not combine with nitrogen).

(*e*) The oxides Li_2O and MgO do not react with excess oxygen to give a peroxide or a superoxide, both of which are produced readily from the other alkali metal oxides.

(*f*) The lithium and magnesium ions distort or polarize the carbonate ion, CO_3^{2-}, to such an extent that a C—O bond is easily broken on heating, liberating carbon dioxide. The other alkali metal carbonates are stable to heat.

(*g*) Solid bicarbonates are known for sodium, potassium, rubidium, and caesium but not for lithium or magnesium.

Occurrence

Salt beds contain both sodium and potassium chlorides. Sea water contains about 2·6 per cent sodium chloride and 0·8 per cent potassium chloride. Double chlorides and sulphates of potassium also occur as carnallite, $KCl.MgCl_2.6H_2O$, and kainite, $K_2SO_4.MgSO_4.MgCl_2.6H_2O$. Lithium, rubidium, and caesium only occur in a few rare aluminosilicate minerals.

Extraction of the Metals

All the elements possess high electrode potentials and electrolysis of aqueous solutions produces the hydroxide. Therefore electrolysis of the fused halides is employed, although vacuum distillation of the anhydrous chlorides of rubidium and caesium with excess calcium is also used. A section of the Downs cell, used for the production of sodium by the electrolysis of fused sodium chloride, is shown in Figure 11.4.

Since sodium chloride fuses at 800 °C, which is very close to the boiling point of sodium, an electrolyte of sodium chloride plus 58 per cent calcium chloride with a little potassium fluoride, is used to lower the fusion point to 600 °C. At 600 °C sodium has a lower discharge potential than calcium.

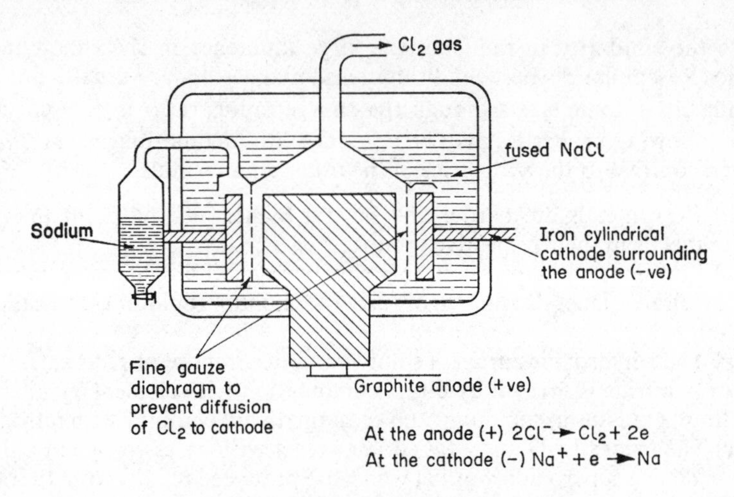

FIGURE 11.4. Section of the Downs cell

Properties

The alkali metals are extremely reactive chemically; a summary of the principal chemical reactions is given in Table 11.4.

TABLE 11.4

Reagent	*Reaction*	*Comments*
Regulated quantity of air or oxygen at 180 °C	$4M + O_2 \rightarrow 2M_2O$ white	Cs_2O is unusual in being red
Regulated quantity of oxygen at 300 °C	$2M + O_2 \rightarrow M_2O_2$ peroxide	No reaction with lithium
Excess oxygen 300 °C	$M + O_2 \rightarrow MO_2$ superoxide	No reaction with lithium; sodium needs high pressure
Heating with non-metallic elements, halogens, N_2, S, P, and H_2	$2M + Cl_2 \rightarrow 2MCl$ $6M + N_2 \rightarrow 2M_3N$ $2M + S \rightarrow M_2S$ $3M + P \rightarrow M_3P$ $2M + H_2 \rightarrow 2MH$	Only lithium reacts with nitrogen

TABLE 11.4—*continued*

Water	$2M + 2H_2O \rightarrow 2MOH + H_2$	Lithium reacts slowly, sodium fuses, and the other metals ignite
Dilute acids	$2M + 2H^+ \rightarrow 2M^+ + H_2$	Explosive
Ammonia gas (400 °C)	$2M + 2NH_3 \rightarrow 2MNH_2 + H_2$ metallic amides, e.g. sodamide $NaNH_2$	
Aluminium chloride with heat	$3M + AlCl_3 \rightarrow 3MCl + Al$	

Uses

The alkali metals are an important source of numerous alkali metal compounds. Important alloys include:

Na/Pb for the production of lead tetra-ethyl (used as an antiknock petrol additive);

Li/Pb for flexible cable sheaths;

Li/Al for increasing the corrosion resistance of aluminium;

Na/K (liquid) as a cooling liquid and heat-exchanger in high-temperature nuclear reactors.

Rubidium and caesium, which readily lose electrons, form alloys with sodium and potassium and are used in the production of photoelectric cells and radio valves.

Caesium may find a use as a rocket fuel in the 'ion engine' where caesium is vaporized, ionized, and the ions accelerated in an electric field to high speeds and then ejected through the rocket nozzles.

The Compounds of the Alkali Metals

The Hydrides

These are prepared from the elements at 400 °C (lithium at 800 °C); their stability decreases from lithium to caesium. They are all ionic solids with a sodium chloride type lattice. They are good electrolytes when molten but are hydrolysed by water.

$$2H^- + 2H_2O \rightarrow 2OH^- + 2H_2$$

The Oxides

All the oxides are ionic; the method of preparation is indicated in Table 11.4. The monoxides $M_2{}^+O^{2-}$ contain the oxide ion O^{2-}; the peroxides $M_2{}^+O_2{}^{2-}$

the peroxide ion $O—O^{2-}$; and the highly coloured superoxides $M^+O_2^-$ (sodium and potassium are yellow, rubidium orange, and caesium red) contain the paramagnetic superoxide ion $O—O^-$, which has one unpaired electron. All these oxides are decomposed by cold water.

$$O^{2-} + H_2O \rightarrow 2OH^-$$
$$O - O^{2-} + 2H_2O \rightarrow 2OH^- + H_2O_2$$
$$2O - O^- + 2H_2O \rightarrow 2OH^- + H_2O_2 + O_2$$

The Hydroxides

Sodium and potassium hydroxides are the most important alkalis and are made commercially by the electrolysis of chloride solutions. Two kinds of cell are used.

(*a*) The mercury cathode cell (Castner–Kellner cell). Mercury flows along the bottom of the cell and is made the cathode (Figure 11.5). The brine electrolyses and, since hydrogen has a high overvoltage at a mercury cathode (+0·78 V), sodium is preferentially discharged and forms an amalgam with the mercury.

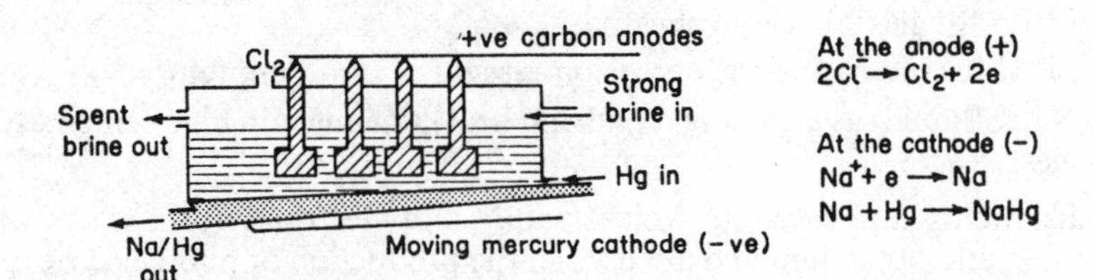

FIGURE 11.5. The Castner–Kellner cell

The sodium amalgam flows out and is reacted with water.

$$2NaHg + 2H_2O \rightarrow 2NaOH + 2Hg + H_2$$

The mercury is recirculated to the cell. Hydrogen and chlorine are two important by-products (p. 325) from a method which gives caustic soda of high purity but requires a large capital outlay for the mercury ($7{\cdot}3 \times 10^4$ kg per cell).

(*b*) Diaphragm cells. In this type of cell (Figure 11.6) alkali and chlorine are kept apart by means of a porous diaphragm, usually of asbestos. In the Billiter cell, brine soaks through an asbestos diaphragm and, on contact with a negative wire gauze, electrolysis begins. Chlorine is liberated at graphite anodes and sodium hydroxide solution is formed at the outside edges of the cathode. Sodium chloride separates out from the concentrated solution which is then evaporated to give solid caustic soda.

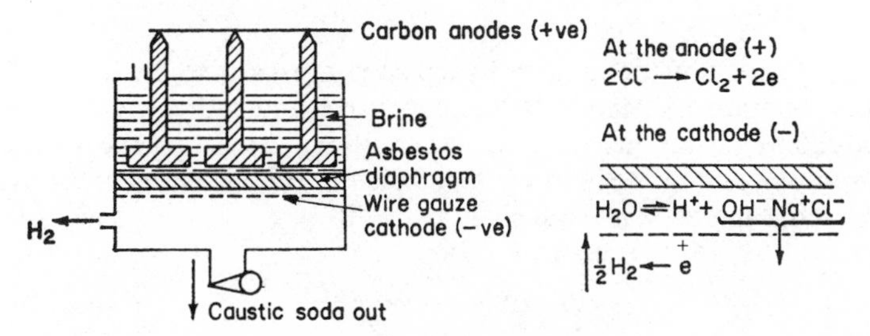

FIGURE 11.6. A diaphragm cell

Properties

All the hydroxides are white, soluble, crystalline solids. Hydroxide stability increases with increasing cation size. The small lithium ion polarizes the electron clouds of the hydroxide ion to such an extent that decomposition occurs on heating. As the cation size increases, the deformation becomes less and the hydroxides more stable. This deformation also imparts covalent character to the hydroxide and again as cation size increases covalent character decreases and the hydroxide become more ionic. Thus ionic character and hence alkali strength increase from lithium to caesium.

All are deliquescent apart from lithium hydroxide and the resulting solutions absorb carbon dioxide from the atmosphere forming carbonate solutions. Sodium hydroxide is unusual in absorbing moisture and carbon dioxide, finally setting to a solid hydrated carbonate.

Aqueous solutions of the hydroxides contain a large concentration of hydroxyl ion (OH^-) and generally precipitate insoluble metallic hydroxides from solutions containing metallic cations. The hydroxides of aluminium, zinc, lead, and tin, however, dissolve in excess of alkali giving clear solutions which can also be obtained by the action of strong alkali on the metal.

$$M^{2+} + 2OH^- \rightarrow M(OH)_2\downarrow \; (M^{2+} = Cu^{2+}, Zn^{2+}, Pb^{2+}, Fe^{2+})$$

$$M^{3+} + 3OH^- \rightarrow M(OH)_3\downarrow \; (M^{3+} = Al^{3+}, Fe^{3+})$$

$$Zn(OH)_2 + 2OH^- \rightarrow Zn(OH)_4^{2-} \text{ (zincate ion as in sodium zincate } Na_2Zn(OH)_4)$$

$$Al(OH)_3 + OH^- \rightarrow Al(OH)_4^- \text{ (aluminate ion as in sodium aluminate } NaAl(OH)_4)$$

$$Zn + 2OH^- + 2H_2O \rightarrow Zn(OH)_4^{2-} + H_2$$

$$2Al + 2OH^- + 6H_2O \rightarrow 2Al(OH)_4^- + 3H_2$$

The action of caustic alkali on non-metals is discussed under the appropriate element.

Uses

Both sodium and potassium hydroxides are used to hydrolyse fats into soaps. Sodium hydroxide is less expensive and more widely used in the production of rayon, textiles, paper, dyes, drugs, in the extraction of aluminium, and in the refining of petroleum. It is also used in the extraction of phenols and cresols from coal tar.

The Carbonates

Again the most important carbonates are the sodium and potassium compounds. In the Solvay or ammonia-soda process for the manufacture of sodium carbonate, the two main stages are:

(*a*) Brine is saturated with ammonia gas and then with carbon dioxide. The equilibrium

$$CO_2 + H_2O \rightleftharpoons HCO_3^- + H^+$$

is moved to the right-hand side due to the high proton affinity of ammonia:

$$NH_3 + H^+ \rightleftharpoons NH_4^+$$

The least soluble salt capable of formation by the four ions present (NH_4^+, HCO_3^-, Na^+, Cl^-) is sodium bicarbonate.

$$NH_4^+ + HCO_3^- + Na^+ + Cl^- \rightarrow Na^+HCO_3^-\downarrow + NH_4^+Cl^-$$

(*b*) The bicarbonate is vacuum-filtered and calcined, the carbon dioxide produced being used once again to saturate the brine.

$$2NaHCO_3 \rightarrow Na_2CO_3 + H_2O + CO_2$$

The ammonium chloride produced in (*a*) is heated with lime to produce ammonia for saturating the brine.

$$Ca(OH)_2 + 2NH_4Cl \rightarrow CaCl_2 + 2H_2O + 2NH_3$$

The raw materials for the process are brine, which is pumped from salt mines, and limestone which is used as a source of both carbon dioxide and quicklime.

$$CaCO_3 \rightleftharpoons CaO + CO_2$$

The only waste product in the process is calcium chloride for which no large-scale use has been found.

Pearl ash (K_2CO_3) cannot be made by the Solvay process because of the high solubility of potassium bicarbonate. Two processes used to produce potassium carbonate are summarized below:

(*a*) The Engel-Precht process. Saturated potassium chloride solution plus hydrated magnesium carbonate is treated with carbon dioxide, precipitating a double salt which is filtered off and decomposed by hot water.

$$3MgCO_3.3H_2O + 2KCl + CO_2 \rightarrow 2MgCO_3.KHCO_3.4H_2O + MgCl_2$$

$$2MgCO_3.KHCO_3.4H_2O \rightarrow 2MgCO_3 + K_2CO_3 + CO_2 + 9H_2O$$

(*b*) Purified producer gas (30 per cent CO) is bubbled through hot concentrated potassium sulphate solution and milk of lime under pressure.

$$K_2SO_4 + Ca(OH)_2 + 2CO \rightarrow \underset{\text{potassium formate}}{2HCOOK} + CaSO_4\downarrow$$

The calcium sulphate is filtered off and the formate solution is evaporated with free access of air.

$$2HCOOK + O_2 \rightarrow K_2CO_3 + CO_2 + H_2O$$

Properties

Except for lithium carbonate, which resembles the alkaline earth carbonates in being insoluble in water and readily decomposed, the alkali metal carbonates are white solids, soluble in water, and stable to heat. The potassium, rubidium, and caesium salts are deliquescent. A variety of hydrates is known for the sodium and potassium salts, of which washing soda ($Na_2CO_3.10H_2O$) is the most important. This effloresces in air to a white powder, the monohydrate. Acids liberate carbon dioxide from the carbonates and aqueous solutions are all alkaline because of hydrolysis.

$$CO_3^{2-} + 2H^+ \rightarrow CO_2 + H_2O$$

$$CO_3^{2-} + H\text{—}OH \rightleftharpoons HCO_3^- + OH^-$$

Both sodium and potassium carbonates are used in the manufacture of glass, soap, and caustic alkalis; considerable quantities are also consumed in the paper, water-softening, and petroleum industries.

The Bicarbonates

Solid bicarbonates are known for all the alkali metals (apart from lithium) but these readily decompose on heating.

$$2NaHCO_3 \rightarrow Na_2CO_3 + H_2O + CO_2$$

Solutions are slightly alkaline and acids liberate carbon dioxide.

$$HCO_3^- + H\text{—}OH \rightleftharpoons H_2CO_3 + OH^-$$

$$HCO_3^- + H^+ \rightarrow H_2O + CO_2$$

If equimolecular amounts of the normal carbonate and the bicarbonate are dissolved in hot water, crystals of a sesquicarbonate are deposited on cooling: $Na_2CO_3.NaHCO_3.2H_2O$.

The Nitrates

Sodium nitrate (plus 60 per cent clay) existing in vast quantities in Chile is known as Chile saltpetre and is probably formed by the decay of large masses of seaweed. It is extracted by hot water and crystallized. Potassium nitrate, known as nitre or saltpetre, is usually made from cheaper sodium nitrate by adding a hot concentrated solution of potassium chloride. The least soluble salt capable of formation by the four ions in solution (K^+, Cl^-, Na^+, NO_3^-) is sodium chloride which crystallizes out. On cooling, potassium nitrate crystallizes. The nitrates are all soluble in water, lithium and sodium being deliquescent. Lithium nitrate is the only member to crystallize out as a hydrate, $LiNO_3.3H_2O$. All the nitrates on heating decompose into the nitrite and oxygen,

$$2NO_3^- \rightarrow 2NO_2^- + O_2$$

the ease of decomposition decreasing from lithium to caesium. Sodium nitrate finds uses as a fertilizer; potassium nitrate in the manufacture of gunpowder and fireworks.

The Sulphates

Sodium sulphate occurs as salt cake in Canada, U.S.A., and the U.S.S.R. but is usually manufactured from common salt and concentrated sulphuric acid.

$$NaCl + H_2SO_4 \rightarrow NaHSO_4 + HCl$$
$$NaCl + NaHSO_4 \rightarrow Na_2SO_4 + HCl$$

Potassium sulphate may be prepared by a similar method but is usually extracted from the naturally occurring double salt, kainite ($K_2SO_4.MgSO_4.MgCl_2.6H_2O$). Potassium, rubidium, and caesium sulphates crystallize anhydrous, lithium as the monohydrate, sodium as the decahydrate below 32·35 °C and anhydrous above this temperature. Sodium sulphate finds uses in the glass, paper, rayon, soap, and coal-tar dye industries. Potassium sulphate is used as a fertilizer and for making potash alum.

The Bisulphates

These are known in the solid state, aqueous solutions of which are slightly acidic.

$$HSO_4^- + H\text{—}OH \rightarrow SO_4^{2-} + H_3O^+$$

On heating the solid bisulphates of sodium and potassium, decomposition occurs, in two stages.

$$2HSO_4^- \xrightarrow{300\,°C} S_2O_7^{2-} + H_2O$$

pyro-
sulphate
ion (as in
sodium
pyrosulphate
$Na_2S_2O_7$)

$$S_2O_7^{2-} \xrightarrow{800\,°C} SO_4^{2-} + SO_3$$

sulphate
ion

The Halides

Sodium chloride occurs in sea water (2·8 per cent). Dead Sea water contains as much as 23·2 per cent sodium chloride. It also occurs in vast rock salt deposits 150 m thick in Europe and the U.S.A. Potassium chloride occurs naturally as sylvine (KCl) and as the double salt carnallite ($KCl.MgCl_2.6H_2O$). The chlorides of lithium, rubidium, and caesium only occur in minute traces. Lithium chloride is the only deliquescent chloride in Group IA. The dampening or caking of table salt is due to calcium and magnesium chloride impurities. These chlorides may be converted into non-deliquescent carbonates or phosphates by the addition of sodium bicarbonate or phosphate. Sodium chloride is essential to diet. It is also used in the salting-out of soap, the preservation of food, the regeneration of water-softeners, and as a source of chlorine, sodium, caustic soda, and sodium carbonate. Ice and salt is a common freezing mixture, giving a eutectic point at −21·2 °C containing 23·6 per cent of salt.

The bromides and iodides are much more readily oxidized to free halogen than the chlorides. The alkali metal fluorides have high melting points, boiling points and heats of vaporization. The solubilities in water increase markedly from lithium fluoride (sparingly soluble) to caesium fluoride (very soluble).

Complexes

The alkali metal ions show very little tendency to form complex compounds. In view of their large radii (apart from lithium) and small charge, this is not surprising. As might be expected, the ability to form complexes is greatest with lithium and least with caesium.

A number of chelate complexes are known for the alkali metals with ligands such as salicylaldehyde and benzoyl acetone. The complexes are soluble in non-polar organic solvents indicating considerable covalent bonding in these complexes.

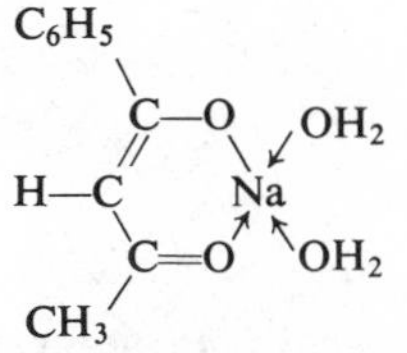

```
   C6H5
      \
       C—O    OH2
      //   \ ↙
  H—C       Na
      \    ↗ ↖
       C=O    OH2
      /
   CH3
```

FIGURE 11.7. The neutral benzoyl acetone dihydrate chelate complex of sodium

12

The Alkaline Earth Metals Group IIA

Physical Properties

	Beryllium Be	Magnesium Mg	Calcium Ca	Strontium Sr	Barium Ba
Atomic mass	9·0122	24·305	40·08	87·62	137·34
Atomic number	4	12	20	38	56
Isotopes	9	24, 25, 26	40, 42, 43, 44, 46, 48	84, 86, 87, 88	130, 132, 134, 138
Electronic structure	$1s^2 2s^2$	$2p^6 3s^2$	$3p^6 4s^2$	$4p^6 5s^2$	$5p^6 6s^2$
Atomic radius (nm)	0·089	0·136	0·174	0·191	0·198
Ionic radius (nm) (M^{2+})	0·031	0·065	0·099	0·113	0·135
Ionization potential (I) ($kJ\ mol^{-1}$)	906	742	596	554	508
Ionization potential (II) ($kJ\ mol^{-1}$)	1766	1456	1156	1066	972
Electronegativity	1·5	1·2	1·0	1·0	0·9
Density ($g\ cm^{-3}$)	1·86	1·74	1·55	2·54	3·59
Melting point (°C)	1,280	650	842	774	850
Boiling point (°C)	2477	1,100	1487	1,366	1640
Electrode potential (V)	−1·69	−2·37	−2·87	−2·89	−2·90
Heat of hydration ($kJ\ mol^{-1}$)	—	−1920	−1650	−1480	−1360

General Characteristics

All the atoms of the alkaline earth metals, except beryllium, have the configuration $(n-1)p^6 ns^2$ in their outermost orbitals. The increase in atomic radius down the group gives a decrease in the ionization potentials, which are higher than the corresponding values for the Group I metals, and thus the chemistry of the alkaline earth metals is not entirely the chemistry of cations. The combined effects of the increase in charge and decrease in size of the alkaline earth metal cations make them good polarizers or distorters. The polarizing ability decreases down the group as the ions become larger. Thus, although the basicity of the oxides and strength of the hydroxides increases down the group, the oxides are less basic and the hydroxides are weaker alkalis, than their counterparts in Group I. The electrode potentials are only slightly lower than the values for the alkali metals and, since the ionization potentials are higher, this indicates considerable energy recovery as hydration energy. This is illustrated by the action of water on the metals: with beryllium there is no action; magnesium reacts with boiling water or steam, whereas calcium, strontium, and barium react with increasing vigour on cold water.

Since the atoms are smaller than those of the alkali metals, the metals have a higher density. However, two electrons per metallic atom are available for metallic bonding, resulting in a harder metal with a higher melting and boiling point. The melting and boiling points show considerable irregularities down the group due to the different crystalline structures taken up by the metal atoms. Beryllium and magnesium are hexagonal close packed; calcium is face-centred cubic or hexagonal close packed. Strontium is face-centred cubic and barium is a body-centred cubic structure.

The small doubly charged cations exert a bigger attraction for water than the larger corresponding alkali metal ions: crystalline compounds tend to be more heavily hydrated, e.g. $MgCl_2.6H_2O$, $CaCl_2.6H_2O$, $BaCl_2.2H_2O$.

Solubility

(*a*) The solubilities of the sulphates, nitrates, and chlorides decrease from beryllium to barium.

(*b*) The solubilities of the fluorides and hydroxides increase from magnesium to barium. Beryllium fluoride is anomalous in being very soluble.

As already stated (Chapter 11), the energy of hydration must be greater than the lattice energy of a crystal for a compound to be soluble in water. However, increase in ionic size down the group causes decreases in both the lattice energy and the heat of hydration. In (*a*) the hydration energy decreases more rapidly than the lattice energy and in (*b*) the lattice energy decreases more rapidly than the hydration energy.

Melting Points

As mentioned above, the alkaline earth cations are fairly good polarizers; Be^{2+} and Mg^{2+} in particular have high charge/radius ratios and all values are larger than the corresponding Group I metals (Table 12.1).

The higher charge/radius ratios give rise to considerable covalent character in the halides (p. 164) and, in contrast to the alkali metal halides, which were regarded as being ionic, all halides of the alkaline earths have covalent character which decreases from beryllium to barium. The smaller the percentage of covalent bonding, the less volatile the halide and, therefore, the melting points of the chlorides increase as the charge/radius ratio decreases.

TABLE 12.1

	Be^{2+}	Mg^{2+}	Ca^{2+}	Sr^{2+}	Ba^{2+}
Ionic radius (nm)	0·031	0·065	0·095	0·113	0·135
Charge	+2	+2	+2	+2	+2
Charge/radius	66·6	30·8	20·2	18·2	15·5
Melting point of chlorides (°C)	440	708	772	873	962

An examination of the halides of calcium shows a decrease in the melting point with increasing atomic number of the halogen. The anion radius increases from F^- to I^-, the iodide ion being most susceptible to deformation. Hence covalent character increases and melting points decrease from calcium fluoride to calcium iodide (Table 12.2).

TABLE 12.2

	CaF_2	$CaCl_2$	$CaBr_2$	CaI_2
Radius of anion X^- (nm)	0·136	0·181	0·195	0·216
Melting point (°C)	1,360	772	730	575

Beryllium is somewhat anomalous in the group and shows a strong diagonal similarity to aluminium. The ionic radii of these two elements differ but the charge/radius ratios are very similar, indicating similar field strengths at the ionic surfaces. The following properties may be noted for beryllium and aluminium which are in contrast to the properties of the other alkaline earth metals.

(*a*) The metals are fairly resistant to acids unless amalgamated or finely divided. The resistance to acid attack is due to a protective surface film of oxide

$$Be(s) \rightarrow Be^{2+}(aq) + 2e \qquad E^{\ominus} = -1{\cdot}69 \text{ V}$$

$$Al(s) \rightarrow Al^{3+}(aq) + 3e \qquad E^{\ominus} = -1{\cdot}67 \text{ V}$$

(*b*) The metals dissolve in strong alkalis to give soluble complex beryllates, $Be(OH)_4{}^{2-}$, and aluminates, $Al(OH)_4{}^-$.

(*c*) The chlorides are 'bridged' in the vapour state

```
         /Cl\              Cl\    /Cl\    /Cl
Cl—Be        Be—Cl            Al      Al
         \Cl/              Cl/    \Cl/    \Cl
```

(*d*) Due to the excessive distortion of the carbonate ion which would occur with the small highly charged cations (the ionic radius of Al^{3+} is 0·050 nm, Be^{2+} is 0·031 nm), the carbonates do not exist under normal conditions.

(*e*) Both ions (Be^{2+}, Al^{3+}) have a strong tendency to form complexes, e.g. BeF_4^{2-} (tetrahedral), AlF_6^{3-} (octahedral). The structures of the oxalato complexes of beryllium and aluminium are shown in Figure 12.1.

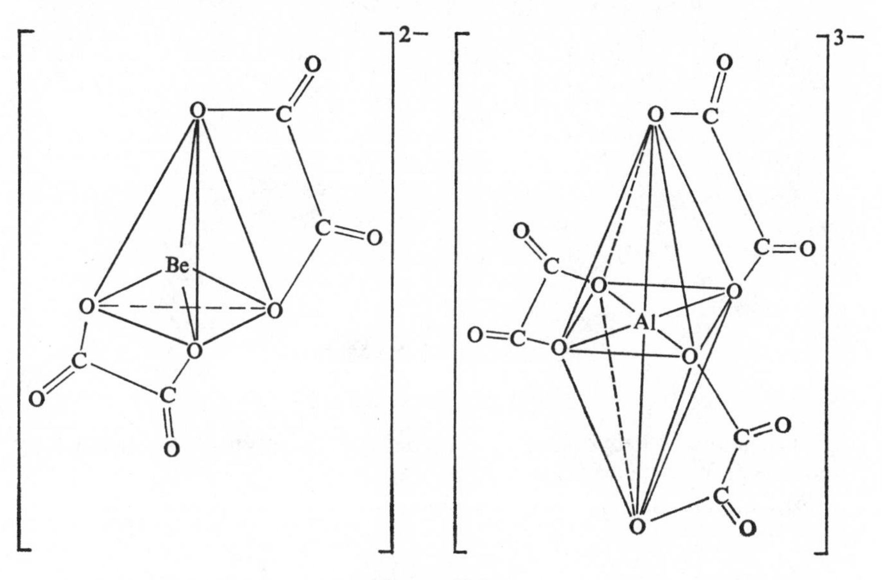

FIGURE 12.1

(*f*) Salts of both metals (nitrates, sulphates, and chlorides) furnish hydrated ions in aqueous solution, $Be(H_2O)_4^{2+}$, $Al(H_2O)_6^{3+}$. Both hydrated cations are Brönsted acids and hydrolyse in water.

$$Be(H_2O)_4^{2+} + H_2O \rightleftharpoons Be(H_2O)_3OH^+ + H_3O^+$$

Occurrence

The chlorides and sulphates of magnesium occur in sea water. The Stassfurt salt beds contain carnallite ($KCl.MgCl_2.6H_2O$). Insoluble minerals include magnesite ($MgCO_3$) and dolomite ($MgCO_3.CaCO_3$) of which whole mountain ranges are composed. Calcium is also widely distributed, the common rock being limestone ($CaCO_3$), other forms of which are chalk, calcite, aragonite, and marble. The sulphate occurs as anhydrite ($CaSO_4$) and gypsum ($CaSO_4.2H_2O$). Strontium and barium are much more rare and occur as the sulphates celestine ($SrSO_4$) and barytes ($BaSO_4$) and the carbonates strontianite ($SrCO_3$) and witherite ($BaCO_3$).

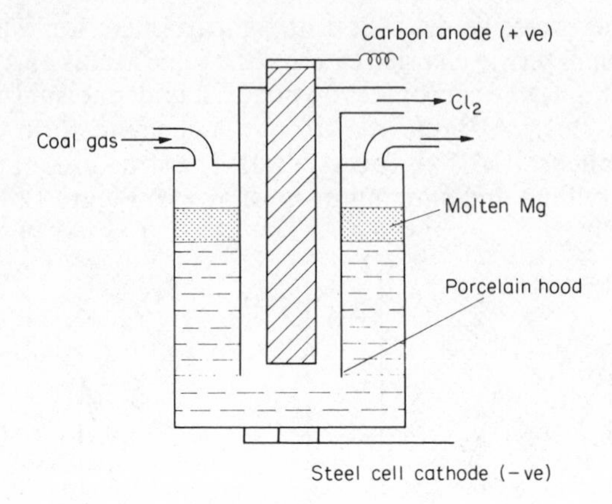

FIGURE 12.2

Extraction of the Metals

Magnesium

(*a*) Carnallite is dehydrated in a current of hydrogen chloride and the mixture of fused chlorides is electrolysed.

(*b*) Dolomite or magnesite is calcined

$$MgCO_3.CaCO_3 \rightarrow MgO.CaO + 2CO_2$$

and either (i) the MgO is mixed with carbon and heated in chlorine gas and the molten chloride is electrolysed,

$$MgO + C + Cl_2 \rightarrow MgCl_2 + CO$$

or (ii) oxides are reduced by ferrosilicon under reduced pressure above 1000 °C (Pidgeon process).

$$2MgO.CaO + FeSi \rightarrow 2Mg + Fe + (CaO)_2SiO_2$$

(*c*) Sea water, from which bromine has been extracted (p. 326), is treated in giant tanks with slaked lime.

$$MgCl_2 + Ca(OH)_2 \rightarrow Mg(OH)_2 + CaCl_2$$

The hydroxide is filtered off, heated to give the oxide, treated as in (*b*) (i) and then electrolysed in a cell through which coal gas is blown to prevent oxidation of magnesium metal (Figure 12.2). Approximately 10^6 kg of sea water must be treated to produce 10^3 kg of magnesium.

Calcium

Calcium is prepared by the electrolysis of the fused chloride mixed with 10 per cent of calcium fluoride to lower the melting point to 700 °C (Figure 12.3).

The cathode is a water-cooled iron rod which, at the beginning of the process, dips just below the surface of the fused electrolyte but, as the calcium is deposited, the cathode is slowly raised forming a rod of the metal

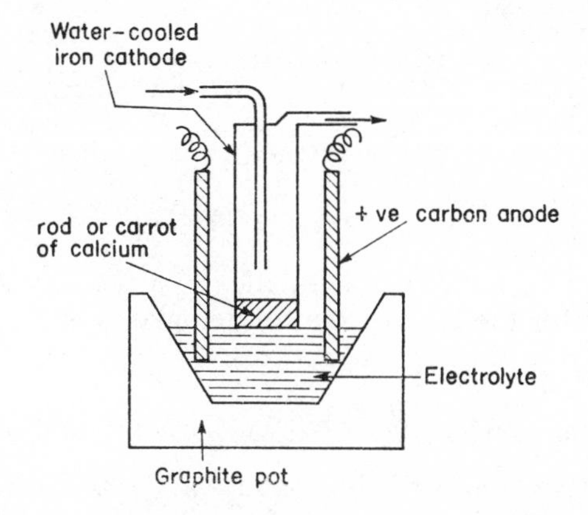

FIGURE 12.3. Electrolytic cell for the production of calcium

which, for the major part of the electrolysis, serves as the cathode. The layer of electrolyte which is withdrawn with the calcium solidifies and prevents oxidation of the metal. The technique cannot be employed for magnesium as the melting point of the metal is much lower than that of calcium.

Strontium and Barium

Electrolysis of fused chlorides is also used for preparing strontium and barium. Another method for barium involves the reduction of a barium oxide and peroxide mixture with aluminium powder in a vacuum furnace.

$$3BaO + 2Al \rightarrow Al_2O_3 + 3Ba$$

All the metals may be purified by vacuum distillation.

Properties

These are summarized in Table 12.3 in which M = Mg, Ca, Sr, or Ba.

TABLE 12.3

Reagent	*Reaction*	*Comments*
Air or oxygen	$2M + O_2 \rightarrow 2MO$	All burn if heated. Ba gives some BaO_2
Halogens	$M + X_2 \rightarrow MX_2$	
Nitrogen	$3M + N_2 \rightarrow M_3N_2$	High temp.
Ammonia	$3M + 2NH_3 \rightarrow M_3N_2 + 3H_2$	Low temp.
Carbon	$M + 2C \rightarrow MC_2$	Carbides ionic with NaCl type structure
Hydrogen	$M + H_2 \rightarrow MH_2$	High temp. Mg reacts under pressure
Water	$M + 2H_2O \rightarrow M(OH)_2 + H_2$	Mg rapidly in steam. Ca, Sr, Ba in cold water
Dilute acids	$M + 2H^+ \rightarrow M^{2+} + H_2$	

Uses

Magnesium forms a number of light alloys, particularly with aluminium, which are used for the construction of aeroplanes and automobiles, e.g.

Magnalium (Mg 15%, Al 83%, and Ca 2%)

Duralumin (Mg 0·5%, Al 95%, Cu 4%, and Mn 0·5%)

Magnesium is also used with alkyl halides in the preparation of valuable synthetic reagents (Grignard reagents) in organic chemistry.

Calcium is used as a dehydrating agent and a hardening agent for lead. Both magnesium and calcium are used as reducing agents for the production of uranium, titanium, and thorium from their fluorides and chlorides. Strontium has no large-scale uses. Barium has been employed as a 'degasser' for radio valves and Ba–Ni alloys are used to make vacuum tubes and sparking plugs.

The Compounds of the Alkaline Earth Metals

The Oxides

All are basic and prepared by heating the carbonate or nitrate. The decomposition of limestone ($CaCO_3$) is applied commercially in the production of quicklime, which is carried out in lime kilns.

$$CaCO_3 \rightleftharpoons CaO + CO_2$$

Magnesium oxide is slightly soluble in water producing an insoluble hydroxide, but all the other oxides slake rapidly with considerable evolution of heat, e.g.

$$CaO + H_2O \rightleftharpoons Ca(OH)_2 \qquad \Delta H = -64{\cdot}5 \text{ kJ}$$

Magnesium oxide is used in toilet powders and toothpastes. Quicklime slaked with caustic soda gives a solid, soda lime, used for removing water and carbon dioxide from gases. All the elements form peroxides but only barium peroxide is formed by heating the oxide in excess oxygen.

$$2BaO + O_2 \underset{700\,°C}{\overset{500\,°C}{\rightleftharpoons}} 2BaO_2$$

Magnesium, calcium, strontium, and barium peroxides are precipitated from solutions of salts as octahydrates ($MO_2.8H_2O$) by the addition of hydrogen peroxide

The Hydroxides

Magnesium oxide does not slake well and the hydroxide (milk of magnesia) is best prepared by double decomposition.

$$Mg^{2+} + 2OH^- \rightarrow Mg(OH)_2\downarrow$$

The hydroxide dissolves in acids and in ammonium chloride.

$$NH_4^+ + H_2O \rightleftharpoons H_3O^+ + NH_3$$

$$Mg(OH)_2 + 2H_3O^+ \rightarrow Mg^{2+} + 4H_2O$$

Calcium, strontium, and barium hydroxides are prepared by slaking the oxides and strontium and barium hydroxides can be crystallized from the resulting solutions as octahydrates $M(OH)_2.8H_2O$.

$$CaO + H_2O \rightleftharpoons Ca(OH)_2 \qquad \Delta H = -64{\cdot}5\ kJ$$

$$SrO + H_2O \rightleftharpoons Sr(OH)_2 \qquad \Delta H = -83{\cdot}2\ kJ$$

$$BaO + H_2O \rightleftharpoons Ba(OH)_2 \qquad \Delta H = -102{\cdot}5\ kJ$$

The solubility in water, the thermal stability, and the alkali strength increase from calcium to barium. Solid calcium hydroxide is known as slaked lime, a dilute solution as limewater, and the milky suspension as milk of lime—a useful cheap alkali. Mortar consists of slaked lime, sand, and water, the mixture drying and then hardening in air due to the formation of crystalline calcium carbonate.

The Carbonates

The stability to heat and solubility in water increases from calcium to barium. Calcium carbonate is polymorphic and exists in two crystalline forms:

(i) hexagonal calcite as in limestone, Iceland spar, chalk, and marble, and (ii) the rarer rhombic aragonite. Magnesium carbonate crystallizes with the calcite structure, and strontium and barium carbonates in the aragonite form.

Magnesium salts with sodium carbonate solution precipitate a basic salt $Mg_2(OH)_2CO_3.3H_2O$, the preparation probably involving the following hydrolytic reaction.

$$CO_3^{2-} + H_2O \rightleftharpoons HCO_3^- + OH^-$$

$$2Mg^{2+} + CO_3^{2-} + 2OH^- \rightarrow Mg_2(OH)_2CO_3$$

Magnesium carbonate is prepared by the addition of a bicarbonate to a magnesium salt.

$$Mg^{2+} + 2HCO_3^- \rightarrow MgCO_3 + CO_2 + H_2O$$

Although hydrolysis of magnesium salts is not as extensive as that of beryllium salts, the magnesium cation is small (ionic radius 0·065 nm) and some hydrolysis does occur particularly at high temperatures (*see below*). Mineral acid causes the evolution of carbon dioxide from all the carbonates.

$$CO_3^{2-} + 2H^+ \rightarrow H_2O + CO_2$$

The Halides

The chlorides are prepared by crystallization of a solution obtained by the action of dilute hydrochloric acid on a carbonate. Barium chloride crystallizes as a dihydrate and magnesium, calcium, and strontium as hexahydrates. Barium chloride is made technically by the reduction of the insoluble naturally occurring barium sulphate with carbon and calcium chloride.

$$BaSO_4 + CaCl_2 + 4C \rightarrow BaCl_2 + CaS + 4CO$$

The fluorides (except BeF_2) are all insoluble due to their high lattice energies. The chlorides are all soluble and magnesium and calcium chlorides are deliquescent. Anhydrous salts are made by heating crystals of the hexahydrate in hydrogen chloride gas at 180°C to prevent hydrolysis.

$$MgCl_2 + H_2O \rightleftharpoons MgO + 2HCl$$

A concentrated solution of magnesium chloride added to magnesium oxide rapidly sets to a hard white mass of a basic salt of indefinite composition known as Sorel's cement.

The Hypohalites

These are all known but the most important is calcium hypochlorite which is called bleaching powder and has the following approximate composition:

$$Ca(OCl)_2.CaCl_2.Ca(OH)_2.H_2O$$

It is prepared by passing chlorine over cold dry slaked lime. Dilute acids, even carbonic acid, liberate chlorine.

$$Ca(OCl)_2.CaCl_2.Ca(OH)_2.H_2O + 6H^+ \rightarrow 3Ca^{2+} + 5H_2O + 2Cl_2$$

Bleaching powder is used as a bleaching agent and disinfectant. The powder slowly decomposes on exposure to the atmosphere forming the chloride and some chlorate.

$$3OCl^- \rightarrow 2Cl^- + ClO_3^-$$

The decomposition of bleaching powder is catalysed by cobalt salts, when oxygen is produced.

$$2OCl^- \rightarrow 2Cl^- + O_2$$

The Sulphates

The solubility of the sulphates decreases markedly from magnesium to barium, the latter being insoluble. Gypsum, $CaSO_4.2H_2O$, is unusual with regard to its decomposition,

$$2CaSO_4.2H_2O \xrightarrow{\geqslant 100\,^{\circ}C} (CaSO_4)_2.H_2O + 3H_2O$$

forming the hemihydrate (plaster of Paris) which, when powdered and added to water, sets with expansion to a hard mass of interlocking crystals of gypsum. Plaster of Paris is used as an interior plaster and for statuary.

Barium sulphate and zinc sulphide mixed together in equimolecular proportions form an important white pigment, lithopone.

The Nitrates

The nitrates are made by dissolution of the carbonates in dilute nitric acid. Magnesium nitrate crystallizes with six, and calcium and strontium nitrates with four, molecules of water. Barium nitrate crystallizes anhydrous. All decompose to the oxide on heating.

$$2M(NO_3)_2 \rightarrow 2MO + 4NO_2 + O_2$$

Strontium and barium nitrates are used in pyrotechny giving red and green flames.

Complexes

The beryllium ion is most ready to form complexes (p. 177) but complexes of the other alkaline earth metal ions with larger radii are mainly restricted to ringed systems or chelates, e.g. calcium EDTA (p. 102).

13

Boron and Aluminium Group IIIB

Physical Properties

		Boron B	Aluminium Al
Atomic mass		10·811	26·9815
Atomic number		5	13
Isotopes		10 (19·57 %) 11 (80·43 %)	27
Electronic structure		$1s^2 2s^2 2p^1$	$1s^2 2s^2 2p^6 3s^2 3p^1$
Atomic radius (nm)		0·080	0·125
Ionic radius (nm) (M^{3+})		0·020	0·050
Ionization potential (eV)	(I)	805	583
	(II)	2426	1826
	(III)	3666	2746
Electrode potential (V)			
$M(H_2O)_6^{3+} + 3e \rightarrow M(s)$		—	−1·67
Electronegativity		2·0	1·5
Density (g cm^{-3})		2·3	2·7
Melting point (°C)		2,300	660
Boiling point (°C)		3930	2470

General Characteristics

Boron and aluminium atoms have the electronic configuration $ns^2 np^1$ in their valency shell. Thus in the majority of their compounds these elements have an oxidation state of +3. The atoms have small radii and high ionization potentials indicating that the valency electrons are strongly bound to the nucleus. It is interesting to note from the values of ionization potential that the *p* electron is more easily removed than the *s* electrons. The B^{3+} ion is not known and nearly all compounds of both elements are predominantly covalent as is supported by the intermediate value of electronegativity. The covalent character of the compounds is illustrated by the melting points of the chlorides:

BCl_3 −107 °C $AlCl_3$ 193 °C (under pressure)

cf. Na^+Cl^- 801 °C

However the Al^{3+} ion is found in a few compounds, e.g. aluminium fluoride (anhydrous) and aluminium perchlorate.

The high electrode potential of aluminium, normally associated with metallic character, is surprising in view of the high values of the ionization potentials. This is because the aluminium ion with its high charge and small size can polarize water molecules which then hydrate the cation (Figure 13.1).

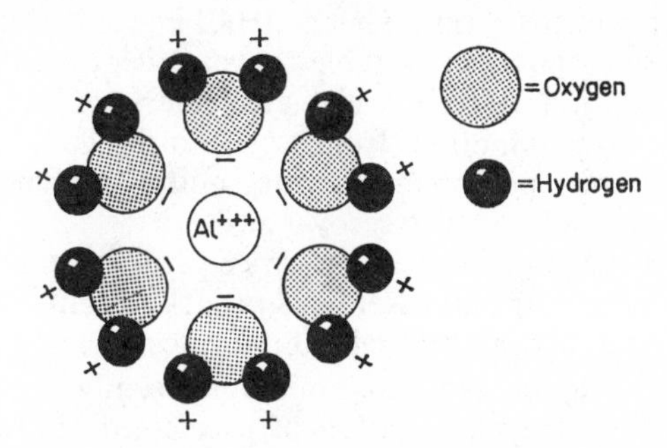

FIGURE 13.1. Polar water molecules grouped round an aluminium cation and held by electrostatic attraction, giving $Al(H_2O)_6^{3+}$

The process of hydration results in the liberation of a considerable amount of energy in the form of heat of hydration.

$$Al^{3+}(g) + 6H_2O \rightarrow Al(H_2O)_6^{3+}(aq) \qquad \Delta H = -4690 \text{ kJ mol}^{-1}$$

Thus the hydrated aluminium ion is readily produced in solution resulting in the unexpectedly high value obtained for the electrode potential. The much smaller B^{3+} ions are unable to accommodate a sufficient number of water molecules around themselves to enable heats of hydration to compensate for the energy lost in ionization. Hence the hydrated B^{3+} ion does not exist.

The covalent compounds of these elements contain only three electron pairs and hence they tend to form electron-deficient compounds, e.g. BF_3, AlH_3. A stable configuration (ns^2np^6) may be produced by accepting electrons from other donor molecules, atoms or ions, e.g. $K^+BF_4^-$, $Li^+AlH_4^-$.

Boron resembles silicon in many respects, showing a diagonal relationship already discussed for lithium and magnesium, beryllium and aluminium.

This resemblance is due to similarity in atomic radii and electronegativity. Boron and silicon can each form a series of hydrides by indirect methods; each gives a readily hydrolysed chloride; the oxides are acidic, giving glasses when fused and cooled, and borides and silicides are produced by fusion with metals at high temperature.

Occurrence

Principal ores	*Location*
(*a*) Boron:	
kernite $Na_2B_4O_7.4H_2O$ borax $Na_2B_4O_7.10H_2O$ colemanite $Ca_2B_6O_{11}.5H_2O$	U.S.A., California, South America, U.S.S.R., and Turkey
(*b*) Aluminium:	
bauxites–gibbsite $Al_2O_3.3H_2O$ diaspore or bohmite $Al_2O_3.H_2O$ cryolite Na_3AlF_6	South America, Ghana, Indonesia, Europe, and U.S.A., Greenland, Australia

Boron is a comparatively rare element (0·001 per cent of the earth's crust) and its chief ores are shown in the table above. Other less important sources are volcanic emanations in Italy, which are a source of boric acid. Borax is also obtained in large quantities from brines in the U.S.A. (Searles Lake, California) from which potassium chloride, sodium sulphate and carbonate, and lithium compounds are also recovered.

Aluminium is the third element in order of abundance and occurs in great quantities in the earth's crust (7·5 per cent). Aluminium occurs widely combined in clays as hydrated aluminium silicates. Its large-scale extraction from these has so far been uneconomic but recently a process has been developed involving the chlorination of a calcined mixture of clay and bauxite. The product, $AlCl_3$, is reduced to the metal with manganese.

Extraction

Boron

Boron is extremely difficult to obtain in the pure state; the following methods are available for its extraction:

(*a*) Reduction of boric oxide or a borate with an electropositive metal, e.g. the oxide is mixed with excess magnesium powder and ignited.

$$B_2O_3 + 3Mg \rightarrow 3MgO + 2B$$

The product is a brown amorphous powder (amorphous boron) and is contaminated with some magnesium boride.

$$3Mg + 2B \rightarrow Mg_3B_2$$

Boiling with dilute hydrochloric acid removes this and the product is heated to 1,200°C in an electric furnace giving pure amorphous boron.

(*b*) Reduction of boron tribromide or trichloride with hydrogen by passing a mixture of the gases over a heated tungsten or tantalum filament. At 1300 °C almost pure crystalline boron is obtained.

$$2BCl_3 + 3H_2 \rightarrow 6HCl + 2B \text{ (99 per cent pure)}$$

Aluminium

Crude bauxite contains iron(III) oxide and silica as impurities. The ore is pulverized and digested with caustic soda under pressure.

$$Al_2O_3 + 2OH^- + 3H_2O \rightarrow 2Al(OH)_4^-$$

$$SiO_2 + 4OH^- \rightarrow SiO_4^{4-} + 2H_2O$$

The aluminium oxide and silica dissolve and the insoluble residue containing iron(III) oxide is filtered off. The filtrate is seeded with freshly

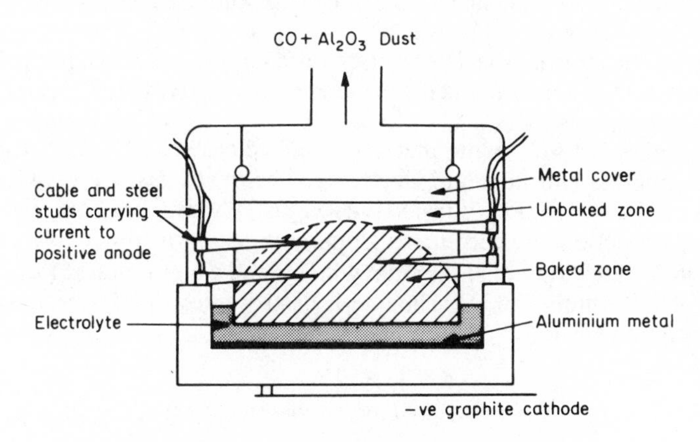

FIGURE 13.2. Cell for the electrolytic extraction of aluminium

precipitated aluminium hydroxide which precipitates approximately 60 per cent of the alumina from solution $Al(OH)_4^- \rightleftharpoons Al(OH)_3 + OH^-$. The silicate is not decomposed and remains in solution. The precipitate is filtered off, washed, and calcined in rotary kilns. Reduction of the oxide is difficult due to its extreme stability:

$$4Al + 3O_2 \rightarrow 2Al_2O_3 \qquad \Delta H = -3370 \text{ kJ}$$

and the energy required to liberate aluminium can only be supplied electrolytically. Since the oxide itself is a non-electrolyte it is added to a bath of molten cryolite and electrolysed in the cell shown (Figure 13.2). The electrolytic reduction is carried out at 900 °C using a Söderberg self-baking electrode (anode) and a carbon cathode. The anode consists of a container, open at the bottom, filled with ground coke and pitch which acts as a binder. The anode becomes baked by the heat from the cell and, as it is slowly eroded by the liberated oxygen, the studs carrying current to the anode can be withdrawn and inserted at a higher level. The mechanism of electrolysis is:

$$\text{At the cathode } Al^{3+} + 3e \rightarrow Al$$

$$\text{At the anode } 6F^- \rightarrow 3F_2 + 6e$$

$$Al_2O_3 + 3F_2 \rightarrow 2Al^{3+} + 6F^- + \tfrac{3}{2}O_2$$

Properties

Crystalline boron has a black metallic appearance, high melting point, is very hard, and is a poor conductor of electricity. X-ray data indicate a near perfect tetragonal crystal lattice in which the boron atoms are covalently bound to form a giant molecule (p. 210). The physical properties mentioned above are in agreement with this structure. Crystalline boron is very unreactive chemically, its only reaction being with fused caustic soda to give borates.

Amorphous boron is a chestnut-coloured powder which is probably micro-crystalline boron and hence is much more reactive than the crystalline form.

Aluminium is a soft, white metal with a face-centred cubic lattice and is a good conductor of heat and electricity. The metal has a brilliant lustre when scraped but rapidly dulls owing to atmospheric oxidation. The more important reactions of boron and aluminium are shown in Table 13.1.

The lack of reaction of aluminium with air and water is due to a coherent and protective film of oxide. The film is destroyed by amalgamation when rapid reaction ensues.

TABLE 13.1

The Reactions of Boron and Aluminium

Reagent	*Boron*	*Aluminium*
Air at room temp.	No action	No action unless amalgamated when rapid oxidation to Al_2O_3 occurs
Cold water	No action	No action unless amalgamated when rapid action giving hydrogen
Heated in N_2, S, O_2, Cl_2	$BN, B_2S_3, B_2O_3, BCl_3$	$AlN, Al_2S_3, Al_2O_3, AlCl_3$
Conc. HCl	No action	$2Al + 6H^+ \rightarrow 2Al^{3+} + 3H_2$ (slow)
Conc. H_2SO_4	Oxidation to orthoboric acid H_3BO_3	$2Al + 12H^+ + 3SO_4^{2-} \rightarrow 2Al^{3+} + 6H_2O + 3SO_2$
Conc. HNO_3	Oxidation to orthoboric acid	No action
Hot. conc. alkali	$2B + 2OH^- + 2H_2O \rightarrow 2BO_2^- + 3H_2$ metaborate	$2Al + 2OH^- + 6H_2O \rightarrow 2Al(OH)_4^- + 3H_2$ aluminate

Uses

Amorphous boron is used for its burning characteristics in flares, and igniters for explosive charges and propellants. Boron is a good neutron absorber and is used in shields and control rods in nuclear reactor vessels. Minute quantities (<0·003 per cent) of boron are used in hardening steels. Crystalline boron is also used in the construction of transistors.

Aluminium can be rolled, pressed or extruded to any shape and is extensively used in building, aircraft construction, shipbuilding, and as a wrapping material in the food industry. Reinforced with steel it is used for overhead power transmission cables. A very recent application is as a core sheath to protect uranium fuel rods in atomic piles from chemical attack. Aluminium finds application in the production of numerous metals by the thermite process. Owing to its high affinity for oxygen (page 187), it reduces the oxides of iron, chromium, manganese, and tungsten to the metal.

$$Fe_2O_3 + 2Al \rightarrow Al_2O_3 + 2Fe \quad (\Delta H = -834 \text{ kJ})$$

Aluminium readily alloys with many metals; some important alloys are:

Duralumin	Al(95%)	Cu(4%)	Mg(0·5%)	Mn(0·5%)
Magnalium	Al(83%)	Mg(15%)	Ca(2%)	
Aluminium bronze		Al(10%)	Cu(90%)	
Alnico	Al(20%)	Fe(50%)	Ni(20%)	Co(10%)

Alloys are usually superior to the metal in strength and hardness but less resistant to corrosion due to the film of oxide which is less tenaciously held. Thus aluminium alloys are frequently coated with a layer of aluminium and are then known as Alclads or Aldurals.

Anodizing is another process used to increase corrosion resistance of aluminium and its alloys. The object is made the anode in an electrolytic bath containing 10 per cent chromic acid, oxalic acid or sulphuric acid. Oxygen is discharged at the anode and forms a transparent but impervious coating of oxide (0·00025 cm thick) on the metal surface which is very absorbent and can be dipped into a dye to give a brightly coloured finish.

Compounds of Boron and Hydrogen

Hydrides

Several hydrides of boron have been prepared and analysis shows them to have the general formula B_nH_{n+4} or B_nH_{n+6}. The following *boranes* have been isolated.

B_2H_6 diborane—gas

B_4H_{10} tetraborane—gas

B_5H_9 pentaborane (stable)—liquid

B_5H_{11} pentaborane (unstable)—liquid
B_6H_{10} hexaborane—liquid
B_9H_{15} enneaborane—liquid
$B_{10}H_{14}$ decaborane—solid

Preparation

(*a*) Hydrolysis of magnesium boride with dilute acid in vacuo.

$$Mg_3B_2 + 6H_2O \rightarrow 3Mg(OH)_2 + B_2H_6 \text{ (+ other hydrides)}$$

Very little diborane is formed since it is readily decomposed by water and air to boric acid and hydrogen.

(*b*) By interaction of boron trichloride and lithium aluminium hydride in dry ether. This gives pure diborane.

$$3LiAlH_4 + 4BCl_3 \rightarrow 3LiCl + 3AlCl_3 + 2B_2H_6$$

(*c*) The higher boranes are made by heating B_2H_6 with hydrogen under specific conditions.

Structure

The structure of the boron hydrides is a difficult problem due to the electron-deficient nature of the molecules. Much evidence has been accumulated for a bridged configuration and this is supported by infrared spectra and electron diffraction. It may be represented simply as shown in Figure 13.3*a*.

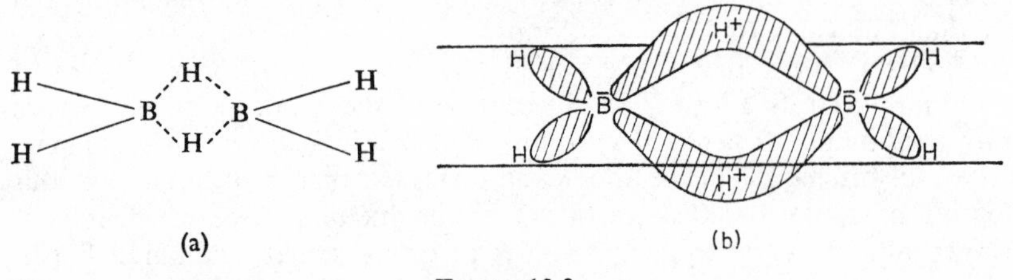

FIGURE 13.3

An alternative suggestion is that the electrons are removed from the hydrogen atoms composing the bridge and that the two protons are embedded in the electron cloud between the boron atoms which are sp^3 hybridized (Figure 13.3*b*).

Properties

Some reactions of diborane are as follows:

$$B_2H_6 + 6H_2O \rightarrow 2B(OH)_3 + 6H_2$$
$$B_2H_6 + 6Cl_2 \rightarrow 2BCl_3 + 6HCl$$

$$B_2H_6 + Br_2 \rightarrow B_2H_5Br + HBr$$

$$B_2H_6 + 2NaH \xrightarrow{\text{ether}} 2NaBH_4$$

$$B_2H_6 + 2(CH_3)_3N \rightarrow 2H_3B \leftarrow N(CH_3)_3$$

All the boron hydrides contain both bridging and terminal hydrogen atoms. In a number of the higher boron hydrides, the boron atoms form part of the icosahedral framework characteristic of elementary boron, e.g. B_9H_{15}, $B_{10}H_{14}$. Other hydrides exhibit various structures for the boron framework. Thus in B_5H_9 the boron atoms have a tetragonal pyramidal arrangement, whereas in B_4H_{10} the boron atoms form two triangles with one side in common and an angle of 120° between the planes of these triangles.

Metalloboranes such as $Na_2{}^+B_{12}H_{12}{}^{2-}$, $K_2{}^+B_{12}H_{12}{}^{2-}$ in which the boron atoms are situated at the corners of an icosahedron have been prepared by a reaction between diborane, triethylamine, and sodium or potassium borohydride at 100–180 °C.

$$2Na^+BH_4^- + 5B_2H_6 \xrightarrow{(C_2H_5)_3N} Na_2{}^+B_{12}H_{12}{}^{2-} + 13H_2$$

These compounds are stable up to 600 °C and are not affected by acids, alkalis, or oxidizing agents. However, chlorine, bromine or iodine readily replace hydrogen atoms to give substituted species such as $B_{12}H_6Br_6{}^{2-}$, $B_{12}Br_{12}{}^{2-}$. A polyhedral boron and carbon hydride (carborane) containing two carbon and ten boron atoms at the corners of an icosahedron has also been prepared from decaborane.

$$B_{10}H_{14} + 2(CH_3)_2S \xrightarrow{40^\circ} B_{10}H_{12}\,[(CH_3)_2S]_2 + H_2$$

$$B_{10}H_{12}\,[(CH_3)_2S]_2 + C_2H_2 \xrightarrow{90^\circ} B_{10}C_2H_{12} + H_2 + 2(CH_3)_2S$$

Treatment of carborane with hot concentrated sodium hydroxide yields an anion $B_9C_2H_{12}{}^-$, but replacement of a hydrogen atom by halogen is not possible below 100 °C.

The boron hydrides are used as rocket propellants and as high-energy fuels.

Borohydrides

Boron is also associated with hydrogen in the complex borohydride anion $(BH_4)^-$. The lithium and sodium salts are important commercial reducing agents. Lithium borohydride is prepared by the action of boron trifluoride on lithium hydride.

$$4LiH + BF_3 \rightarrow LiBH_4 + 3LiF$$

Sodium borohydride can be prepared by heating sodium hydride with boric oxide.

$$2B_2O_3 + 4NaH \rightarrow 3NaBO_2 + NaBH_4$$

The products are both salt-like compounds with a face-centred cubic lattice of sodium ions and tetrahedral borohydride ions.

Borazole (Borazine)

On heating diborane with excess ammonia at 200 °C, borazole, $B_3N_3H_6$, is produced (Figure 13.4). The structure is a flat hexagonal ring consisting

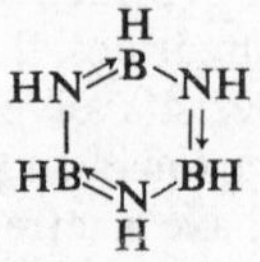

$$\begin{array}{ccc} & H & \\ HN & \overset{\displaystyle B}{} & NH \\ | & & \| \\ HB & N & BH \\ & H & \end{array}$$

FIGURE 13.4

of alternate BH and NH groups. It is isoelectronic with benzene, i.e. has the same number of electrons, but shows a higher degree of unsaturation than benzene. For example,

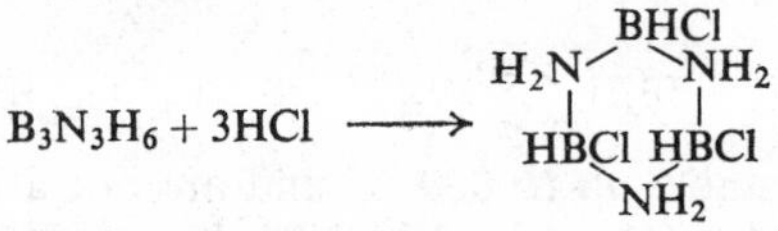

$$B_3N_3H_6 + 3HCl \longrightarrow \begin{array}{ccc} & BHCl & \\ H_2N & & NH_2 \\ | & & | \\ HBCl & & HBCl \\ & NH_2 & \end{array}$$

borazole trihydrochloride

Compounds of Aluminium and Hydrogen

Aluminium and hydrogen do not react directly and no stable monomeric hydride exists. A non-volatile polymerized aluminium hydride is produced by the action of lithium hydride on aluminium chloride in dry ether.

$$AlCl_3 + 3LiH \rightarrow \underset{\displaystyle (AlH_3)_x}{\overset{}{AlH_3 \downarrow}} + 3LiCl$$

A greater proportion of lithium hydride gives the white solid lithium aluminium hydride.

$$4LiH + AlCl_3 \rightarrow LiAlH_4 + 3LiCl$$

This is a valuable reducing agent in organic chemistry. It is readily hydrolysed by water.

$$LiAlH_4 + 4H_2O \rightarrow LiOH + Al(OH)_3 + 4H_2$$

The structure is similar to that of lithium borohydride.

Compounds with Oxygen

Boric oxide (B_2O_3) exists in two forms, a crystalline and a glassy supercooled liquid form (Figure 13.5). The glassy form is obtained by igniting amorphous boron or dehydrating boric acid.

$$4H_3BO_3 \rightarrow 2B_2O_3 + 6H_2O$$

The crystalline form is obtained by carrying out this reaction *in vacuo* over P_2O_5 and consists of spiral chains of triangular BO_3 units.

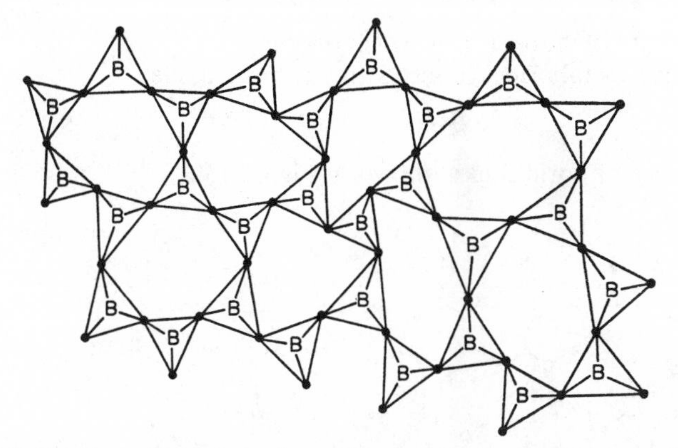

FIGURE 13.5. Structure of the glassy form of boric oxide: a random three-dimensional network of triangular BO_3 groups

It has no basic properties and combines with water to give metaboric and orthoboric acids.

$$B_2O_3 + H_2O \rightarrow 2HBO_2 \text{ (metaboric)}$$

$$B_2O_3 + 3H_2O \rightarrow 2H_3BO_3 \text{ (orthoboric)}$$

Boric oxide combines with bases and alkalis to give metaborates, many of which are brightly coloured and are the basis of the borax bead test.

Orthoboric acid (H_3BO_3) is prepared in the laboratory by the action of concentrated hydrochloric acid on a concentrated solution of borax. Boric acid separates on cooling as white flaky crystals.

$$B_4O_7^{2-} + 2H^+ + 5H_2O \rightarrow 4H_3BO_3$$

It is a very weak monobasic acid which functions not as a proton donor, but as a Lewis acid.

$$B(OH)_3 + H_2O \rightleftharpoons B(OH)_4^- + H^+ \qquad (K = 6{\cdot}5 \times 10^{-10} \text{ at } 25\ ^\circ C)$$

It forms a strong monobasic acid by complexing with glycerol and can then be titrated with alkali using phenolphthalein indicator.

$$B(OH)_3 + \begin{array}{l} HOCH_2CHOHCH_2OH \\ HOCH_2CHOHCH_2OH \end{array} \longrightarrow \left[\begin{array}{ccc} CH_2{-}O & & O{-}CH_2 \\ | & B & | \\ CH{-}O & & O{-}CH \\ | & & | \\ CH_2OH & & CH_2OH \end{array} \right]^- H_3O^+ + 2H_2O$$

On heating, orthoboric acid loses water at 100 °C giving metaboric acid:

$$H_3BO_3 \rightarrow HBO_2 + H_2O$$

This reverts to orthoboric acid when dissolved in water. Further dehydration converts metaboric acid into vitreous boric oxide.

$$2HBO_2 \rightarrow B_2O_3 + H_2O$$

The structure of boric acid is shown in Figure 13.6.

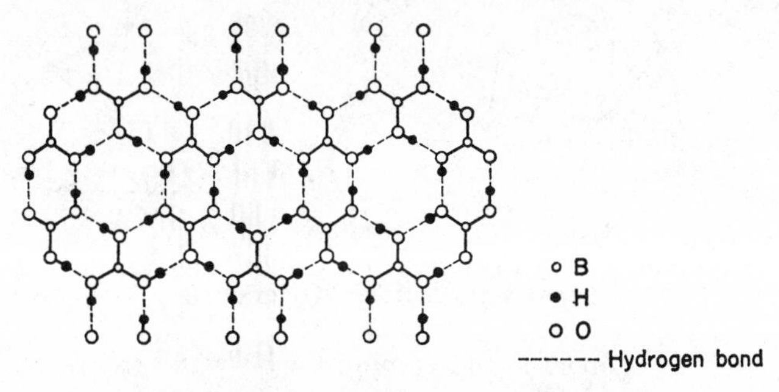

FIGURE 13.6. Layer structure of boric acid

Borax and the Borates

Borax is the sodium salt of the theoretical tetraboric acid, $H_2B_4O_7$. It occurs as the decahydrate, $Na_2B_4O_7.10H_2O$, which crystallizes as colourless, efflorescent, monoclinic crystals. On heating it loses water giving a white spongy mass of anhydrous borax, which contracts to a colourless glass of sodium metaborate (Figure 13.7(*a*)).

A large number of complex borates containing trigonal BO_3 and tetrahedral BO_4 groups have been isolated. The relevant equilibrium is:

$$B(OH)_3 + 2H_2O \rightleftharpoons B(OH)_4^- + H_3O^+$$

The main species present depends upon the pH and the concentration. Thus strong alkali at high concentration shifts the equilibrium to the right producing tetrahedral $B(OH)_4^-$ which polymerizes to the trigonal metaborates (Figure 13.7(*a*) and (*b*)).

$$nB(OH)_4^- \rightarrow nBO_2^- + 2nH_2O$$

Strong acid at high concentration shifts the equilibrium to the left producing trigonal boric acid. At intermediate pH values and concentrations numerous condensations can occur producing complex borates containing both trigonal BO_3 and tetrahedral BO_4 groups, e.g. borax (Figure 13.7(*c*)).

$$\underset{\text{trigonal}}{2B(OH)_3} + \underset{\text{tetrahedral}}{2B(OH)_4^-} \rightleftharpoons \underset{\text{borax}}{B_4O_5(OH)_4^{2-}} + 5H_2O$$

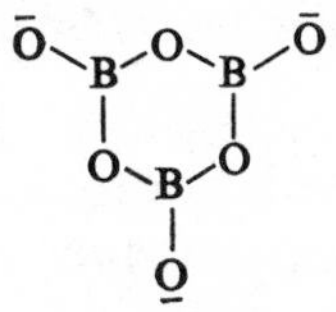

(a) Sodium and potassium metaborates contain the cyclic $B_3O_6^{3-}$ ion and should therefore be written $Na_3B_3O_6$ and $K_3B_3O_6$.

(b) 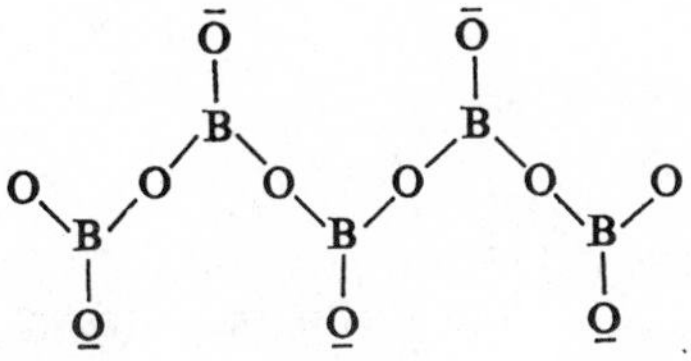 Calcium metaborate ($Ca(BO_2)_2$) consists of long chains of anions held together by calcium ions.

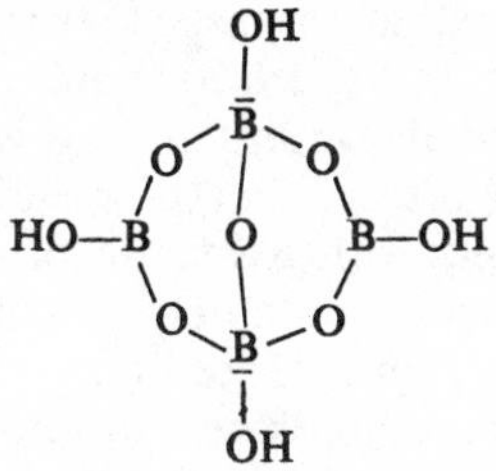

(c) The hydrated anion $B_4O_7^{2-}\cdot 2H_2O$ or $B_4O_5(OH)_4^{2-}$ present in borax containing both trigonal and tetrahedral boron.

FIGURE 13.7

Many dyeing and bleaching processes utilize borax as an alkaline buffer. Borax and many metallic borates, e.g. zinc borate, are good flame-proofing agents for textile fabrics.

Sodium perborate

Sodium perborate, $Na_2B_2(O_2)_2(OH)_4\cdot 6H_2O$ (sodium peroxoborate), is prepared by the action of alkaline hydrogen peroxide on borax at 0 °C. It is a stable compound, sparingly soluble in water, the solution having strong oxidizing and alkaline properties. Hence it is a constituent of domestic washing powders (Persil, Daz, Tide, etc.). The anion $B_2(O_2)_2(OH)_4^{2-}$ contains tetrahedral boron atoms with two bridging peroxo groups.

Aluminium oxide

Aluminium oxide, Al_2O_3, occurs naturally as corundum or white sapphire and, coloured by traces of other metal oxides, as ruby (Cr), amethyst (Mn), etc., which are valuable as gemstones on account of their beauty and hardness. Artificial precious stones can be prepared by fusing alumina with the colouring oxide in an oxyhydrogen flame.

The oxide may be prepared in the laboratory by heating the hydroxide or ammonium alum.

$$2Al(OH)_3 \rightarrow Al_2O_3 + 3H_2O$$

$$(NH_4)_2SO_4.Al_2(SO_4)_3.24H_2O \rightarrow (NH_4)_2SO_4 + 3SO_3 + Al_2O_3 + 24H_2O$$

The oxide is amphoteric:

$$Al_2O_3 + 6H^+ \rightarrow 2Al^{3+} + 3H_2O$$

$$Al_2O_3 + 2OH^- + 3H_2O \rightarrow 2Al(OH)_4^-$$

The anhydrous oxide exists in two principal forms, γ-alumina, prepared by heating the hydroxide to 500 °C *in vacuo*, and α-alumina, formed by heating γ-alumina above 1000 °C. α-Alumina, present in an impure form in corundum, sapphire, ruby, etc., has a hexagonal close-packed structure in which each aluminium ion is surrounded by six oxygen ions and each oxygen by four aluminium ions. This gives a hard, dense structure which is insoluble in acids. The γ-form has a badly organized micro-crystalline spinel structure and hence has a smaller particle size with greater absorptive power. It is less dense and readily soluble in acids. Aluminium oxide is used, together with silica, as a refractory material for furnace linings. The oxidation of amalgamated aluminium gives aluminium oxide in a surface-active form (activated alumina). This is used as a dehydrating agent, a catalyst, and as a packing material for chromatographic columns.

Aluminium hydroxide

Aluminium hydroxide, $Al(OH)_3$, is prepared as a white gelatinous precipitate by addition of alkali to an aluminium salt.

$$2Al^{3+} + 6OH^- \rightarrow 2Al(OH)_3$$

The precipitate is soluble in excess sodium hydroxide giving an aluminate solution.

$$Al(OH)_3 + OH^- \rightarrow Al(OH)_4^-$$

Aluminium hydroxide is amphoteric due to its ability to ionize in two ways.

$$Al^{3+} + 3OH^- \rightleftharpoons Al(OH)_3 + H_2O \rightleftharpoons Al(OH)_4^- + H^+$$

Aluminium hydroxide is a good example of a layer structure. Layers of the type shown (Figure 13.8) are held together by hydrogen bonds.

Alkali metal aluminates furnish the $Al(OH)_4^-$ ion in solution. This anion appears to be hydrated with two molecules of water giving an octahedral structure (Figure 13.9). Solutions are viscous due to hydrogen bonding with water molecules. This tendency for aluminium to form octahedral complexes accounts for the lack of resemblance between borates and aluminates.

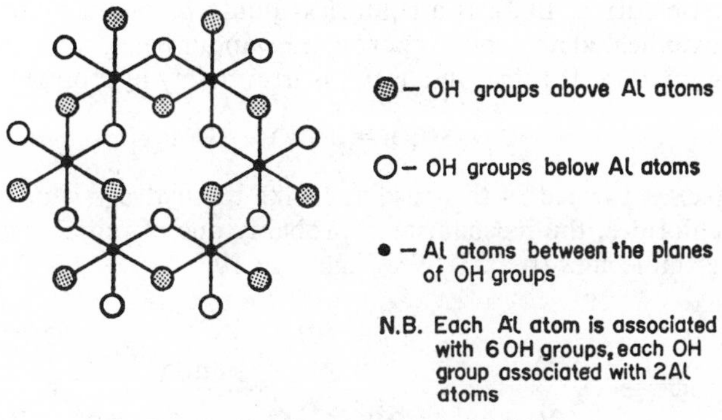

FIGURE 13.8. Part of a layer of $Al(OH)_3$

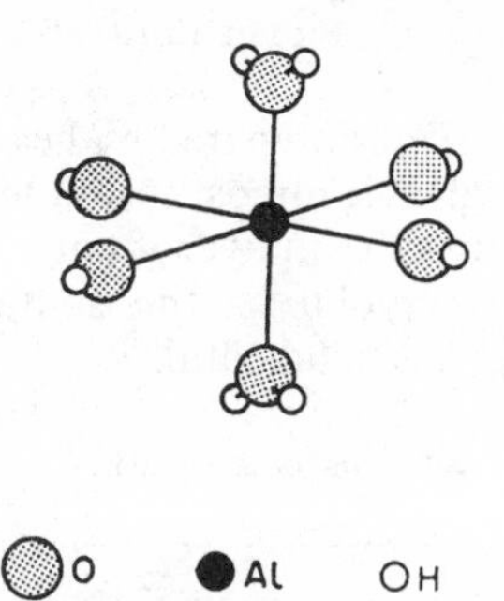

FIGURE 13.9. The octahedral structure of the hydrated aluminate ion

Halides

Boron trifluoride, BF_3, is a colourless fuming gas, prepared by the action of concentrated sulphuric acid on a mixture of calcium fluoride and boric oxide.

$$CaF_2 + H_2SO_4 \rightarrow CaSO_4 + 2HF$$

$$B_2O_3 + 6HF \rightarrow 3H_2O + 2BF_3$$

The gas is collected over mercury since it reacts vigorously with water giving fluoboric acid.

$$4BF_3 + 3H_2O \rightarrow H_3BO_3 + 3HBF_4.$$

This acid gives rise to salts, the fluoborates, e.g. $K^+BF_4^-$, in which the anion is tetrahedral. Surprisingly BF_3 is a weaker Lewis acid than BCl_3.

The short B—F bond allows for π overlap between full fluorine and empty boron $2p$ orbitals thus reducing the electron-attracting power of boron. Nevertheless BF_3 forms many adducts with Lewis bases, e.g. the liquid etherate $F_3B \leftarrow (OC_2H_5)_2$ is used as a source of gaseous BF_3.

Boron trichloride, BCl_3, is a colourless liquid prepared by passing dry chlorine over heated amorphous boron, the vapour being condensed out in a freezing mixture. It fumes in air and is irreversibly hydrolysed by water.

$$BCl_3 + 3H_2O \rightarrow H_3BO_3 + 3HCl$$

No complex is formed in this case and, like the hydrolysis of most non-metallic chlorides, the mechanism is probably one of initial coordination with water molecules (p. 333). *See* Figure 13.10.

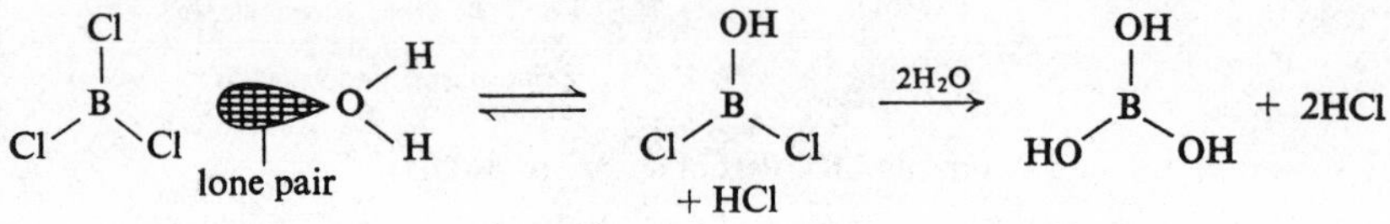

$$\underset{\text{lone pair}}{Cl_3B \; + \; H_2O} \rightleftharpoons B(OH)Cl_2 + HCl \xrightarrow{2H_2O} B(OH)_3 + 2HCl$$

FIGURE 13.10

Aluminium fluoride, AlF_3, is prepared by heating aluminium in fluorine at red heat. It is an anhydrous, almost insoluble, white, ionic solid which sublimes on heating. It dissolves in hydrofluoric acid giving fluoroaluminic acid, H_3AlF_6, of which cryolite is the sodium salt. The hexafluoroaluminate(III) ion, AlF_6^{3-}, is octahedral.

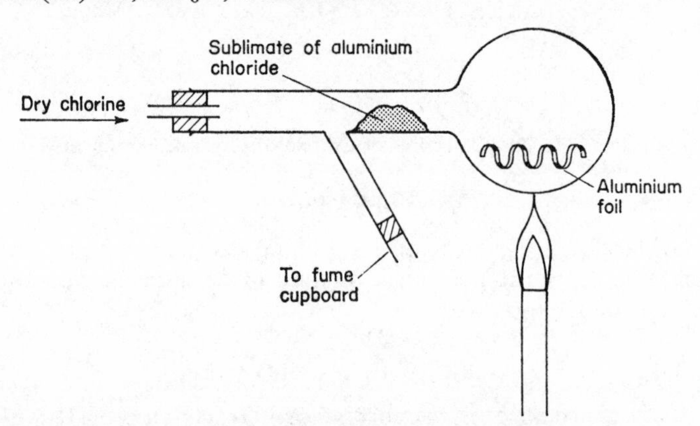

FIGURE 13.11. The preparation of aluminium chloride

Aluminium chloride, $AlCl_3$, is prepared anhydrous by passing dry chlorine over heated aluminium (Figure 13.11).

$$2Al + 3Cl_2 \rightarrow 2AlCl_3$$

It is prepared commercially by passing chlorine over a mixture of bauxite and coke.

$$Al_2O_3 + 3C + 3Cl_2 \rightarrow 2AlCl_3 + 3CO$$

The anhydrous chloride is a white, fibrous solid which fumes in air due to hydrolysis and dissolves in water furnishing $Al(H_2O)_6^{3+}$ and Cl^- ions. The vapour density up to 400 °C corresponds to the formula Al_2Cl_6 and electron diffraction data on the vapour show that the chlorine atoms are tetrahedrally arranged around the aluminium atoms, giving molecules of the form shown in Figure 13.12. The main uses of aluminium chloride are catalytic, as in the Friedel-Crafts reaction and in petroleum cracking.

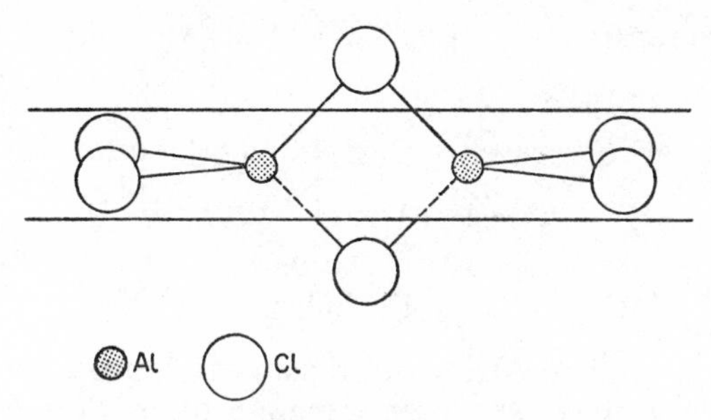

FIGURE 13.12. Structure of the molecule Al_2Cl_6

Nitrides

Nitrides are prepared by heating amorphous boron and aluminium powder in nitrogen at 1000 °C and 700 °C respectively. Boron nitride has a layer lattice similar to that of graphite (*see* Figure 13.13(*a*)) but, in the former, the atoms in the layers are vertically above each other and, in contrast to graphite, it is a white solid and a very good insulator. A diamond form of boron nitride (borazon) has been prepared by subjecting the layer form to high temperatures and pressures. This compound is as hard as diamond

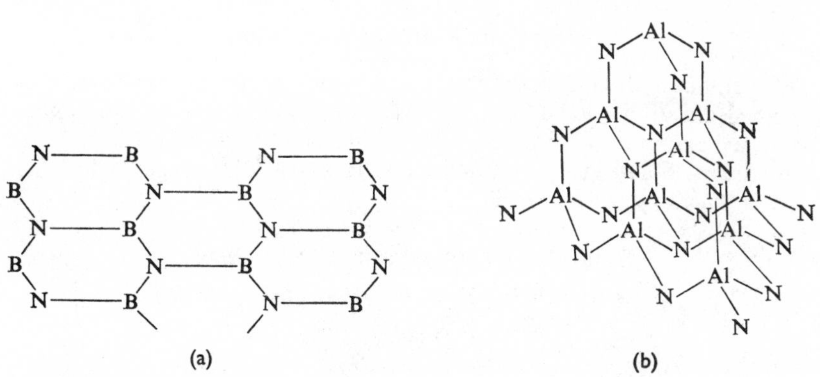

(a) (b)

FIGURE 13.13 (*a*) graphite layer structure of boron nitride; (*b*) diamond structure of aluminium nitride

and is a good electrical insulator. Aluminium nitride has a diamond-type lattice (Figure 13.13(*b*)). It is, however, hydrolysed by cold water indicating that the bonding is not strictly covalent.

$$AlN + 3H_2O \rightarrow Al(OH)_3 + NH_3$$

Sulphides

Sulphides are prepared by heating the elements with sulphur.

$$2Al + 3S \rightarrow Al_2S_3$$

Both sulphides are extensively hydrolysed in solution

$$B_2S_3 + 6H_2O \rightarrow 2B(OH)_3 + 3H_2S$$

Carbides

The carbides $B_{12}C_3$ and Al_4C_3 are prepared by heating the elements with carbon at high temperature in an electric furnace. Both compounds are extremely hard, forming giant molecular complex lattices. Boron carbide is used as an abrasive.

Borides

The borides are prepared by heating a mixture of powdered metal and boron to high temperature; the structure and formula of many of them is uncertain. In the borides, the boron atoms can form chains, layers, or three-dimensional structures, the position of the metal atoms depending on the type of structure (e.g. FeB chain type; AlB_2 layer type; CaB_6 three-dimensional type). These structures cannot be accounted for on the simple valency theory.

Ionic Compounds of Aluminium

These compounds both in the solid state and in solution usually contain the octahedral cation $Al(H_2O)_6^{3+}$. The Al^{3+} ion is found in a few compounds such as aluminium fluoride and perchlorate which can be prepared anhydrous. Solutions of aluminium salts are acidic since the hydrated cation functions as a triprotic acid.

$$Al(H_2O)_6^{3+} + H_2O \rightleftharpoons Al(H_2O)_5OH^{2+} + H_3O^+$$

$$Al(H_2O)_5OH^{2+} + H_2O \rightleftharpoons Al(H_2O)_4(OH)_2^{+} + H_3O^+$$

$$Al(H_2O)_4(OH)_2^{+} + H_2O \rightleftharpoons Al(H_2O)_3(OH)_3\downarrow + H_3O^+$$

Hence salts of weak acids such as carbonates, sulphides, cyanides, etc., cannot be prepared in solution since these anions are strong proton acceptors (strong bases). Thus addition of carbonate or sulphide ions to an aluminium salt precipitates aluminium hydroxide.

$$Al(H_2O)_5OH^{2+} + CO_3^{2-} \rightarrow Al(H_2O)_3(OH)_3\downarrow + H_2CO_3$$

Aluminium nitrate, $Al(NO_3)_3$, is prepared by dissolving aluminium hydroxide in nitric acid and it crystallizes from solution as an enneahydrate $Al(NO_3)_3.9H_2O$. It decomposes on heating.

$$4Al(NO_3)_3 \rightarrow 2Al_2O_3 + 12NO_2 + 3O_2$$

Aluminium sulphate, $Al_2(SO_4)_3.18H_2O$ (hair salt or feather alum), is prepared industrially by treating bauxite or clay with concentrated sulphuric acid.

$$\underset{\text{china clay}}{Al_2O_3.2SiO_2} + 3H_2SO_4 \rightarrow Al_2(SO_4)_3 + 3H_2O + 2SiO_2$$

The residue is extracted with water, filtered, and recrystallized.

Aluminium sulphate finds uses in sizing, papermaking, mordanting, in the precipitation of colloidal matter from sewage, and as a constituent of styptic pencils and foam fire-extinguishers.

Alums

Alums have the general formula $M^I M^{III} (SO_4)_2.12H_2O$ where

M^I may be Li^+, Na^+, K^+, Rb^+, Cs^+, NH_4^+

M^{III} may be Al^{3+}, Cr^{3+}, Fe^{3+}, Mn^{3+}, Co^{3+}

Some common examples are: potash alum, $KAl(SO_4)_2.12H_2O$; chrome alum, $KCr(SO_4)_2.12H_2O$; ferric alum, $NH_4Fe(SO_4)_2.12H_2O$; and ammonium alum, $NH_4Al(SO_4)_2.12H_2O$. These are true double salts giving simple ions only in solution. In the solid state each cation is associated with six water molecules and generally the alums crystallize as octahedra. They may be prepared by mixing saturated solutions containing equimolecular proportions of the component sulphates and allowing to crystallize.

Complexes

Aluminium forms complexes more readily than the *s* block elements because of its smaller size and greater charge. In addition to the halo, hydro, and hydroxo complexes already mentioned, the octahedral chelates are worthy of note. The acetylacetonate $Al(acac)_3$ is a volatile covalent solid, melting point 192 °C. The 8-hydroxyquinolate Al(O, N-quinolinolate)$_3$

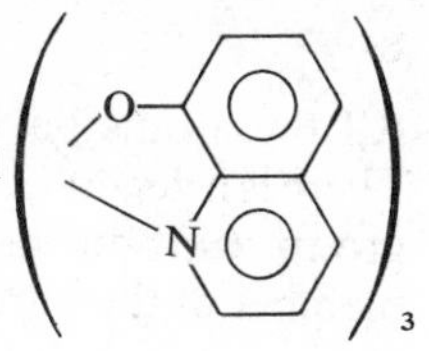

is used for the gravimetric estimation of aluminium.

14

Group IVB

Physical Properties

	Carbon C	Silicon Si	Germanium Ge	Tin Sn	Lead Pb
Atomic mass	12·01115	28·086	72·59	118·69	207·19
Atomic number	6	14	32	50	82
Isotopes	12, 13, 14	28, 29, 30	70, 72, 73, 74, 76	112, 114 to 120, 122, 124	204, 206, 207, 208
Electronic structure	$2s^2 2p^2$	$3s^2 3p^2$	$3d^{10}$ $4s^2 4p^2$	$4d^{10}$ $5s^2 5p^2$	$4f^{14} 5d^{10}$ $6s^2 6p^2$
Atomic radius (nm)	0·077	0·117	0·122	0·140	0·154
Ionic radius (nm) (M^{4+})	0·015	0·041	0·053	0·071	0·084
Ionization potential ($kJ\ mol^{-1}$)					
(I)	1096	792	768	713	722
(II)	2356	1586	1546	1416	1456
Electronegativity	2·5	1·9	1·7	1·7	1·6
Density ($g\ cm^{-3}$)	3·52 (diamond) 2·22 (graphite)	2·33	5·36	7·30	11·34
Melting point (°C)	3730 (sub) (graphite)	1,420	958·5	231·8	327·4
Boiling point (°C)	4830 (graphite)	2,600	2,700	2,260	1,620

General Characteristics

All the atoms have the configuration $ns^2 np^2$ in their outermost orbitals. There is a general, though irregular, trend from non-metal to metal down the group. Carbon, however, shows little chemical resemblance to any other member of the group. Examination of the bond energies (Table 14.1) shows the great strength of the C—C single bond; the tendency to form long chains of carbon atoms (catenation) is a marked characteristic of carbon. Although molecules with chains of up to six silicon atoms have been made, catenation almost disappears with the remaining elements.

All elements of the group show a characteristic oxidation state of +4. This is achieved by promotion of an *s* electron into a *p* orbital. The *s* and *p* orbitals then undergo hybridization (sp^3) giving rise to four strong covalent bonds which are directed towards the corners of a tetrahedron. The energy required for promotion and hybridization in the case of carbon is more than compensated by the establishment of four stable links (p. 98). In many cases, however, the bond energies of a series of compounds MX_4 decrease markedly down the group and may be insufficient to compensate for the promotion energy. Thus the tetra-covalent state becomes less extensive and less stable down the group,

$$\text{e.g. } PbCl_4 \xrightarrow{\text{room temp.}} PbCl_2 + Cl_2$$

There are a few cases in which the bond energies increase down the group (Table 14.1).

TABLE 14.1

Bond energies (kJ mol⁻¹)

C—O	360	C—Cl	338	C—C	348
Si—O	451	Si—Cl	380	Si—Si	176

It will be noted that the Si—O bond is particularly strong and, in contrast to gaseous carbon dioxide, silica has a three-dimensional macromolecular structure in which each silicon atom is covalently bonded to and tetrahedrally surrounded by four oxygen atoms (p. 229). Bond energy considerations would predict $SiCl_4$ to be more stable than CCl_4 but the former is very susceptible to attack by water molecules whereas the latter is not. This is because the carbon atom in CCl_4 is exerting its maximum covalency whereas silicon and other members of the group have vacant *d* orbitals into which water molecules are able to donate a lone pair of electrons facilitating hydrolysis (Figure 14.1). The increasing polarity of the M—Cl bond down the group also facilitates hydrolysis.

Carbon is unique in the group in its ability to undergo:

(*a*) sp^2 hybridization, giving rise to double bonding as in graphite and the carbonate ion (pp. 99, 218);

(*b*) *sp* hybridization, giving rise to double bonding in carbon dioxide and triple bonding in carbon monoxide (pp. 38, 214). There are no corresponding analogues with the remaining elements of the group.

The reason why carbon dioxide forms a multiple-bonded monomer whereas silica forms a single-bonded giant molecule may be partially explained in terms of the respective atomic radii of carbon and silicon.

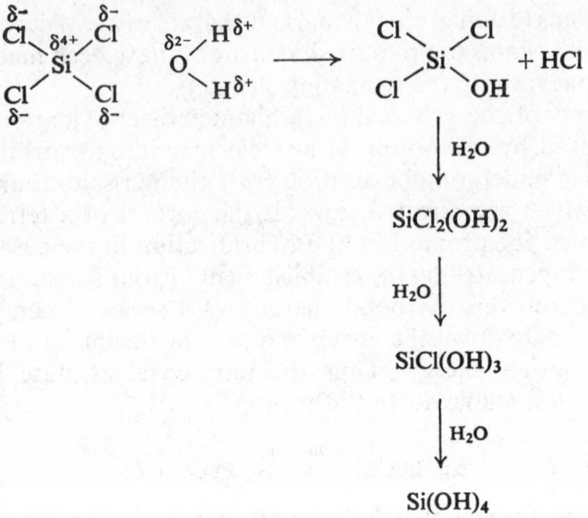

$$\overset{\delta-}{Cl}_3\overset{\delta-}{Cl}\,\overset{\delta 4+}{Si} + \overset{\delta 2-}{O}(\overset{\delta+}{H})_2 \longrightarrow SiCl_3(OH) + HCl$$

$$SiCl_3(OH) \xrightarrow{H_2O} SiCl_2(OH)_2 \xrightarrow{H_2O} SiCl(OH)_3 \xrightarrow{H_2O} Si(OH)_4$$

FIGURE 14.1. Hydrolysis of $SiCl_4$

Silicon has a much larger radius than carbon, giving a greater Si—O bond length with less efficient lateral overlap of *p* orbitals. Thus the formation of a strong π bond, as in carbon dioxide, is precluded and the strong silicon–oxygen single bond is preferred. Since this is highly polar, the silica becomes polymerized and forms a giant molecule.

Germanium, tin, and lead are capable of forming a +2 ionic oxidation state of which Pb^{2+} is the most stable. This was initially ascribed to an inert electron pair but this is not borne out by ionization potential values. For ion formation, the energy required for the process $M \rightarrow M^{2+} + 2e$ must be recoverable as the lattice energy of the resulting ionic structure. Because of the small atomic radii of carbon and silicon, the ionization energy would be high and the lattice energy would be small due to the inability of very small ions to form stable crystalline lattices. Thus carbon and silicon never form dipositive ions, but the stability of the dipositive ions of other members of the group increases with atomic radius. In view of the considerable ionization energy required for the process $M^{2+} \rightarrow M^{4+} + 2e$, none of the elements forms a tetrapositive ion.

Silicon, germanium, tin, and lead possess a fairly strong tendency to form octahedral complexes utilizing *d* orbitals in d^2sp^3 hybridization, e.g. SiF_6^{2-}, $SiCl_6^{2-}$, $PbCl_6^{2-}$, GeF_6^{2-}. Carbon, with no available *d* orbitals, cannot form such complexes.

Occurrence

Carbon occurs in the elementary form as diamond and graphite. The structures of these have already been described (p. 98) but they are included

TABLE 14.2

Diamond	*Graphite*
Occurrence: Brazil as alluvial deposits. South Africa in shafts of ancient volcanoes as a dark rock–blue ground	Naturally in Ceylon, Germany, and U.S.A., but usually manufactured
Manufacture; Very small artificial diamonds (industrial) have been prepared by subjecting graphite to very high temperatures (1920 K) and very high pressures	Made by the Acheson process. Sand and petroleum coke are electrically heated for 24 hr in an electric furnace $SiO_2 + 3C \rightarrow SiC + 2CO$ $SiC \rightarrow C\,(\text{graphite}) + Si\,(\text{distils off})$
Structure; Giant molecule (p. 98). Each carbon atom is tetrahedrally surrounded by and covalently bonded to four other carbon atoms. C—C distance corresponds to a single bond	Layer structure (p. 99) consisting of planes of hexagons. All C—C bonds are 0·142 nm, indicating resonance within the rings and considerable double-bond character. Layers are weakly bound by van der Waals forces
Properties; Extremely hard octahedral crystals (24–48 faces) with high melting point, density, refractive index, and dispersion. Non-conductor	Planes of hexagons can glide freely over one another and graphite suspended in oil or water is an excellent lubricant. Delocalized π-orbitals give high conductivity.
$C\,(\text{diamond}) + O_2 \xrightarrow{900°C} CO_2$	$C\,(\text{graphite}) + O_2 \xrightarrow{700°C} CO_2$
$C\,(\text{diamond}) + 2F_2 \xrightarrow{700°C} CF_4$	$C\,(\text{graphite}) + 2F_2 \xrightarrow{500°C} CF_4$
$C\,(\text{diamond}) + K_2Cr_2O_7 + \text{conc. } H_2SO_4 \xrightarrow{200°C} CO_2$	$C\,(\text{graphite}) + K_2Cr_2O_7 + \text{conc. } H_2SO_4 \xrightarrow{100°C} CO_2$

again here in tabular form, together with their principal properties (Table 14.2).

In the graphite structure, the openings between the planes constitute points of entry for chemical reagents and a number of graphitic compounds have been made. The so-called graphitic oxide has been produced by the action of concentrated sulphuric acid, concentrated nitric acid, and potassium chlorate on graphite. The graphite swells, giving a green product of variable composition approximating to $C_6O(OH)$. The lustre and conductivity of graphite is lost and the distance between the layers increases up to 1·0 or 1·1 nm. It has been suggested that oxygen atoms link the sheets together. The action of concentrated acids (sulphuric, nitric, or pyrophosphoric) and a small quantity of potassium chlorate or chromium trioxide on graphite again causes swelling and usually gives a coloured compound (blue with concentrated sulphuric acid). Separation of the graphite layers occurs up to 0·8 nm and the anions HSO_4^-, NO_3^-, and $P_2O_7^{2-}$ are contained between the sheets. The products (graphite salts) are all unstable and are decomposed by water with the regeneration of graphite. The product obtained by the action of fluorine on graphite at 500 °C is also a compound of this type, where the fluorine is contained in double layers between the sheets of graphite.

Electrical conducting graphitic compounds have also been prepared with potassium, rubidium, caesium, and iron(III) chloride. The preparative methods involve heating the molten alkali metal or iron(III) chloride with graphite. The coloured compounds obtained with potassium and iron(III) chlorides have empirical compositions C_8K—$C_{24}K$ and C_9FeCl_3—$C_{12}FeCl_3$. The conductivity is increased by the removal or addition of electrons from the conducting layers of graphite by the potassium atoms or iron(III) chloride molecules between the sheets. Thus the electrons are free to move from layer to layer as well as within a layer.

A recent achievement has been the development of strong graphite fibres by the controlled decomposition of acrylic textile fibres. These graphite fibres embedded in epoxy resin have been used for the reinforcement of the plastic blades of compressor fans in jet engines.

Amorphous carbon was originally considered as a third allotropic form but X-ray photographs show that all forms of amorphous carbon are micro-crystalline graphite, i.e. all are composed of minute portions of the graphite structure.

The various forms of amorphous carbon include wood charcoal, animal charcoal, lampblack, and carbon black and are prepared respectively by burning wood, bones, oils, and natural gas in a limited supply of air. Wood charcoal is very porous and exhibits an enormous surface area. When freshly prepared it is an excellent gas adsorber. Charcoal cooled in liquid air is 10–15 times more efficient as a gas adsorber and is used in the production of high vacua. Ordinary charcoal, however, adsorbs no great quantity of gas due to the hydrocarbons already adsorbed and it is activated by heating in a current of steam to drive out the adsorbed impurities. It is then suitable as an adsorption material for chromatographic columns. Animal charcoal is used to decolorize sugar-cane juice in making white sugar; lampblack is used in ink, paint pigment, and shoe blacking, and carbon black is used as a filler in the rubber industry.

Owing to its fine subdivision, charcoal is the most reactive form of carbon.

$$C + O_2 \rightarrow CO_2 \text{ (continues to burn in oxygen)}$$

$$C + F_2 \rightarrow CF_4 \text{ (ignites)}$$

$$C + \underset{\text{conc.}}{4HNO_3} \rightarrow CO_2 + 2H_2O + 4NO_2$$

$$C + \underset{\text{conc.}}{2H_2SO_4} \rightarrow CO_2 + 2H_2O + 2SO_2$$

Coal

The majority of great coal fields are the remains of vast swamps which flourished some three-hundred million years ago in the Primo-Carboniferous period. Large masses of vegetable material have been compressed by overlying strata and have undergone decay and bacterial decomposition.

The different stages in the conversion of the vegetable material to the final product, anthracitic coal, is usually represented by the series: wood → peat → lignite (brown coal) → bituminous coal → anthracite.

The main elements, carbon, hydrogen, and oxygen in coal show the following variations.

	Wood		*Anthracite*
Carbon	50%	→	95%
Hydrogen	43%	→	2·5%
Oxygen	6%	→	2·5%

Coal is an extremely complex substance, the basic unit of the coal structure consisting of polycyclic aromatic nuclei with two to four fused rings in 'low-ranking' coals but with up to thirty fused rings in the 'high-ranking' anthracite variety. These fused rings are surrounded by carbon and hydrogen in the form of short aliphatic side-chains and alicyclic rings. Low-ranking coals also contain up to 10 per cent oxygen in the form of phenolic —OH groups and quinonoid structures or ether linkages between a side-chain and a nucleus or between two side-chains (Figure 14.2). Sulphur, nitrogen, and phosphorus are also found in various forms as minor components. The stacking of these flat plate-like units is not very orderly but the degree of order does increase slightly with the anthracite coals. Research into the structure of coal is still continuing and, once its structure has been fully elucidated, better methods will no doubt become available for its full utilization both as a fuel and a source of chemicals.

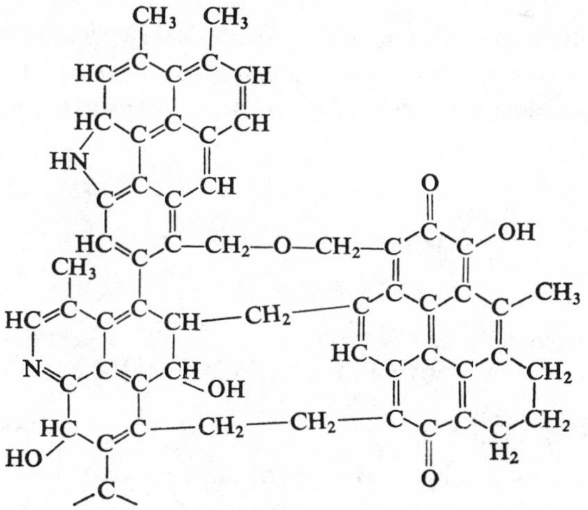

FIGURE 14.2. A possible structural unit of a coal with 80–85 per cent carbon

Coal Gas

The four main products obtained when coal is subjected to high-temperature carbonization, together with their composition and uses, are shown in Table 14.3.

TABLE 14.3

Product	*Composition*	*Uses*
1. Coal gas	50% H_2; 35% CH_4; 15% C_2H_4 and CO	Domestic and industrial fuel gas
2. Ammoniacal liquor	NH_3; $(NH_4)_2SO_4$; $(NH_4)_2CO_3$; $(NH_4)_2S$; NH_4CN, etc.	Source of ammonia and compounds particularly ammonium sulphate, used as a fertilizer
3. Coal tar	Benzene, toluene, phenol, cresol, naphthalene, anthracene, etc.	Road tar, creosote, but more important as a source of organic chemicals for the manufacture of dyes, plastics, drugs, antiseptics, paints, and varnishes
4. Coke	90% carbon	Domestic and industrial fuel (in power stations). Reducing agent. Raw material for production of water gas and producer gas

Low-temperature carbonization of coal (500–700 °C) is now carried out on a large scale and the products are:

(*a*) A smokeless fuel—coalite.

(*b*) Crude light oil which contains a high percentage of phenols.

(*c*) A smaller quantity of coal gas of higher calorific value than that obtained by high-temperature coal distillation.

Many new developments are under way for coal gasification and include:

(*a*) High-pressure gasification with steam and oxygen for low-grade coals,

(*b*) Oxygen–steam gasification of coal dust, which gives a gas of extremely high calorific value.

Other Industrial Gases

Producer gas

Air is passed through a bed of hot coke at 1,000°C. Reactions which take place in the reactor are shown in Figure 14.3. Thus the overall reaction is:

$$2C + O_2 \rightarrow 2CO \text{ and from Hess's law } \Delta H = -394 + 163 = -231 \text{ kJ}$$

The resulting hot gas has a low calorific value and is useless for piping long distances since the initial heat is lost. The gas is used *in situ* for heating coal and in metallurgical and glass furnaces.

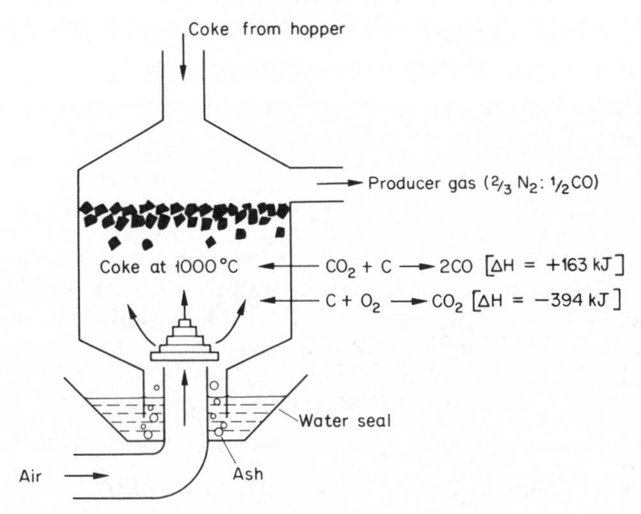

FIGURE 14.3

Water gas

Steam is passed through hot coke at 1,000°C in a reactor which is very similar to that used for the manufacture of producer gas.

$$C + H_2O \rightleftharpoons CO + H_2 \qquad (\Delta H = +121 \text{ kJ})$$

Since the reaction absorbs heat, the temperature falls and eventually carbon dioxide is produced. The methods used to avoid this are: (*a*) to blow air through the coke at intervals, raising the temperature and forming producer gas; (*b*) to blow a mixture of air and steam through the coke, giving a continuous supply of gas called semi-water gas.

Some typical percentage analyses are shown below.

	H_2	CO	CH_4	CO_2, N_2, etc.
Water gas	50	40	1	9
Semi-water gas	17	25	8	50

These gases are used for the manufacture of hydrogen (p. 156), the production of methanol (p. 157), the production of carbon dioxide for the Solvay process (p. 170), and to supplement coal gas.

The Carbides

The metallic carbides are prepared by direct union of the elements at a high temperature (>2000 °C), heating the oxide with carbon or by heating the metal in the vapour of a suitable hydrocarbon, e.g. acetylene.

$$2Na + CH{\equiv}CH \rightarrow Na^{+}{}_2C_2{}^{2-} + H_2$$

The carbides can be classified into three groups, ionic, covalent or interstitial.

Ionic Carbides

These are formed by the representative metals of groups I, II, III—Cu, Ag, Au, Zn, Cd, and the lanthanides. They furnish either methanide, C^{4-}, or acetylide, $(—C{\equiv}C—)^{2-}$, ions. Beryllium carbide, Be_2C, and aluminium carbide, Al_4C_3, are compounds of the former type which give methane on hydrolysis.

$$C_3{}^{4-} + 12H_2O \rightarrow 12OH^- + 3CH_4$$

Carbides of the latter type (Ca, Sr, Ba) yield acetylene on hydrolysis.

$$(—C{\equiv}C—)^{2-} + 2H_2O \rightarrow 2OH^- + HC{\equiv}CH$$

Many carbides have complex structures but most of the acetylides have a structure similar to that of sodium chloride (Figure 14.4(*a*)).

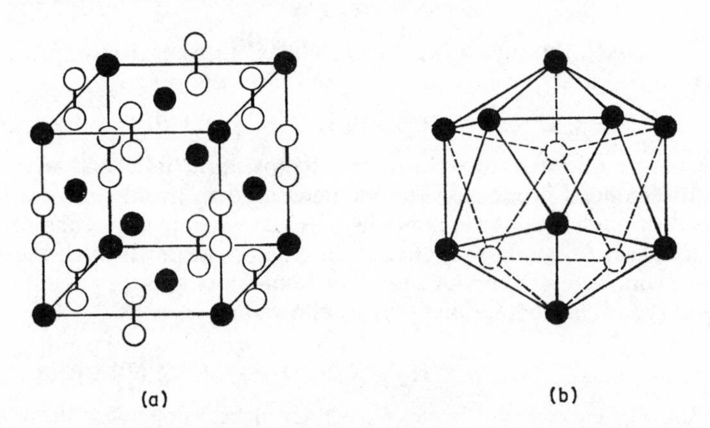

FIGURE 14.4 (*a*) alkaline earth cation M^{2+}, acetylide anion $C{\equiv}C^{2-}$; (*b*) icosahedral B_{12} groups in crystalline boron and boron carbide

Covalent Carbides

These are exemplified by the carbides of silicon (SiC) and boron ($B_{12}C_3$). Both are hard, infusible, and inert. Silicon carbide crystallizes in a large number of different forms but all are giant molecules with a tetrahedral diamond-type arrangement of silicon and carbon atoms. In boron carbide, B_{12} icosahedral groups are linked at some corners to other B_{12} groups and at others to a linear carbon chain containing three carbon atoms. The giant covalent structure thus contains B—B and B—C covalent bonds (Figure 14.4(*b*)).

Interstitial Carbides

In these carbides the small carbon atoms enter the spaces between the metallic atoms giving a slightly expanded metallic lattice. Examples include TiC, ZrC, HfC, UC, VC, NbC, TaC, MoC, WC. The lattice is usually of the rock-salt type with neutral atoms in place of the ions and has a higher stability than the pure metal lattice. These carbides are not attacked by water or dilute acids, have high melting points (3000–3500 °C), are extremely hard, lustrous, and good electrical conductors.

Iron, manganese, cobalt, nickel, and chromium have atomic radii less than 0·13 nm and, in these carbides (M_3C), the carbon atoms cannot fit into the metallic lattice without a great deal of distortion. The carbon atoms appear to link up with one another forming long chains which run through a very distorted metallic lattice. Consistent with this structure is the production of a variety of hydrocarbons on hydrolysis with water or dilute acids.

Oxides of Carbon

Carbon Monoxide

Carbon monoxide is prepared by the dehydration of formic or oxalic acid with concentrated sulphuric acid.

$$\mathrm{HCOOH} \xrightarrow[\mathrm{H_2SO_4}]{\text{cold conc.}} \mathrm{C \equiv O + H_2O}$$

$$\mathrm{(COOH)_2} \xrightarrow[\mathrm{H_2SO_4}]{\text{hot conc.}} \mathrm{O{=}C{=}O + C \equiv O + H_2O}$$

The reduction of carbon dioxide by passage over red hot charcoal or the action of concentrated sulphuric acid on potassium hexacyanoferrate(II) also produces carbon monoxide.

$$\mathrm{Fe(CN)_6^{4-} + 6H_2SO_4 + 6H_2O \rightarrow 2SO_4^{2-} + FeSO_4 + 6CO + 3(NH_4)_2SO_4}$$

Carbon dioxide may be removed from the gas with potassium hydroxide solution and the carbon monoxide dried by concentrated sulphuric acid.

Properties

Carbon monoxide is a colourless, odourless, water-insoluble gas which is extremely poisonous. The red pigment in the blood, haemoglobin, combines with carbon monoxide to give a pink complex, carboxyhaemoglobin, preventing the normal oxygenation of the blood.

Gaseous carbon monoxide burns in air with a blue flame and combines directly with many non-metallic elements.

$$2CO + O_2 \rightarrow 2CO_2$$

$$CO + Cl_2 \rightarrow COCl_2 \quad \text{(carbonyl chloride: charcoal catalyst required)}$$

$$CO + S \rightarrow COS \quad \text{(carbonyl sulphide: hot tube)}$$

The gas will also reduce the heated oxides of copper, lead, and iron to the metal. Carbon monoxide is a constituent of producer and water gas and can be used in many important synthetic reactions such as:

$$CO + 2H_2 \xrightarrow[300°C\ Al_2O_3\ \text{catalyst}]{\text{pressure}} CH_3OH \text{ (methanol)}$$

$$CO + \underset{\text{conc.}}{NaOH} \xrightarrow{200°C} HCOONa \text{ (sodium formate)}$$

Carbon monoxide may be absorbed in ammoniacal copper(I) chloride or formate when complexes are formed:

$$\begin{matrix} H_2O \searrow & & \diagup Cl \diagdown & & \swarrow OH_2 \\ & Cu & & Cu & \\ OC \nearrow & & \diagdown Cl \diagup & & \nwarrow CO \end{matrix}$$

The *carbonyls* are another series of complex covalent compounds which can be made by the action of carbon monoxide on

(*a*) the finely divided transition metal or compound under pressure.

(*b*) the anhydrous transition metal chloride in an organic solvent (ether or tetrahydrofuran) with a reducing agent (sodium, magnesium, or aluminium) under pressure. The structure of the mononuclear carbonyls of iron, nickel, and chromium is shown in Figure 14.5.

The mononuclear carbonyls are fairly stable liquids and solids, decomposition occurring at about 100 °C to carbon monoxide and the metal. The properties are those typical of a covalent compound, i.e. soluble in organic solvents and volatile. These carbonyls are all spin paired complexes (p. 351) and the central metallic atoms all have an effective atomic number of 36 (as for krypton).

A large number of polynuclear carbonyls, often formed as intermediates in the decomposition of mononuclear carbonyls, are also known, e.g. $Fe_2(CO)_9$, $Co_2(CO)_8$, $Mn_2(CO)_{10}$, $Fe_3(CO)_{12}$, $Co_4(CO)_{12}$, etc. The structures of some binuclear carbonyls are shown in Figure 14.5. The binuclear

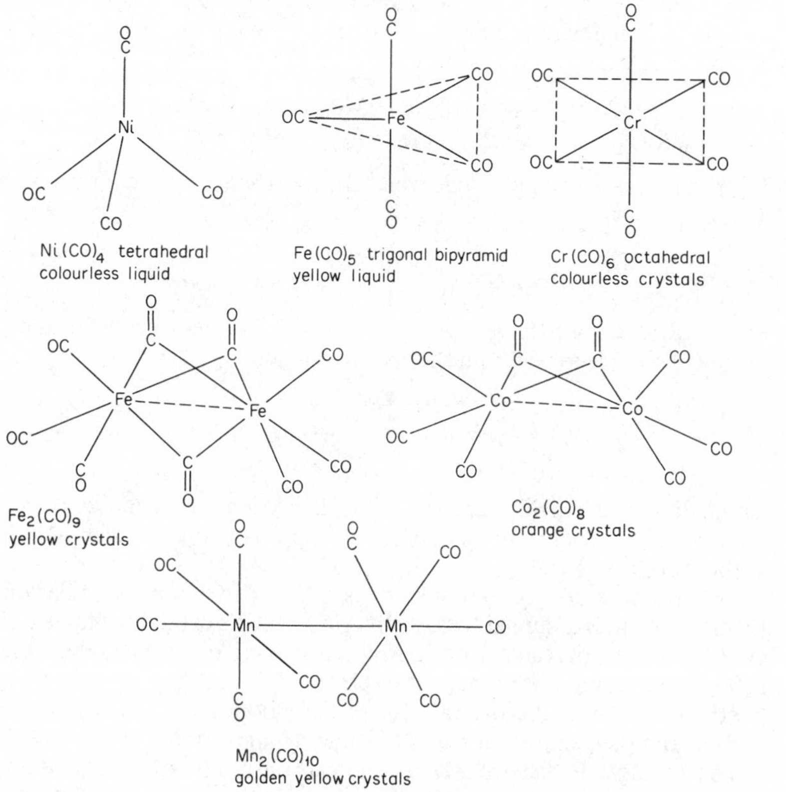

$Ni(CO)_4$ tetrahedral
colourless liquid

$Fe(CO)_5$ trigonal bipyramid
yellow liquid

$Cr(CO)_6$ octahedral
colourless crystals

$Fe_2(CO)_9$
yellow crystals

$Co_2(CO)_8$
orange crystals

$Mn_2(CO)_{10}$
golden yellow crystals

FIG. 14.5

carbonyls of iron and cobalt contain both terminal CO groups and bridging keto groups,

where two sigma bonds are formed, one to each metallic atom. Since there are no unpaired electrons in these complexes, and the metal-metal distances are extremely short, metal-metal bonds have been postulated. In contrast, $Mn_2(CO)_{10}$ possesses no bridging keto groups and there is no simple explanation to offer for this anomaly. Infrared spectra of the carbonyls provides additional evidence for the structures of these complexes. The bridging keto groups show absorption near 1800 cm^{-1} (ketones absorb near 1700 cm^{-1}) whereas the terminal CO groups absorb near 2000 cm^{-1} (carbon monoxide absorbs at 2143 cm^{-1}).

Carbonylate anions have been obtained by the treatment of a mononuclear carbonyl with sodium hydroxide or cleaving the metal–metal bond of a polynuclear carbonyl with sodium amalgam.

$$Fe(CO)_5 + 3NaOH \rightarrow Na[HFe(CO)_4](aq) + Na_2CO_3(aq) + H_2O$$

$$Co_2(CO)_8 + 2Na/Hg \xrightarrow{THF} 2NaCo(CO)_4 + 2Hg$$

The products are yellow liquids which yield carbonyl hydrides on acidification.

$$NaHFe(CO)_4 + H^+ \rightarrow H_2Fe(CO)_4 + Na^+$$

$$NaCo(CO)_4 + H^+ \rightarrow HCo(CO)_4 + Na^+$$

Both hydrides are pale yellow liquids at very low temperatures, they are sparingly soluble in water and function as acids:

$$HCo(CO)_4 \rightleftharpoons H^+ + Co(CO)_4^-$$

$$H_2Fe(CO)_4 \rightleftharpoons H^+ + HFe(CO)_4^-$$

$$HFe(CO)_4^- \rightleftharpoons H^+ + Fe(CO)_4^{2-}$$

On heating, the hydrides decompose to hydrogen and metallic carbonyls. The hydrogen atom in these compounds appears to be covalently bonded to the metal.

The structure of carbon monoxide is: $^-C{\equiv}O^+$. On the molecular orbital theory, carbon is assumed to be *sp* hybridized. One orbital contains a single electron which overlaps with a single *p* orbital of oxygen, the other *sp* orbital containing a lone pair of electrons. The full $2p_z$ orbital of oxygen overlaps the empty $2p_z$ orbital of carbon (π-bond) and the $2p_y$ orbitals of carbon and oxygen form a second π-bond (Figure 14.6).

The negative formal charge on the carbon atom is unusual, the bond polarity being opposite in the majority of C—O compounds, but this fact is not inconsistent with the properties of carbon monoxide, particularly with regard to its donor properties in the carbonyls. However, apart from the coordinate link from the lone pair on the carbon atom, there is considerable π-bonding between the *d* orbitals on the metal and the empty π^* antibonding orbital on the carbon.

Carbon Dioxide

Occurs to 0·03–0·04 per cent by volume in the atmosphere although larger concentrations may exist in volcanic regions and caverns. Carbon dioxide absorbs solar radiation and significantly reduces the earth's temperature. The gas may be prepared in the laboratory by the action of 50 per cent hydrochloric acid on marble chips or any other carbonate or bicarbonate. Purification from acid spray is effected by passage through potassium bicarbonate solution and the gas may be dried by concentrated sulphuric acid and collected by downward delivery.

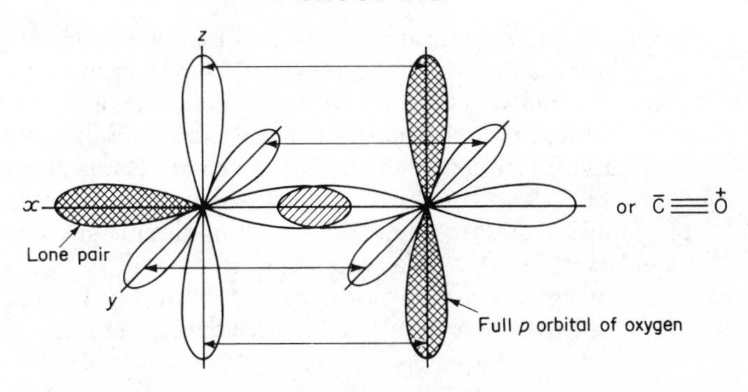

FIGURE 14.6

$$CO_3^{2-} + 2H^+ \rightarrow H_2O + CO_2$$

$$HCO_3^- + H^+ \rightarrow H_2O + CO_2$$

On the industrial scale carbon dioxide is a by-product in (*a*) the production of quicklime

$$CaCO_3 \rightleftharpoons CaO + CO_2$$

(*b*) the fermentation of monosaccharides, e.g. glucose, in the presence of yeast at 35 °C.

$$C_6H_{12}O_6 \rightarrow 2C_2H_5OH + 2CO_2$$

The purified gas is compressed at 60 times atmospheric pressure in cylinders as liquid carbon dioxide.

Properties

Carbon dioxide is a colourless, odourless gas which is heavier than air. It stimulates the respiratory centres and is used with oxygen for revival in cases of drowning and gas poisoning. The gas dissolves slightly in water giving a weakly acid solution.

$$CO_2 + H_2O \rightleftharpoons H_2CO_3 + H_2O \rightleftharpoons H_3O^+ + HCO_3^- \quad (K_1 = 4.2 \times 10^{-7})$$

$$H_2CO_3 + 2H_2O \rightleftharpoons 2H_3O^+ + CO_3^{2-} \quad (K_2 = 4.8 \times 10^{-11})$$

The dissociation constants are surprisingly low since carbonic acid

$$(HO)_2C{=}O$$

might be expected to be a strong acid. In fact less than 1 per cent of the dissolved carbon dioxide is converted into carbonic acid and the first dissociation constant K_1 is incorrect since it assumes that all the dissolved carbon dioxide is present as H_2CO_3. The true value of K_1 is about 2×10^{-4} and the acid is stronger than acetic acid. The anhydrous acid H_2CO_3

exists in ether at –30 °C. A saturated solution of carbon dioxide at 8 times atmospheric pressure is known as soda water. If liquid carbon dioxide is allowed to expand rapidly, a solid, 'dry ice' ('Drikold') is obtained which, when mixed with ether or amyl acetate, forms an excellent freezing mixture particularly valuable for ice cream and perishable goods. Rapid evaporation of the solid does not occur, because of the high latent heat of vaporization (364 kJ mol^{-1}) and because the blanket of carbon dioxide gas which surrounds the solid exerts a substantial vapour pressure.

The gas does not support combustion except when the burning substance is able to break the two double bonds and release the oxygen, e.g.

$$2Mg + CO_2 \rightarrow 2MgO + C$$

When three volumes of ammonia and one volume of carbon dioxide are mixed, a white solid, ammonium carbamate, is deposited.

$$O{=}C{=}O + 2NH_3 \longrightarrow \begin{matrix} NH_4O \diagdown \\ \quad C{=}O \\ NH_2 \diagup \end{matrix}$$

ammonium carbamate

This is the ammonium salt of carbamic acid which has never been isolated in the free state.

$$\begin{matrix} HO \diagdown \\ \quad C{=}O \\ NH_2 \diagup \end{matrix}$$

carbamic acid

The decomposition of ammonium carbamate under pressure with water is used for the commercial production of urea.

$$\begin{matrix} NH_4O \diagdown \\ \quad C{=}O \\ NH_2 \diagup \end{matrix} \longrightarrow H_2O + \begin{matrix} NH_2 \diagdown \\ \quad C{=}O \\ NH_2 \diagup \end{matrix}$$

urea

The structure of carbon dioxide has been described on p. 38.

The Carbonates

The salts of carbonic acid, the carbonates, are usually well-defined crystalline compounds. Those of the alkali metals (apart from lithium carbonate) and ammonium are soluble, whereas the carbonates of all the other metallic elements are insoluble. The carbonate ion is hydrolysed in water giving an alkaline solution.

$$CO_3^{2-} + H_2O \rightleftharpoons HCO_3^- + OH^-$$

Acidification of carbonate yields carbon dioxide.

$$CO_3^{2-} + 2H^+ \rightarrow H_2O + CO_2$$

Structures

In $Na_2CO_3.H_2O$ the packing of the Na^+, CO_3^{2-} ions, and water molecules is extremely complex and will not be discussed further. In the decahydrate $Na_2CO_3.10H_2O$ the extra water molecules are loosely held in the interstices of the crystal structure as lattice water and are easily driven off by heat. Sodium bicarbonate contains infinite hydrogen-bonded chains linked together laterally by Na^+ ions (p. 51).

The calcite form of calcium carbonate has a deformed rock-salt structure with the Ca^{2+} replacing the Na^+ ions and the CO_3^{2-} replacing the Cl^- ions but, due to the large size of the carbonate ion compared to chloride, there is some expansion and deformation of the rock-salt structure. The other form of calcium carbonate, aragonite, is related in a similar manner to the nickel arsenide structure. Bivalent metal carbonates whose cationic radii are within the range 0·06–0·106 nm, e.g. $ZnCO_3$ and $MgCO_3$, crystallize with the calcite form, whereas bivalent carbonates with cationic radii within the range 0·106–0·143 nm, e.g. $SrCO_3$, $BaCO_3$, crystallize with the aragonite structure.

Besides the normal carbonates, there are many basic carbonates which are intermediate in composition between the normal carbonate and the hydroxide. Thus zinc, lead, magnesium, and copper all form insoluble basic carbonates, produced by precipitation with sodium carbonate solution, e.g. $MgCO_3.Mg(OH)_2$; $2ZnCO_3.3Zn(OH)_2.H_2O$; $2PbCO_3.Pb(OH)_2$; $CuCO_3.Cu(OH)_2$. Since the carbonate ion is fairly extensively hydrolysed in water the formation of basic carbonates probably proceeds in the following stages:

$$CO_3^{2-} + H_2O \rightleftharpoons HCO_3^- + OH^-$$

$$2M^{2+} + CO_3^{2-} + 2OH^- \rightarrow M_2(OH)_2CO_3\downarrow$$

Since bicarbonate solutions are only slightly hydrolysed and solid bicarbonates are only known for the alkali metals, bicarbonate solutions are used to precipitate the normal carbonates of these metals. The structure of these basic salts is extremely complex but a number of them form hydrogen-bonded layer arrangements.

Stability to Heat

The stability of carbonates to heat depends upon the size and charge of the cation associated with the carbonate ion. A small, highly charged cation causes considerable distortion of the C—O bonds in the carbonate ion (Figure 14.7). The smaller the cation, the greater is the distortion and the more easily the carbonate ion is decomposed.

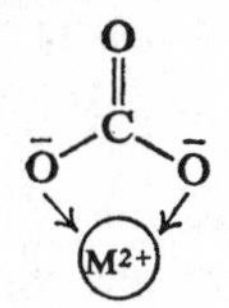

FIGURE 14.7

Thus the carbonates of cations with large radii (e.g. Na^+, K^+, Ba^{2+}, Sr^{2+}, Ca^{2+}) are stable to heat, whereas carbonates of cations with small ionic radii (0·06–0·08 nm e.g. Li^+, Be^{2+}, Mg^{2+}, Zn^{2+}, Cu^{2+}, Mn^{2+}, Ni^{2+}) are decomposed fairly readily on heating. It is interesting to note that with the Fe^{3+} and Al^{3+} ions (ionic radii ≈ 0·05 nm) the extremely small ion and the very high charge produces such a large strain that the bond is broken at ordinary temperatures and the carbonates cannot be prepared.

The above generalizations are somewhat over-simplified and ignore the thermodynamics of the decomposition process. Thus the decomposition temperatures of $CaCO_3$ and $CdCO_3$ are 900 °C and 350 °C respectively even though the ionic radii of Ca^{2+} and Cd^{2+} are very similar ($Ca^{2+} = 0{\cdot}099$ nm, $Cd^{2+} = 0{\cdot}097$ nm). Using the relationship $\Delta G = \Delta H - T\Delta S$ for the carbonate decomposition $MCO_3 \rightarrow MO + CO_2$, at 350 °C, where M = Cd, the free energy change for the decomposition (ΔG) = −3 kJ mol^{-1}. Where M = Ca at this temperature, $\Delta G = +78$ kJ mol^{-1}. Hence the decomposition of calcium carbonate is not feasible at 350 °C. At 900 °C, however, $\Delta G =$ −11 kJ mol^{-1} and the decomposition of calcium carbonate proceeds at this temperature.

Ions

The carbonate ion is planar with a bond angle of 120° and the carbon atom is sp^2 hybridized. Three bonds are formed by overlap with singly occupied *p* orbitals in oxygen (Figure 14.8(*a*)); π overlap also occurs as shown in Figure

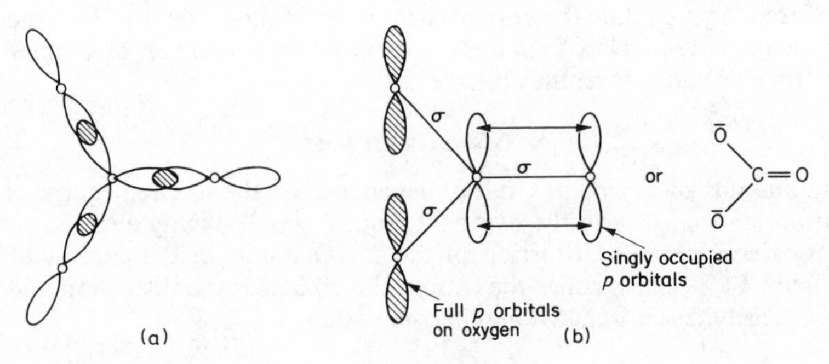

FIGURE 14.8

14.8(*b*). This, however, represents only one possible canonical form since π overlap may occur with any of the oxygen atoms. Thus Figure 14.8(*b*) shows one canonical form of the resonance hybrid.

The Halides

TABLE 14.4

Formula	Melting point (°C)	Boiling point (°C)	Bond energies C—X (kJ mol^{-1})	Properties
CF_4	−185	−128	484	Stable gas
CCl_4	−22·9	76·4	338	Stable, colourless liquid
CBr_4	93·7	decomp.	276	Yellow solid, decomposed on heating
CI_4	171·0	decomp.	238	Red solid, decomposed readily by heat or light

The physical properties of the halides are shown in Table 14.4. They are all prepared by prolonged halogenation of a hydrocarbon. Carbon tetrachloride is important commercially as a solvent and is prepared by passing chlorine into hot carbon disulphide in the presence of an antimony pentachloride or aluminium chloride catalyst.

$$CS_2 + 3Cl_2 \rightarrow CCl_4 + S_2Cl_2$$

The disulphur dichloride is removed by fractional distillation. All the halides are very resistant to hydrolysis (p. 333). The instability of the bromide and iodide to heat is probably due to the relatively large size of the halogen atoms compared to carbon atoms.

The Carbonyl Compounds

These have the formula COX_2 (where X = F, Cl, Br) and are all planar molecules, prepared by the action of carbon monoxide on a halogen. Fluorine reacts explosively, and chlorine and bromine readily, in the presence of light. Carbonyl chloride (phosgene) is an extremely poisonous gas. All the carbonyl halides are hydrolysed by water.

$$O{=}C\begin{matrix}\diagup Cl \\ \diagdown Cl\end{matrix} + O\begin{matrix}\diagup H \\ \diagdown H\end{matrix} \longrightarrow O{=}C{=}O + 2HCl$$

Carbon Disulphide

This is a linear molecule S=C=S. It is prepared by passing sulphur vapour through hot coke in an electric furnace.

$$C + 2S \rightarrow CS_2 \qquad (\Delta H = +88 \text{ kJ mol}^{-1})$$

Carbon disulphide is a colourless, volatile liquid, boiling point 46·2 °C, which is immiscible with water but miscible with alcohol and ether. The liquid usually has an unpleasant smell which can be removed by purification. The liquid dissolves sulphur, white phosphorus, rubber, camphor, etc., and although the vapour is highly inflammable, carbon disulphide is a very useful solvent and is used extensively to make viscose rayon and to vulcanize rubber. Carbon disulphide reacts with alkali sulphide or ammonium pentasulphide to form salts of thiocarbonic acid:

$$(HS)_2C{=}S$$

$$CS_2 + Na_2S \rightarrow Na_2CS_3 \quad \text{(yellow crystals).}$$

Excess concentrated hydrochloric acid liberates the free acid (H_2CS_3) as an unstable red liquid.

Cyanogen, Hydrocyanic Acid, and the Cyanides

Cyanogen, C_2N_2, may be prepared by heating mercury(II) or silver cyanide.

$$Hg(CN)_2 \rightarrow Hg + (CN)_2$$

$$2AgCN \rightarrow 2Ag + (CN)_2$$

The gas is colourless and poisonous and surprisingly quite stable, although $\Delta H_f = +308$ kJ mol^{-1}. Cyanogen burns in air with a purple flame:

$$(CN)_2 + 2O_2 \rightarrow 2CO_2 + N_2$$

On heating to 500 °C, it polymerizes to para-cyanogen, $(CN)_n$, for which a ring structure has been proposed.

Water causes hydrolysis to hydrocyanic and isocyanic acid although some ammonium formate and oxamide are also produced.

$$(CN)_2 + H_2O \rightarrow HCN + HNCO$$

The molecule of cyanogen is linear and the following structure has been proposed: N≡C—C≡N. However, the C—C distance is 0·138 nm indicating considerable π-bonding.

Hydrogen cyanide gas, *which is extremely poisonous*, is produced by distilling potassium cyanide with 50 per cent sulphuric acid.

$$KCN + H_2SO_4 \rightarrow KHSO_4 + HCN$$

A solution of the acid may be prepared by the distillation of potassium hexacyanoferrate(II) with dilute sulphuric acid.

$$2K_4Fe(CN)_6 + 3H_2SO_4 \rightarrow 3K_2SO_4 + K_2Fe^{II}Fe^{II}(CN)_6 + 6HCN$$

potassium iron(II) hexacyanoferrate(II)

A solution of the acid containing 2–3 per cent hydrogen cyanide is called prussic acid and is extremely poisonous. The acid is weak and in water some ammonium formate is obtained.

$$HCN + H_2O \rightleftharpoons H_3O^+ + CN^- \qquad (K = 4 \times 10^{-10})$$
$$CN^- + H_3O^+ + H_2O \rightarrow HCOONH_4$$

In the solid state, HCN molecules are hydrogen-bonded in endless chains.

$$H\cdots N{\equiv}C{-}H\cdots N{\equiv}C{-}H\cdots$$

The hydrogen bonding persists to some extent in the pure liquid (boiling point 26°C) which has a high relative permittivity (116).

The Cyanides

The methods of preparation of these poisonous compounds are as follows:

(*a*) *Ionic cyanides* of the alkali metals, viz. Na^+CN^-, K^+CN^-, are prepared by the reduction of:

(i) sodamide with red hot carbon

$$NaNH_2 + C \rightarrow NaCN + H_2$$

(ii) potassium carbonate with carbon and ammonia

$$K_2CO_3 + C + 2NH_3 \rightarrow 2KCN + 3H_2O$$

(*b*) *Covalent cyanides* are generally only sparingly soluble or insoluble in water and can be prepared by:

(i) double decomposition

$$Ag^+ + CN^- \rightarrow AgCN\downarrow \text{ (white)}$$
$$Ni^{2+} + 2CN^- \rightarrow Ni(CN)_2\downarrow \text{ (green)}$$

(ii) dissolution of the oxide in hydrocyanic acid followed by crystallization

$$HgO + 2HCN \rightarrow Hg(CN)_2 + H_2O$$

Cyanide structures

Ionic cyanides. If the cyanide ion is regarded as a charged sphere with a radius of 0·192 nm, this is intermediate between the radius of the chloride ion (0·181 nm) and that of the bromide ion (0·195 nm). Thus the structures of the alkali metal cyanides are similar to those of the halides: sodium, potassium, and rubidium cyanides have a rock-salt structure and caesium cyanide has a caesium chloride structure. Aqueous solutions of ionic cyanides are strongly alkaline due to hydrolysis.

$$CN^- + H_2O \rightleftharpoons HCN + OH^-$$

Covalent cyanides. The structures may be of four types:

(*a*) Discrete linear covalent molecules as in $Hg(CN)_2$:

$$N{\equiv}C{-}Hg{-}C{\equiv}N$$

(*b*) Infinite chains as in silver and gold cyanides:

$$-Ag-C{\equiv}N-Ag-C{\equiv}N-$$

(*c*) Layer or planar arrangements as in nickel and platinum cyanides:

```
  |          |
—Ni—C≡N—Ni—C≡N—Ni—
  |          |          |
  N          N          N
  |||        |||        |||
  C          C          C
  |          |          |
—Ni—C≡N—Ni—C≡N—Ni—
  |          |          |
```

(*d*) Three-dimensional networks as in zinc and cadmium cyanides where the cross-linking of the chains takes place in more than one plane.

Complex cyanides

The cyanide ion is an extremely efficient ligand and many complexes have been prepared (p. 102) by dissolution of the appropriate covalent cyanide in excess cyanide solution. Examples are:

(*a*) linear $Ag^{I}(CN)_2^-$
(*b*) square planar $Ni^{II}(CN)_4^{2-}$, $Pt^{II}(CN)_4^{2-}$
(*c*) tetrahedral $Zn^{II}(CN)_4^{2-}$, $Cd^{II}(CN)_4^{2-}$
(*d*) octahedral $Fe^{II}(CN)_6^{4-}$, $Fe^{III}(CN)_6^{3-}$
(Pure iron(II) and iron(III) cyanides do not appear to exist.)
(*e*) dodecahedral $W^{IV}(CN)_8^{4-}$, $Mo^{III}(CN)_8^{5-}$.

The Isocyanates and Thiocyanates

The fusion of litharge with an alkali metal cyanide produces the corresponding alkali metal isocyanate which may be extracted with water.

$$KCN + PbO \rightarrow KNCO + Pb$$

Acidification produces isocyanic acid, H—N=C=O, which is rapidly hydrolysed.

$$HNCO + H_2O \rightarrow NH_3 + CO_2$$

The isocyanates contain the linear ion $(N{\equiv}C{-}O)^-$.

The thiocyanates are prepared by heating the cyanides with sulphur.

$$KCN + S \rightarrow KNCS$$

The gaseous acid is produced by heating potassium thiocyanate with potassium bisulphate.

$$KNCS + KHSO_4 \rightarrow K_2SO_4 + HNCS$$

Solutions are strongly acidic and contain the linear anion $(N{\equiv}C{-}S)^-$. Silver salts give a white precipitate of silver thiocyanate with potassium thiocyanate solution and iron(III) salts give a blood-red coloration containing all the complexes from $FeNCS^{2+}$ to $Fe(NCS)_6^{3-}$. The latter reaction is used as a qualitative test for the presence of iron(III) ions and thiocyanate ions.

Silicon, Germanium, Tin, and Lead

Occurrence

The earth's crust contains a high percentage of silica and silicate minerals. Impure forms of silica include sand and sandstone and important silicates are: feldspar, $KAlSi_3O_8$ (found in igneous rocks); clay or kaolin, $Al_2Si_2O_5(OH)_4$; and asbestos, $Ca_2Mg_3Si_2O_9$.

Germanium occurs in bituminous coals and as complex sulphide ores, e.g. germanite (contains Cu, Fe, Zn, Ga, Ge). Tin occurs almost entirely as tinstone or cassiterite, SnO_2, and lead as galena, PbS.

Extraction of Elements

Silicon

(*a*) An amorphous brown powder of elementary silicon may be obtained by: (i) heating silicon tetrafluoride with sodium or potassium,

$$SiF_4 + 4K \rightarrow Si + 4KF$$

or (ii) heating silica with aluminium powder and sulphur (supplies heat).

$$3SiO_2 + 4Al \rightarrow 2Al_2O_3 + 3Si$$

The product is washed with dilute hydrochloric and hydrofluoric acids to remove silica.

(*b*) Crystalline silicon is made by the reduction of silica with the calculated quantity of carbon in an electric furnace. The product is a dark lustrous mass of metallic appearance.

$$SiO_2 + 2C \rightarrow Si + 2CO$$

Germanium

Coke from bituminous coals contains germanium and gallium oxides and sulphides. When the coke is used to make producer gas these oxides and sulphides, together with many other metallic impurities, are deposited in the flue systems. The purification of the flue dust to recover germanium and gallium as the dioxide is complex but, once this is done, the dioxides can be reduced to the metal with hydrogen or carbon at red heat. Extraction from the sulphides is also a complex process involving volatilization of germanium as $GeCl_4$, followed by hydrolysis to the dioxide and reduction to the metal with carbon or hydrogen.

Tin

The powdered ore is washed, concentrated magnetically, and then roasted to remove arsenic and sulphur as their volatile oxides. Reduction with anthracite or with coke is then carried out at 1300–1400 °C in a reverberatory furnace.

$$SnO_2 + 2C \rightarrow Sn + 2CO$$

Molten tin sinks to the bottom of the furnace and is drawn off and cast into blocks.

Lead

The sulphide ore is contaminated with variable quantities of zinc sulphide and the ore is first concentrated by oil flotation. After roasting and sintering on a Dwight-Lloyd sintering machine (p. 125):

$$2PbS + 3O_2 \rightarrow 2PbO + 2SO_2$$

the sintered ore is transferred to a blast furnace and reduced with coke, scrap iron, and limestone as a flux. Crude lead is tapped off from the furnace.

$$PbO + C \rightarrow Pb + CO$$

$$PbO + CO \rightarrow Pb + CO_2$$

$$PbS + Fe \rightarrow Pb + FeS$$

Purification

Both silicon and germanium are purified by zone refining (p. 127). The increasing importance of silicon and germanium in semiconductor rectifiers and transistors has led to the demand for large, pure, single crystals of the elements. These are prepared by the very slow withdrawal of a small seed crystal of silicon or germanium from a prepared elemental melt. Silicon or germanium freeze on to the seed crystal and take on the same crystalline shape.

Crude tin is purified by 'liquation' (p. 127), although smaller quantities may be zone refined.

The melting and stirring of crude lead removes arsenic, antimony, bismuth, and copper as an oxide scum. Further refining may be carried out for the removal of silver and gold by Parkes's process (p. 422). Electrolytic purification may be carried out for both tin and lead from an electrolytic bath of hexafluosilicic acid (H_2SiF_6) using impure metal anodes and pure metal cathodes.

Considerable quantities of tin are recovered from scrap tin plate and cans by treatment with caustic soda to give a solution of stannite which is electrolytically reduced to the metal.

Uses

Apart from the uses mentioned above, silicon imparts hardness and strength to bronze and renders iron acid-resistant. Tin forms many useful alloys, e.g. bronze (Cu–Sn), solder, type metal, and bearing metal (Sn–Pb plus a little Sb). The tinning of iron and copper surfaces increases the resistance to corrosion. Lead is used in the foundation of buildings to reduce vibration and is also used in X-ray shields, piping, roofing, cable sheaths, etc.

Allotropy

The amorphous and crystalline varieties of silicon are not true allotropic forms. The amorphous form actually consists of minute octahedral crystals which have the diamond structure, each silicon atom being covalently bonded to four other silicon atoms at the corners of a regular tetrahedron. 'Crystalline' silicon is a polycrystalline mass with the same structure. No graphitic silicon has ever been prepared.

Germanium also has the diamond lattice but tin has three allotropic modifications with the following transition temperatures:

$$\text{grey tin } (\alpha) \underset{}{\overset{13\,^\circ\text{C}}{\rightleftharpoons}} \text{white tin } (\beta) \underset{}{\overset{161\,^\circ\text{C}}{\rightleftharpoons}} \text{rhombic tin } (\gamma)$$

Grey tin appears amorphous but, in fact, consists of minute octahedral crystals with the diamond lattice; β- and γ-tin are metallic but both have a distorted packing of the metallic atoms; β-tin (normal metallic tin) is malleable and γ-tin is more brittle. At very low temperatures, white tin (β) becomes slowly converted into grey tin and once the white tin has been 'inoculated' with a small quantity of the grey tin the change rapidly accelerates and the phenomenon is known as tin plague.

Lead has only one metallic form which is cubic close packed.

Chemical Properties

A summary of the principal reactions of silicon, germanium, tin, and lead is shown in Table 14.5. The reactions quoted for silicon refer to the amorphous form, since the crystalline variety is comparatively inert chemically.

TABLE 14.5

Reagent	*Reaction*	*Remarks*
Dilute HCl or H_2SO_4	$M + 2H^+ \rightarrow M^{2+} + H_2$	No reaction with Si, Ge. Very slow with Sn, Pb.
Conc. HNO_3	$3M + 4H^+ + 4NO_3^- \rightarrow 3MO_2 + 4NO + 2H_2O$ $3Pb + 8H^+ + 2NO_3^- \rightarrow 3Pb^{2+} + 4H_2O + 2NO$	Si is insoluble. Ge and Sn give hydrated oxide. Lead gives nitrate.
Alkali (conc. or molten)	$M + 2OH^- + 2H_2O \rightarrow M(OH)_4^{2-} + H_2$	Si and Ge give silicates and germanates Na_4XO_4. Sn and Pb give stannites and plumbites slowly.
Oxygen or air, heated	$M + O_2 \rightarrow MO_2$	Pb gives PbO and some Pb_3O_4.
Sulphur	$M + 2S \rightarrow MS_2$	Lead forms PbS.
Chlorine	$M + 2Cl_2 \rightarrow MCl_4$	Lead forms $PbCl_2$. Ge ignites if finely divided.
Water	$2Pb + 2H_2O + O_2 \xrightarrow{\text{air}} 2Pb(OH)_2$ $\left.\begin{matrix}Si\\Sn\end{matrix}\right\} + 2H_2O \xrightarrow[\text{steam}]{} \left.\begin{matrix}Si\\Sn\end{matrix}\right\}O_2 + 2H_2$	Slowly for Pb.

The Hydrides

A number of hydrides of germanium and silicon have been prepared (Si_nH_{2n+2}, Ge_nH_{2n+2}, where $n = 1$ to 6) but only SnH_4, Sn_2H_6 and PbH_4 are known for tin and lead. The methods available for their preparation include:

(*a*) The action of hydrochloric acid on magnesium silicide or germanide (Mg_2Si or Mg_2Ge) or on Mg–Sn, Mg–Pb alloys in an atmosphere of hydrogen. In the case of silicon and germanium, the gases are cooled in liquid nitrogen and fractionated to give individual hydrides.

(*b*) The action of lithium aluminium hydride on the appropriate halide in ether, e.g.

$$GeCl_4 + LiAlH_4 \rightarrow \underset{\text{mono-germane}}{GeH_4} + LiAlCl_4$$

$$2Si_2Cl_6 + 3LiAlH_4 \rightarrow \underset{\text{disilane}}{2Si_2H_6} + 3LiAlCl_4$$

In the case of silicon and germanium, the stability of the hydrides decreases with increasing relative molecular mass, e.g. SiH_4 decomposes at 400 °C and Si_3H_8 decomposes spontaneously at room temperature. Considering the tetrahydrides MH_4 of germanium, silicon, tin, and lead, the stability decreases and the volatility decreases with increasing atomic number, e.g. PbH_4 decomposes into the elements at room temperature.

The structures of the hydrides mentioned are similar to the structures of the paraffins. There are, however, no unsaturated hydrides corresponding to the olefines or acetylenes and no isomers have been separated for compounds above the third member of the series.

The Halides

The tetrahalides are all known apart from $PbBr_4$ and PbI_4, which would not be expected to be stable due to the strong oxidizing power of Pb^{4+} and the reducing power of Br^- and I^-. The tetrahalides may be prepared directly by heating the element in a stream of halogen gas. Sometimes the dioxides are a more convenient source for their preparation, e.g. silicon and germanium tetrafluorides can be made by heating the dioxides with calcium fluoride and concentrated sulphuric acid.

$$CaF_2 + H_2SO_4 \rightarrow CaSO_4 + 2HF$$

$$SiO_2 + 4HF \rightarrow SiF_4 + 2H_2O$$

Lead tetrachloride is prepared by dissolving lead dioxide in cold concentrated hydrochloric acid.

$$PbO_2 + 4HCl \rightarrow PbCl_4 + 2H_2O$$

The stability of the tetrahalides decreases from silicon to lead.

$$PbCl_4 \xrightarrow{\text{room temp}} PbCl_2 + Cl_2$$

All the tetrahalides are hydrolysed by water giving dioxides or hydrated dioxides.

$$MCl_4 + 4H_2O \rightarrow MO_2.2H_2O + 4HCl$$

The tetrafluorides also dissolve in excess hydrofluoric acid giving complexes (H_2MF_6) containing octahedral MF_6^{2-} ions. Salts of these complex acids have been made by neutralization, e.g. $BaSiF_6$ (barium hexafluorosilicate(IV)) and K_2SnF_6 (potassium hexafluorostannate(IV)).

Excess of concentrated hydrochloric acid on the tetrachlorides of tin and lead also yields complex anions of the type $SnCl_6^{2-}$, $PbCl_6^{2-}$ and addition of ammonium chloride precipitates salts $(NH_4)_2SnCl_6$ and $(NH_4)_2PbCl_6$. Ammonium hexachlorostannate(IV) is known as 'pink salt' and is used as a mordant in dyeing.

Besides the tetrahalides mentioned there are other halides of silicon such as Si_2X_6, Si_3X_8, Si_4X_{10}, which are formed during the preparation of the tetrahalides and may be separated by fractional distillation. These compounds contain chains of quadrivalent silicon atoms, i.e.

```
    Cl  Cl  Cl
    |   |   |
Cl—Si—Si—Si—Cl
    |   |   |
    Cl  Cl  Cl
```

Compounds analogous to chloroform, $SiHCl_3$ and $GeHCl_3$, have been made by the action of hydrogen chloride gas on heated silicon or germanium.

The dihalides of germanium, tin, and lead may be obtained conveniently by the methods given below. No dihalides of silicon are known.

(*a*) $\underset{\text{vapour}}{GeX_4} + Ge \xrightarrow{400\,^{\circ}C} 2GeX_2$

(*b*) $Sn + \underset{\text{gas}}{2HX} \rightarrow SnX_2 + H_2$

(*c*) $Pb^{2+} + 2X^- \rightarrow PbX_2\downarrow$ ($X = F, Cl, Br, I$)

All the chlorides form complexes with excess concentrated hydrochloric acid, $GeCl_3^-$, $SnCl_4^{2-}$, $PbCl_4^{2-}$, for which complex salts have been isolated, e.g. $CsGeCl_3$, $K_2SnCl_4.2H_2O$, K_2PbCl_4.

Germanium and tin dihalides are hydrolysed by water, unless strong hydrochloric acid is present. Germanium gives the unstable hydroxide $Ge(OH)_2$ and tin gives the oxochloride.

$$4SnCl_2 + 6H_2O \rightarrow \underset{\substack{\text{tin(II)} \\ \text{oxochloride (polymeric)}}}{Sn_4(OH)_6Cl_2} + 6HCl$$

The dihalides are much less volatile than the tetrahalides and become increasingly powerful as reducing agents on passing from lead to germanium. The tin(II) chloride molecule is angular in the vapour state with a bond angle of 95°:

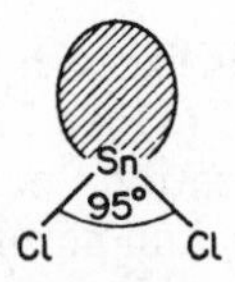

FIGURE 14.9

sp^2 hybridization of the tin atom occurs, one of the hybrid orbitals having a lone pair of electrons. Repulsion between lone pair and bond pair accounts for the contraction in bond angle from 120° to 95° (Figure 14.9). Germanium dichloride probably has a bridged structure in the solid.

Ge—Cl→Ge—Cl→Ge—Cl→Ge (each Ge bearing a terminal Cl)

Lead(II) chloride is mainly ionic but the packing of the ions is rather complex.

The Dioxides

Silica

Silica, SiO_2, occurs in an amorphous form as flint and opal and may be prepared by:

(*a*) heating hydrated silica gel, $SiO_2.2H_2O$, from the hydrolysis of $SiCl_4$, or

(*b*) adding hydrochloric acid to sodium silicate solution and evaporating on a water bath.

$$Na_4SiO_4 + 4HCl \rightarrow 4NaCl + SiO_2.2H_2O$$

Silica exists in three crystalline forms which are stable in the following temperature ranges:

$$\text{quartz} \xrightarrow{870°} \text{tridymite} \xrightarrow{1,470°} \text{crystobalite} \xrightarrow[\text{m.p.}]{1,710°} \text{silica-quartz glass}$$

All three forms are built up from SiO_4 tetrahedra so that each oxygen atom is common to two tetrahedra forming a giant macromolecule. In crystobalite the silicon atoms have the diamond structure with oxygen atoms at the mid-points of each Si—Si bond. The arrangement of the linked tetrahedra is rather different for quartz and tridymite. All the varieties of silica begin to soften below 1600 °C and melt at 1710 °C to give a viscous liquid which, on cooling, does not crystallize but supercools, forming silica or quartz glass. The tetrahedral SiO_4 groups are still present in this glass but, whereas in the crystal the SiO_4 units are built up in a regular manner, in the glass the arrangement is completely random (Figure 14.10).

Silica glass possesses a small coefficient of expansion and a high refractive index. It is transparent to ultraviolet and infrared radiation and does not crack if suddenly heated or cooled. Apart from hydrogen fluoride and fused alkali, the glass is extremely inert toward chemical reagents and hence is used in optical and special laboratory glass apparatus. With hydrogen fluoride, silicon tetrafluoride is produced.

$$SiO_2 + 4HF \rightarrow SiF_4 + 2H_2O$$

Excess fused alkali converts silica into sodium silicate, a solution of which is known as water glass.

$$4NaOH + SiO_2 \rightarrow Na_4SiO_4 + 2H_2O$$

The reaction is probably not as simple as outlined by the equation and the solution may contain a mixture of anions, SiO_3^{2-}, $(SiO_3)_n^{2n-}$, SiO_4^{4-}, and $Si_2O_5^{2-}$, etc.

Other dioxides of the group

White dioxides of germanium and tin are prepared by evaporation of the metals with concentrated nitric acid. The hydrated oxides first deposited

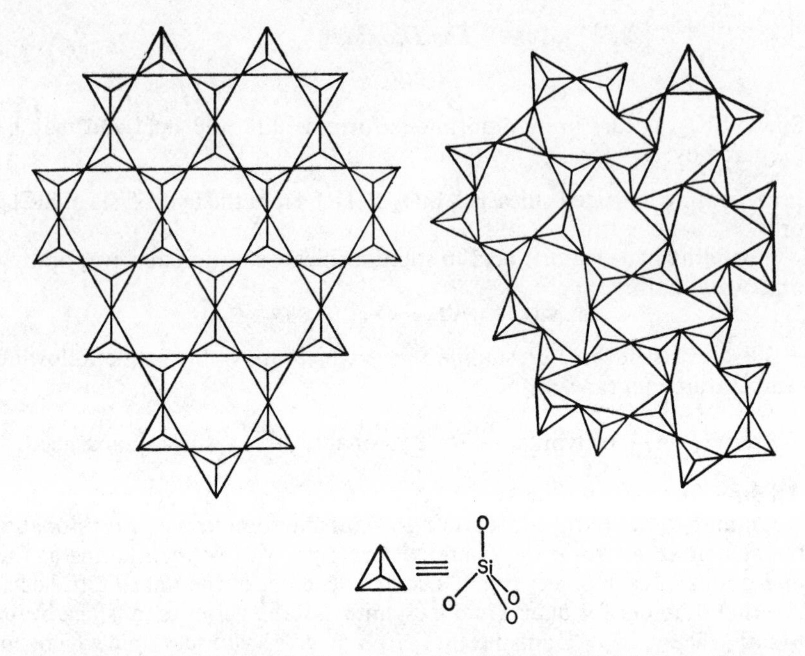

FIGURE 14.10. Two-dimensional representation of crystobalite and quartz glass

yield the anhydrous oxide on ignition. Both oxides are predominantly acidic, yielding germanates (GeO_4^{4-}) and stannates ($Sn(OH)_6^{2-}$) with molten or concentrated alkali.

$$SnO_2 + 2OH^- + 2H_2O \rightarrow Sn(OH)_6^{2-}$$

Brown PbO_2 is somewhat anomalous. Evaporation of lead with concentrated nitric acid produces lead nitrate, which on ignition produces lead monoxide (PbO). The preparation of the dioxide involves the oxidation of the monoxide in alkali (plumbite solution) with hypochlorite.

$$PbO + OCl^- \rightarrow PbO_2 + Cl^-$$

Although molten alkali dissolves the dioxide forming plumbates $Pb(OH)_6^{2-}$, it is also soluble in cold concentrated hydrochloric acid giving lead tetrachloride which, on warming, liberates chlorine and forms $Pb^{II}Cl_2$.

$$PbO_2 + 4HCl \rightarrow PbCl_4 + 2H_2O$$

The dioxide also evolves oxygen on heating.

$$2PbO_2 \xrightarrow{300\,°C} 2PbO + O_2$$

All the dioxides possess a rutile lattice, although germanium dioxide exhibits a quartz structure at 1033 °C.

The Monoxides

Silicon monoxide is obtained as a brown glassy sublimate by heating silicon and silica to 1250 °C *in vacuo*. It is the only divalent silicon compound which has been isolated. Germanium, tin, and lead form monoxides MO, which are made by heating the nitrate or hydroxide, although PbO may be made by blowing air over molten lead. GeO is black, SnO is olive green, and PbO exists in two polymorphic forms, a yellow rhombic form (massicot) which, on fusion, crystallizes as red tetragonal litharge, identical in structure with tetragonal SnO (Figure 14.11).

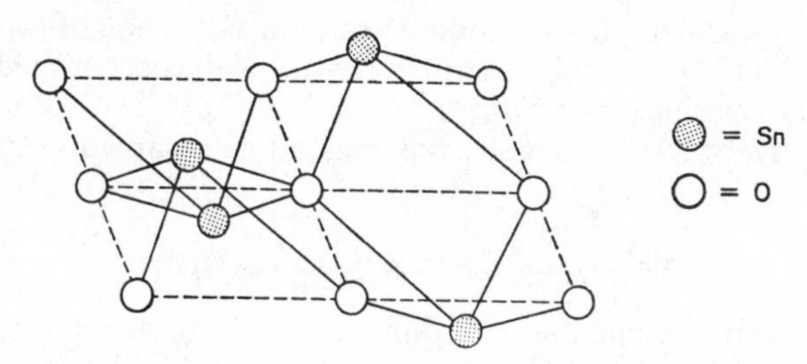

FIGURE 14.11. Part of the layer structure of SnO or PbO: metal–metal bonds link the layers together

All the oxides are amphoteric, dissolving in acids to give salts and in alkalis giving solutions of germanites ($Ge(OH)_3^-$ or $Ge(OH)_4^{2-}$), stannites ($Sn(OH)_3^-$ or $Sn(OH)_4^{2-}$), and plumbites ($Pb(OH)_3^-$ or $Pb(OH)_4^{2-}$).

$$SnO + 2OH^- + H_2O \rightarrow Sn(OH)_4^{2-}$$

Solutions of stannites are powerful reducing agents.

$$Sn(OH)_4^{2-} + 2OH^- \rightarrow Sn(OH)_6^{2-} + 2e \qquad (E^\circ = -0{\cdot}91\ V)$$

Red lead

Red lead, Pb_3O_4, is a red powder prepared by heating litharge in air to 400 °C.

$$6PbO + O_2 \underset{470\,°C}{\overset{400\,°C}{\rightleftharpoons}} 2Pb_3O_4$$

It is often classified as a compound oxide containing lead in the divalent and tetravalent state. With hot dilute nitric acid it behaves as such forming $Pb^{II}(NO_3)_2$ and depositing brown $Pb^{IV}O_2$.

$$Pb_3O_4 + 4HNO_3 \rightarrow 2Pb(NO_3)_2 + 2H_2O + PbO_2$$

The structure consists of $Pb^{IV}O_6$ octahedra attached together in chains which are linked together by pyramidal $Pb^{II}O_3$ groups.

Silica Gel, Silicic and Stannic Acids

The addition of concentrated hydrochloric acid to a concentrated solution of a silicate produces a gel of hydrated silica (57 per cent water) called silica gel. This gel has an open porous structure with a large surface area and is used to adsorb gases and as a catalyst or catalyst support for reactions in the gaseous phase. It is also used for absorbing moisture, e.g. in desiccators and optical instruments. The gel was sometimes called orthosilicic acid (*see* note on nomenclature, p. 271).

$$SiO_4^{4-} + 4H^+ \rightarrow H_4SiO_4 \qquad (SiO_2.2H_2O)$$

Fusion of silica with alkalis produces sodium salts containing the anions quoted. Acidification of solutions of these salts with concentrated sulphuric acid is said to produce definite acids such as H_2SiO_3, $H_2Si_2O_5$, etc., but this is not conclusively proven. These acids may all be regarded as dehydration products of H_4SiO_4, e.g.

$$2H_4SiO_4 \xrightarrow{-2H_2O} 2H_2SiO_3 \xrightarrow{-H_2O} H_2Si_2O_5$$

Acidification of stannates with dilute acid also produces α-stannic acid which, when dried at 100°C, has the composition H_2SnO_3 or $SnO_2.H_2O$. β-stannic acid, prepared by the action of concentrated nitric acid on tin, when dried *in vacuo* has the formula $(SnO_2.H_2O)_5$ and appears to be polymerized α-stannic acid. Salts of α-stannic acid are the stannates, e.g. $Na_2Sn(OH)_6$ or $Na_2SnO_3.3H_2O$, the water of which cannot be driven off without decomposing the salt. Salts of β-meta-stannic acid, e.g. $Na_2Sn_5O_{11}.4H_2O$, have also been isolated as white powders.

The Silicates

Although a great variety of silicates is known, the fundamental unit is SiO_4 in which the silicon atom lies at the centre of a tetrahedron with oxygen atoms at the corners. These tetrahedra may exist as discrete units or they may share corners, edges or, in some cases, faces in a variety of ways:

Simple orthosilicates and pyrosilicates

The negative charges on the anions are balanced by those on cations which pack into the vacant spaces between the SiO_4 tetraheda and bind them together (Figure 14.12(*a*) and (*b*)).

Cyclic silicate ions

The rings are arranged in sheets held together by the cations (Figure 14.12(*c*) and (*d*)).

Infinite chain ions

These may contain single chains of composition $(SiO_3^{2-})_n$, the pyroxenes (Figure 14.12(*e*)), or double chains or bands of composition $(Si_4O_{11}^{6-})_n$, the amphiboles (Figure 14.12(*f*)). The chains or bands are held together by the cations which interleave them but, as these forces are weaker than the SiO bonds in the chains, these compounds are of a fibrous nature.

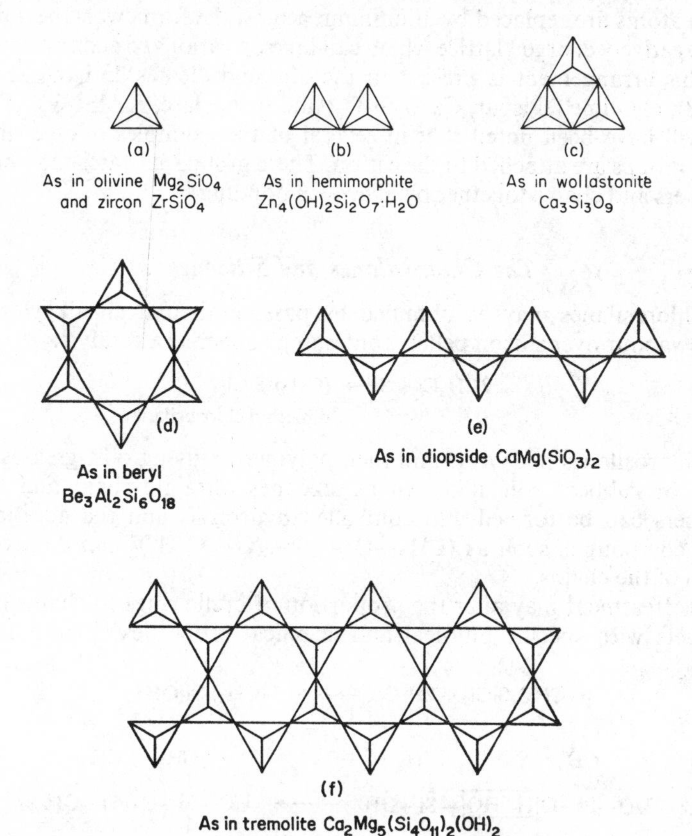

FIGURE 14.12(*a*) SiO_4^{4-} orthosilicate ion; (*b*) $Si_2O_7^{6-}$ pyrosilicate ion; (*c*) $Si_3O_9^{6-}$ ion; (*d*) $Si_6O_{18}^{12-}$ ion; (*e*) $(SiO_3^{2-})_n$ ion; (*f*) $(Si_4O_{11}^{6-})_n$ ion

Infinite sheets

These are formed by the extension of the double chains. The sheets of composition $(Si_2O_5^{2-})_n$ are bound together by the cations and the minerals again tend to cleave, e.g. talc $Mg_3(OH)_2Si_4O_{10}$ and chrysotile $Mg_3(OH)_4Si_2O_5$ (90 per cent chrysotile plus 10 per cent asbestos is commercial asbestos). Replacement of one-quarter of the silicon atoms in talc by potassium and aluminium atoms gives rise to negatively charged layers interleaved with potassium ions characteristic of the micas, $KAl_2(OH)_2Si_3AlO_{10}$.

Three-dimensional lattices

This type of lattice has already been seen with silica. If, however, some silicon atoms are replaced by aluminium atoms, the framework becomes a vast negatively charged lattice which can take up cations to acquire neutrality. This arrangement is present in the aluminosilicates and the zeolites (p. 288), e.g. lime feldspar, $CaAl_2Si_2O_8$, and orthoclase, $KAlSi_3O_8$.

It will have been noted that in several of the examples quoted above, —OH groups are attached to the cation. These groups also assist in holding the layers and chains together by hydrogen bonding.

The Chlorosilanes and Silicones

The chlorosilanes may be obtained by passing air and an alkyl or aryl halide vapour over silicon pellets containing copper as a catalyst.

$$2CH_3Cl + Si \rightarrow (CH_3)_2SiCl_2$$

dimethyl chlorosilane

The chlorosilanes hydrolyse and then polymerize giving oils, greases, and resins or rubbery solids known as silicones. Straight-chain and cyclic polymers can be formed and controlled hydrolysis and the addition of other compounds such as $(CH_3—O)_2—Si—(O - C_2H_5)_2$ can control the length of the chains.

This treatment may alter the proportion of cyclic rings to chains giving products with specific physical and chemical properties (Figure 14.13).

$$(CH_3)_2SiCl_2 + 2H_2O \longrightarrow (CH_3)_2Si(OH)_2$$

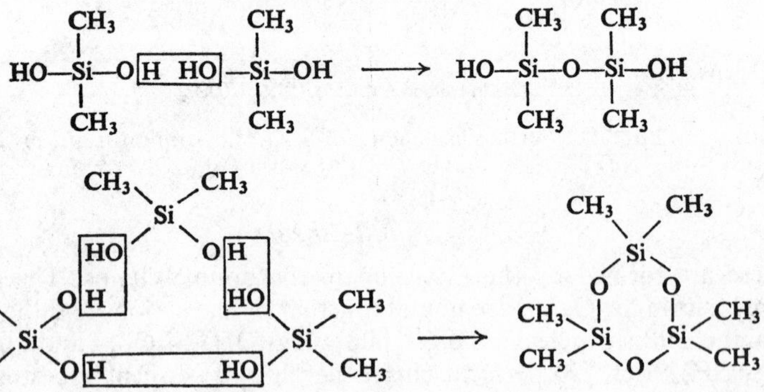

$$HO—Si(CH_3)_2—O\boxed{H\ HO}—Si(CH_3)_2—OH \longrightarrow HO—Si(CH_3)_2—O—Si(CH_3)_2—OH$$

$$3\,(CH_3)_2Si(OH)_2 \longrightarrow [(CH_3)_2SiO]_3 \text{ (cyclic)}$$

FIGURE 14.13

The hydrolysis of certain chlorosilanes (e.g. phenyl trichlorosilane) produces cross-linked silicones (Figure 14.14).

```
   C6H5   C6H5   C6H5   C6H5
    |      |      |      |
 —Si—O—Si—O—Si—O—Si—
    |      |      |      |
    O      O      O      O
    |      |      |      |
 —Si—O—Si—O—Si—O—Si—
    |      |      |      |
   C6H5   C6H5   C6H5   C6H5
```

FIGURE 14.14

The silicones are stable to heat, oxidation, and chemical reagents and are extremely good water repellents. They are used for water-proofing materials, as vacuum oils, lubricants, varnishes, paints, and polishes.

Glasses

As already mentioned in connection with silica (p. 229), glasses are super-cooled liquids and, in fact, very few oxides (e.g. B_2O_3, SiO_2, GeO_2, P_2O_5) form glasses. In the glassy condition there is the same structural unit and environment as in the crystalline state but the structural units are linked together in a random manner.

For the formation of glasses:

(*a*) The element must have a coordination number of three or four with respect to oxygen (boron has coordination number 3 and silicon 4). Larger coordination numbers give more rigid structures which do not permit the random linking of the basic structural units.

(*b*) The sharing of edges or faces of the basic units is not permitted as this again confers too great a rigidity on the system for the formation of a glass. Thus in silica glass, the SiO_4 tetrahedra share corners only.

Commercial soda glass is produced from sand, soda ash, limestone, salt cake, and broken cullet (scrap glass) with a decolorizer of manganese dioxide, arsenious oxide or selenium. This effectively adds basic oxides Na_2O, CaO to the silica. The cations Na^+, Ca^{2+} occupy the interstices in the structure but excess oxygen is also introduced causing gaps between the bonds since a number of oxygen atoms are only attached to one silicon atom (Figure 14.15). Too much distortion of the basic silica structure must not occur, otherwise a glass will not be formed. The rather more open structure of soda glasses also explains the lower melting point range compared with silica glass.

If sodium carbonate is replaced by potassium carbonate and calcium carbonate by red lead, a flint glass is obtained. Pyrex glass is a borosilicate glass containing silica, boric oxide, and aluminium oxide.

In making fibre glass, molten glass is dropped on to a refractory rotating disc from which it flies off in the form of fibres. This is used to make vehicle panels, reinforced plastics, aircraft components, furniture, etc.

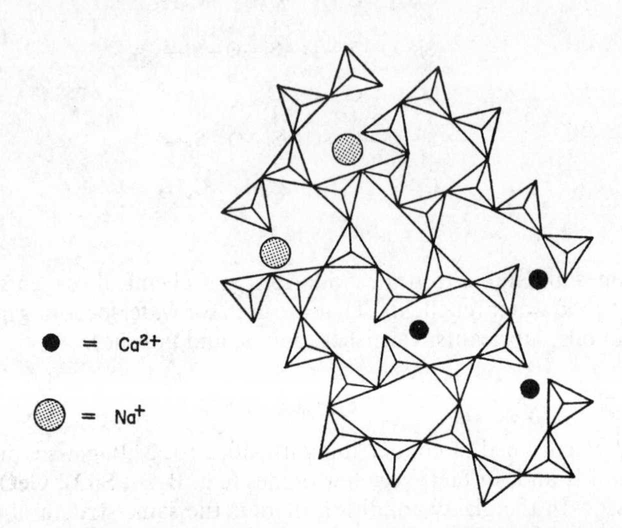

FIGURE 14.15. Glass containing excess oxygen by addition of basic oxides

Spectroscopic studies carried out in the visible region for glasses correlated with energy difference (Δ) calculations for transition metal complexes (p. 351) have revealed the coordination numbers with respect to oxygen of a large number of transition metal ions in glasses. Thus in glasses with a low basic oxide content cobalt(II) oxide gives a pink colour, due to octahedral coordination, which changes to blue (tetrahedral coordination) if the basic oxide content reaches 20 per cent. Iron compounds in sand impart a green colour to glass, due to octahedrally coordinated iron with respect to oxygen. The colour is removed by the addition of manganese(IV) oxide, arsenic(III) oxide, or sodium nitrate, which oxidize the octahedral iron(II) oxygen complex to a tetrahedral iron(III) complex of much paler colour.

This colour is completely neutralized by the addition of small amounts of selenium or cobalt. In cheap amber glasses (beer bottles, etc.) sand containing iron is used. Sulphur and carbon (to prevent oxidation of sulphur to sulphate) are also added to the glass melt during manufacture. The species responsible for the amber colour is an iron(III) cation surrounded tetrahedrally by three oxygen atoms and one sulphur atom.

Another type of coloured glass is the copper and gold ruby glass and the cadmium sulphide selenium glass. The latter varies in colour from yellow to red depending upon the selenium content. Here the reducing conditions of the melt are adjusted so that the substances concerned are precipitated in very minute crystals which are responsible for the colour.

X-ray, NMR, and infrared techniques have been used to establish the coordination numbers in borosilicate glasses which contain BO_4 and SiO_4 tetrahedra in low alkali oxide concentrations. If the alkali oxide concentration is high, the boron becomes three coordinate with respect to oxygen. The same effect is observed if aluminium oxide is incorporated into borosilicate glass where the aluminium becomes tetrahedrally coordinated at the expense of the boron which again becomes three coordinate. Alumina glasses, e.g. pyrex, are chemical- and temperature-resistant.

Other developments is glass technology include:

(*a*) Semi-conducting glasses made from phosphorus(V) oxide and vanadium(V) oxide containing V^{5+} and V^{4+} ions, conduction taking place between the ions.

(*b*) Red arsenic(III) sulphide and arsenic(III) telluride glasses which contain no oxygen or silicon and are very soft. These glasses are opaque to visible light but transmit infrared, and are used as prisms and windows in spectrometers.

(*c*) Glasses containing neodymium ions (Nd^{2+}) which are used to make glass lasers.

(*d*) Anti-glare aluminium phosphate glasses containing trapped silver ions which darken when exposed to bright sunlight but become clear when placed in the dark.

$$Ag^+ + e \rightleftharpoons Ag$$

The Sulphides

The disulphides

Of the disulphides, MS_2, lead disulphide is not known. The other disulphides may be made by heating the elements with sulphur. The germanium and tin disulphides are soluble in lithium hydroxide or yellow ammonium sulphide giving thiogermanates, GeS_4^{4-}, and thiostannates, SnS_3^{2-}. Acidification reprecipitates the disulphides. Silicon disulphide is hydrolysed by water.

$$SiS_2 + 2H_2O \rightarrow SiO_2 + 2H_2S$$

Silicon and germanium disulphides consist of long chains of tetrahedra.

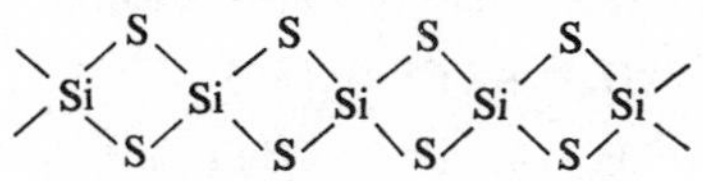

```
 \  /S\    /S\    /S\    /S\    /
  Si    Si     Si     Si     Si
 /  \S/    \S/    \S/    \S/    \
```

Tin disulphide has the cadmium iodide lattice.

The monosulphides

The monosulphides, MS, of tin (brown) and lead (black) may be prepared by passing hydrogen sulphide through acidified solutions of metallic salts. Since germanium(II) salts are unstable in solution, germanium(II) sulphide is prepared by the reduction of germanium(IV) sulphide with hydrogen. Unlike the monosulphides of tin and lead it gives metallic germanium on heating.

Tin monosulphide, unlike lead, is soluble in yellow ammonium sulphide or lithium hydroxide solution producing a solution containing thiostannates SnS_3^{2-}. Reacidification precipitates the tetravalent sulphide.

$$SnS_3^{2-} + 2H^+ \rightarrow SnS_2 + H_2S$$

Oxo salts and Aqueous Chemistry

The tetravalent salts of all four elements are almost unknown. Tin(IV) sulphate, $Sn(SO_4)_2$, has been prepared by the evaporation of tin(IV) oxide in sulphuric acid and a preparation of plumbic sulphate, $Pb(SO_4)_2$, can be carried out electrolytically. Both compounds are rapidly hydrolysed by water.

The tetravalent states are predominantly covalent as seen in the oxides, sulphides, and halides. Silicon does not exhibit a divalent state and has no cationic chemistry. The divalent ions Sn^{2+} and Pb^{2+} exist in a number of crystalline compounds such as SnF_2, $SnSO_4$, $Pb(NO_3)_2$, etc. In aqueous solution, however, the stability of the divalent ion increases from tin to lead. There is no evidence for the existence of a Ge^{2+} ion.

$$3Sn^{2+} + 4H_2O \rightarrow Sn_3(OH)_4^{2+} + 4H^+$$

Hydrolysis is extensive with soluble tin oxosalts. This causes precipitation of basic salts, e.g. $Sn_3(OH)_4(NO_3)_2$. Hydrolysis of lead salts is very slight.

Tin(II) compounds are powerful reducing agents in acid solution.

$$Sn^{2+} \rightarrow Sn^{4+} + 2e \qquad (E^\ominus = +0{\cdot}15\ V)$$

$$Pb^{2+} \rightarrow Pb^{4+} + 2e \qquad (E^\ominus = +1{\cdot}70\ V)$$

They will reduce iron(III) and mercury(II) salts.

$$2Fe^{3+} + Sn^{2+} \rightarrow 2Fe^{2+} + Sn^{4+}$$

$$2Hg^{2+} + Sn^{2+} \rightarrow Hg_2^{2+} + Sn^{4+}$$

$$Hg_2^{2+} + Sn^{2+} \rightarrow 2Hg + Sn^{4+}$$

Oxo salts of germanium(II) are unknown. Tin(II) sulphate and nitrate are prepared by standard methods but are very rapidly hydrolysed to basic salts (*see* above). Sodium carbonate precipitates the basic carbonate $SnCO_3 . SnO$ from solutions of tin(II) salts.

The sulphate and carbonate of lead are insoluble and are, therefore, prepared by double decomposition but again, with sodium carbonate, a basic carbonate $2PbCO_3 . Pb(OH)_2$ (white lead) is precipitated. This is used as a paint pigment and is manufactured by the action of acetic acid and carbon dioxide vapour on lead. The nitrate and acetate $(CH_3COO)_2Pb . 3H_2O$ (sugar of lead) are two important soluble lead salts prepared by dissolving the oxide or carbonate in the respective acid and crystallizing.

15

Group VB

Physical Properties

	Nitrogen N	Phosphorus P	Arsenic As	Antimony Sb	Bismuth Bi
Atomic mass	14·0067	30·9738	74·9216	121·75	208·980
Atomic number	7	15	33	51	83
Isotopes	14, 15	31	75	121, 123	209
Electronic structure	$2s^2 2p^3$	$3s^2 3p^3$	$3d^{10}$ $4s^2 4p^3$	$4d^{10}$ $5s^2 5p^3$	$4f^{14}$ $5d^{10}$ $6s^2 6p^3$
Atomic radius (nm)	0·074	0·110	0·121	0·141	0·152
Ionic radius (nm)	0·171(N^{3-})	0·212(P^{3-})	0·222(As^{3-})	0·245(Sb^{3-})	0·120(Bi^{3+})
Electronegativity	3·07	2·06	2·00	1·82	1·67
Ionization potential (I) (kJ mol^{-1})	1406	1066	972	840	780
Density (g cm^{-3})	0·96 (solid)	1·82 (white)	5·72 (grey)	6·69	9·80
Melting point (°C)	–210	44·1 (white)	817 (grey)	630·5	271
Boiling point (°C)	–195·8	280	633	1,375	1,560
Latent heat of fusion (kJ mol^{-1})	0·36	0·63	27·7	19·8	11·0

General Characteristics

All the atoms have the configuration $ns^2 np^3$ in their outermost orbitals and there is a regular transition in properties from non-metal to metal down the group. Nitrogen may complete its outermost *p* orbital by electron gain as in the ionic nitrides, e.g. $(Li^+)_3N^{3-}$. Phosphorus, arsenic, antimony, and bismuth need considerable energy for electron gain and only phosphorus forms a trivalent anion P^{3-} (phosphide ion) as in $(Na^+)_3P^{3-}$.

In covalent bonding the atoms utilize hybrid orbitals: sp^3 hybridization produces pyramidal molecules with one lone pair directed towards the fourth corner of a tetrahedron. Overlap may occur with a half-filled *s* orbital of hydrogen to give hydrides or with a half-filled *p* orbital of a halogen atom to give halides (Figure 15.1).

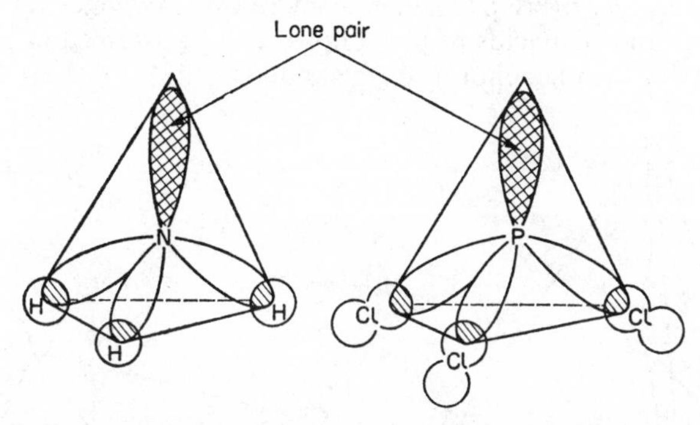

FIGURE 15.1. Tetrahedral structures of Group V hydrides and halides

The lone pair may also be used to form a coordinate link, overlap occurring with the empty *s* orbitals in the hydrogen ion H^+ giving the tetrahedral ammonium ion (NH_4^+) and phosphonium ion (PH_4^+). The lone pair of electrons on the phosphorus atom in the trichlorides may overlap with an empty *p* orbital of oxygen to give diamagnetic oxochlorides (Figure 15.2).

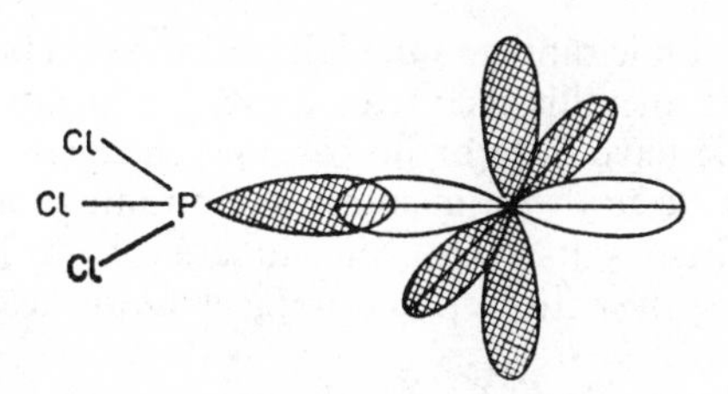

FIGURE 15.2

There is, however, some overlap of the full *p* orbitals of oxygen with empty *d* orbitals of phosphorus giving double-bond character and phosphorus oxochloride is usually written as shown in Figure 15.3(*a*).

d-Orbitals are also used in gaseous phosphorus pentachloride, arsenic pentafluoride, etc., where sp^3d hybridization gives a trigonal bipyramidal structure (Figure 15.3(*b*)). Nitrogen, with no available *d* orbitals, cannot form pentahalides but, like carbon, nitrogen does have a strong tendency to form multiple bonds with oxygen as in the oxides and in the nitrite and nitrate ions. In some oxides and oxo anions sp^2 trigonal hybridization occurs with double bonding due to $p\pi$–$p\pi$ overlap. There is much less tendency to form multiple bonds involving $p\pi$–$p\pi$ overlap with phosphorus since the phosphorus atom is too large for efficient overlap of this kind. As mentioned previously, any multiple bonding between phosphorus and oxygen involves $d\pi$–$p\pi$ overlap. The tendency towards polymerization shown by the oxides and oxo acids of phosphorus can be ascribed in part to the high polarity of the phosphorus–oxygen bond.

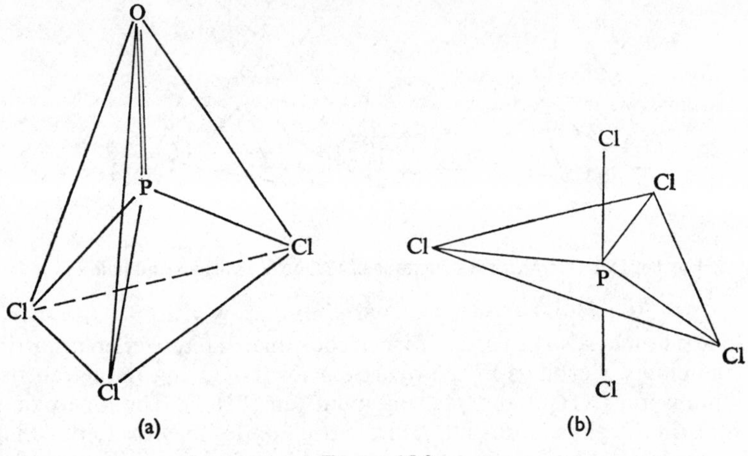

FIGURE 15.3

The increase in atomic radii and the fall in ionization potential correlates with the increase in metallic character down the group. Nitrogen, phosphorus, and arsenic have little or no cationic chemistry but considerable cationic behaviour occurs with bismuth and antimony as in $Bi^{3+}F_3^-$ and $Sb_2^{3+}(SO_4^{2-})_3$. These salts, however, undergo hydrolysis in solution, depositing oxo salts, the cations functioning as Lewis acids.

$$Sb^{3+} + O\begin{matrix}H\\H\end{matrix} \longrightarrow \underset{\text{antimonyl ion}}{SbO^+ + 2H^+}$$

Nitrogen

Occurrence

Nitrogen occurs free in the atmosphere (78 per cent by volume) which may also contain traces of ammonia from the decay of nitrogenous substances or nitric acid, particularly after thunderstorms. It also exists, combined in vast quantities, as sodium and potassium nitrates. The tissues of all living organisms contain combined nitrogen in the form of protein.

Preparation

(*a*) By the action of heat on certain ammonium oxo salts, e.g. ammonium nitrite and ammonium dichromate.

$$NH_4NO_2 \rightarrow N_2 + 2H_2O$$

$$(NH_4)_2Cr_2O_7 \rightarrow N_2 + 4H_2O + Cr_2O_3$$

Since ammonium nitrite is unstable at ordinary temperatures, ammonium chloride and sodium nitrite are intimately mixed, water is added, and the resulting solution of ammonium nitrite is heated.

(*b*) The industrial method involves the fractional distillation of liquid air, which gives nitrogen gas containing argon and a small amount of oxygen (p. 281).

Properties

The gas is colourless, odourless, and tasteless with a boiling point of −196 °C. Nitrogen gas exists as diatomic molecules with triple bonds (one

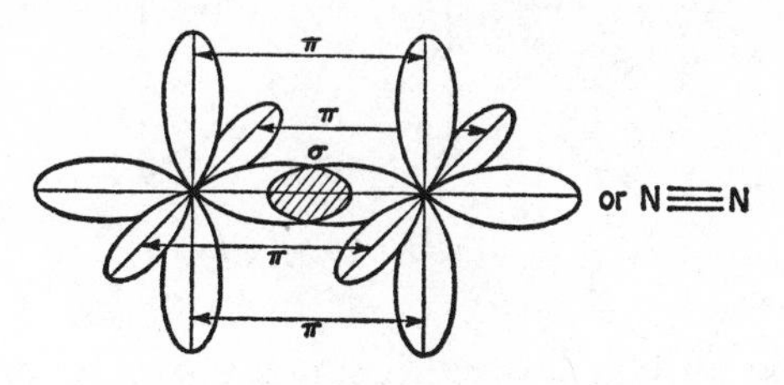

FIGURE 15.4

σ- and two π-bonds) between the atoms (Figure 15.4). The high chemical stability of nitrogen compared to a molecule such as ethyne (CH≡CH) can be mainly ascribed to the difference in bond enthalpies between the triple- and single-bonded systems:

$N{\equiv}N$	944 kJ mol^{-1}	$>N{-}N<$	163 kJ mol^{-1}
$-C{\equiv}C-$	836 kJ mol^{-1}	$\gtrless C{-}C \lessgtr$	348 kJ mol^{-1}

Thus the energy requirement for the change from a triple-bonded to a single-bonded system is much greater for nitrogen, accounting for the greater chemical stability.

When subjected to a continuous electrical discharge, nitrogen combines with oxygen above 500 °C to give nitric oxide. It also combines with a number of metals to give nitrides, which may be prepared by heating the metal in nitrogen or ammonia or heating the metallic amides, e.g.

$$3Ba(NH_2)_2 \rightarrow Ba_3N_2 + 4NH_3$$

Nitrides in general may be classified into three types:

(*a*) Ionic nitrides of lithium and group IIA metals. These all contain the nitride ion N^{3-} which has an ionic radius of 0·171 nm, e.g.

$$\underset{\text{red}}{Li^+{}_3N^{3-}}, \quad \underset{\text{colourless}}{Mg^{2+}{}_3N^{3-}{}_2, Ca^{2+}{}_3N^{3-}{}_2}$$

All these nitrides have disordered and complex modes of ionic packing and, with water, readily yield ammonia.

$$N^{3-} + 3H_2O \rightarrow 3OH^- + NH_3$$

(*b*) Covalent nitrides of the non-metals. Some examples are BN, S_4N_4, P_3N_3, Si_3N_4, C_2N_2. The methods of preparation are very specific and

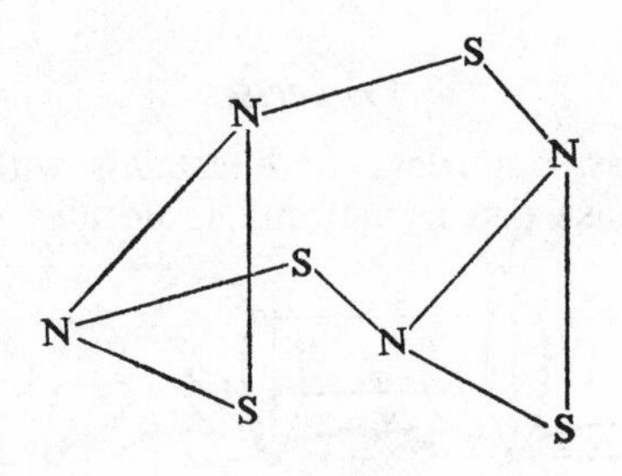

FIGURE 15.5

structurally these nitrides have either simple molecular or giant molecular lattices. Thus boron nitride possesses a graphite lattice whereas sulphur nitride exists as discrete puckered rings (Figure 15.5).

(*c*) The nitrides of the transition metals. Here the nitrogen atoms occupy vacant sites between the close-packed metallic atoms. These interstitial nitrides may have simple formulae as in TiN, VN, ZrN, TaN, but many others have formulae which do not correspond to any normal oxidation state of the metal, e.g. Fe_4N, Mn_4N. Many of these nitrides are not exactly stoichiometric, being deficient in nitrogen. Since the metallic lattice is generally not disturbed during the inclusion of the nitrogen atoms, these compounds all have the high melting point, hardness, lustre, and conducting power of the parent metal. The surfaces of many metals are treated with nitrogen to increase the hardness of the metal and its resistance to impact. The process is known as *nitriding*.

Active Nitrogen

When nitrogen is subjected to an electric discharge at low pressure, a yellow glowing gas is obtained consisting of up to 30 per cent of excited nitrogen atoms which undergo slow recombination. This is a very reactive form of nitrogen which readily combines even with the normally unreactive metals.

$$3Hg + 2N \rightarrow Hg_3N_2$$

The Hydrides

Ammonia

Laboratory preparative methods for ammonia include:

(*a*) The action of heat on any ammonium salt with any alkali, slaked lime being the usual cheap alkali employed.

$$NH_4^+ + OH^- \rightarrow NH_3 + H_2O$$

(*b*) The hydrolysis of ionic nitrides (*see* above)

$$N^{3-} + 3H_2O \rightarrow NH_3 + 3OH^-$$

On the industrial scale ammonia is obtained from:

(*a*) The Haber synthesis from nitrogen (obtained from liquid air) and hydrogen (obtained electrolytically).

$$\underset{1\ \text{vol}}{N_2} + \underset{3\ \text{vol}}{3H_2} \rightleftharpoons \underset{2\ \text{vol}}{2NH_3} \qquad (\Delta H_f^{\ominus} = -46{\cdot}2\ \text{kJ mol}^{-1})$$

Using Le Chatelier's principle, the best yield would be obtained at high pressures and low temperatures. However, a low temperature gives a slow reaction rate and, to obtain the maximum yield in the minimum time a temperature of 500 °C is chosen with a catalyst of finely divided iron. This, together with a pressure of 200 times atmospheric pressure, gives a 15 per cent yield. The Claude process employs pressures of up to 1000 times atmospheric pressure, giving yields of up to 40 per cent, but this is offset by the engineering difficulties involved in working at high pressures and the increase in plant cost.

(*b*) The cyanamide process produces ammonia by the following sequence of reactions.

$$Ca^{2+}C_2^{2-} + N_2 \xrightarrow{1100\,^{\circ}C} \underset{\text{calcium cyanamide}}{CaNCN} + C$$

$$Ca^{2+}NCN^{2-} + 3H_2O \longrightarrow CaCO_3 + 2NH_3$$

Calcium cyanamide is the salt of cyanamide $NH_2{-}C{\equiv}N$. The compound is ionic and the cyanamide ion is linear: $Ca^{2+}({}^{-}N{=}C{=}N^{-})$.

Properties of ammonia

It is a colourless, pungent-smelling gas, lighter than air. It is readily liquefied by cooling and compressing (latent heat of vaporization is 23 kJ mol^{-1}). Although, at one time, ammonia was valuable as a refrigerant, it is now being replaced by less toxic and unreactive fluorocarbons (freons, arctons, etc.). The ammonia molecule is pyramidal and highly polar; liquid ammonia shows extensive hydrogen bonding (Figure 15.6).

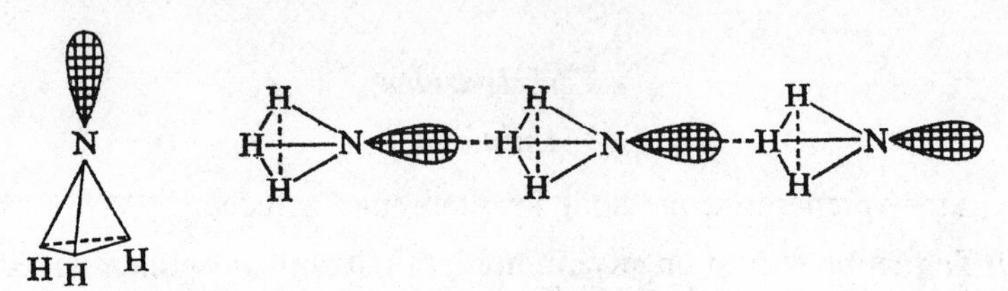

FIGURE 15.6

Ammonia is a base and accepts protons from water, although the equilibrium is well over to the left-hand side.

$$NH_3 + H_2O \rightleftharpoons NH_4^+ + OH^-$$

The so-called ammonium hydroxide (NH_4OH) cannot exist as a discrete molecule as there are no vacant orbitals available for bonding and, in solution, the ammonia molecule is probably hydrogen-bonded to a water molecule (Figure 15.7).

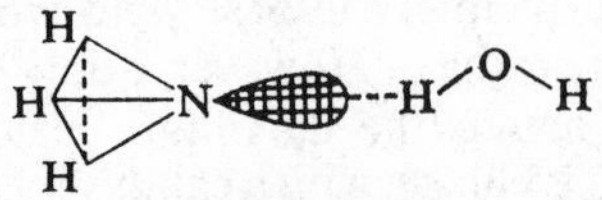

FIGURE 15.7

Supporting evidence for this comes from structural studies on the two crystalline hydrates, $NH_3.H_2O$ and $2NH_3.H_2O$, obtained by cooling ammonia solutions. Both these hydrates have three-dimensional networks containing long chains of hydrogen-bonded water molecules, cross-linked by more hydrogen bonds through ammonia molecules.

Although it is a weak base, ammonia readily accepts protons from good proton donors, e.g. mineral acids.

$$NH_3 + HCl \rightarrow NH_4^+ + Cl^-$$

The ammonia molecule, with the lone pair on the nitrogen atom, functions as an excellent ligand and forms numerous complex ammines (p. 103).

$$Ag^+ + 2NH_3 \longrightarrow [H_3N\rightarrow Ag\leftarrow NH_3]^+ \quad \text{linear}$$

$$Cu^{2+} + 4NH_3 \longrightarrow \left[\begin{array}{ccc} H_3N & & NH_3 \\ & \searrow \; \swarrow & \\ & Cu & \\ & \nearrow \; \nwarrow & \\ H_3N & & NH_3 \end{array}\right]^{2+} \quad \text{square planar}$$

$$Cr^{3+} + 6NH_3 \longrightarrow \left[\begin{array}{ccc} & NH_3 & \\ H_3N & \downarrow & NH_3 \\ & \searrow \; Cr \; \swarrow & \\ & \nearrow \; \uparrow \; \nwarrow & \\ H_3N & & NH_3 \\ & NH_3 & \end{array}\right]^{3+} \quad \text{octahedral}$$

Apart from reducing many heated oxides (CuO, PbO), the ammonia molecule is stable and not easily decomposed. High-temperature electric sparks cause slow decomposition to the elements. In air, ammonia burns with a yellow flame to nitrogen but a platinum catalyst causes oxidation to nitric oxide. This is the basis of the Ostwald process for the manufacture of nitric acid (p. 257).

$$4NH_3 + 3O_2 \longrightarrow 2N_2 + 6H_2O$$

$$4NH_3 + 5O_2 \xrightarrow[600\,°C]{Pt} 4NO + 6H_2O$$

Liquid ammonia

This very closely resembles water in acting as a solvent for many substances, producing solutions which readily act as electrical conductors. Nitrogen, however, is less electronegative than oxygen (nitrogen = 3·0, oxygen = 3·5) and association due to hydrogen bonding is weaker in the case of nitrogen. This, combined with the lower relative permittivity gives a much weaker solvating power. Liquid ammonia and water are both poor conductors of electricity, although slight ionization does occur.

$$2NH_3 \rightleftharpoons NH_4^+ + NH_2^- \qquad (K = 10^{-30} \text{ at } -50\,°C)$$

$$2H_2O \rightleftharpoons H_3O^+ + OH^- \qquad (K = 10^{-14} \text{ at } 25\,°C)$$

Substances which dissolve in ammonia to give NH_2^- ions are bases and those giving NH_4^+ ions are acids.

The equations below illustrate some of the reactions of liquid ammonia and the corresponding aqueous reactions.

(*a*) Sodium, potassium, barium, and calcium dissolve in ammonia and the free metals are recovered by evaporation. Concentrated solutions are bronze in colour and dilute solutions are bright blue, e.g.

$$Na \rightarrow Na^+ + e^-$$

$$Na^+ + nNH_3 \rightarrow Na(NH_3)_n^+$$

$$e^- + mNH_3 \rightarrow e^-(NH_3)_m \text{ (solvated electron)}$$

The solvated electron is responsible for the colour and strong reducing action of such solutions which are used in organic and inorganic reductions. On standing, the solutions liberate hydrogen.

$$e^-(NH_3)_m \rightarrow mNH_2^- + \frac{m}{2} H_2$$

(*b*) Neutralization in liquid ammonia can be carried out using indicators (e.g. phenolphthalein) just as in aqueous media.

In liquid ammonia

$$Na^+NH_2^- + NH_4^+Cl^- \rightarrow Na^+Cl^- + 2NH_3$$

In water

$$Na^+OH^- + H_3O^+Cl^- \rightarrow Na^+Cl^- + 2H_2O$$

(*c*) Double decomposition reactions proceed in a different manner, i.e.

In liquid ammonia

$$Ca^{2+}(NO_3^-)_2 + 2Ag^+I^- \rightarrow Ca^{2+}I^-_2\downarrow + 2Ag^+NO_3^- \quad (1)$$

In water

$$Ca^{2+}I^-_2 + 2Ag^+NO_3^- \rightarrow 2Ag^+I^-\downarrow + Ca^{2+}(NO_3^-)_2 \quad (2)$$

Reaction (1) proceeds because of the preferential ammoniation of the silver ion.

$$Ag^+ + 2NH_3 \rightarrow Ag(NH_3)_2^+$$

Reaction (2) proceeds because of the preferential hydration of the calcium ion.

$$Ca^{2+} + nH_2O \rightarrow Ca(H_2O)_n^{2+}$$

Ammonium salts

The ammonium, potassium, and rubidium ions have the same charge and approximately the same ionic radius:

$$K^+ = 0{\cdot}133 \text{ nm} \qquad NH_4^+ = 0{\cdot}143 \text{ nm} \qquad Rb^+ = 0{\cdot}148 \text{ nm}$$

Thus a close resemblance exists in the solubility of salts, in general chemistry, and in ionic character of the salts. Solutions of ammonium salts are, however, slightly acidic due to hydrolysis.

$$NH_4^+ + H_2O \rightleftharpoons NH_3 + H_3O^+ \qquad (K = 5{\cdot}5 \times 10^{-10} \text{ at } 25\,^\circ\text{C})$$

Thus the ammonium ion is a proton donor and, according to Lowry and Brönsted (p. 145), a cationic acid. The free ammonium radical NH_4 has never been isolated as attempts at reduction of the ammonium ion always produce ammonia and hydrogen.

$$2NH_4^+ + 2e \rightarrow 2NH_3 + H_2$$

Many ammonium salts sublime with dissociation at about 300 °C.

$$NH_4Cl \rightleftharpoons NH_3 + HCl$$

$$NH_4NO_3 \rightleftharpoons NH_3 + HNO_3 \text{ (higher temperatures give } N_2O)$$

$$(NH_4)_2SO_4 \rightleftharpoons 2NH_3 + H_2SO_4$$

Other oxo acid salts decompose on heating.

$$NH_4NO_2 \rightarrow N_2 + 2H_2O$$

$$(NH_4)_2Cr_2O_7 \rightarrow N_2 + 4H_2O + Cr_2O_3$$

Ammonium chloride (sal ammoniac) is prepared by the action of ammonia on hydrochloric acid and is used in dry cells and as a fertilizer. Ammonium nitrate and sulphate are prepared in a similar manner and find uses as an explosive and fertilizer respectively. Ammonium sulphide is prepared by saturating ammonia solution with hydrogen sulphide, giving the bisulphide, and then adding an equal volume of ammonia.

$$NH_4HS + NH_3 \rightarrow (NH_4)_2S$$

Although colourless when freshly prepared, oxidation to yellow polysulphides soon occurs. Yellow ammonium sulphide, which seems to be mainly $(NH_4)_2S_5$, is produced by digesting sulphur with ammonium bisulphide.

Commercial ammonium carbonate (sal volatile) is made from ammonia, carbon dioxide, and steam. The white crystalline product is, however, a double compound of ammonium bicarbonate and carbamate.

$$NH_4^+ \; \bar{O}\text{—}\overset{\overset{\large OH}{|}}{C}\text{=}O \qquad NH_4^+ \; \bar{O}\text{—}\overset{\overset{\large NH_2}{|}}{C}\text{=}O$$

ammonium bicarbonate — ammonium carbamate

In ammonium carbamate, the —OH group of the bicarbonate has been replaced by an —NH_2 group. In water, the commercial carbamate produces the normal salt.

$$NH_4^+ \; \bar{O}\text{—}\overset{\overset{\large NH_2}{|}}{C}\text{=}O + H_2O \rightarrow NH_4^+ \; \bar{O}\text{—}\overset{\overset{\large \bar{O}NH_4^+}{|}}{C}\text{=}O$$

Hydrazine

Hydrazine, N_2H_4, is theoretically derived from ammonia by the replacement of one hydrogen atom by an —NH_2 group. It is prepared commercially and on the laboratory scale by the action of hypochlorite on ammonia solution in the presence of glue or gelatin (Raschig process).

$$NH_3 + Na^+OCl^- \rightarrow \underset{\text{chlor-amine}}{NH_2Cl} + Na^+OH^-$$

$$NH_3 + NH_2Cl \rightarrow N_2H_4 + H^+Cl^-$$

The gelatin catalyses the latter reaction and also sequesters minute traces of cations such as Cu^{2+} and Fe^{2+} which catalyse the reaction:

$$N_2H_4 + 2NH_2Cl \rightarrow 2NH_4Cl + N_2$$

Hydrazine, a colourless liquid, is a very weak base but forms two series of hydrazinium salts, e.g. $N_2H_5^+Cl^-$ and $N_2H_6^{2+}Cl^-{}_2$.

$$N_2H_4 + H_2O \rightleftharpoons N_2H_5^+OH^- \qquad (K = 8{\cdot}5 \times 10^{-7} \text{ at } 25\ ^\circ C)$$

$$N_2H_5^+OH^- + H_2O \rightleftharpoons N_2H_6^{2+}(OH^-)_2 \qquad (K = 8{\cdot}8 \times 10^{-16} \text{ at } 25\ ^\circ C)$$

The latter series of salts are, however, extensively hydrolysed in aqueous solution. Hydrazine is an endothermic compound ($\Delta H_f^\ominus = +50$ kJ mol^{-1}) which decomposes to nitrogen and ammonia on heating, inflames in dry oxygen, and burns in air. It is a very powerful reducing agent in alkaline solution.

$$N_2H_4 + 4OH^- \rightarrow N_2 + 4H_2O + 4e \qquad E^\ominus = -1{\cdot}16 \text{ V}$$

The structure of hydrazine (Figure 15.8) shows some resemblance to hydrogen peroxide. The nitrogen atoms are sp^3 hybridized with one hybrid orbital occupied by a lone pair of electrons.

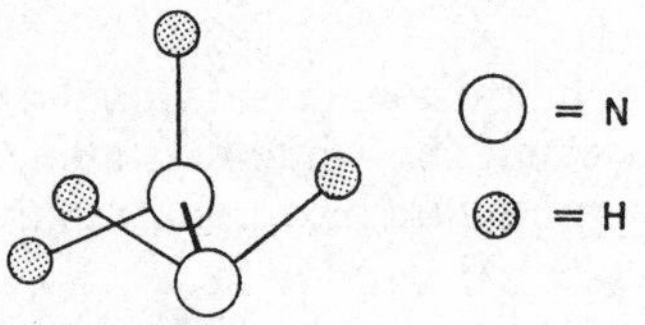

○ = N

● = H

FIGURE 15.8

Hydrazoic acid or azoimide

Hydrazoic acid, HN_3, or azoimide, $N_2.NH$, may be prepared by oxidizing hydrazine with hydrogen peroxide or nitric acid or by heating sodamide in a stream of nitrous oxide.

$$3N_2H_4 + 5H_2O_2 \rightarrow 2HN_3 + 10H_2O$$

$$NaNH_2 + N_2O \rightarrow NaN_3 + H_2O$$

Distillation of the sodium salt with excess dilute sulphuric acid gives a solution of hydrazoic acid. The pure acid is a colourless, poisonous liquid which explodes on heating. The heavy metal salts (AgN_3, PbN_3) are dangerously explosive compounds employed as detonators. The alkali metal azides are not explosive and decompose smoothly on heating. Although the acid is weak, it dissolves many metals.

$$HN_3 + H_2O \rightleftharpoons H_3O^+ + N_3^- \qquad (K = 1{\cdot}8 \times 10^{-5})$$

$$Cu + 3HN_3 \rightarrow Cu(N_3)_2 + N_2 + NH_3$$

The azide ion is linear and is best represented as a resonance hybrid of the following.

$$\overset{+}{N}\equiv\overset{-}{N}-N \leftrightarrow \overset{-}{N}-\overset{+}{N}\equiv N \leftrightarrow \overset{-}{N}=\overset{+}{N}=\overset{-}{N}$$

Hydroxylamine

Hydroxylamine, NH_2OH, is theoretically derived by the replacement of one of the hydrogen atoms in ammonia by an —OH group. The compound may be prepared by the electrolytic reduction of ice-cold 50 per cent nitric acid using lead electrodes or by the reduction of nitrite solution in the cold (–2 °C) by sulphur dioxide.

$$NO_3^- + 7H^+ + 6e \rightarrow NH_2OH + 2H_2O$$

$$HNO_2 + 2HSO_3^- + H_2O \rightarrow NH_2OH + 2HSO_4^-$$

Pure hydroxylamine forms colourless deliquescent crystals which decompose above 15 °C to give nitrogen, ammonia, and nitrous oxide. The compound is a weaker base than ammonia, but readily forms hydroxylaminium salts which are soluble stable solids.

$$NH_2OH + H_2O \rightleftharpoons \underset{\text{hydroxylaminium cation}}{NH_3OH^+} + OH^- \quad (K = 6{\cdot}6 \times 10^{-9} \text{ at } 25° \text{ C})$$

$$NH_2OH + HCl \rightarrow \underset{\text{hydroxylaminium chloride}}{NH_3OH^+Cl^-}$$

$$2NH_2OH + H_2SO_4 \rightarrow \underset{\text{hydroxylaminium sulphate}}{(NH_3OH)^+{}_2SO_4^{2-}}$$

Hydroxylamine and its salts are powerful reducing agents.

$$2NH_2OH \rightarrow N_2O + 4H^+ + H_2O + 4e \quad (E^\ominus = -1{\cdot}05 \text{ V})$$

Oxides of Nitrogen

Table 15.1 lists the oxides of nitrogen with some of their physical properties.

TABLE 15.1

Oxide	N_2O	NO	N_2O_3	NO_2	N_2O_5
Oxidation state of nitrogen	+1	+2	+3	+4	+5
Properties	Colourless gas	Colourless gas	Blue liquid	Brown gas (NO_2) Yellow liquid (N_2O_4) Colourless solid (N_2O_4)	Colourless gas White unstable solid
	m.p. –90·8°	m.p. –163·6°	Decomposes 27 °C	m.p. –9·04 °C b.p. 21·9 °C	Solid unstable 0 °C

The oxide structures are very unusual in many ways and are best interpreted in terms of the valency bond theory as resonance hybrids.

Nitrogen(I) oxide, nitrous oxide (N_2O), is linear with a very small dipole moment (0·16 D), suggesting resonance between the forms:

$$\overset{-}{N}{=}\overset{+}{N}{=}O \qquad N{\equiv}\overset{+}{N}{-}\overset{-}{O}$$

The nitrogen(II) oxide, nitric oxide (NO), molecule contains an odd number of electrons and is paramagnetic. It has a low dipole moment and the N—O distance of 0·114 nm is intermediate between that of a double bond and that estimated for a triple bond. The NO molecule possesses one more electron than the nitrogen molecule (one σ- and two π-bonds). On the molecular orbital theory, the additional electron occupies a higher energy orbital termed a π-antibonding orbital (p. 40). The electron in the π-antibonding orbital effectively reduces the strength of one of the π-bonds and this correlates with the observed bond length.

Nitrogen(IV) oxide, nitrogen dioxide (NO_2), is also an odd-electron paramagnetic molecule. The molecule is non-linear with an O—N—O angle of 134° and the structure is a hybrid of:

$$O{=}N^{+}{-}O^{-} \qquad {}^{-}O{-}{}^{+}N{=}O$$

Dinitrogen tetroxide, N_2O_4, is the dimer of NO_2 and, since nitrogen dioxide is paramagnetic, N_2O_4 is diamagnetic with a planar structure, having four N—O bonds of equal length. The N—N bond is, however, extraordinarily long (0·175 nm) compared to the N—N bond in hydrazine (0·075 nm). The structure is a hybrid of:

$$(O{=})(O^{-}{-})\overset{+}{N}{-}\overset{+}{N}({-}O^{-})({=}O) \qquad ({}^{-}O{-})(O{=})\overset{+}{N}{-}\overset{+}{N}({=}O)({-}O^{-})$$

The structures of nitrogen(III) oxide, (nitrogen trioxide (N_2O_3)), and nitrogen(V) oxide, (nitrogen pentoxide (N_2O_5)), in the vapour state are shown below. That of nitrogen pentoxide is rather uncertain and in each case only one canonical form is shown.

$$(O^{-}{-})(O{=})N^{+}{-}O{-}{}^{+}N({-}O^{-})({=}O) \qquad O{=}N{-}O{-}N{=}O$$

Solid nitrogen pentoxide is an ionic compound $NO_2^{+}NO_3^{-}$ (nitronium nitrate) containing linear nitronium ions $(O{=}N{=}O)^+$ and plane triangular nitrate ions:

$$\begin{array}{c} {}^{-}O \quad\;\; O^{-} \\ \diagdown \overset{+}{N} \diagup \\ \| \\ O \end{array}$$

All the oxides may be prepared from nitric acid or one of its salts.

Nitrous oxide is obtained by heating ammonium nitrate or reducing dilute nitric acid with a fairly strong reducing agent such as zinc or tin(II) chloride.

$$NH_4NO_3 \rightarrow N_2O + 2H_2O$$

$$4Zn \rightarrow 4Zn^{2+} + 8e$$

$$2NO_3^- + 10H^+ + 8e \rightarrow 5H_2O + N_2O$$

Nitrous oxide, relative density 22, can be collected by downward delivery.

Nitric oxide is prepared by reducing 50 per cent nitric acid with a mild reducing agent, e.g. copper or an acidified iron(II) salt solution.

$$6Fe^{2+} \rightarrow 6Fe^{3+} + 6e$$

$$3Cu \rightarrow 3Cu^{2+} + 6e$$

$$2NO_3^- + 8H^+ + 6e \rightarrow 4H_2O + 2NO$$

The gas may be collected over water.

Nitrogen trioxide may be prepared by cooling nitric oxide and nitrogen dioxide in the ratio 1:1 in liquid air or by distilling arsenious oxide with nitric acid. The red gas obtained condenses to a blue liquid in a freezing mixture.

$$2NO_3^- + 6H^+ + 4e \rightarrow N_2O_3 + 3H_2O$$
$$AsO_2^- + 2H_2O \rightarrow AsO_4^{3-} + 4H^+ + 2e$$

Nitrogen dioxide is prepared by heating lead nitrate or reducing concentrated nitric acid with a mild reducing agent such as copper.

$$2Pb(NO_3)_2 \rightarrow 2PbO + 4NO_2 + O_2$$

$$Cu \rightarrow Cu^{2+} + 2e$$

$$2NO_3^- + 4H^+ + 2e \rightarrow 2H_2O + 2NO_2$$

Nitrogen pentoxide is prepared by the low temperature dehydration of concentrated nitric acid with phosphorus pentoxide.

$$2HNO_3 \xrightarrow{P_2O_5} N_2O_5 + H_2O$$

White deliquescent crystals are obtained by cooling the product in solid carbon dioxide and ether.

Chemical properties

Thermal stability

There is no direct correlation between stability and oxidation state. Nitrous oxide is most readily decomposed. A glowing splint ignites and weakly burning carbon, sulphur, and phosphorus continue to burn when placed in the gas.

Nitric oxide is the most stable gas (stable up to 1000 °C) and only strongly burning phosphorus decomposes the oxide. Strongly burning carbon, sulphur, and phosphorus continue to burn in, and therefore decompose, nitrogen dioxide. Dinitrogen tetroxide undergoes thermal dissociation as shown below.

N_2O_4	$\rightleftharpoons$	N_2O_4	$\rightleftharpoons$	$2NO_2$	$\rightleftharpoons$	$2NO + O_2$
—10 °C		2.1 °C		140 °C		600 °C
colourless solid		pale yellow liquid		brown gas		colourless gases

Both nitrogen trioxide and nitrogen pentoxide decompose in the vapour state.

$$N_2O_3 \rightarrow NO + NO_2$$

$$2N_2O_5 \rightarrow 4NO_2 + O_2$$

Solubility

Nitrous oxide is soluble in water but hyponitrous acid is not formed, although nitrous oxide is produced by thermal decomposition of the acid $H_2N_2O_2$. Nitric oxide is insoluble in water. Nitrogen trioxide, which is a stable blue liquid below −27 °C, can be regarded as the anhydride of nitrous acid since equimolecular mixtures of nitric oxide and nitrogen dioxide in alkali give pure nitrite.

$$\underbrace{NO + NO_2}_{N_2O_3} + 2KOH \rightarrow 2KNO_2 + H_2O$$

Nitrogen dioxide and dinitrogen tetroxide are the anhydrides of mixed acids and dissolve in both water and alkalis.

$$N_2O_4 + H_2O \rightarrow \underbrace{HNO_2 + HNO_3}_{\text{nitrous and nitric acids}}$$

Nitrogen pentoxide is the anhydride of nitric acid.

$$N_2O_5 + H_2O \rightarrow 2HNO_3$$

Oxidizing and reducing action

Nitrous oxide is relatively unreactive. Nitric oxide has a strong affinity for oxygen and immediately combines with the element on exposure to the air, giving brown fumes of nitrogen dioxide.

$$2NO + O_2 \rightleftharpoons 2NO_2$$

It acts as a typical reducing agent:

$$NO_3^- + 4H^+ + 3e \rightarrow NO + 2H_2O \quad (E^\ominus = +0{\cdot}96\ V)$$

Both nitrogen dioxide and dinitrogen tetroxide can function as oxidising and reducing agents:

$$N_2O_4 + 4H^+ + 4e \rightarrow 2NO + 2H_2O \quad (E^\ominus = +1{\cdot}03\ V)$$

$$2NO_3^- + 4H^+ + 2e \rightarrow N_2O_4 + 2H_2O \quad (E^\ominus = +0{\cdot}8\ V)$$

Unique properties

Nitrous oxide shows unique physiological activity and is used as an anaesthetic (laughing gas).

Nitric oxide combines with hydrated iron(II) ions to give a brown complex responsible for the brown ring in the nitrate test.

$$Fe(H_2O)_6^{2+} + NO \rightarrow Fe(H_2O)_5NO^{2+} + H_2O$$

As might be expected, the single unpaired electron in the antibonding orbital of the nitric oxide molecule is readily lost, giving rise to the nitrosonium ion (NO^+) produced when nitrogen trioxide or dinitrogen tetroxide is dissolved in concentrated sulphuric acid.

$$NO + NO_2 + 2H_2SO_4 \rightarrow \underset{\text{nitrosyl bisulphate}}{2NO^+HSO_4^-} + H_2O$$

Nitric oxide reacts with halogens, except iodine, to give gaseous nitrosyl halides which are covalent molecules.

$$2NO + Cl_2 \longrightarrow 2\left[\begin{array}{c} N \\ O{/\!\!/}\quad\backslash Cl \end{array}\right]$$

Solid nitrogen pentoxide is an ionic compound, nitronium nitrate $NO_2^+NO_3^-$. The nitronium ion is, in fact, nitrogen dioxide without its unpaired electron. Dissociation occurs in anhydrous sulphuric and perchloric acids to produce the nitronium ion and many nitronium compounds have been prepared (p. 260).

$$NO_2^+NO_3^- + HClO_4 \rightarrow \underset{\text{nitronium perchlorate}}{NO_2^+ClO_4^-} + HNO_3$$

Oxo Acids of Nitrogen

Hyponitrous acid

The solution of the free acid $H_2N_2O_2$ may be obtained by the action of hydroxylamine on nitrous acid.

$$HO{-}N{=}O + H_2N{-}OH \rightarrow HO{-}N{=}N{-}OH + H_2O$$

Soluble salts may be prepared by the reduction of nitrite solutions.

$$2NO_2^- + 4H^+ + 4e \rightarrow N_2O_2^{2-} + 2H_2O \quad (E^\ominus = -0{\cdot}86\ V)$$

The addition of silver nitrate to a solution of the free acid gives insoluble yellow silver hyponitrite which, on adding to an ethereal solution of hydrogen chloride and evaporating in vacuum, gives white crystals of the free acid. The acid, however, decomposes both in the solid state and in solution.

$$H_2N_2O_2 \rightarrow H_2O + N_2O$$

Hyponitrites are strong reducing agents in acidic solution.

$$5N_2O_2^{2-} + 8MnO_4^- + 24H^+ \rightarrow 10NO_3^- + 8Mn^{2+} + 12H_2O$$

Infrared spectra of the salts and the low dipole moment of solutions of organic esters, e.g. $(C_2H_5)_2N_2O_2$, indicate that the molecule of free acid has a *trans* configuration, i.e. the two —OH groups lie on opposite sides of —N═N—:

HO╲N
‖
N╲OH

Nitrous acid

Solutions of the free acid HNO_2 are obtained by the acidification of nitrites. Alkali metal nitrites are made by heating the nitrate or fusing the nitrate with lead and extracting with water.

$$NaNO_3 + Pb \rightarrow PbO + NaNO_2$$

The insoluble nitrites of silver, lead, and barium are prepared by the addition of a solution of sodium nitrite to a soluble silver, lead or barium salt.

$$Ba^{2+} + 2NO_2^- \rightarrow Ba(NO_2)_2\downarrow$$

The acid is extremely weak.

$$HONO + H_2O \rightleftharpoons H_3O^+ + ONO^- \quad (K = 6 \times 10^{-4} \text{ at } 30\ ^\circ C)$$

Solutions, which have a pale blue colour due to nitrogen trioxide, rapidly decompose.

$$3HNO_2 \rightarrow HNO_3 + 2NO + H_2O$$

Acidified solutions of nitrous acid and nitrites function as oxidizing agents and will liberate iodine from potassium iodide and oxidize iron(II) salts to iron(III).

$$2NO_2^- + 4H^+ + 2e \rightarrow 2NO + 2H_2O \quad (E^\ominus = +0{\cdot}99\ V)$$

Strong oxidizing agents such as potassium permanganate, potassium dichromate, and bromine water oxidize the nitrite ion to nitrate

$$NO_2^- + H_2O \rightarrow NO_3^- + 2H^+ + 2e \quad (E^\ominus = -0{\cdot}94\ V)$$

The nitrite ion is angular with an O—N—O angle of 114° and an N—O bond length of 0·123 nm. The nitrogen atom undergoes sp^2 hybridization, the actual structure being a resonance hybrid, the canonical forms of which are shown in Figure 15.9.

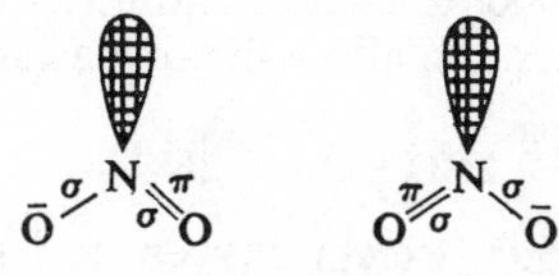

FIGURE 15.9

Contraction of the bond angle from 120° to 114° is due to lone pair–bond pair repulsion being greater than bond pair–bond pair repulsion. Silver nitrite is probably covalent and has the following structure:

$$Ag-\overset{+}{N}\begin{matrix} \nearrow O^- \\ \searrow\!\!= O \end{matrix}$$

Nitric acid

This is the most important oxo acid and may be prepared by distilling sodium nitrate with concentrated sulphuric acid. The reaction is reversible but nitric acid, being more volatile (boiling point = 86 °C), distils off.

$$NaNO_3 + H_2SO_4 \rightleftharpoons NaHSO_4 + HNO_3$$

The modern method of manufacturing the acid is by the Ostwald process which involves the catalytic oxidation of ammonia. Ammonia (1 volume)

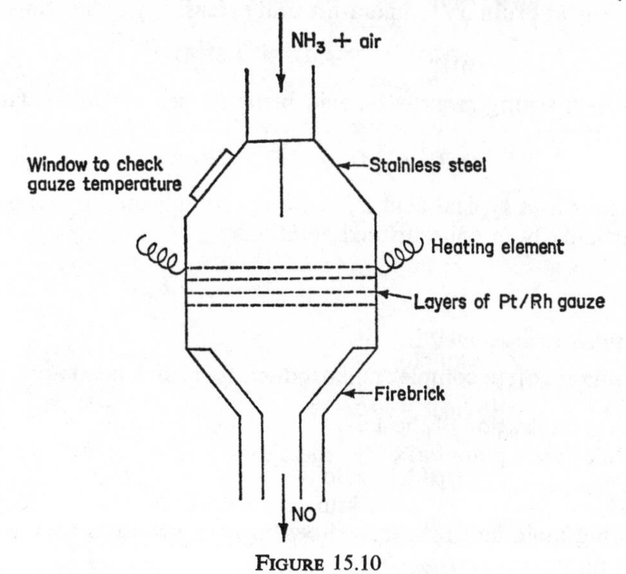

FIGURE 15.10

and air (10 volumes), preheated to 600 °C, are brought into contact with a platinum–rhodium gauze catalyst (Figure 15.10).

$$4NH_3 + 5O_2 \rightarrow 4NO + 6H_2O \qquad (\Delta H = -900 \text{ kJ})$$

The temperature rises to 1000 °C and air is admitted to the reaction chamber to cool the mixture of gases and allow the oxidation of nitric oxide to take place.

$$2NO + O_2 \rightleftharpoons 2NO_2$$

The gases (nitrogen dioxide, excess oxygen and nitrogen) pass through sprays of water in steel absorption towers.

$$3NO_2 + H_2O \rightarrow 2HNO_3 + NO$$

The nitric oxide combines with more oxygen and is recirculated to the absorption towers. Fifty per cent nitric acid collects in the towers and, on distillation, this yields a constant boiling mixture of 68 per cent nitric acid, which is the commercial concentrated acid. In the U.S.A. absorption is carried out under 8 times atmospheric pressure, giving 75 per cent acid.

Properties

Pure nitric acid is a colourless liquid, boiling point 86 °C, which freezes to a white solid at –42 °C. Commercial concentrated nitric acid is a 68 per cent constant boiling mixture which may be further concentrated by distillation with concentrated sulphuric acid. Both the pure acid and the vapour dissociate slightly at ordinary temperature and extensively on heating.

$$4HNO_3 \rightleftharpoons 2N_2O_4 + O_2 + 2H_2O$$

Nitric acid is a strong monobasic acid being 93 per cent dissociated in a normal solution.

$$HNO_3 + H_2O \rightleftharpoons H_3O^+ + NO_3^-$$

The behaviour as a typical acid is, however, complicated by its oxidizing power, particularly in concentrated solutions.

Action of nitric acid on metals

The reactions are often complex and products formed depend upon:

(*a*) The concentration of the acid.
(*b*) The electrode potential of the metal.
(*c*) The temperature.

The following ionic half reactions show how the products vary with acid concentration.

(1) $NO_3^- + 2H^+ + e \rightarrow H_2O + NO_2$		concentrated acid
(2) $NO_3^- + 4H^+ + 3e \rightarrow 2H_2O + NO$		↓ dilution increasing
(3) $NO_3^- + 6H^+ + 5e \rightarrow 3H_2O + \frac{1}{2}N_2$		
(4) $NO_3^- + 7H^+ + 6e \rightarrow 2H_2O + NH_2OH$		
(5) $NO_3^- + 10H^+ + 8e \rightarrow 3H_2O + NH_4^+$		dilute acid

Reactions (1) and (2) occur with mild reducing metals, (copper, silver, mercury, and lead) on concentrated and 8M nitric acid respectively. These metals do not react with dilute nitric acid. Reactions (3), (4), and (5) occur with more dilute nitric acid on strongly reducing metals (iron, tin, magnesium, zinc, cadmium, and manganese). In all the above reactions, the metals supply the necessary electrons and form salts:

$$Cu \rightarrow Cu^{2+} + 2e$$

$$2NO_3^- + 4H^+ + 2e \rightarrow 2H_2O + 2NO_2$$

$$\overline{Cu + 2NO_3^- + 4H^+ \rightarrow Cu^{2+} + 2NO_2 + 2H_2O}$$

It must be noted that the metalloids tin and antimony give hydrated oxides with concentrated nitric acid.

$$Sn + 4NO_3^- + 4H^+ \rightarrow SnO_2.H_2O + 4NO_2 + H_2O$$

Aqua regia consists of concentrated hydrochloric and nitric acids in the ratio 3:1.

$$3HCl + HNO_3 \rightarrow 2H_2O + \underset{\text{nitrosyl chloride}}{NOCl} + Cl_2$$

Aqua regia reacts vigorously with metals and even dissolves gold and platinum as chloro complexes $AuCl_4^-$ and $PtCl_6^{2-}$. Iron, aluminium, and chromium are rendered passive by concentrated nitric acid, probably due to the formation of a protective film of oxide.

The oxidizing action of nitric acid on non-metals and compounds is illustrated by the following half equations.

(*a*) In concentrated acid, sulphur and iodine may be oxidized.

$$NO_3^- + 2H^+ + e \rightarrow H_2O + NO_2$$

$$\overline{S + 4H_2O \rightarrow SO_4^{2-} + 8H^+ + 6e}$$

$$\tfrac{1}{2}I_2 + 3H_2O \rightarrow IO_3^- + 6H^+ + 5e$$

(*b*) In dilute nitric acid, sulphide, iodide, and iron(II) ions are oxidized.

$$NO_3^- + 4H^+ + 3e \rightarrow 2H_2O + NO$$

$$\overline{3S^{2-} \rightarrow 3S + 6e}$$

$$6I^- \rightarrow 3I_2 + 6e$$

$$3Fe^{2+} \rightarrow 3Fe^{3+} + 3e$$

Although aqueous solutions of nitric acid ionize, producing oxonium and nitrate ions, the pure liquid ionizes producing nitronium ions NO_2^+.

$$2HNO_3 \rightleftharpoons H_2O + NO_2^+ + NO_3^-$$

Electrolytic data and Raman spectra indicate that this ion is also produced when concentrated nitric acid is dissolved in concentrated sulphuric acid.

$$2H_2SO_4 + HONO_2 \rightarrow NO_2^+ + 2HSO_4^- + H_3O^+$$

Confirmation of the existence of the nitronium ion has come with the isolation of crystalline nitronium salts from reactions such as those outlined below.

$$N_2O_5 + \underset{\text{anhydrous}}{HClO_4} \rightarrow \underset{\text{nitronium perchlorate}}{NO_2^+ClO_4^-} + HNO_3$$

$$2N_2O_5 + SiF_4 + 2HF \rightarrow \underset{\text{nitronium hexafluorsilicate(IV)}}{(NO_2^+)_2SiF_6^{2-}} + 2HNO_3$$

Nitric acid has many important applications in the manufacture of conventional explosives (dynamite, T.N.T., cyclonite, etc.), dyes, plastics, and drugs. Electron diffraction measurements on nitric acid vapour indicate that the molecule is planar. The bond lengths and bond angles are indicated in Figure 15.11.

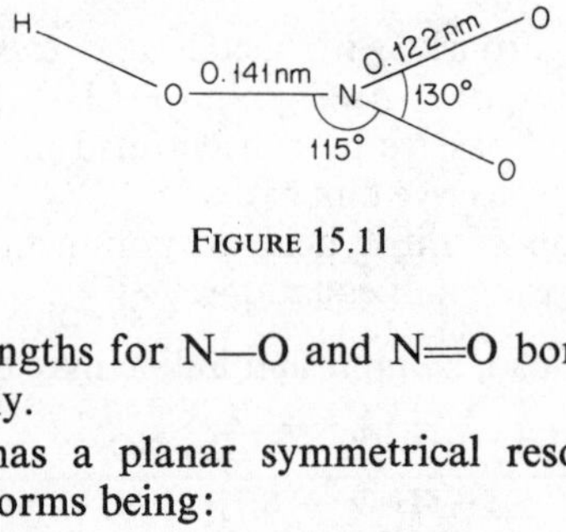

FIGURE 15.11

The normal bond lengths for N—O and N═O bonds are 0·136 nm and 0·118 nm, respectively.

The nitrate ion has a planar symmetrical resonance structure, the principal canonical forms being:

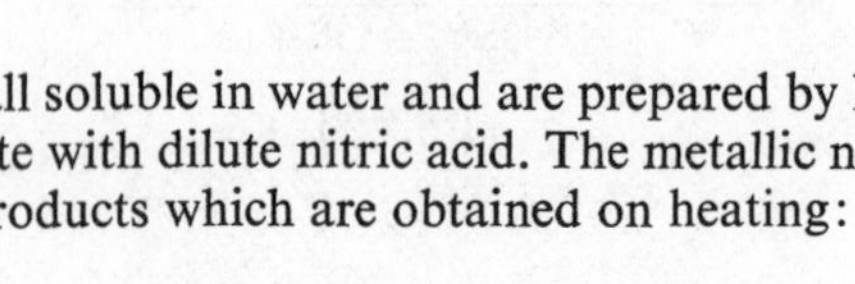

The nitrates are all soluble in water and are prepared by heating the metal, oxide or carbonate with dilute nitric acid. The metallic nitrates vary in the decomposition products which are obtained on heating:

$$2XNO_3 \rightarrow 2XNO_2 + O_2 \quad (X = \text{alkali metal})$$

$$2X(NO_3)_2 \rightarrow 2XO + 4NO_2 + O_2 \quad (X = \text{alkaline earth metals and lead, copper, zinc, aluminium or iron})$$

or

$$2X(NO_3)_3 \rightarrow X_2O_3 + 6NO_2 + \tfrac{3}{2}O_2$$

$$2XNO_3 \rightarrow 2X + 2NO_2 + O_2 \quad (X = \text{silver or mercury})$$

The Halides

Nitrogen trifluoride is a colourless gas obtained by the action of fluorine on ammonia with a copper catalyst. The molecule is pyramidal with a very low dipole moment (0·22 D) and is not affected by water. One would expect the withdrawal of electrons by the fluorine atoms to make the lone pair of electrons on the nitrogen atom unavailable for coordination. However, a gaseous NOF_3 has recently been prepared by the action of fluorine on nitrogen(II) oxide at an elevated temperature and pressure. The gas is resistant to hydrolysis and is thermally stable up to 300 °C.

The trichloride is a yellow oil prepared by the action of chlorine on ammonia, when the former is in excess, or by inverting a flask of chlorine over a saturated solution of ammonium chloride.

$$NH_3 + 3Cl_2 \rightarrow NCl_3 + 3HCl$$

$$NH_4Cl + 3Cl_2 \rightleftharpoons NCl_3 + 4HCl$$

The compound is endothermic (heat of formation = +230 kJ mol^{-1}) and serious explosions have occurred during the preparation, which is extremely dangerous. The oil is slowly hydrolysed by water.

$$Cl_3N\cdots H\text{—}O\text{—}H \longrightarrow Cl_2N\text{—}H + HOCl$$

(H-bond indicated between N and H)

$$Cl_2N\text{—}H \xrightarrow{2H_2O} NH_3 + 2HOCl$$

The bromides and iodides are actually double compounds: $NBr_3.6NH_3$ (purple) and $NI_3.NH_3$ (black). Both are explosive solids prepared from the halogen and ammonia.

Phosphorus, Arsenic, Antimony, and Bismuth

Occurrence

The chief minerals of phosphorus are phosphorite, $3Ca_3(PO_4)_2.Ca(OH)_2$, chloroapatite and fluoroapatite, $3Ca_3(PO_4)_2.CaX_2$, where X is chlorine or fluorine. Arsenic, antimony, and bismuth occur mainly as the sulphides, orpiment (As_2S_3), stibnite (Sb_2S_3), and bismuth glance (Bi_2S_3). Arsenic also occurs fairly widely as mispickel (FeAsS) and bismuth may occur native with lead, silver, arsenic or antimony ores.

Extraction of Elements

Phosphorus

The element is produced by the high-temperature reduction of phosphates in an electric-arc furnace (Figure 15.12). The charge of crushed phosphate,

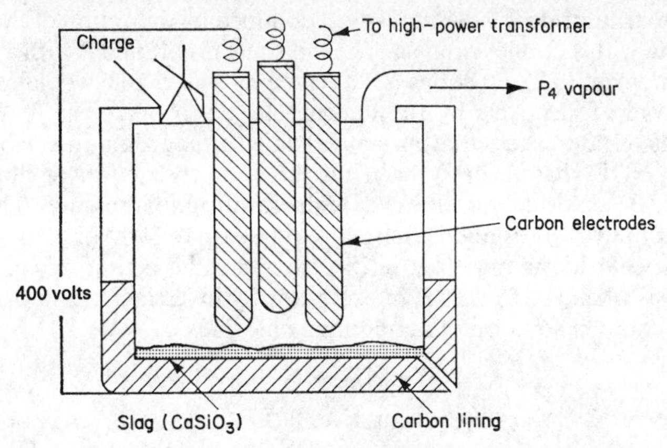

FIGURE 15.12

coke, and sand or crushed quartz is fed into the furnace where electric arcs are struck between the electrodes at the base of the furnace. The high temperature produced (1400 °C) causes the following reactions to occur:

$$2Ca_3(PO_4)_2 + 6SiO_2 \rightarrow \underset{\text{calcium silicate}}{6CaSiO_3} + P_4O_{10}$$

$$P_4O_{10} + 10C \rightarrow 10CO + \underset{\text{vapour}}{P_4}$$

The phosphorus vapour may be condensed under water to give white phosphorus or immediately converted into phosphorus pentoxide and phosphoric acid (p. 268).

Arsenic, Antimony, and Bismuth

These may be obtained by roasting the sulphides in air and reducing the oxides formed with carbon at red heat.

$$2M_2S_3 + 9O_2 \rightarrow 2M_2O_3 + 6SO_2$$

$$M_2O_3 + 3C \rightarrow 2M + 3CO$$

Other methods involve: (*a*) the distillation of mispickel

$$FeAsS \rightarrow FeS + As \text{ (sublimes)}$$

(*b*) the direct reduction of stibnite with iron in a reverberatory furnace

$$Sb_2S_3 + 3Fe \rightarrow 2Sb \text{ (molten)} + 3FeS$$

Native bismuth may be purified by liquation (p. 127).

Uses

Vast quantities of phosphorus are converted into the acids and then salts for use as detergents and fertilizers. Small quantities are consumed for fireworks, phosphor bombs, rat poisons, and matches. Arsenic and its compounds find uses in glass making, the manufacture of weed killers, pest-killing sprays, and alloys. Antimony and bismuth form useful alloys and are also used in certain pharmaceutical preparations.

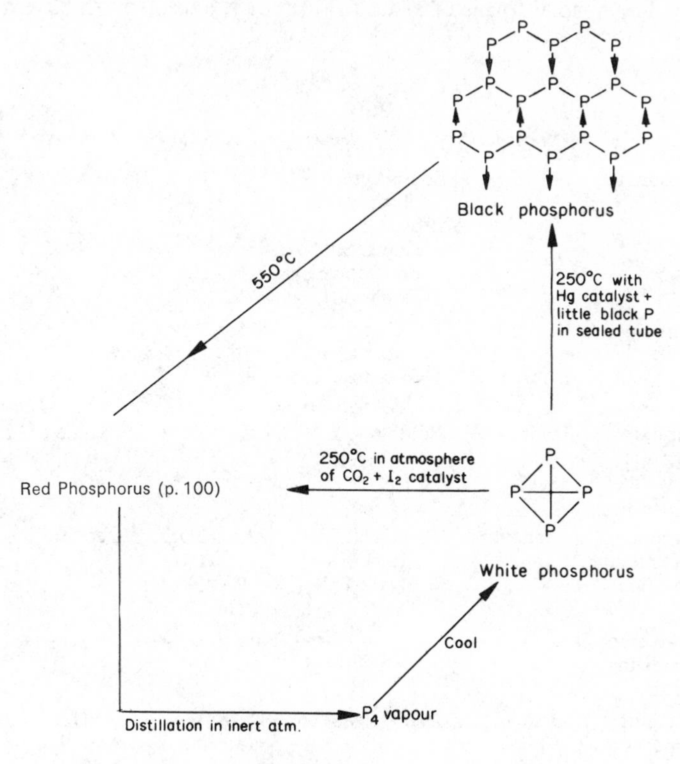

FIGURE 15.13

Allotropy

The structure of the three main allotropic forms of phosphorus and the conditions necessary for their interconversion are outlined in Figure 15.13.

In white phosphorus the P—P—P bond angle is 60° and this small value indicates considerable strain, hence the high reactivity of this form. In the production of red phosphorus, one of the P—P bonds of white phosphorus is ruptured and the free bonds link up to form chains of tetrahedra which possess less strain. In black phosphorus further bond breakage and cross-linking occurs giving almost strainless hexagonal rings with bond angles of 100°.

Although black phosphorus is the most difficult allotrope to prepare, it is the most stable form. White phosphorus is the most unstable and reactive form and reverts to red phosphorus at all temperatures.

$$\text{P(white)} \rightarrow \text{P(red)} \qquad (\Delta H = -18 \text{ kJ mol}^{-1})$$

Arsenic and antimony have tetratomic yellow modifications but these are of little importance since they are extremely unstable even at room temperature. The normal forms of arsenic, antimony, and bismuth have a metallic

TABLE 15.2

	Red phosphorus	*White phosphorus*	*Arsenic*	*Antimony*	*Bismuth*
Exposure to air	No oxidation	Emits a green phosphorescence. Complex photochemical oxidation process	← No oxidation →		
Melting point	Sublimes at 416 °C	44 °C under water. Ignites in air to give P_4O_6 and P_4O_{10}	816 °C (under pressure)	630 °C	273 °C
Action of organic solvents, C_6H_6, CS_2	Insoluble	Soluble	Insoluble	Insoluble	Insoluble
Hot conc. caustic soda solution	No action	$P_4 + 3NaOH + 3H_2O = 3NaH_2PO_2 + PH_3$	$2As + 2OH^- + 2H_2O = 2AsO_2^- + 3H_2$	No action	No action
Dilute, non-oxidizing acids	← No action →				
Conc. nitric acid	Ortho-phosphoric acid H_3PO_4	H_3PO_4 (slowly)	H_3AsO_4 arsenic acid	H_3SbO_4 antimonic acid	$Bi(NO_3)_3$ bismuth nitrate

appearance and lustre and the structures are very similar to that of black phosphorus, the bond distances increasing from phosphorus to bismuth.

Table 15.2 summarizes the properties of red and white phosphorus and the metallic forms of arsenic, antimony and bismuth.

The hydrides

All the elements form pyramidal hydrides MH_3. Phosphine, PH_3, may be prepared by heating white phosphorus with concentrated sodium hydroxide solution or by the action of water on calcium phosphide.

$$P_4 + 3NaOH + 3H_2O \rightarrow \underset{\text{sodium hypophosphite}}{3NaH_2PO_2} + PH_3$$

$$Ca_3P_2 + 6H_2O \rightarrow 3Ca(OH)_2 + 2PH_3$$

Arsine (AsH_3), stibine (SbH_3), and bismuthine (BiH_3) may be produced by reducing the trichlorides with zinc and dilute hydrochloric acid or by the action of hydrochloric acid on compounds of the elements with zinc or magnesium.

$$Zn_3As_2 + 6HCl \rightarrow 3ZnCl_2 + 2AsH_3$$

All the hydrides are gases, their thermal stability decreasing with the decrease in bond energies (the bond energy of the P—H bond is 322 and that of the Sb—H bond is 254 kJ mol^{-1}). Thus arsine and phosphine

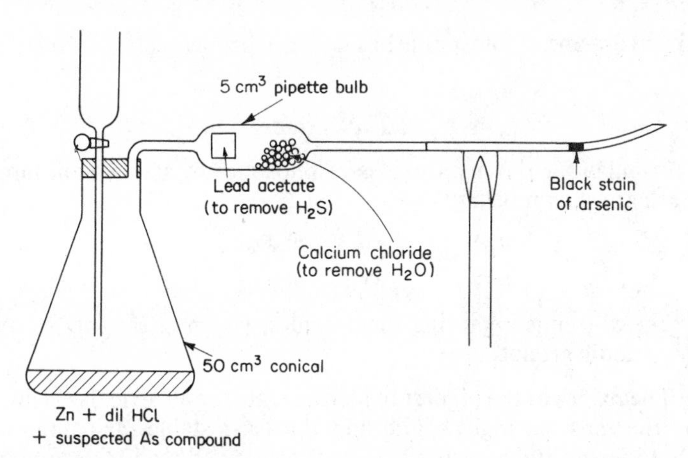

FIGURE 15.14. The Marsh–Berzelius test for arsenic and antimony

decompose to the elements on gently heating, whereas antimony and bismuth hydrides are both unstable at room temperature.

The decomposition of arsine is made the basis of the Marsh–Berzelius test for arsenic. The suspected arsenic-containing compound is reduced by zinc and dilute hydrochloric acid or electrolytically to arsine in the apparatus shown diagrammatically in Figure 15.14.

The arsine is decomposed in the hot tube and a black stain is deposited beyond the heated portion. By a comparison of the stain with standard sets of stains from known weights of arsenic, it is possible to detect as little as 10^{-8} g. Antimony forms a similar mirror but this appears on both sides of the heated portion and is insoluble in hypochlorite solution whereas the arsenic stain is soluble.

All the hydrides are fairly strong reducing agents.

$$PH_3 + 3OH^- \rightarrow P + 3H_2O + 3e \qquad (E^\ominus = -0{\cdot}89\ V)$$

The basic properties shown by ammonia are not exhibited by phosphine although it can unite with halogen hydrides to give phosphonium compounds.

$$PH_3\ (gas) + HI\ (gas) \rightarrow PH_4^+I^-$$

The salts are, however, decomposed by water.

$$PH_4^+ + H_2O \rightarrow PH_3 + H_3O^+$$

A hydride, diphosphine P_2H_4, analogous to hydrazine is also formed during the preparation of phosphine and renders the latter spontaneously inflammable. It may be frozen out from the crude phosphine as a colourless liquid. Both the liquid and the vapour are unstable.

$$3P_2H_4 \rightarrow 2P + 4PH_3$$

Unlike hydrazine, diphosphine has no basic properties.

The Oxides

The formulae for the main oxides of phosphorus, arsenic, antimony, and bismuth are shown below:

P_4O_6 As_4O_6 Sb_4O_6 Bi_2O_3

P_4O_{10} As_4O_{10} Sb_4O_{10}

The general trends regarding these oxides, which also apply to oxides in other periodic groups, are:

(*a*) The oxides of the highest oxidation state for a particular element tend to be the most strongly acidic and the least stable (arsenious oxide is amphoteric and stable; arsenic oxide is acidic and readily decomposed).

(*b*) The stability of the highest oxidation state decreases with increasing atomic number (Bi_2O_5 has never been prepared).

(*c*) For any given oxidation state of an oxide, the basic character increases (acidic character decreases) with increasing atomic number. Thus phosphorous oxide is acidic, arsenious oxide amphoteric, antimonous oxide amphoteric, and bismuth trioxide is basic.

The structures of the two main oxides of phosphorus are both related to the structure of white phosphorus and are shown in Figure 15.15.

The oxides of arsenic and antimony probably adopt similar structures in their main forms, although different crystalline modifications are known (p. 285). The structure of bismuth(III) oxide is that of a distorted giant molecule with bismuth in a distorted octahedral environment.

Phosphorous oxide (phosphorus trioxide, P_4O_6 or phosphorus(III) oxide) is prepared by drawing a stream of air slowly over white phosphorus in a condenser jacketed by water at 60 °C.

$$P_4 + 3O_2 \rightarrow P_4O_6$$

The phosphorous oxide melts (24 °C) and runs into a flask surrounded by a freezing mixture. Any phosphoric oxide (P_4O_{10}) which is formed sublimes

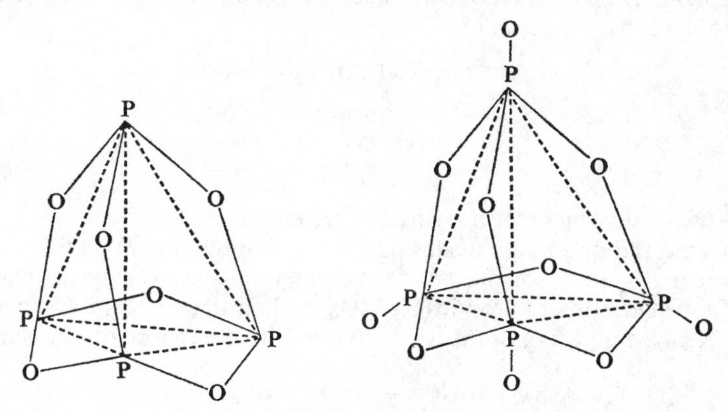

FIGURE 15.15. The four P—O bond distances at the apices of the tetrahedron in P_4O_{10} indicate considerable $d\pi$–$p\pi$ double bonding

(50 °C) and is held back in the condenser by a plug of glass wool. Since the oxides of arsenic and antimony in the pentavalent state are unstable, arsenious and antimonous oxides are prepared by freely heating the elements in air. For bismuth trioxide, the action of heat on the nitrate may be employed.

Phosphorous oxide is a strongly acidic oxide,

$$\underset{\text{cold}}{P_4O_6 + 6H_2O} \rightarrow \underset{\substack{\text{phosphorous}\\ \text{acid}}}{4H_3PO_3} \xrightarrow{\text{warm}} \underset{\substack{\text{ortho-}\\ \text{phosphoric}\\ \text{acid}}}{3H_3PO_4} + PH_3$$

Arsenious and antimonous oxides (arsenic and antimony(III) oxides) are only sparingly soluble in water and behave as amphoteric oxides, dissolving in strong acids and giving arsenite (AsO_2^-) and antimonite (SbO_2^-) solutions with strong alkali.

$$As_4O_6 + 4OH^- \rightarrow 4AsO_2^- + 2H_2O$$

Bismuth trioxide is predominantly basic, dissolving readily in acid and being unattacked by alkali solutions although, with fused alkalis, a brown mass of sodium bismuthate is formed ($NaBiO_3$).

$$Bi_2O_3 + 2OH^- + O_2 \rightarrow 2BiO_3^- + H_2O$$

Phosphoric oxide (phosphorus pentoxide, P_4O_{10}, or phosphorus(V) oxide) is a white powder obtained by burning white phosphorus in air or oxygen.

$$P_4 + 5O_2 \rightarrow P_4O_{10}$$

The solid may be purified by sublimation in a current of dry air. The oxide reacts vigorously with water.

$$P_4O_{10} + 2H_2O \rightarrow \underset{\text{meta-phosphoric acid}}{4HPO_3} \xrightarrow[H_2O]{\text{warm}} \underset{\text{ortho-phosphoric acid}}{H_3PO_4}$$

Thus the oxide is an excellent dehydrating agent.

Arsenic and antimonic oxides (arsenic and antimony(V) oxides) cannot be prepared by direct oxidation of the elements (*see* above) and are made by careful dehydration of arsenic (H_3AsO_4) and antimonic (H_3SbO_4) acids or by careful evaporation of the lower oxides with concentrated nitric acid.

$$As_4O_6 + 8HNO_3 \rightarrow As_4O_{10} + 4H_2O + 8NO_2$$

Although only sparingly soluble in water, the oxides dissolve in alkali hydroxide to give arsenate and antimonate solutions.

$$As_4O_{10} + 12OH^- \rightarrow 4AsO_4^{3-} + 6H_2O$$

The Sulphides

Red phosphorus and sulphur combine directly to give a number of sulphides, the composition of which varies according to the proportions taken, i.e. P_4S_3, P_4S_5, P_4S_7, P_4S_{10}. All are yellow solids, hydrolysed by water to hydrogen sulphide and acids of phosphorus. The structures have been determined and are based on tetrahedral phosphorus. Thus P_4S_{10} has the phosphoric oxide structure and that of P_4S_3, used in the manufacture of matches, is shown in Figure 15.16.

The trisulphides of arsenic, antimony, and bismuth (M_2S_3) may be prepared directly from the elements by passing hydrogen sulphide through acidified solutions of the trichlorides or, in the case of arsenic, by passing hydrogen sulphide through acidified arsenite solution.

$$2AsO_2^- + 2H^+ + 3H_2S \rightarrow As_2S_3\downarrow + 4H_2O$$

The trisulphide of arsenic is yellow, that of antimony is orange (black on heating), and that of bismuth is brown.

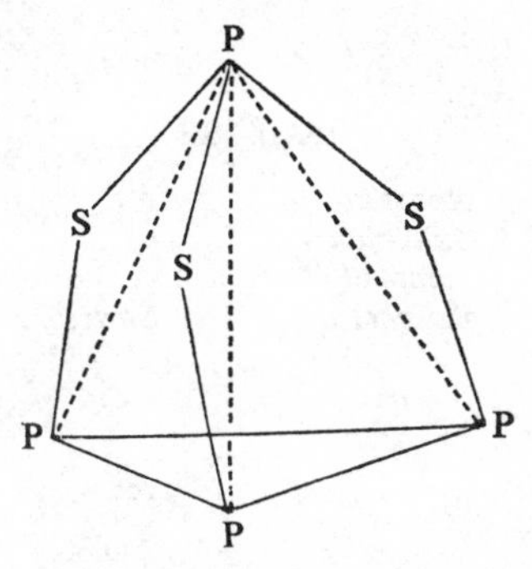

FIGURE 15.16

The trisulphides of arsenic and antimony are soluble in lithium hydroxide solution or in yellow ammonium sulphide.

(*a*) $2As_2S_3 + 4OH^- \rightarrow \underset{\text{arsenite}}{AsO_2^-} + \underset{\text{thio-arsenite}}{3AsS_2^-} + 2H_2O$

(*b*) $As_2S_3 + \underbrace{3S^{2-} + 2S}_{\text{yellow } (NH_4)_2S_x} \rightarrow \underset{\text{thio(tetra)-arsenate}}{2AsS_4^{3-}}$

Antimony trisulphide reacts similarly. Reacidification of (*a*) deposits the trisulphide and of (*b*) the pentasulphide.

$$3AsS_2^- + AsO_2^- + 4H^+ \rightarrow 2As_2S_3\downarrow + 2H_2O$$

$$2AsS_4^{3-} + 6H^+ \rightarrow \underset{\text{bright yellow}}{As_2S_5\downarrow} + 3H_2S \qquad (Sb_2S_5 \text{ is red})$$

Bismuth trisulphide is unaffected by either alkali or yellow ammonium sulphide and the pentasulphide is not known. The pentasulphides of arsenic and antimony are also soluble in alkali and sulphide solutions to give thio-complexes.

$$As_2S_5 + 6OH^- \rightarrow AsS_4^{3-} + AsO_3S^{3-} + 3H_2O$$

$$As_2S_5 + 3S^{2-} \rightarrow 2AsS_4^{3-}$$

Arsenic, like phosphorus, forms a number of other sulphides As_4S_3, As_4S_4, made by direct combination of the elements.

The Oxo Acids of Phosphorus

The principal oxo acids of phosphorus and their structures are shown in Table 15.3.

TABLE 15.3

Formula	*Name*	*Apparent oxidation state of phosphorus*	*Structure*	*Comments*
H_3PO_2	HYPO-phosphorous	+1	O ‖ H—P—H | OH	Distorted tetrahedron. Monoprotic acid.
H_3PO_3	ORTHO-phosphorous	+3	O ‖ HO—P—H | OH	Tetrahedral. Diprotic.
$(HPO_3)_n$	META-phosphoric	+5	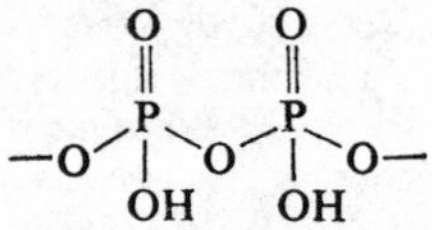 O O ‖ ‖ —O—P—O—P—O— | | OH OH	Not isolated in solid state, only in glassy condition. Structure consists of long chains and rings of PO_4 tetrahedra.
H_3PO_4	ORTHO-phosphoric	+5	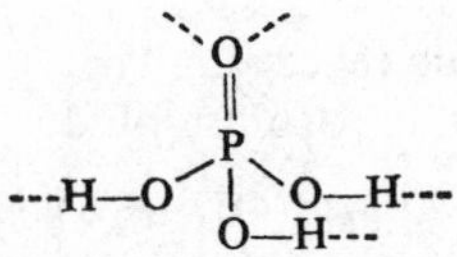 ---O--- ‖ P ---H—O—P—O—H--- O—H---	Tetrahedral molecules linked by hydrogen bonds to give a three-dimensional structure. Triprotic acid.
$H_4P_2O_7$	PYRO-phosphoric	+5	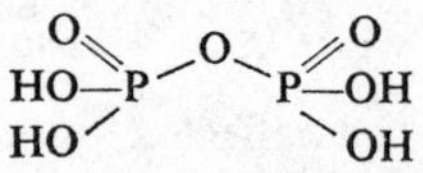 O O O HO—P—O—P—OH HO OH	Two PO_4 tetrahedra linked by a common oxygen atom. Tetraprotic acid.
$H_4P_2O_6$	HYPO-phosphoric	+4	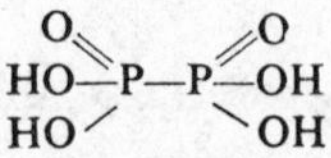 O O HO—P—P—OH HO OH	Probable structure. Tetraprotic acid.

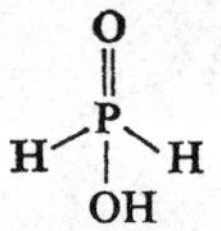

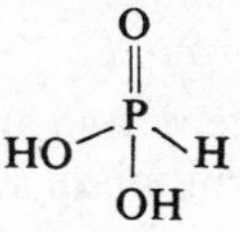

Note on nomenclature

The most highly hydroxylated acid known in a particular oxidation state is the *ortho* acid. The *meta* acid is obtained by the loss of one molecule of water from the ortho acid:

$$H_3PO_4 \xrightarrow{-H_2O} HPO_3$$

The *pyro* acid is an intermediate produced by the action of heat on the ortho acid:

$$2H_3PO_4 \xrightarrow{-H_2O} H_4P_2O_7$$

The term *hypo* indicates a lower oxygen content but is not used systematically.

Chemical properties

Hypophosphorous and phosphorous acids

If, in the preparation of phosphine, sodium hydroxide is replaced by barium hydroxide, barium hypophosphite $Ba(H_2PO_2)_2.H_2O$ can be crystallized from the solution. After treatment with the calculated quantity of dilute sulphuric acid and filtering off the insoluble barium sulphate, the solution is evaporated and cooled in a freezing mixture when colourless crystals of hypophosphorous acid (H_3PO_2) separate.

$$2P_4 + 3Ba(OH)_2 + 6H_2O \rightarrow 2PH_3 + 3Ba(H_2PO_2)_2$$

Phosphorous acid (H_3PO_3) is prepared by careful evaporation and cooling of solutions obtained by the action of cold water on phosphorous oxide or phosphorus trichloride. White deliquescent crystals are obtained.

Both acids are moderately strong, hypophosphorous acid being monoprotic ($K = 10^{-2}$ at 18 °C) and phosphorous acid being diprotic ($K_1 = 10^{-2}$; $K_2 = 2 \times 10^{-7}$ at 18 °C). Salts are known for both acids, namely:

$NaH_2PO_2.H_2O$	sodium hypophosphite
$Ca(H_2PO_2)_2$	calcium hypophosphite
$Na_2HPO_3.5H_2O$	sodium phosphite
$2NaH_2PO_3.5H_2O$	acid sodium phosphite.

Both acids are unstable and decompose on heating evolving phosphine.

$$3H_3PO_2 \rightarrow 2H_3PO_3 + PH_3$$

$$4H_3PO_3 \rightarrow 3H_3PO_4 + PH_3$$

The acids and their salts are good reducing agents.

$$H_3PO_2 + H_2O \rightleftharpoons H_3PO_3 + 2H^+ + 2e \qquad (E^\ominus = -0{\cdot}59\ V)$$

$$H_3PO_3 + H_2O \rightleftharpoons H_3PO_4 + 2H^+ + 2e \qquad (E^\ominus = -0{\cdot}20\ V)$$

Thus although phosphorous acid reduces copper sulphate to copper, hypophosphorous acid reduces it to a red unstable copper(I) hydride.

$$CuSO_4 + H_2O + H_3PO_3 \rightarrow Cu + H_2SO_4 + H_3PO_4$$

$$2CuSO_4 + 3H_2O + 3H_3PO_2 \rightarrow 2CuH + 3H_3PO_3 + 2H_2SO_4$$

Orthophosphoric acid

Orthophosphoric acid, H_3PO_4, is manufactured by burning white phosphorus in a plentiful supply of air and dissolving the product (phosphoric oxide) in hot water.

$$P_4O_{10} + 6H_2O \rightarrow 4H_3PO_4$$

Pure orthophosphoric acid may be obtained in the laboratory by evaporating red phosphorus with concentrated nitric acid.

$$P_4 + 10HNO_3 + H_2O \rightarrow 4H_3PO_4 + 5NO + 5NO_2$$

Evaporation of the resulting syrupy liquid over concentrated sulphuric acid and cooling in a freezing mixture deposits prisms of orthophosphoric acid (melting point 42 °C). The solid possesses a three-dimensional hydrogen-bonded structure and a great deal of this hydrogen bonding persists in the thick viscous solutions. The acid is triprotic and three series of salts are known.

$$H_3PO_4 + H_2O \rightleftharpoons H_2PO_4^- + H_3O^+ \qquad (K_1 = 7{\cdot}5 \times 10^{-3})$$

$$H_2PO_4^- + H_2O \rightleftharpoons HPO_4^{2-} + H_3O^+ \qquad (K_2 = 6{\cdot}2 \times 10^{-8})$$

$$HPO_4^{2-} + H_2O \rightleftharpoons PO_4^{3-} + H_3O^+ \qquad (K_3 = 1{\cdot}1 \times 10^{-12})$$

Some typical salts are:

$NaH_2PO_4.H_2O$	sodium dihydrogen phosphate
$Na_2HPO_4.12H_2O$	disodium hydrogen phosphate
$Na_3PO_4.12H_2O$	trisodium phosphate

Solutions containing the ion $H_2PO_4^-$ are weakly acidic, those containing the ion HPO_4^{2-} weakly alkaline and those containing the ion PO_4^{3-} are strongly alkaline due to hydrolysis.

$$H_2PO_4^- + H_2O \rightleftharpoons HPO_4^{2-} + H_3O^+ \qquad (pH = 3\ to\ 5)$$

$$HPO_4^{2-} + H_2O \rightleftharpoons H_2PO_4^- + OH^- \qquad (pH = 8\ to\ 10)$$

$$PO_4^{3-} + H_2O \rightleftharpoons HPO_4^{2-} + OH^- \qquad (pH = 12)$$

Thus, on titrating phosphoric acid with caustic soda, methyl orange changes colour when sodium dihydrogen phosphate is formed, and phenolphthalein when disodium hydrogen phosphate is formed. On heating orthophosphoric acid to 250 °C, pyrophosphoric acid is formed which, on further heating, yields metaphosphoric acid as a final product.

$$2H_3PO_4 \rightarrow \underset{\text{pyrophosphoric acid}}{H_4P_2O_7} + H_2O$$

$$H_4P_2O_7 \rightarrow \underset{\text{metaphosphoric acid}}{2HPO_3} + H_2O$$

Sodium dihydrogen phosphate and trisodium phosphate are used as water-softeners. Soluble calcium dihydrogen phosphate, $Ca(H_2PO_4)_2$, finds extensive use as a fertilizer. Treatment of the insoluble naturally occurring rock phosphate $Ca_3(PO_4)_2$ with crude sulphuric or phosphoric acids produces the soluble dihydrogen salt.

$$Ca_3(PO_4)_2 + 2H_2SO_4 \rightarrow Ca(H_2PO_4)_2 + 2CaSO_4$$

$$Ca_3(PO_4)_2 + 4H_3PO_4 \rightarrow 3Ca(H_2PO_4)_2$$

The products are known respectively as superphosphate and triple superphosphate of lime and are valuable fertilizers.

A double phosphate, magnesium ammonium phosphate ($MgNH_4PO_4.6H_2O$), is precipitated in the test for magnesium ions in qualitative analysis.

$$Mg^{2+} + NH_4^+ + HPO_4^{2-} \rightarrow MgNH_4PO_4\downarrow + H^+$$

Solutions of phosphates with excess nitric acid and ammonium molybdate solution give a bright yellow precipitate of ammonium phosphomolybdate (p. 384).

Pyrophosphoric acid

Pyrophosphoric acid, $H_4P_2O_7$, is prepared by heating orthophosphoric acid alone or with phosphorus oxochloride.

$$5H_3PO_4 + POCl_3 \rightarrow 3H_4P_2O_7 + 3HCl$$

Strong heating converts pyrophosphoric acid into a sticky mass of metaphosphoric acid. Pyrophosphoric acid is tetraprotic and is a stronger acid than orthophosphoric acid.

$$H_4P_2O_7 + H_2O \rightleftharpoons H_3O^+ + H_3P_2O_7^- \qquad (K = 1{\cdot}4 \times 10^{-1})$$

Salts of the acid are known, e.g.: $Na_4P_2O_7.10H_2O$, $Na_2H_2P_2O_7$, $Mg_2P_2O_7$ (obtained by heating magnesium ammonium phosphate). In contrast to the orthophosphates, which give a yellow precipitate (Ag_3PO_4), pyrophosphates give a white precipitate ($Ag_4P_2O_7$) with silver nitrate solution.

Metaphosphoric acid

Metaphosphoric acid, $(HPO_3)_n$, is obtained as a syrupy mass when ortho or pyrophosphoric acid is heated above 300 °C. The acid is a polymer consisting of a mixture of linear and cyclic chains (Figure 15.17(*a*)).

The sodium salt is formed as a clear glass by heating sodium dihydrogen phosphate.

$$NaH_2PO_4 \rightarrow H_2O + NaPO_3$$

The product was originally known as sodium hexametaphosphate $(NaPO_3)_6$, but it is now known to consist of long chains of cross-linked PO_4 tetrahedra (Figure 15.17(*b*)). The salt is known commercially as Calgon and sequesters

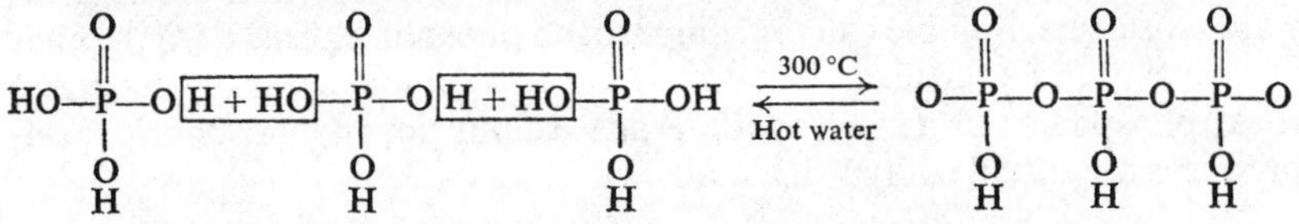

300 °C

Hot water

(a)

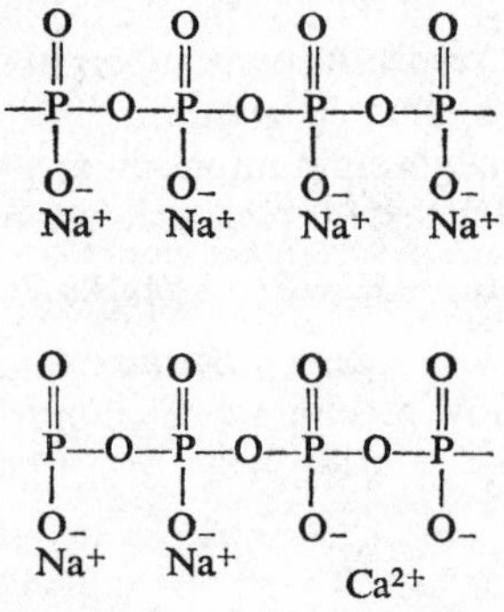

(b)

$$NaH_2PO_4 + 2Na_2HPO_4 \longrightarrow Na_5P_3O_{10} + 2H_2O$$

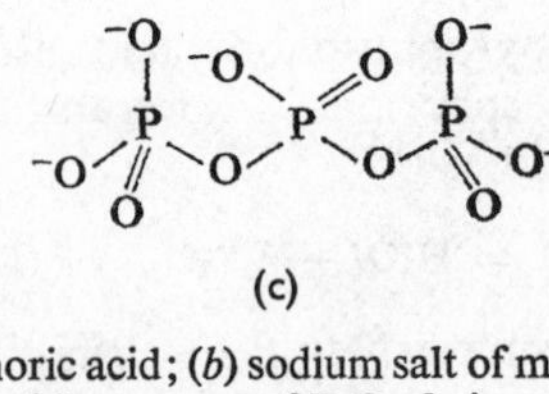

(c)

FIGURE 15.17(*a*) metaphosphoric acid; (*b*) sodium salt of metaphosphoric acid (Calgon); (*c*) structure of $P_3O_{10}^{5-}$ ion

magnesium and calcium ions from hard water, these ions replacing the sodium ions in the original structure (p. 289).

Sodium triphosphate, $Na_5P_3O_{10}$ ($Na_4P_2O_7.NaPO_3$), is mixed with detergents and soaps in domestic washing powders and is prepared by heating together the mono and disodium hydrogen phosphates. The structure of the $P_3O_{10}^{5-}$ ion is shown in Figure 15.17(*c*).

Hypophosphoric acid

Hypophosphoric acid, $H_4P_2O_6$, can be prepared by oxidizing phosphorous acid with iodine.

$$2H_3PO_3 + I_2 \rightarrow H_4P_2O_6 + 2HI$$

Evaporation in vacuum over concentrated sulphuric acid yields crystals of the dihydrate. The acid has no reducing action and decomposes on heating.

$$H_4P_2O_6 \rightarrow HPO_3 + H_3PO_3 \quad (\xrightarrow{180\,°C} PH_3)$$

Salts are known, e.g. $Na_2H_2P_2O_6.6H_2O$, disodium dihydrogen hypophosphate.

Oxo Acids of Arsenic, Antimony, and Bismuth

A solution of arsenious acid ($HAsO_2$ or H_3AsO_3), which is produced by dissolving arsenious oxide in water, is a very weak acid ($K = 5 \times 10^{-10}$). Evaporation of such solutions yields crystalline precipitates of the hydrated oxide and the pure acid has never been isolated. Solutions of arsenious acid and arsenites are readily oxidized to arsenates.

$$AsO_4^{3-} + 2H_2O + 2e \rightarrow AsO_2^- + 4OH^- \; (E^\ominus = 0{\cdot}67 \text{ V})$$

Heating the trioxide or metal with concentrated nitric acid and cooling deposits crystals of arsenic acid (H_3AsO_4). The acid is tribasic and, although weaker than orthophosphoric acid ($K = 5 \times 10^{-3}$), it is a stronger oxidizing agent and will liberate iodine from potassium iodide solution.

$$AsO_4^{3-} + 2I^- + 4H^+ \rightarrow AsO_2^- + I_2 + 2H_2O$$

A number of polyarsenates exist but these are less stable than the corresponding phosphorus compounds and all rapidly hydrolyse with water. No lower acid is known for antimony, only the hydrated oxide, $Sb_2O_3.xH_2O$. Salts, the antimonites, are, however, well characterized, e.g. $NaSbO_2.3H_2O$. The antimonic acids are also indefinite in composition but again crystalline antimonates exist which do not contain SbO_4^{3-} ions but octahedral $Sb(OH)_6^-$ ions (cf. tin and lead). No oxo acids of bismuth exist but salts, the bismuthates ($NaBiO_3$), can be made by fusion of bismuth trioxide with sodium hydroxide in air (p. 268). They are powerful oxidizing agents and will oxidize manganese salts to the purple permanganate. Sodium bismuthate has the ilmenite structure (p. 364).

Halides

The trihalides

All the trihalides have pyramidal structures in the gaseous state and form molecular lattices in the solid state. Apart from phosphorus trifluoride (gas), all the trihalides can be prepared by direct halogenation, e.g. phosphorus trichloride can be prepared by passing chlorine over molten white phosphorus in a distillation apparatus from which all the air has been displaced by carbon dioxide, when the trichloride (boiling point 76 °C) distils over.

$$P_4 + 6Cl_2 \rightarrow 4PCl_3$$

Solutions of the trichlorides of antimony and bismuth may also be made by dissolving the trioxides in dilute hydrochloric acid. The trifluoride of phosphorus is prepared indirectly by the reaction:

$$PCl_3 + AsF_3 \rightarrow \underset{\text{gas}}{PF_3} + \underset{\text{liquid}}{AsCl_3}$$

All the trihalides undergo hydrolysis with water. The ease of hydrolysis decreases with increase in atomic number of the element and increases as the electronegativity of the halogen decreases from fluorine to iodine.

$$PX_3 + 3H_2O \rightarrow H_3PO_3 + 3HX$$

$$4AsX_3 + 6H_2O \rightarrow As_4O_6 + 12HX \quad \text{(arsenious oxide)}$$

$$SbX_3 + H_2O \rightarrow SbOX + 2HX \quad \text{(antimony oxohalide)}$$

$$BiX_3 + H_2O \rightarrow BiOX + 2HX \quad \text{(bismuth oxohalide)}$$

The pentahalides

All the pentafluorides are known, together with phosphorus pentachloride (phosphorus(V)chloride) and pentabromide and antimony pentachloride. The non-existence of pentaiodides is probably due to a steric factor (p. 26). The pentafluorides can be prepared from the elements with fluorine (in excess) or by the fluorination of the trifluorides. Similarly, the pentachlorides and pentabromides are prepared by the action of halogen on the trihalide. The apparatus for preparing phosphorus pentachloride is shown diagrammatically in Figure 15.18.

$$PCl_3 + Cl_2 \rightleftharpoons PCl_5 \text{ (solid)}$$

The bonding in the vapour state of the pentahalides involves 3*d* orbitals with dsp^3 hybridization, giving the trigonal bipyramidal structure. In the solid state, however, phosphorus pentachloride and pentabromide have ionic structures $[PCl_4^+][PCl_6^-]$, $[PBr_4^+][Br^-]$. The PCl_4^+ cation is tetrahedral and the PCl_6^- anion is octahedral. The non-existence of the PBr_6^- anion is probably due to the low electronegativity of bromine and a steric factor, i.e. the difficulty of surrounding the relatively small phosphorus

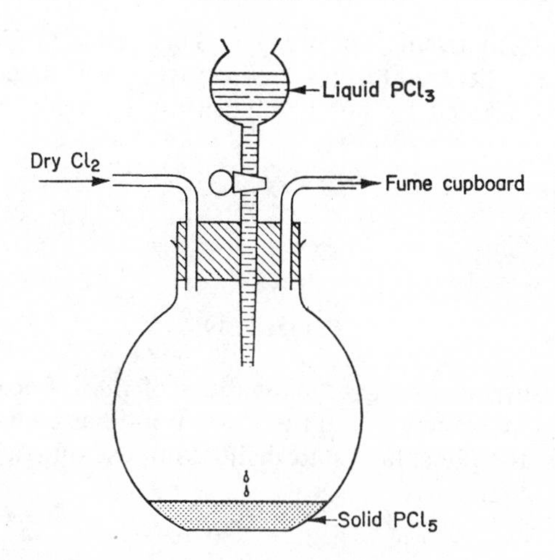

FIGURE 15.18

atoms by six larger bromine atoms. Phosphorus pentachloride also forms a number of addition compounds with certain halides:

$$2PCl_5 + 2AsF_3 \xrightarrow{\text{in } AsCl_3} [PCl_4^+][PF_6^-] + 2AsCl_3$$

$$PCl_5 + ICl \xrightarrow{\text{in } CS_2} [PCl_4^+][ICl_2^-]$$

The pentafluorides are fairly stable to heat but the pentachlorides and pentabromides undergo thermal dissociation.

$$PCl_5 \rightleftharpoons PCl_3 + Cl_2$$
$$SbCl_5 \rightleftharpoons SbCl_3 + Cl_2$$

The pentahalides also hydrolyse readily, the reaction with phosphorus pentahalides involving two stages.

$$PX_5 + H_2O \rightarrow \underset{\text{phosphorus oxohalide}}{POX_3} + 2HX$$

$$POX_3 + \underset{\text{excess}}{3H_2O} \rightarrow H_3PO_4 + 3HX$$

Antimony pentachloride deposits a hydrated pentoxide, $Sb_2O_5.2H_2O$.

Oxohalides

The oxochlorides of phosphorus may be prepared by the action of a little water on the pentachloride or by the distillation of the pentahalides with phosphoric oxide.

$$6PX_5 + P_4O_{10} \rightarrow 10POX_3 \qquad (X = F, Cl, Br)$$

Solid oxochlorides of arsenic, antimony, and bismuth (MOCl) are deposited on hydrolysis of the trichlorides (*see* above). The arsenic compound, however, is decomposed by water, depositing the oxide As_4O_6. Electron

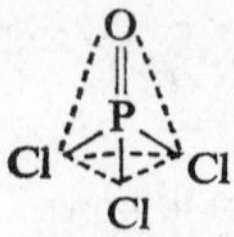

FIGURE 15.19

diffraction measurements on the oxohalides of phosphorus indicate that the molecules are tetrahedral and the P—O bond has considerable double bond character, the phosphorus atoms making use of available *d* orbitals (Figure 15.19).

Oxo Salts of Arsenic, Antimony, and Bismuth

There are no oxo salts of arsenic and an As^{3+} cation is unknown. Antimony also shows little tendency to form the Sb^{3+} cation although antimony sulphate (from antimonious oxide and concentrated sulphuric acid) is a fairly well-defined salt. The hydrated Sb^{3+} ion in water readily hydrolyses, precipitating basic salts containing the antimonyl cation SbO^+.

$$Sb^{3+} + H_2O \rightarrow SbO^+ + 2H^+$$

Bismuth has fairly extensive cation chemistry and salts such as $Bi(NO_3)_3.5H_2O$, $Bi_2(SO_4)_3$, $Bi(ClO_4)_3.5H_2O$ have been crystallized from strongly acidic solutions. Again, although the hydrated bismuth cation is fairly stable in acid solution, dilution and warming initiates hydrolysis and causes precipitation of bismuthyl salts, e.g. $(BiO^+)_2SO_4^{2-}$, $BiO^+NO_3^-$.

16

Group VIB

Physical Properties

	Oxygen O	Sulphur S	Selenium Se	Tellurium Te
Atomic mass	15·9994	32·064	78·96	127·60
Atomic number	8	16	34	52
Isotopes	16, 17, 18	32, 34, 33, 36	80, 78, 82, 76, 77, 74	130, 128, 126, 125, 124, 123, 122, 120
Electronic structure	$2s^2 2p^4$	$3s^2 3p^4$	$3d^{10} 4s^2 4p^4$	$4d^{10} 5s^2 5p^4$
Atomic radius (nm)	0·074	0·103	0·117	0·137
Ionic radius (nm) X^{2-}	0·140	0·184	0·198	0·221
Electronegativity	3·5	2·5	2·4	2·1
Ionization potential ($kJmol^{-1}$) (I)	1316	1006	947	876
(II)	3396	2266	2086	1806
Density (gcm^{-3})	1·27 (solid)	2·06 (rhombic)	4·81 (grey)	6·25
Melting point (°C)	−218·9	119 (monoclinic)	220·2 (grey)	450
Boiling point (°C)	−182·9	444·6	684·8	990
Electrode potential (V) $X_2(g) + 2e + (aq) \rightarrow 2X^-(aq)$	+0·401	−0·48	−0·92	−1·14

General Characteristics

All the atoms of these elements have the electronic configuration ns^2np^4 in their outermost orbitals. They are two electrons short of a noble gas configuration ns^2np^6 and thus can function as electron acceptors.

The bigger the lattice energy, the greater is the stability of an ionic compound. Lattice energies are considerably higher for metallic oxides than for sulphides, selenides, and tellurides. Hence, although ionic oxides exist for a large number of metals, sulphur, selenium, and tellurium only form ionic solids with the alkali metals. Also, since metallic character increases down the group, the metallic selenides and tellurides tend to be more alloy-like. The existence of a large number of ionic oxides as compared to sulphides, selenides, and tellurides can also be accounted for by the higher electronegativity of oxygen (p. 64).

The electrode potentials show that the tendency to form X^{2-} ions in solution decreases markedly from oxygen to tellurium. $E^{\ominus}$ values for oxygen refer to a molar solution of OH^- ions, whereas $E^{\ominus}$ values for sulphur, selenium, and tellurium refer to molar solutions of S^{2-}, Se^{2-}, Te^{2-}.

The high ionization potentials prohibit the formation of cations for all atoms but other methods of completing the outer *p* orbitals involve the formation of:

(*a*) Two single covalent bonds as in water, hydrogen sulphide, and sulphur and tellurium dihalides.

H–O–H   H–S–H   Cl–S–Cl   Cl–Te–Cl

(*b*) A double bond in the case of oxygen and sulphur, e.g.

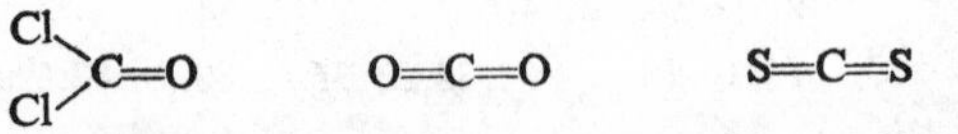

$Cl_2C{=}O$ phosgene   O=C=O carbon dioxide   S=C=S carbon disulphide

Predominantly covalent higher oxidation states of +4 and +6 are shown by sulphur, selenium, and tellurium, e.g. SO_2, SeO_2, TeO_2, SO_3, TeO_3, SF_6, TeF_6. In these compounds promotion of electrons into empty *d* orbitals occurs and these become available for bonding. Thus in TeF_6, sp^3d^2 hybridization occurs giving an octahedral structure. Oxygen, however, has no *d* orbitals and promotion of electrons from the 2*p* to the 3*s* orbital requires energy (1674 kJ mol^{-1}) far in excess of that which would be recovered during bond formation. Oxygen, therefore, has little tendency to form higher oxidation states but is tricovalent in the hydroxonium ion and certain hydroxonium salts, prepared by the action of halogen acid on aldehydes, ethers, and alcohols (p. 154).

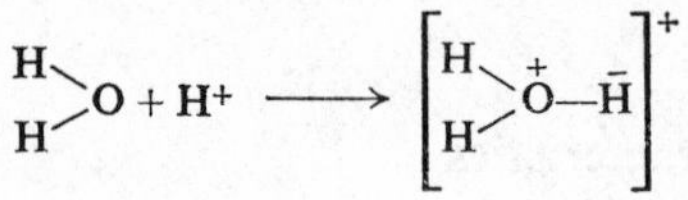

$$H_2O + H^+ \longrightarrow [H_2\overset{+}{O}{-}H]^+ \quad \text{hydroxonium ion}$$

$$(CH_3)_2O + H{-}Cl \longrightarrow [(CH_3)_2\overset{+}{O}{-}H]^+\,Cl^- \quad \text{dimethylhydroxonium chloride}$$

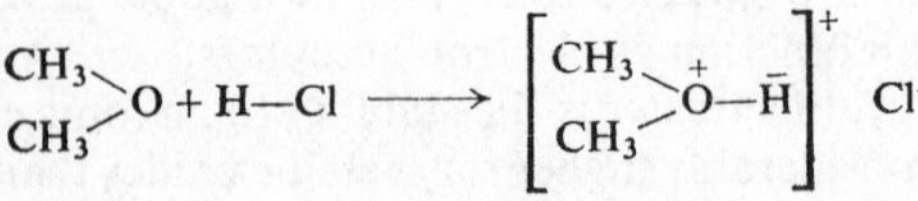

The tendency to form long chains of atoms is evident with elemental sulphur, selenium, and tellurium, and the hydrogen and alkali metal polysulphides. This is accounted for by the fairly high bond energy of the S—S single bond (264 kJ mol^{-1}) and, although the bond energies of Se—Se and Te—Te bonds are smaller, the tendency to form chains persists in the crystalline forms of the elements.

Oxygen

Occurrence

Oxygen comprises 21 per cent by volume of the atmosphere and is found combined as water in the oceans and lakes and as silicates and oxides in the earth's crust.

Preparation

Oxygen may be obtained by heating potassium chlorate, certain higher oxides or the alkali and alkaline earth nitrates.

$$2KClO_3 \xrightarrow{MnO_2 \text{ catalyst}} 2KCl + 3O_2$$

$$2PbO_2 \rightarrow 2PbO + O_2$$

$$2KNO_3 \rightarrow 2KNO_2 + O_2$$

A very convenient method of preparation is to drop water on to sodium peroxide ($Na^+{}_2O—O^{2-}$)

$$2Na^+{}_2(O—O)^{2-} + 2H_2O \rightarrow 4Na^+OH^- + O_2$$

The gas may be dried with calcium chloride or phosphoric oxide. The industrial manufacture of oxygen involves the liquefaction of air and its subsequent fractional distillation. The diagram (Figure 16.1) and the associated notes explain the essentials of this important process.

(*a*) Air at 150 times atmospheric pressure is allowed to expand freely by means of the expansion engine, which also does external work by driving a generator or compressor. Air leaves the engine at 5 times atmospheric pressure and −160 °C.

(*b*) The air liquefies at *A* and is then pumped to *B* at atmospheric pressure. Nitrogen vaporizes and oxygen condenses. Liquid oxygen collects in *C* and gaseous nitrogen passes out at the top of the rectifier.

(*c*) Nitrogen vapour also passes up *D* where it is condensed by the condensers cooled in liquid oxygen. Some collects in the containers at *E* and some passes down *D* condensing the oxygen rising up the column.

(*d*) The nitrogen in *E* is pumped to the top of the column where it vaporizes and cools and condenses the gaseous oxygen which collects in *C*.

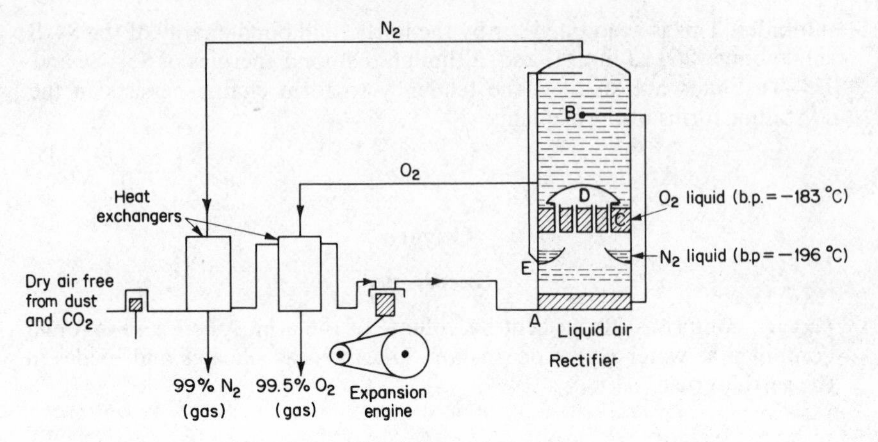

FIGURE 16.1. The Linde process

Properties

Oxygen is a colourless gas, slightly soluble in water (30 cm³ per 1000 cm³ at 20 °C). Many elements and compounds combine with and sometimes burn in oxygen giving corresponding oxides. For example,

$$(a)\ 4Na + O_2 \rightarrow 2Na_2O \qquad (\Delta H = -832 \text{ kJ})$$

The large amount of heat evolved causes ignition.

$$(b)\ C\ (\text{charcoal}) + O_2 \rightarrow CO_2 \qquad (\Delta H = -394 \text{ kJ})$$

Here the large surface area of the charcoal facilitates reaction. Finely divided metals, such as lead, may be so reactive with oxygen as to ignite spontaneously. They are described as pyrophoric.

(*c*) In many cases the products are volatile or gaseous.

$$S + O_2 \rightarrow SO_2 \qquad (\Delta H = -297 \text{ kJ})$$

$$2H_2 + O_2 \rightarrow 2H_2O(l) \qquad (\Delta H = -286 \text{ kJ})$$

A few elements and compounds react with oxygen in the cold, e.g.

$$P_4\ (\text{white}) + 3O_2 \rightarrow P_4O_6$$

$$2NO + O_2 \rightarrow 2NO_2$$

The structure of the oxygen molecule has been discussed previously (p. 40).

Ozone

This is the only allotropic form of oxygen and is an important constituent of the stratosphere where it is formed by the action of ultraviolet light on oxygen. Ozone also occurs to a small extent near working electrical machines and is prepared by subjecting oxygen to an electrical strain using a silent electrical discharge.

$$3O_2 \rightleftharpoons 2O_3 \qquad (\Delta H = +284 \text{ kJ})$$

The mechanism is probably one involving dissociation of molecular oxygen into atoms thus:

$$O_2 \rightarrow O + O$$

$$O + O_2 \rightarrow O_3$$

The equilibrium at ordinary temperatures is not a true one as the ozone very slowly reverts back to oxygen. Catalysts such as manganese dioxide and silver oxide increase the rate of decomposition so that the true equilibrium (with scarcely any ozone) is obtained at room temperature. At 300 °C, however, although a better yield of ozone is obtained as predicted by Le Chatelier's principle applied to endothermic reactions, the true equilibrium is attained rapidly and little ozone is produced. Hence in the ozonizer, no sparking must occur, otherwise ozone will be converted into oxygen. A typical laboratory ozonizer is shown in Figure 16.2, which gives

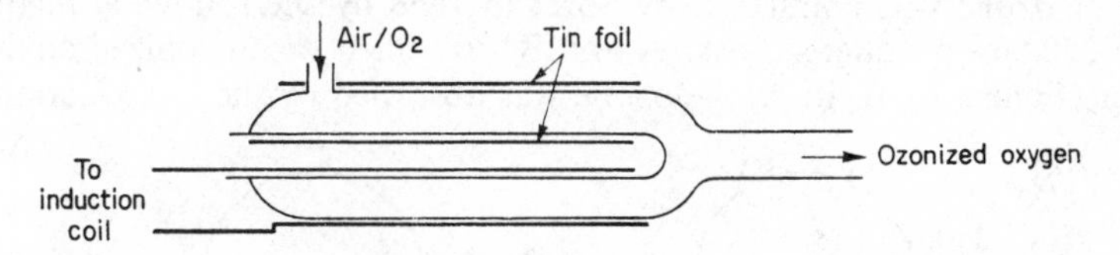

FIGURE 16.2. The Siemens ozonizer consists of two coaxial glass tubes with tinfoil on the inside of the inner and outside of the outer tube. Insulating layer of glass prevents sparking across the air space

a yield of 8–10 per cent of ozone. Higher yields of up to 20 per cent may be obtained with commercial ozonizers.

Properties

Ozone is a pale blue coloured gas, boiling point −112 °C, which is slightly water-soluble. It is a strong oxidizing agent.

$$O_3 + 2H^+ + 2e \rightarrow O_2 + H_2O \qquad (E^\ominus = +2{\cdot}1 \text{ V})$$

Thus ozone liberates iodine from acidified potassium iodide, oxidizes iron(II) salts to iron(III), sulphides to sulphates, and metals to oxides.

$$2I^- + 2H^+ + O_3 \rightarrow H_2O + O_2 + I_2$$

This reaction is used in the quantitative estimation of ozone.

$$2Fe^{2+} + 2H^+ + O_3 \rightarrow 2Fe^{3+} + H_2O + O_2$$

$$S^{2-} + 4O_3 \rightarrow SO_4^{2-} + 4O_2$$

$$2Hg + O_3 \rightarrow (Hg + HgO) + O_2$$

The last reaction listed causes mercury to tail (or stick to glass) and is used as a test for the presence of ozone. Ozone also adds on to unsaturated organic compounds forming ozonides, e.g. turpentine is used to absorb ozone (Figure 16.3).

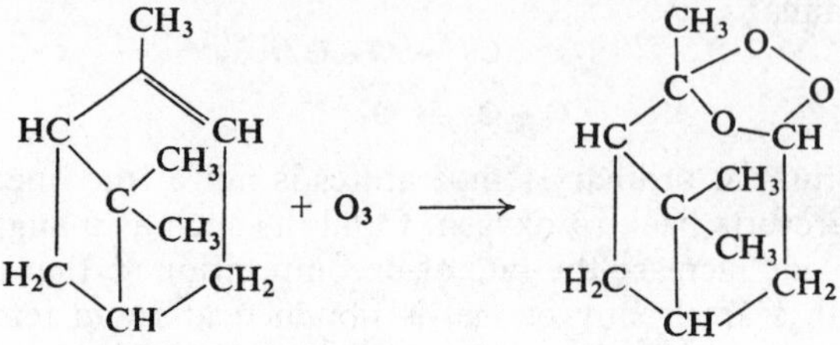

CH₃ ... HC, CH, CH₃, C, CH₃, H₂C, CH₂, CH $+ O_3 \longrightarrow$ CH₃, C, O, O, O—CH, HC, CH₃, C, CH₃, H₂C, CH₂, CH

α-pinene (main constituent of turpentine)

FIGURE 16.3

Ozone is used to sterilize drinking water and to bleach wax. The constitution of ozone was confirmed by Soret in 1868 by the following method. Two 250 cm³ graduated flasks A and B were filled with ozonized air over water (Figure 16.4). In A, the ozone was absorbed by the introduction of

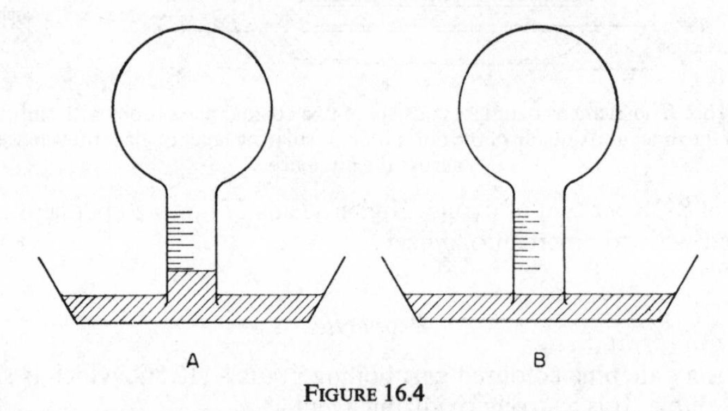

FIGURE 16.4

turpentine and the volume contracted by v cm³. Flask B was then heated and the volume was found to expand by $v/2$ cm³.

Therefore, v cm³ of ozone give $3/2 \times v$ cm³ of oxygen

$$\therefore \text{1 vol of ozone gives 3/2 vol of oxygen}$$

or 2 vol of ozone give 3 vol of oxygen

$$\therefore 2O_n = 3O_2$$

$$\text{and } n = 3$$

Thus the formula for ozone is O_3.

Infrared spectra and electron diffraction measurements have shown that the molecule of ozone is angular with a bond angle of 117°. The three oxygen atoms are equidistant, with a bond length of 0·126 nm. The actual structure is a resonance hybrid of the canonical forms shown below:

$$O{=}O^{+}{-}O^{-} \qquad {}^{-}O{-}O^{+}{=}O \qquad {}^{-}O{-}O{-}O^{+} \qquad {}^{+}O{-}O{-}O^{-}$$

Oxides

The structures of some common oxides are summarized in Table 16.1. More details concerning the individual structures are to be found in the appropriate part of the text.

TABLE 16.1

Structure	*Examples*
1. Antifluorite	Li_2O, Na_2O, K_2O, Rb_2O
2. Rock salt	MgO, CaO, SrO, BaO, CdO, FeO
3. Rutile	TiO_2, SnO_2, MnO_2, PbO_2, VO_2, TeO_2
4. Zinc blende	ZnO, BeO
5. Wurtzite	ZnO
6. Long chains	Sb_2O_3(rhombic), SeO_2, HgO
7. Layers	As_2O_3(monoclinic), SnO, PbO
8. Molecular oxides	P_4O_6, As_4O_6, Sb_4O_6, CO_2, etc.
	P_4O_{10}, As_4O_{10}, Sb_4O_{10}

Unfortunately, no rigid relationship of chemical properties to structure exists but some useful generalizations may be made. The oxides with the antifluorite structure are ionic, readily soluble in water, releasing the oxide ion O^{2-} which reacts rapidly with water producing alkaline solutions.

$$O^{2-} + H_2O \rightarrow 2OH^-$$

Ionic lattices of the rock-salt type have higher lattice energies than those of the antifluorite type (decrease in cationic size). Thus the former are less soluble in and react more slowly with water. The resulting hydroxides also have low solubility, although the solubility of both oxides and hydroxides increases with cation size. Thus magnesium and cadmium oxides are unattacked by cold water owing to the high lattice energies (cationic radii Mg^{2+} 0·065 nm, Cd^{2+} 0·097 nm) and high insolubility of the hydroxides. Barium oxide, however, is soluble and barium hydroxide is the strongest alkali of the group (cationic radius Ba^{2+} 0·135 nm). Oxides of both classes readily dissolve in dilute acids.

Oxides with the rutile-type structures represent the higher oxides of metals, are acidic in nature, and often function as oxidizing agents.

$$PbO_2 + 2NaOH + 2H_2O \rightarrow Na_2Pb(OH)_6$$

$$PbO_2 + 4HCl \xrightarrow{\text{warm}} PbCl_2 + 2H_2O + Cl_2$$

Zinc, beryllium, and aluminium oxides and a large number of layer and chain-type oxides behave amphoterically.

$$SnO + 2H^+ \rightarrow Sn^{2+} + H_2O$$

$$SnO + 2OH^- + H_2O \rightarrow Sn(OH)_4^{2-}$$

Covalent molecular oxides of non-metals are usually strongly acidic oxides and dissolve in water producing acidic solutions.

$$P_4O_{10} + 2H_2O \rightarrow 4HPO_3$$

$$CO_2 + H_2O \rightleftharpoons H_2CO_3$$

However, oxides of the transition metals in their higher oxidation states are also strongly acidic (p. 359), e.g.

$$CrO_3 + H_2O \rightleftharpoons 2H^+ + CrO_4^{2-}$$

A class of oxides not included in the table is the compound oxides which contain the metal in two different oxidation states. Thus red lead, Pb_3O_4 (or $2PbO.PbO_2$), is a macromolecular covalent oxide containing lead in oxidation states +2 and +4. Iron(II) diiron(III) oxide, Fe_3O_4, has a spinel structure with iron(II) and iron(III) ions distributed in the holes between close-packed oxide ions. The chemical properties of these oxides confirm the different oxidation states:

$$Fe_3O_4 + 8HCl \rightarrow Fe^{II}Cl_2 + 2Fe^{III}Cl_3 + 4H_2O$$

$$Pb_3O_4 + 4HNO_3 \rightarrow 2Pb^{II}(NO_3)_2 + Pb^{IV}O_2 + 2H_2O$$

Water

The boiling point of water is 100 °C and the freezing point is 0 °C at s.t.p. The density varies with temperature and is 1 g cm^{-3} at 4 °C. Water is a universal solvent and dissolves most compounds which are essentially ionic but it is difficult to make accurate predictions regarding solubility (p. 162). Besides being a universal solvent and reaction medium, water reacts chemically with many salts (p. 146), halides (p. 331), oxides (p. 285), sulphides (p. 300), nitrides (p. 244), hydrides (p. 157), etc. These reactions are all termed hydrolyses and are discussed under the references given (*hydro*—water; *lysis*—breakdown).

Water is a very weak electrolyte.

$$2H_2O \rightleftharpoons H_3O^+ + OH^- \quad (K = 10^{-14} \text{ at } 25\ °C)$$

Although the value of the dissociation constant K is small, it does account for the amphiprotic character of water and a large number of its reactions.

Water can function both as an oxidizing and reducing agent.

$$2H_2O + 2e \rightarrow H_2 + 2OH^- \quad (E^\ominus = -0{\cdot}83\ V)$$

$$2H_2O \rightarrow O_2 + 4H^+ + 4e \quad (E^\ominus = +1{\cdot}23\ V)$$

Thus sodium, potassium, and calcium liberate hydrogen from cold water.

$$2K \rightarrow 2K^+ + 2e \qquad (E^\circ = -2{\cdot}92\ V)$$

$$2H_2O + 2e \rightarrow H_2 + 2OH^- \qquad (E^\circ = -0{\cdot}83\ V)$$

$$2K + 2H_2O \rightarrow 2K^+OH^- + H_2$$

Zinc ($E^\ominus = -0{\cdot}76$ V) and iron ($E^\ominus = -0{\cdot}44$ V) only react with steam at elevated temperatures. Magnesium ($E^\ominus = -2{\cdot}37$ V) does not react with cold water due to the formation of an insoluble film of magnesium hydroxide (p. 141).

Non-metals such as fluorine ($E^\ominus = +2{\cdot}87$ V) and chlorine ($E^\ominus = +1{\cdot}36$ V) liberate oxygen from water.

$$2H_2O \rightarrow O_2 + 4H^+ + 4e$$

$$2F_2 + 4e \rightarrow 4F^-$$

$$2F_2 + 2H_2O \rightarrow O_2 + 4HF$$

With chlorine, hypochlorous acid is first formed which, in the presence of sunlight, decomposes, liberating oxygen.

$$Cl_2 + H_2O \rightleftharpoons HCl + HOCl$$

$$2HOCl \rightarrow 2HCl + O_2$$

Sulphur ($E^\ominus = -0{\cdot}48$ V), bromine ($E^\ominus = +1{\cdot}06$ V), and iodine ($E^\ominus = +0{\cdot}54$ V) scarcely react with water (p. 327).

Compared with the other gaseous hydrides of Group VI (hydrogen sulphide, hydrogen selenide, and hydrogen telluride), water is anomalous with its high boiling point, latent heat of vaporization, and low vapour pressure. This has already been explained (p. 49) on the basis that liquid water consists of small, hydrogen-bonded units $(H_2O)_n$. Solid ice consists of a three-dimensional open network of water molecules, held together by hydrogen bonds (p. 51). The network of ice is a very open one so that, when it melts, the majority of the hydrogen bonds break and the structure collapses producing liquid water of a higher density than the solid ice.

As the temperature of liquid water rises, the density increases due to the more compact structure arising from hydrogen bond breakage and, at the same time, decreases due to the thermal expansion of the liquid. At 4 °C these two opposing effects balance and water exhibits its maximum density of 1 g cm^{-3}. Above 4 °C, thermal expansion predominates and the density decreases up to the boiling point.

A material known as 'polywater' or 'anomalous water' has been recently prepared by the condensation of water vapour in very fine quartz capillaries suspended in an evacuated chamber. The liquid obtained has a high density, 1·4 g cm^{-3}, a boiling point above 200 °C and a freezing point of –40 °C. The structure of polywater has not been completely resolved but spectroscopic and other studies indicate a three-dimensional network of rings of water molecules. However, recent studies indicate significant amounts of hydrated silica in the water.

Hardness in water

Water is said to be hard when it will not give a lather with soap. Naturally occurring water dissolves magnesium and calcium sulphates by flowing through deposits of gypsum, causing *permanent hardness* (unable to be removed by boiling). The presence of carbon dioxide in rain water causes limestone and magnesium carbonates to dissolve, giving rise to *temporary hardness*.

$$\underset{\text{insoluble}}{CaCO_3} + H_2O + CO_2 \rightarrow \underset{\text{soluble}}{Ca(HCO_3)_2}$$

The cations (Ca^{2+} and Mg^{2+}), which are held in solution, form an insoluble stearate scum with soap.

$$\underset{\substack{\text{sodium stearate}\\\text{(soap)}}}{2C_{17}H_{35}COONa} + Ca^{2+} \rightarrow \underset{\substack{\text{calcium stearate}\\\text{(scum)}}}{(C_{17}H_{35}COO)_2Ca} + 2Na^+$$

The hardness in water causes wastage of soap, damage to fine fabrics, and causes the formation of thermally insulating layers in boilers and pipes due to the decomposition of the bicarbonate.

$$Ca(HCO_3)_2 \rightarrow CaCO_3\downarrow + H_2O + CO_2$$

Temporary hardness may be removed by boiling or by adding the calculated quantity of slaked lime.

$$HCO_3^- + OH^- + Ca^{2+} \rightarrow CaCO_3\downarrow + H_2O$$

Both methods remove calcium ions as insoluble calcium carbonate.

Temporary and permanent hardness may be removed by :

(*a*) Adding the calculated quantity of washing soda ($Na_2CO_3.10H_2O$).

$$Ca^{2+} + CO_3^{2-} \rightarrow CaCO_3\downarrow$$

(*b*) The use of zeolites which are complex hydrated aluminosilicates of sodium and potassium, e.g. $Na_2Al_2Si_3O_{10}.2H_2O$. These occur naturally but are now usually made synthetically under various trade names, e.g. Permutit. Zeolite structures have been elucidated from X-ray analysis. In the zeolites, some of the silicon atoms in the three-dimensional structure of silica are replaced by aluminium and the framework acquires a negative charge which is balanced by positive ions of sodium and potassium which occupy cavities in the open structure (Figure 16.5). Zeolites can take up, and lose, varying amounts of water without any change in structure. Also, the exchange of positive cations may be carried out by running a solution containing the cation through a packed column of zeolite. Thus, in water-softening, the zeolites take up calcium and magnesium ions from the hard water, releasing sodium ions. The zeolites can eventually be regenerated by flushing out with strong brine when the reverse exchange of ions takes place.

Dehydration *in vacuo* removes all the water from the cavities of a zeolite which then functions as a molecular sieve, the cavities trapping atoms or molecules of a certain size. Thus the separation of the individual noble gases and the separation of straight chain from branched chain hydrocarbons has been effected.

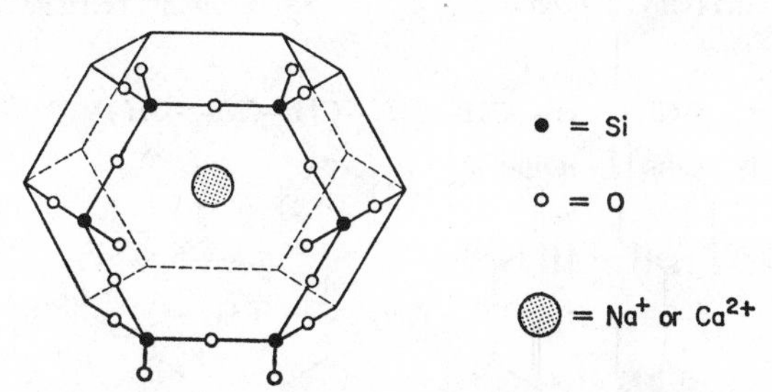

FIGURE 16.5. The three-dimensional framework of $SiAlO_4$ tetrahedra containing four- and six-membered rings which form cavities containing K^+, Na^+, or Ca^{2+} ions (only one such cavity is shown here and the tetrahedra are only drawn for the front face)

(*c*) The use of sodium hexametaphosphate (Calgon), $(NaPO_3)_n$, which consists of rings or long chains of PO_4 tetrahedra.

```
   O     O     O     O
   ||    ||    ||    ||
—P—O—P—O—P—O—P—
   |     |     |     |
   O-    O-    O-    O-
  Na+   Na+      Ca2+
```

When added to hard water, Calgon removes calcium and magnesium ions releasing sodium ions. Calgon is used in the treatment of industrial boiler feed water, in domestic softeners, and in washing powders.

(*d*) The use of synthetic ion-exchange resins which may completely demineralize water, removing both cations and anions. The majority of such resins are made by polymerization of styrene (vinyl benzene) to form long chains of polystyrene which can be cross-linked by copolymerization with divinyl benzene (Figure 16.6).

Groups such as $—SO_3H$, $—COOH$, $—OH$, can be substituted in the aromatic rings to give a cation exchange resin and groups such as $—N(C_2H_5)_3OH$ give anion exchange resins. A mixture of anion and cation exchange resins demineralizes water by reactions such as:

$$2RCOO^-H^+ + Ca^{2+} \rightarrow (RCOO^-)_2Ca^{2+} + 2H^+ \quad \text{cation exchange}$$

$$RN(C_2H_5)_3{}^+OH^- + Cl^- \rightarrow RN(C_2H_5)_3{}^+Cl^- + OH^- \quad \text{anion exchange}$$

$$H^+ + OH^- \rightleftharpoons H_2O$$

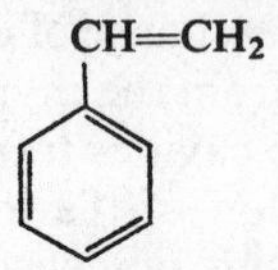

styrene (vinyl benzene)

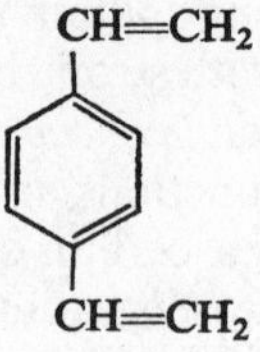

divinyl benzene

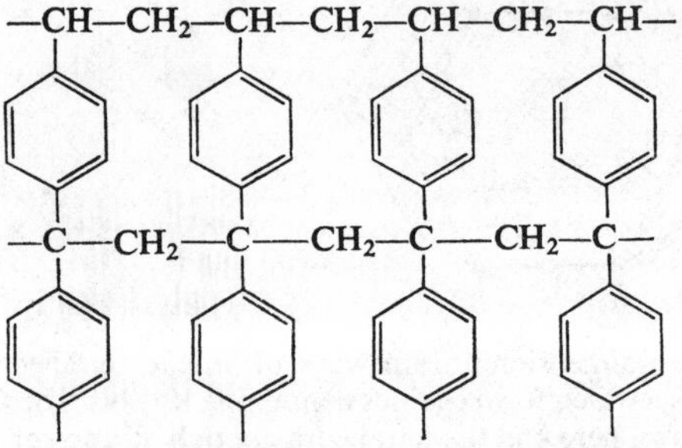

FIGURE 16.6

For regeneration, the cation and anion resins are separated by flotation methods and then treated with dilute acid and alkali respectively.

$$(RCOO)_2Ca + 2H^+ \rightarrow 2RCOOH + Ca^{2+}$$

$$RN(C_2H_5)_3Cl + OH^- \rightarrow RN(C_2H_5)_3OH + Cl^-$$

Other uses of ion exchange resins include the separation of the lanthanides and actinides, the purification of antibiotics and pharmaceutical products, and the removal of salts from sugars and wines.

Water of crystallization

Solutions of many salts, on evaporation and subsequent cooling, deposit crystals containing the salt and water in definite proportions by weight. Such compounds are known as crystalline hydrates and the bound water is known as water of hydration or crystallization. In the majority of crystalline hydrates, the cation is surrounded by a fixed number of molecules of water, usually four or six. The size of the hydrated cation is usually approximately the same as the size of the anion, thus facilitating efficient packing of the two ions. Also, the hydrated cation has less polarizing power than the free ion, enabling the hydrated salt to crystallize with a simple ionic lattice. Thus $Be(H_2O)_4SO_4$ crystallizes with the caesium chloride structure. Some examples of hydrated ions are shown in Figure 16.7.

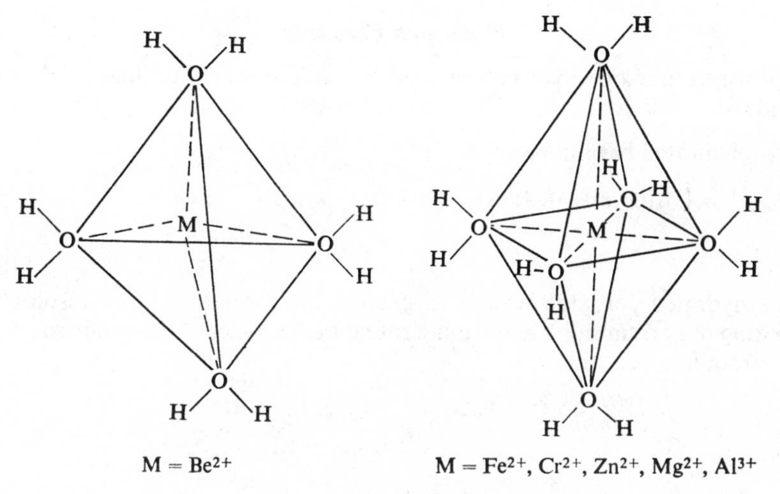

FIGURE 16.7

Although coordinate bonding has been suggested, this probably occurs only in a few aquo-complexes in which full *p* orbitals on oxygen atoms in the water molecules may overlap empty *d* orbitals on the cation, e.g. $Cr(H_2O)_6^{3+}$ $Cl^-{}_3$. In most cases, the bonding is predominantly electrostatic between the charged cation and the polarized water molecule. Note that in $CuSO_4.5H_2O$ and $NiSO_4.7H_2O$, one water molecule is unaccounted for in the given arrangement. The extra water molecule is held in the crystal lattice between the cation water and the anion by hydrogen bonds.

Hydration is most probable with a small, highly charged cation which enhances polarization in the water molecules and attracts the negative ends of the latter. Thus, the tendency of salts to hydrate increases as the ionic radius decreases across a period. Passing down a group, the tendency to hydrate decreases with increasing ionic radius.

Highly hydrated salts, e.g.

$$Na_2SO_4.10H_2O,\ Na_2CO_3.10H_2O,\ Na_2HPO_4.12H_2O$$

do not fit into this class of hydrate because the sodium ion is relatively large and has a small charge. Here water occupies the interstices of the crystal structure and probably some of these water molecules are held together by hydrogen bonds which may also be responsible for binding the ions together in the crystal structure. An extreme form of this is shown by the zeolites where water molecules, contained in open cavities, can be taken in or expelled without any change in structure (p. 289). The inert gases and halogens also form another rather rarer type of hydrate discussed under clathrates (p. 153).

Hydrogen Peroxide

Hydrogen peroxide may be prepared by adding ice-cold dilute sulphuric acid to:

(*a*) hydrated barium peroxide

$$BaO_2.8H_2O + H_2SO_4 \rightarrow BaSO_4\downarrow + 8H_2O + \underset{\text{15 per cent solution}}{H_2O_2}$$

The hydrated peroxide is used to prevent the formation of an insoluble coating of barium sulphate forming round the particles of barium peroxide.

(*b*) sodium peroxide

$$Na_2O_2 + H_2SO_4 \rightarrow Na_2SO_4 + \underset{\text{30 per cent solution—perhydrol}}{H_2O_2}$$

The sodium sulphate separates out as Glauber's salt ($Na_2SO_4.10H_2O$).

Industrial methods include:

(*a*) Electrolysis of concentrated solutions of sulphuric acid or ammonium sulphate at 5 °C, using a platinum anode. Anodic oxidation occurs, producing perdisulphuric acid which, on fractional distillation, produces a 30 per cent solution of hydrogen peroxide.

$$H_2SO_4 \rightleftharpoons H^+ + HSO_4^-$$

$$2HSO_4^- \rightleftharpoons \underset{\text{perdisulphate anion}}{S_2O_8^{2-}} + 2H^+ + 2e$$

$$S_2O_8^{2-} + 2H_2O \rightarrow H_2O_2 + 2HSO_4^-$$

(*b*) The autoxidation of an anthraquinol. Air is blown through a solution of 2-ethyl anthraquinol dissolved in benzene and long-chain alcohols. The hydrogen peroxide formed is extracted by water, giving a product containing 18–20 per cent hydrogen peroxide. The 2-ethyl anthraquinone produced is reduced back to 2-ethyl anthraquinol with hydrogen using a palladium catalyst.

$$\text{2-ethyl anthraquinol (OH, OH, } C_2H_5) \underset{H_2/\text{Pd catalyst}}{\overset{O_2}{\rightleftharpoons}} \text{2-ethyl anthraquinone (O, O, } C_2H_5) + H_2O_2$$

The concentration of hydrogen peroxide solutions is effected by fractional distillation under reduced pressure at 60 °C. By this means, a 90 per cent solution may be obtained which, on fractional crystallization at low temperature, deposits pure crystals of hydrogen peroxide.

Properties

The pure liquid is faintly blue and viscous. The molecule is polar and, like water, is associated due to hydrogen bonding. The boiling point is 84·85 °C at 68 torr and the freezing point is −1·7 °C. Hydrogen peroxide is more acidic than water.

$$H_2O_2 + H_2O \rightleftharpoons H_3O^+ + OOH^- \qquad (K = 1{\cdot}5 \times 10^{-12} \text{ at } 20\ °C)$$

It forms salts, the peroxides, e.g.

$$Na_2{}^+CO_3{}^{2-} + H_2O_2 \rightarrow Na_2{}^+O_2{}^{2-} + H_2O + CO_2$$

Solutions of hydrogen peroxide have a strong tendency to decompose.

$$2H_2O_2 \rightarrow 2H_2O + O_2 \qquad (\Delta H = -196 \text{ kJ})$$

The decomposition is catalysed by the heavy metals, silver, gold, and platinum and also by manganese dioxide. Stabilizers, such as orthophosphoric acid and acetanilide, which complex with traces of heavy-metal ions are added to prevent this decomposition.

The strength of hydrogen peroxide solutions is quoted in volumes, e.g. '20-volume' hydrogen peroxide. For a 20-vol solution, 1 cm^3 of hydrogen peroxide produces 20 cm^3 of oxygen at s.t.p. on decomposition.

$$2H_2O_2 \rightarrow 2H_2O + O_2$$

68 g of hydrogen peroxide give 22·4 l of oxygen at s.t.p.

∴ 1 g of hydrogen peroxide gives 22·4/68 l of oxygen at s.t.p.

1 l of 20-vol hydrogen peroxide gives 20 l of oxygen at s.t.p.

∴ 1 l of 20-vol hydrogen peroxide contains

$$\frac{20}{22{\cdot}4} \times 68 = 60{\cdot}8 \text{ g of } H_2O_2$$

Molarity of 20-vol hydrogen peroxide

$$= \frac{\text{concentration}}{\text{relative molecular mass}} = \frac{60{\cdot}8}{34} = 1{\cdot}78 \text{ M}$$

In aqueous solution, hydrogen peroxide can function both as an oxidizing and as a reducing agent.

$$H_2O_2 + 2H^+ + 2e \rightarrow 2H_2O \qquad (E^\ominus = +1{\cdot}77 \text{ V})$$

$$H_2O_2 \rightarrow O_2 + 2H^+ + 2e \qquad (E^\ominus = +0{\cdot}68 \text{ V})$$

Thus hydrogen peroxide liberates iodine from potassium iodide, oxidizes iron(II) salts to iron(III), and sulphites and sulphides to sulphates:

$$2I^- + 2H^+ + H_2O_2 \rightarrow 2H_2O + I_2$$

$$2Fe^{2+} + 2H^+ + H_2O_2 \rightarrow 2Fe^{3+} + 2H_2O$$

$$SO_3^{2-} + H_2O_2 \rightarrow SO_4^{2-} + H_2O$$

$$S^{2-} + 4H_2O_2 \rightarrow SO_4^{2-} + 4H_2O$$

The latter reaction is used in the restoration of old paintings where the white lead paint has been blackened by traces of hydrogen sulphide in the atmosphere. Hydrogen peroxide also oxidizes acidified potassium dichromate solution to blue chromium peroxide which is stable in ether. The reaction is used both as a test for hydrogen peroxide and chromates.

$$Cr_2O_7^{2-} + 2H^+ + 4H_2O_2 \rightarrow 2CrO_5 + 5H_2O$$

Owing to its powerful oxidizing action, hydrogen peroxide is used as a bleaching agent for hair and silk and in toothpastes, etc.

As a reducing agent, it will reduce ammoniacal silver nitrate to silver, acidified potassium permanganate to a manganese(II) salt, hypohalites to halides, and ozone to oxygen.

$$2Ag(NH_3)_2^+ + H_2O_2 \rightarrow 2Ag + 4NH_3 + 2H^+ + O_2$$

$$2MnO_4^- + 5H_2O_2 + 6H^+ \rightarrow 2Mn^{2+} + 8H_2O + 5O_2$$

$$OBr^- + H_2O_2 \rightarrow Br^- + H_2O + O_2$$

$$O_3 + H_2O_2 \rightarrow 2O_2 + H_2O$$

Recent studies using the ${}^{18}_{8}O$ isotope in hydrogen peroxide show that in the catalytic decomposition of hydrogen peroxide and in the action of oxidizing agents, the evolved oxygen comes entirely from the hydrogen peroxide. Thus no breakage of an O—O bond occurs, indicating the mechanism shown in Figure 16.8.

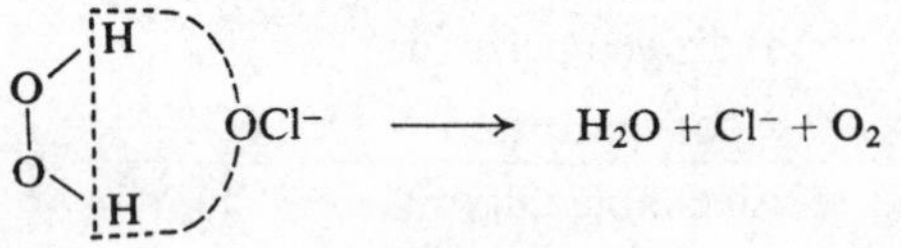

$$\text{H–O–O–H} \cdots OCl^- \longrightarrow H_2O + Cl^- + O_2$$

FIGURE 16.8

Dipole moments and infrared spectral analysis indicate that hydrogen peroxide has a skew structure (Figure 16.9).

Pure crystalline hydrogen peroxide is an infinite, three-dimensional, hydrogen-bonded structure and some of this hydrogen bonding persists in the liquid state giving the viscosity characteristic of concentrated solutions.

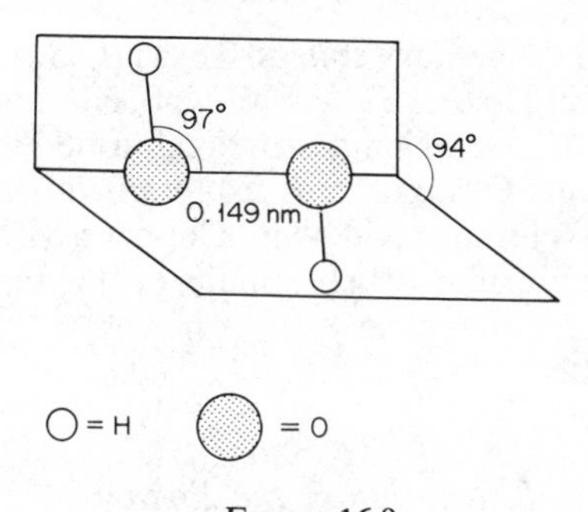

FIGURE 16.9

Peroxides

These are the salts of hydrogen peroxide. The alkali metal and alkaline earth metal peroxides are ionic, containing the peroxide ion $(O—O)^{2-}$. The structures of a large number of peroxides is still not certain, although barium peroxide has a structure of the calcium carbide type (p. 210). *See* Figure 16.10.

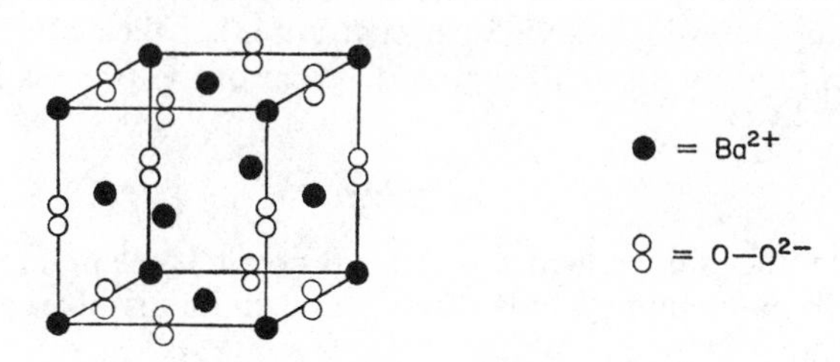

FIGURE 16.10

The peroxides all give hydrogen peroxide on acidification or hydrolysis.

$$(O—O)^{2-} + 2H_2O \rightarrow 2OH^- + H_2O_2$$

$$(O—O)^{2-} + 2H^+ \rightarrow H_2O_2$$

The final oxidation products of the alkali and alkaline earth metals are dark-coloured, crystalline solids termed superoxides, e.g. KO_2, $Ba(O_2)_2$, RbO_2, CsO_2. X-ray studies have shown the presence of the paramagnetic ion $(O—O)^-$. The superoxides readily yield oxygen on heating or with water.

$$2M^+(O—O)^- \rightarrow M^+{}_2(O—O)^{2-} + O_2$$

Some transition metal peroxides are discussed on pp. 371, 379, 448.

Sulphur, Selenium, and Tellurium

Occurrence

Sulphur occurs in an elementary state in Texas (U.S.A.), Italy, and Japan. It is also found as sulphide ores FeS_2, PbS, ZnS, etc., and sulphates $CaSO_4 . 2H_2O$ and $MgSO_4 . 7H_2O$. Selenides and tellurides occur admixed with certain sulphides. Thus FeS_2 contains selenium which accumulates in the lead chambers of the sulphuric acid plant. Copper pyrites, $CuFeS_2$, contains both selenium and tellurium which remain in the anode slimes after the electrolytic purification of copper.

Extraction of the Elements

Sulphur

Elementary sulphur often occurs in extensive beds 150 m beneath rock, quicksand, and clay so that conventional mining is impossible. The first successful extraction was made by Frasch in 1904. A 15 cm diameter metal pipe containing two smaller concentric pipes is sunk into the sulphur bed. Superheated steam, forced through the outer pipe, melts the sulphur; compressed air, pumped through the smallest diameter pipe, causes a light froth of sulphur to be formed which is forced upwards through the third pipe. The sulphur, which is 99·5 per cent pure, is collected at the surface in large vats holding up to 10^8 kg. A diagram of the process is shown in Figure 16.11.

Selenium

The lead chamber sludge, which is reddish brown if selenium is present, is digested with potassium cyanide when the selenium dissolves as potassium selenocyanide.

$$KCN + Se \rightarrow KCNSe$$

After filtering, hydrochloric acid is added to the filtrate, depositing selenium as a red powder.

$$KCNSe + HCl \rightarrow KCl + HCN + Se$$

Selenium and tellurium

The anode slimes from copper refining are fused with sodium nitrate, giving sodium selenite and tellurite (Na_2SeO_3, Na_2TeO_3), which are extracted with water. Acidification with sulphuric acid precipitates tellurium dioxide which is filtered and reduced with carbon. Sulphur dioxide precipitates selenium.

$$TeO_2 + 2C \rightarrow Te + 2CO$$

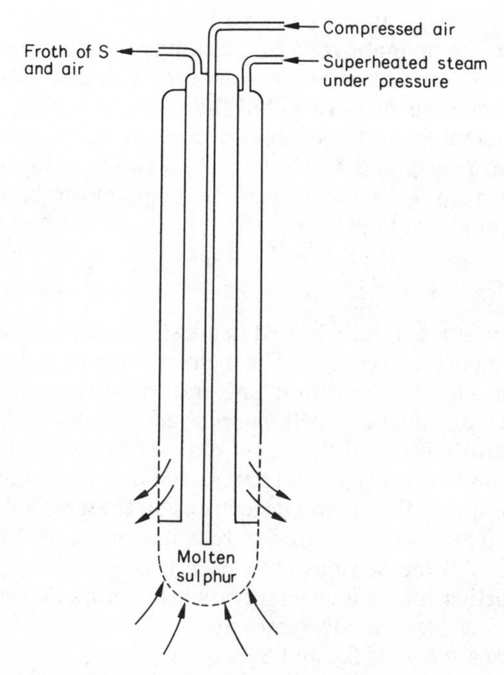

FIGURE 16.11. Frasch pump for the extraction of sulphur

Uses

Vast quantities of sulphur are used for the manufacture of sulphuric acid, required for making fertilizers, etc. Sulphur is also used in the vulcanization of rubber, the manufacture of gunpowder and matches, 'sulpha' drugs, and fungicidal fruit and flower sprays. Selenium and tellurium have also both been used in the vulcanization of rubber, increasing its resistance to oxidation and heat. Selenium is used in the manufacture of red glass, photoelectric cells, and a.c. rectifiers. Certain alloys have also been prepared with selenium and tellurium, e.g. with lead and magnesium, the corrosion resistance of which is increased by alloying.

Allotropy

Sulphur has two solid allotropic modifications:

(*a*) α-Sulphur (rhombic sulphur). This is the main constituent of roll sulphur and consists of S_8 molecules in the form of puckered rings (p. 101). It is soluble in alcohol, ether, and carbon disulphide and it separates from these solvents on slow evaporation as octahedral crystals.

(*b*) β-Sulphur (monoclinic or prismatic sulphur). This form crystallizes from hot molten sulphur above 95·6 °C as long prismatic needles and is also obtained by crystallizing sulphur from hot toluene. The molecule of β-sulphur also consists of S_8 rings but the mode of packing of the rings in the solid is different from that of solid α-sulphur. Above 95·6 °C α-sulphur passes into the β-form and below 95·6 °C β-sulphur slowly reverts to α-sulphur. Thus 95·6 °C is the transition temperature between the two enantiotropic modifications.

$$S\alpha \overset{95.6\,°C}{\rightleftharpoons} S\beta \qquad (\Delta H = +0{\cdot}3 \text{ kJ mol}^{-1})$$

When roll sulphur is carefully heated in a test tube it melts to a pale yellow liquid consisting of S_8 molecules. The melting point of S_α is 113 °C, and of S_β, 119 °C, and since the transition temperature between the two modifications is 95·6 °C, the observed melting point depends upon the heating rate. As the temperature is raised the liquid darkens in colour, becoming more viscous since the S_8 rings begin to rupture and link up forming chains. The colour darkens due to the unpaired electrons on the terminal sulphur atoms of the chains. The viscosity increases to a maximum at 200 °C when the liquid is black. All the S_8 rings have been broken, forming chains which intertwine. Further increase of temperature increases thermal agitation of the chains and the viscosity decreases up the the boiling point at 444·6 °C. The vapour consists of S_6, S_4, and S_2 molecules.

On pouring boiling liquid sulphur into cold water, a grey-brown plastic form (γ-sulphur) is obtained which is, in fact, the supercooled liquid consisting of long intertwined spiral chains (p. 101). On standing, the chains revert to S_8 rings characteristic of rhombic sulphur but some of the chains persist.

Amorphous sulphur (δ-sulphur) may be prepared by: (*a*) cooling sulphur vapour to give flowers of sulphur; or (*b*) acidification of calcium pentasulphide (CaS_5) to give white milk of sulphur.

Colloidal sulphur may be obtained by:

(*a*) Mixing cold solutions of hydrogen sulphide and sulphur dioxide

$$2H_2S + SO_2 \rightarrow 2H_2O + 3S$$

(*b*) Acidifying a solution of sodium thiosulphate with concentrated hydrochloric acid.

$$S_2O_3^{2-} + 2H^+ \rightarrow H_2O + SO_2 + S$$

The stable modifications of selenium and tellurium are the metallic forms, namely grey selenium and silver-white tellurium. Both have similar structures consisting of spiral chains of atoms. Selenium has a very small conductivity which increases 500 times on exposure to light but falls again when the light is cut off. This is termed the photoelectric effect.

Amorphous forms of both elements (red for selenium and brown for tellurium) are obtained by reduction of selenites and tellurites with sulphur dioxide. Both forms revert to the stable metallic form on heating. A monoclinic form is also known for selenium and is prepared by the slow evaporation of a solution of amorphous selenium in carbon disulphide. The red crystals obtained consist of Se_8 molecules (cf. sulphur).

Properties of Sulphur, Selenium, and Tellurium

All the elements burn in air with a blue flame to the dioxides: SO_2 (gas), SeO_2 and TeO_2 (solids). They combine directly with halogens to give halides (p. 316) and with many metals giving the corresponding sulphides, selenides, and tellurides, e.g. ZnS, Na_2Se, Al_2Te_3, etc. Although all the elements are insoluble in non-oxidizing acids, dissolution occurs in concentrated nitric acid, e.g.

$$Se + 4HNO_3 \rightarrow H_2SeO_3 + H_2O + 4NO_2$$

$$S + 6HNO_3 \rightarrow H_2SO_4 + 2H_2O + 6NO_2$$

Sulphur is oxidized to sulphuric and selenium and tellurium to selenious and tellurous acids respectively. On evaporation of the solution containing tellurous acid, a basic nitrate, $(2TeO_2).HNO_3$, is deposited.

With hot concentrated alkali, sulphides (selenides and tellurides) and sulphites (selenites and tellurites) are produced, e.g.

$$3Se + 6OH^- \rightarrow \underset{\text{selenide}}{2Se^{2-}} + \underset{\text{selenite}}{SeO_3^{2-}} + 3H_2O$$

The reaction is complicated for sulphur because of other reactions such as:

$$\underset{\text{sulphite}}{SO_3^{2-}} + \underset{\text{sulphur}}{S} \rightarrow \underset{\text{thiosulphate}}{S_2O_3^{2-}}$$

$$\underset{\text{sulphur}}{4S} + \underset{\text{sulphide}}{S^{2-}} \rightarrow \underset{\text{pentasulphide}}{S_5^{2-}}$$

The metallic sulphides

The structure and properties of the metallic selenides and tellurides are, when known, very similar to the sulphides and will not be discussed further. The structures of some common sulphides are summarized in Table 16.2.

TABLE 16.2

Structure	*Examples*
1. Antifluorite	Na_2S, K_2S
2. Rock salt	MgS, CaS, SrS, BaS
3. Zinc blende	ZnS, CdS, HgS
4. Wurtzite	ZnS, CdS, MnS
5. Pyrites	FeS_2
6. Nickel arsenide	FeS, CoS, NiS
7. Chain structures	Sb_2S_3, Bi_2S_3, SiS_2
8. Layer structures (cadmium iodide type)	SnS_2, PtS_2, TiS_2
9. Molecular sulphides	P_4S_3, P_4S_{10}, N_4S_4

It will be noted that there are often considerable differences between oxide and sulphide structures. The main reason for this is that the bonding in the sulphides has much more covalent character. This is because sulphur is

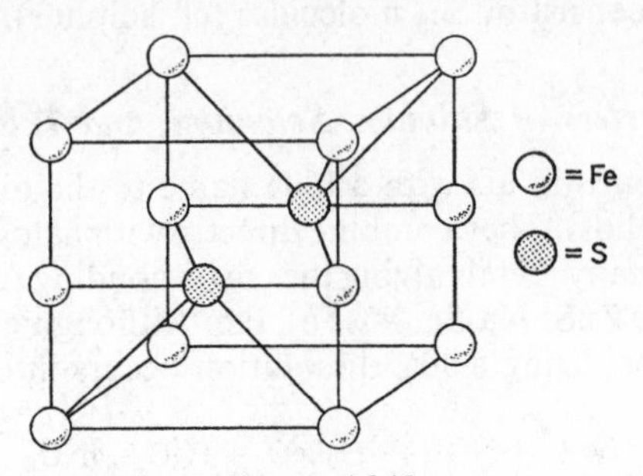

FIGURE 16.12

much less electronegative than oxygen and hence ionic sulphides are formed only with strongly electropositive metals. They are rapidly hydrolysed by water

$$S^{2-} + 2H_2O \rightleftharpoons H_2S + 2OH^-$$

Apart from the nickel arsenide type, the structural types listed have been discussed in Chapter 5 or under the appropriate element. In the nickel arsenide structure (Figure 16.12), exemplified by iron(II) sulphide, each iron atom is surrounded octahedrally by six sulphur atoms and each sulphur atom by six iron atoms.

The metallic atoms are fairly close ($\approx$ 0·26 nm) giving some metallic bonding, probably accounting for the metallic appearance and lustre of these sulphides. Iron(II) sulphide is also an example of a non-stoichiometric sulphide with all the sulphur atoms present in the nickel arsenide type framework but with some of the iron atoms missing giving a composition varying from $Fe_{0\cdot86}S$ to $Fe_{0\cdot89}S$.

The hydrides

Hydrogen sulphide (H_2S), hydrogen selenide (H_2Se), and hydrogen telluride (H_2Te) are all colourless gases with extremely unpleasant smells and are prepared by the acidification of metallic sulphides, selenides, and tellurides, e.g.

$$FeS + 2H^+ \rightarrow Fe^{2+} + H_2S$$

$$Al_2Se_3 + 6H^+ \rightarrow 2Al^{3+} + 3H_2Se$$

The gases all burn in a plentiful supply of air with a blue flame, forming the dioxides.

$$2H_2S + 3O_2 \rightarrow 2H_2O + 2SO_2$$

In a limited supply of air the free elements are deposited. The stability of the hydrides decreases from sulphur to tellurium and solutions of hydrogen selenide and telluride rapidly deposit selenium and tellurium respectively on standing. Acid strength, however, increases from sulphur to tellurium and the first dissociation constants are:

$$K_1(H_2S) = 9 \times 10^{-8}$$
$$K_1(H_2Se) = 1{\cdot}3 \times 10^{-4}$$
$$K_1(H_2Te) = 2{\cdot}7 \times 10^{-3} \quad \text{at} \quad 25^\circ C$$

All the hydrides function as reducing agents.

$$H_2S \rightarrow S + 2H^+ + 2e \qquad (E^\ominus = +0{\cdot}14\ V)$$
$$H_2Se \rightarrow Se + 2H^+ + 2e \qquad (E^\ominus = -0{\cdot}40\ V)$$
$$H_2Te \rightarrow Te + 2H^+ + 2e \qquad (E^\ominus = -0{\cdot}72\ V)$$

Thus iron(III) salts and potassium permanganate are readily reduced.

$$2Fe^{3+} + H_2S \rightarrow 2Fe^{2+} + 2H^+ + S$$
$$2MnO_4^- + 6H^+ + 5H_2S \rightarrow 2Mn^{2+} + 8H_2O + 5S$$

The hydrogen persulphides (sulphanes) and polysulphides

Acidification of the polysulphides of alkali and alkaline earth metals gives a yellow oil which, on careful fractionation under reduced pressure, produces the hydrides H_2S_6, H_2S_5, H_2S_4, and H_2S_2.

$$CaS_x + 2HCl \rightarrow CaCl_2 + H_2S_x$$

Alkali and alkaline earth metal polysulphides are prepared by dissolving sulphur in a hot concentrated solution of the appropriate alkali and have formulae such as Na_2S_n ($n = 2$ to 6) and CaS_n ($n = 2, 3, 4$ or 5). In the formation of polysulphides, the S_8 rings of sulphur are attacked and broken by the sulphide ions (S^{2-}) which function as Lewis bases. The rings may be broken at various points giving chains with different numbers of sulphur atoms (Figure 16.13).

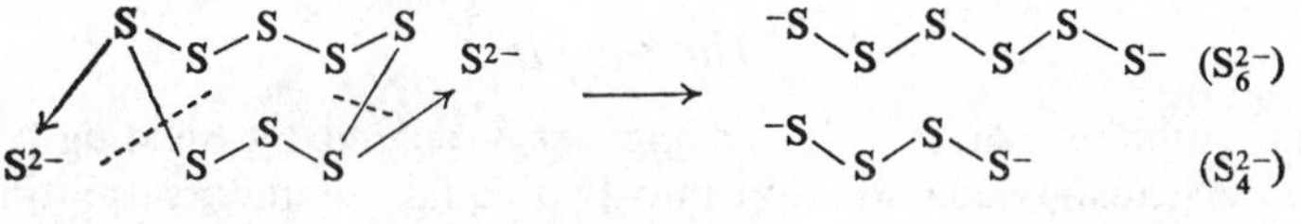

S_8 ring $+ S^{2-}$ $\longrightarrow$ $^-S-S-S-S-S-S^-$ (S_6^{2-}); $^-S-S-S-S^-$ (S_4^{2-})

FIGURE 16.13

The oxides

Sulphur dioxide

Sulphur dioxide, SO_2, may be prepared in the laboratory by the action of:

(*a*) Concentrated sulphuric acid on copper, silver, or mercury.

$$Cu + 2H_2SO_4 \rightarrow CuSO_4(+CuS) + 2H_2O + SO_2$$

(*b*) Dilute acids on sulphites, metabisulphites or thiosulphates.

$$\underset{\text{sulphite}}{SO_3^{2-}} + 2H^+ \rightarrow H_2O + SO_2$$

$$\underset{\text{metabisulphite}}{S_2O_5^{2-}} + 2H^+ \rightarrow H_2O + 2SO_2$$

$$\underset{\text{thiosulphate}}{S_2O_3^{2-}} + 2H^+ \rightarrow H_2O + SO_2 + S$$

This heavy colourless gas with a pungent smell is easily liquefied by compression and cooling in a freezing mixture. The liquid may be used as a medium for non-aqueous reactions, e.g. preparation of thionyl bromide.

$$SOCl_2 + 2KBr \xrightarrow{SO_2} 2KCl + SOBr_2$$

The gas is very soluble in water, the aqueous solution functioning as a reducing agent. The free acid H_2SO_3 has never been isolated.

$$SO_2 + 2H_2O \rightleftharpoons HSO_3^- + H_3O^+ \qquad (K_1 = 1{\cdot}7 \times 10^{-2} \text{ at } 25\ ^\circ\text{C})$$

$$HSO_3^- + H_2O \rightleftharpoons SO_3^{2-} + H_3O^+ \qquad (K_2 = 6{\cdot}2 \times 10^{-8} \text{ at } 25\ ^\circ\text{C})$$

$$H_2SO_3 + H_2O \rightleftharpoons SO_4^{2-} + 4H^+ + 2e \qquad (E^\ominus = +0{\cdot}20 \text{ V})$$

Thus the aqueous solution will reduce iron(III) salts and iodine.

$$2Fe^{3+} + H_2SO_3 + H_2O \rightarrow 2Fe^{2+} + SO_4^{2-} + 4H^+$$

$$I_2 + H_2SO_3 + H_2O \rightarrow 2I^- + SO_4^{2-} + 4H^+$$

With moist hydrogen sulphide, sulphur is precipitated, the sulphite functioning as an oxidizing agent.

$$2H_2S + SO_3^{2-} + 2H^+ \rightarrow 3H_2O + 3S\downarrow$$

Sulphur dioxide bleaches straw, wool, paper, etc., but, on exposure to the air, the original colours may return.

The sulphites

Sodium sulphite, $Na_2SO_3.7H_2O$, may be prepared by dividing a given volume of caustic soda solution into two equal volumes, saturating one with sulphur dioxide, adding the other volume of caustic soda solution, and evaporating.

$$SO_2 + OH^- \rightarrow HSO_3^-$$

$$HSO_3^- + OH^- \rightarrow SO_3^{2-} + H_2O$$

Other sulphites such as potassium and calcium sulphites may be made in a similar manner.

Solid bisulphites are unknown and spectra indicate that, even in solution, the bisulphite ion HSO_3^- does not exist. Evaporation of solutions of so-called bisulphites gives metabisulphites and this latter ion is probably the one which exists in solution.

$$2HSO_3^- \rightarrow \underset{\text{metabi-sulphite}}{S_2O_5^{2-}} + H_2O$$

Structure of sulphur dioxide and the sulphites

Sulphur dioxide is an angular molecule with an O—S—O angle of 119° and a high dipole moment of 1·6 D. The sulphur atom undergoes sp^2 hybridization, resonance occurring between canonical forms (Figure 16.14(*a*)).

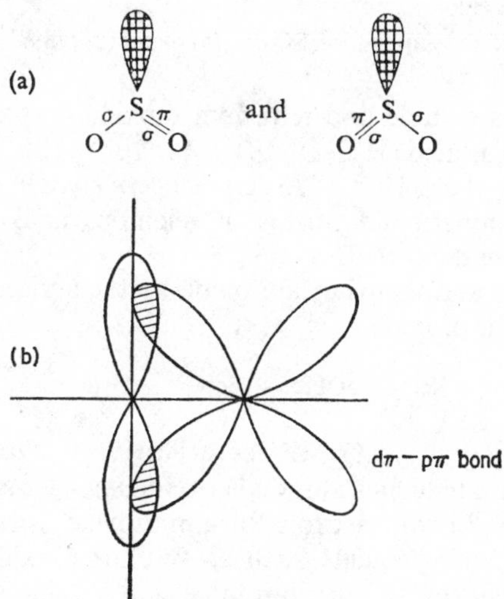

FIGURE 16.14

However, in order to account for the abnormally short S—O bond length of 0·143 nm, some *dπ–pπ* overlap between full oxygen *p* orbitals and empty sulphur *d* orbitals has been postulated (Figure 16.14(*b*)).

The sulphite ion SO_3^{2-} is tetrahedral, indicating sp^3 hybridization (Figure 16.15(*a*)). The metabisulphite ion $S_2O_5^{2-}$ contains an S—S bond and has an S—O bond length of 0·146 nm. All the bonds in the metabisulphite ion are equalized by resonance (Figure 16.15(*b*)).

Selenium and tellurium dioxides

Selenium and tellurium dioxides, SeO_2 and TeO_2, may be prepared by acidifying selenites and tellurites and heating to decompose the resulting selenious and tellurous acids.

$$SeO_3^{2-} + 2H^+ \rightarrow H_2SeO_3 \rightarrow H_2O + SeO_2$$

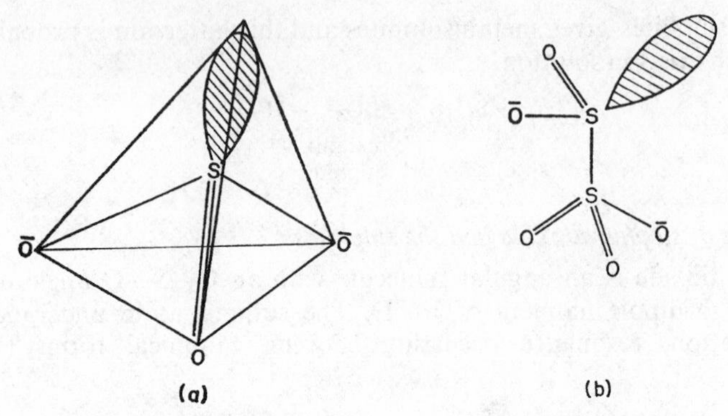

FIGURE 16.15 (*a*) sulphite ion SO_3^{2-}; (*b*) metabisulphite ion $S_2O_5^{2-}$

Selenium dioxide is soluble and tellurium dioxide is sparingly soluble in water giving selenious acid (H_2SeO_3; $K_1 = 4 \times 10^{-3}$ at 25 °C) and tellurous acid (H_2TeO_3; $K_1 = 0{\cdot}6 \times 10^{-5}$ at 25 °C) respectively. On careful evaporation of these solutions, solid prisms of selenious acid and a hydrated tellurium dioxide are deposited.

Metallic selenites and tellurites are made by the action of concentrated or fused alkali on the dioxide.

$$SeO_2 + 2OH^- \rightarrow SeO_3^{2-} + H_2O$$

Alkali metal salts, e.g. Na_2SeO_3, K_2TeO_3, $NaHSeO_3$, $NaHTeO_3$, are well known. Selenites and tellurites are weaker reducing agents than sulphites.

Selenium dioxide, in contrast to tellurium dioxide, is readily reduced to the element by reducing agents such as sulphur dioxide and hydrogen iodide. It is also reduced to selenium in certain specific organic reactions involving the oxidation of —CH_2—CO— to —CO—CO—.

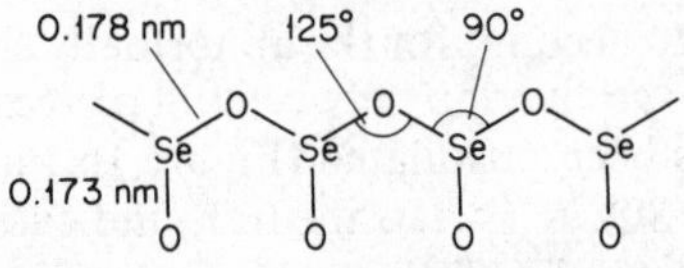

0.178 nm 125° 90°

0.173 nm

FIGURE 16.16

X-ray examination of the solids shows that selenium dioxide consists of non-planar, zig-zag chains (Figure 16.16) and tellurium dioxide has a rutile structure.

Sulphur trioxide

Sulphur trioxide, SO_3, is prepared from sulphur dioxide and oxygen or air.

$$2SO_2 + O_2 \rightleftharpoons 2SO_3 \qquad (\Delta G_f^\ominus = -370 \text{ kJ mol}^{-1})$$

The reaction is thermodynamically feasible but very slow and a platinized asbestos catalyst and a temperature of 500 °C are necessary. The sulphur trioxide condenses as white crystals when cooled in a freezing mixture. Other methods of preparation involve the dehydration and thermal decomposition of sulphates, e.g.

$$H_2SO_4 - H_2O \xrightarrow{P_4O_{10}} SO_3$$

$$2NaHSO_4 \longrightarrow Na_2S_2O_7 + H_2O$$

$$\hookrightarrow Na_2SO_4 + SO_3$$

$$Fe_2(SO_4)_3 \longrightarrow Fe_2O_3 + 3SO_3$$

$$2FeSO_4 \longrightarrow Fe_2O_3 + SO_2 + SO_3$$

Sulphur trioxide is polymorphic and, on cooling the vapour, transparent crystals of α-SO_3 are produced which, in the presence of moisture, are converted to silky asbestos-like needles of β-SO_3. The α-SO_3 structure is built up of S_3O_9 rings but the β-form consists of SO_4 tetrahedra linked together to form long chains (Figure 16.17).

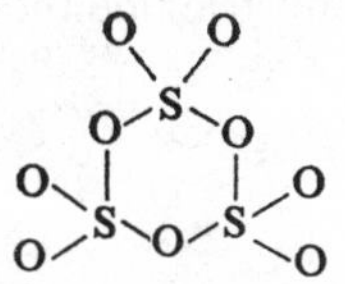

α-SO_3 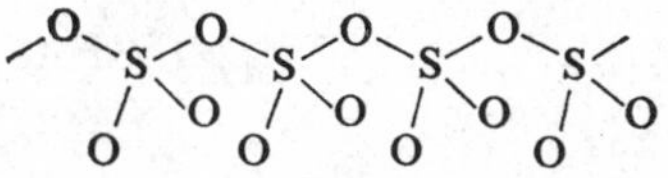 β-SO_3

FIGURE 16.17

In addition to these two well-defined forms, other solid and liquid polymorphs have been described but their structures are not known. Although β-SO_3 dissolves slowly in water, α-SO_3 dissolves vigorously with evolution of heat forming sulphuric acid

$$\alpha\text{-}SO_3\,(s) + H_2O \rightarrow H_2SO_4 \qquad (\Delta H = -156\ \text{kJ})$$

Dissolution also occurs smoothly in concentrated sulphuric acid producing oleum or pyrosulphuric acid.

$$SO_3 + H_2SO_4 \rightarrow H_2S_2O_7$$

The product deposits as colourless crystals on cooling.

The oxide is strongly acidic and reacts readily with basic oxides on heating.

$$CaO + SO_3 \rightarrow CaSO_4$$

Structure. In sulphur trioxide, the sulphur atom is sp^2 hybridized and the molecule is flat and planar with a bond angle of 120° and an S—O bond length of 0·153 nm. Resonance exists between the canonical forms shown.

$$\begin{array}{c}O^-\\|\\S^{2+}\\{}^-O\diagup\ \diagdown\!\!=O\end{array}\qquad\begin{array}{c}O^-\\|\\S^{2+}\\O=\!\!\diagup\ \diagdown O^-\end{array}\qquad\begin{array}{c}O\\\|\\S^{2+}\\{}^-O\diagup\ \diagdown O^-\end{array}$$

In addition to $p\pi$–$p\pi$ overlap, some $d\pi$–$p\pi$ overlap occurs, accounting for the short S—O bond length.

Selenium and tellurium trioxides

Both trioxides, SeO_3 and TeO_3, decompose on heating to the dioxide and oxygen and are prepared by careful dehydration of higher oxo acids.

$$\underset{\text{selenic acid}}{H_2SeO_4} \xrightarrow{\text{vacuum}} H_2O + SeO_3$$

$$\underset{\text{telluric acid}}{Te(OH)_6} \xrightarrow{300\,°C} 3H_2O + TeO_3$$

Tellurium trioxide is insoluble in water but selenium trioxide dissolves in both water and selenic acid.

$$SeO_3 + H_2O \rightarrow \underset{\text{selenic acid}}{H_2SeO_4}$$

$$SeO_3 + H_2SeO_4 \rightarrow H_2Se_2O_7$$

Two crystalline forms of selenium trioxide are known, α-SeO_3 consisting of trimers $(SeO_3)_3$ and a polymerized β-form (cf. sulphur). Tellurium trioxide has a three-dimensional structure consisting of TeO_6 octahedra sharing all corners.

The oxo acids of sulphur

Sulphuric acid

Sulphuric acid is a chemical of vast importance and the prosperity of any industrial community is fairly closely related to its sulphuric acid production.

The manufacture of sulphuric acid requires the production of sulphur dioxide which may be obtained from:

(*a*) Sulphur by the reaction $S + O_2 \rightarrow SO_2$.

(*b*) Iron pyrites and zinc blende (U.K. and Europe). On roasting in air, these ores give a small yield of sulphur dioxide but, in the case of zinc blende, zinc oxide is used to produce zinc and the sulphur dioxide is only a by-product.

$$4FeS_2 + 11O_2 \rightarrow 2Fe_2O_3 + 8SO_2$$

$$2ZnS + 3O_2 \rightarrow 2ZnO + 2SO_2$$

(*c*) Anhydrite, $CaSO_4$ (U.K. and Europe). Crushed anhydrite or gypsum ($CaSO_4 . 2H_2O$), coke, shale, and sometimes pyrites, is charged into a furnace fired by a pulverized-coal burner when the following reactions occur:

$$CaSO_4 + 2C \rightarrow 2CO_2 + CaS$$

$$CaS + 3CaSO_4 \rightarrow 4CaO + 4SO_2$$

$$3CaS + CaSO_4 \rightarrow 4CaO + 4S$$

The lime produced reacts with the shale or clay, producing calcium silicates and aluminates (cement). One kilogramme of cement is produced for every kilogramme of acid obtained from anhydrite.

$$4CaO + Al_2Si_2O_7 \rightarrow \underbrace{2CaSiO_3 + Ca_2Al_2O_5}_{\text{cement}}$$

The impure gas from sources (*b*) and (*c*) only contains up to 9 per cent sulphur dioxide and is purified by electrostatic precipitators, washed, dried, and passed into converters for the production of sulphuric acid. There are two main industrial processes for the production of sulphuric acid from sulphur dioxide.

(*a*) *The contact process.* This requires the formation of sulphur trioxide from sulphur dioxide and oxygen from air.

$$2SO_2 + O_2 \rightleftharpoons 2SO_3 \qquad (\Delta H = -192 \text{ kJ})$$

From Le Chatelier's principle:

(i) As the temperature rises, the percentage of sulphur trioxide in the equilibrium mixture falls. For example, at 400 °C there is 99 per cent conversion to sulphur trioxide, at 600 °C there is 70 per cent conversion, and at 1000 °C, no conversion.

(ii) Since there is a contraction in volume during the gaseous reaction (3 volumes of reactants producing 2 volumes of products), increase in pressure will increase the yield of sulphur trioxide. At constant pressure, for a given temperature, the equilibrium constant (K) may be written:

$$K = \frac{[SO_3]^2}{[SO_2]^2[O_2]}$$

An increase in the concentration of oxygen will therefore cause an increase in the concentration of sulphur trioxide.

Thus the best yield of sulphur trioxide is obtained at low temperature and high pressure with excess oxygen. However, although a good yield is obtained at 400 °C, equilibrium is only established after a considerable time and, to obtain the maximum yield in the minimum of time, a higher temperature of 500 °C is used, together with a catalyst of platinized asbestos or vanadium pentoxide. Excess air is used but increased pressures are not employed since the yield (98 per cent) is sufficiently good at atmospheric pressure. The gaseous mixture of sulphur trioxide and air from the converter is absorbed in concentrated sulphuric acid. Water is unsuitable for absorption since the fine mist of sulphur trioxide passes straight through it.

$$SO_3 + H_2SO_4 \rightarrow \underset{\text{oleum}}{H_2S_2O_7}$$

The acid leaving the absorption tower is diluted with water to give 98–99 per cent sulphuric acid. Some of this acid is recirculated to the absorbers and the rest is run off for sale. An average contact plant can produce up to 100 tonnes of 98 per cent sulphuric acid per day.

(*b*) *The lead chamber process.* Here the oxidation of sulphur dioxide is effected catalytically by means of oxides of nitrogen in the presence of water. Sulphur dioxide and oxygen from air are mixed with nitric oxide from the catalytic oxidation of ammonia (p. 257). The gases pass through the Glover tower into lead chambers which are sprayed with water or steam.

$$2NO + O_2 \rightleftharpoons 2NO_2$$

$$NO_2 + SO_2 + H_2O \rightarrow H_2SO_4 + NO$$

Sulphuric acid (65 per cent) collects in the lead chambers.

The remaining gases, nitric oxide, oxygen, and nitrogen, pass to the Gay-Lussac tower which is packed with coke and meet a descending stream of 78 per cent sulphuric acid. This absorbs the remaining oxides of nitrogen forming nitrosonium bisulphate.

$$2H_2SO_4 + NO_2 + NO \rightleftharpoons 2NO^+HSO_4^- + H_2O$$

This, together with some of the chamber acid, is pumped to the top of the Glover tower where the spray meets an ascending stream of hot gases. In

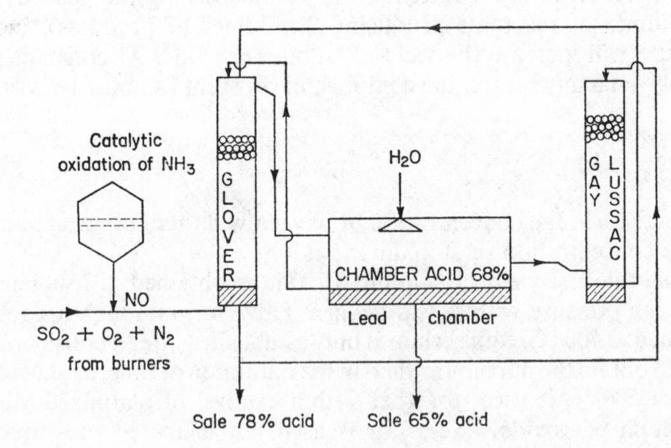

FIGURE 16.18

this tower the ascending gases are cooled and the nitrosonium bisulphate is decomposed, releasing the oxides of nitrogen.

$$2NO^+HSO_4^- + H_2O \rightleftharpoons 2H_2SO_4 + NO_2 + NO$$

The decomposition of the nitrosonium bisulphate concentrates the chamber acid up to 78 per cent, some of which is pumped to the top of the Gay-Lussac tower and the remainder run off.

Chamber acid (68 per cent) is used to make superphosphate of lime and the 78 per cent acid (brown oil of vitriol) may be concentrated to 98 per cent but, since it is usually cheaper to produce highly concentrated acid by the contact method, most of the acid is used directly as produced. A diagram of the lead chamber process is shown in Figure 16.18.

The principal uses of sulphuric acid, together with the percentages of the total production, are summarized below.

	per cent
Superphosphate and fertilizers	23·7
Titanium dioxide	14·2
Ammonium sulphate	13·0
Rayon and transparent paper	10·0
Soaps, detergents	5·0
Iron pickling	5·0
Dyestuffs, drugs, explosives.	4·7
Oil refining, petroleum products	3·0
Miscellaneous	21·4
	100.0

Properties of sulphuric acid

Pure sulphuric acid is a heavy, colourless, oily liquid which freezes when cooled to give colourless crystals of melting point 10·5 °C. These crystals have a hydrogen-bonded layer structure (p. 51), and hydrogen bonding persists to some extent in the liquid state giving rise to the viscosity characteristic of concentrated sulphuric acid. A constant boiling mixture of 98·3 per cent sulphuric acid distils at 338 °C. Pure sulphuric acid contains a number of ionic and molecular species in equilibrium:

$$2H_2SO_4 \rightleftharpoons H_2O + H_2S_2O_7$$
$$H_2SO_4 + H_2O \rightleftharpoons H_3O^+ + HSO_4^-$$
$$H_2SO_4 + H_2S_2O_7 \rightleftharpoons H_3SO_4^+ + HS_2O_7^-$$

The chemical properties are now considered:

(*a*) *Affinity for water*. The addition of concentrated sulphuric acid to water causes the evolution of a great deal of heat and a number of hydrates, e.g. $H_2SO_4.H_2O$ and $H_2SO_4.2H_2O$, have been frozen out from the resulting solution. These hydrates are probably hydroxonium salts, $H_3O^+ HSO_4^-$, $(H_3O^+)_2SO_4^{2-}$, and the heat evolved when concentrated sulphuric acid is added to water is due to the hydration of the proton.

$$H^+ + H_2O \rightleftharpoons H_3O^+ \qquad (\Delta H = -1200 \text{ kJ mol}^{-1})$$

Thus concentrated sulphuric acid can remove water from many compounds, e.g.

$$\underset{\text{blue}}{CuSO_4.5H_2O} \rightarrow \underset{\text{white}}{CuSO_4} + 5H_2O$$

$$\underset{\text{cane sugar}}{C_{12}H_{22}O_{11}} \rightarrow 12C + 11H_2O$$

(*b*) *Acidic properties*. Sulphuric acid is a strong acid. The first-stage dissociation is almost complete in dilute solution.

$$H_2SO_4 + H_2O \rightleftharpoons H_3O^+ + HSO_4^-$$

The second dissociation is less complete.

$$HSO_4^- + H_2O \rightleftharpoons H_3O^+ + SO_4^{2-} \qquad (K_2 = 10^{-2} \text{ at } 25\ °C)$$

Both normal and acid sulphates are known for the majority of metals and many sulphates occur naturally, e.g. $CaSO_4.2H_2O$ (gypsum), $MgSO_4.7H_2O$ (Epsom salts), $Na_2SO_4.10H_2O$ (Glauber's salt). All sulphates are soluble in water apart from those of calcium, strontium, barium, lead, and mercury. The alkali metal bisulphates (p. 172) are stable and anhydrous and decompose at red heat to the pyrosulphate. Sulphuric acid vapour decomposes at 444 °C.

$$H_2SO_4 \rightarrow H_2O + SO_3$$

At higher temperatures sulphur dioxide and oxygen are produced.

Although the sulphates of sodium, potassium, magnesium, and calcium are strongly resistant to heat many other sulphates decompose producing sulphur dioxide and oxygen or sulphur trioxide alone.

$$2Al_2(SO_4)_3 \rightarrow 2Al_2O_3 + 6SO_2 + 3O_2$$

$$CuSO_4 \rightarrow CuO + SO_3$$

(*c*) *Oxidizing properties*. The concentrated acid has some oxidizing action which falls rapidly with dilution.

$$SO_4^{2-} + 4H^+ + 2e \rightarrow H_2SO_3 + H_2O \qquad (E^{\ominus} = +0{\cdot}20 \text{ V})$$

Thus the concentrated acid will oxidize carbon and sulphur to the dioxides and liberate iodine from potassium iodide.

$$SO_4^{2-} + 4H^+ + 2I^- \rightarrow 2H_2O + SO_2 + I_2$$

(*d*) *Action of sulphuric acid on metals*

(i) Metals with $E^{\ominus}$ values above hydrogen liberate hydrogen from the dilute acid, e.g. magnesium, zinc, and iron.

$$Mg + 2H^+ \rightarrow Mg^{2+} + H_2$$

(ii) Copper, silver, and mercury with $E^{\ominus}$ values from +0·34 to +0·80 V are not attacked by dilute sulphuric acid.

(iii) Most metals are attacked by concentrated sulphuric acid, yielding sulphur dioxide.

$$SO_4^{2-} + 4H^+ + 2e \rightarrow SO_2 + 2H_2O$$

$$Cu \rightarrow Cu^{2+} + 2e$$

$$Cu + SO_4^{2-} + 4H^+ \rightarrow Cu^{2+} + SO_2 + 2H_2O$$

Further reduction often occurs producing metallic sulphides.

$$SO_4^{2-} + 8H^+ + 8e \rightarrow S^{2-} + 4H_2O$$

The constitution of sulphuric acid

Phosphorus pentachloride reacts with concentrated sulphuric acid producing chlorosulphonic acid and sulphuryl chloride, where one and two hydroxyl groups respectively are replaced by chlorine.

$$PCl_5 + SO_2(OH)_2 \rightarrow \underset{\text{chlorosulphonic acid}}{SO_2.OH.Cl} + POCl_3 + HCl$$

$$PCl_5 + SO_2OHCl \rightarrow \underset{\text{sulphuryl chloride}}{SO_2Cl_2} + POCl_3 + HCl$$

Both chlorosulphonic acid and sulphuryl chloride are liquids, boiling at 151°C and 60°C respectively, and can be separated by fractional distillation. Only one chlorosulphonic acid is known, indicating that the —OH groups

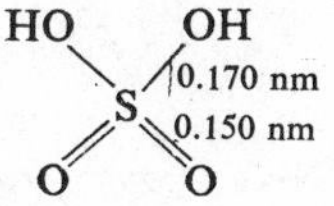

HO OH
S
O O
0.170 nm
0.150 nm

FIGURE 16.19

in sulphuric acid are symmetrically distributed. Sulphuryl chloride hydrolyses back through chlorosulphonic acid to sulphuric acid. Hence the formula for sulphuric acid is $SO_2(OH)_2$. The molecule of sulphuric acid is tetrahedral (Figure 16.19).

In the sulphate ion, which is also tetrahedral, all the bond lengths are 0·15 nm indicating resonance between a number of forms.

Selenic and telluric acids

Both acids have been isolated as colourless crystals by the oxidation of the dioxides or the selenites and tellurites by chloric and perchloric acid. Selenic acid is also obtained by the oxidation of silver selenite with bromine.

$$Ag_2SeO_3 + Br_2 + H_2O \rightarrow H_2SeO_4 + 2AgBr$$

Selenic acid, melting point 57 °C, is soluble in water and the solution is strongly acidic (K_1 complete; $K_2 = 1{\cdot}15 \times 10^{-2}$ at 298 K). The acid chars organic material, oxidizes concentrated hydrochloric acid to chlorine, liberates iodine from potassium iodide, and dissolves copper forming copper selenate ($CuSeO_4$).

$$SeO_4^{2-} + 4H^+ + 2e \rightarrow H_2O + H_2SeO_3 \qquad (E^{\ominus} = +1{\cdot}15\ V)$$

Well-characterized salts, the selenates and biselenates, have been made and, in most cases, these are isomorphous with the sulphates or bisulphates. The solid has a similar structure to that of crystalline sulphuric acid but is less thermally stable and loses oxygen at 210 °C.

Telluric acid ($Te(OH)_6$) is quite anomalous, being sparingly soluble in water, the solution being weakly acidic ($K = 6 \times 10^{-7}$ at 25 °C). Tellurates are known and may be prepared by oxidizing tellurites with hypochlorite.

$$TeO_3^{2-} + OCl^- + 2H_2O \rightarrow TeO_2(OH)_4^{2-} + Cl^-$$

The structure of the acid consists of octahedral $Te(OH)_6$ groups linked by hydrogen bonds. The acid is an oxidising agent.

$$H_6TeO_6 + 2H^+ + 2e \rightarrow 4H_2O + TeO_2 \qquad (E^{\ominus} = +1{\cdot}02\ V)$$

Amide derivatives of sulphuric acid

Sulphamic acid and sulphamide are formed by the action of ammonia gas on chlorosulphonic acid and sulphuryl chloride, respectively.

$$O_2S(OH)Cl + NH_3 \longrightarrow HCl + O_2S(OH)NH_2$$

chlorosulphonic acid → sulphamic acid

$$O_2SCl_2 + 2NH_3 \longrightarrow 2HCl + O_2S(NH_2)_2$$

sulphuryl chloride → sulphamide

The technical production of sulphamic acid is carried out by heating urea with excess of fuming sulphuric acid.

$$HO{-}SO_2{-}O{-}SO_2{-}OH + H_2N{-}CO{-}NH_2 \longrightarrow CO_2 + HO{-}SO_2{-}NH_2 + NH_2{-}SO_2{-}OH$$

Both sulphamic acid and sulphamide are stable white solids. Sulphamic is one of the very few strong solid monobasic acids which can be directly weighed and used as a standard for bases. Solutions of the free acid slowly undergo hydrolysis.

$$O_2S(OH)(NH_2) + H{-}OH \longrightarrow O_2S(OH)(OH) + NH_3 \longrightarrow O_2S(ONH_4)(OH)$$

ammonium bisulphate

Ammonium sulphamate finds uses as a weed killer and fireproofing agent.

$$O_2S(ONH_4)(NH_2)$$

Peroxoacids of sulphur (persulphuric acids)

Two peroxoacids are well known, peroxomonosulphuric acid (H_2SO_5) and peroxodisulphuric acid ($H_2S_2O_8$). Crystals of the latter may be obtained by the low-temperature electrolysis of 50 per cent sulphuric acid using a platinum anode and high current density (p. 292). Both peroxomono- and peroxodisulphuric acid can be made by the action of anhydrous hydrogen peroxide on chlorosulphonic acid, the reactions indicating the structures of both compounds.

$$O_2S(OH)(Cl) + H{-}O{-}O{-}H \longrightarrow HCl + O_2S(OH)(O{-}O{-}H)$$

peroxomonosulphuric acid

$$O_2S(OH)(O{-}OH) + HO(Cl)SO_2 \longrightarrow HCl + O_2S(OH){-}O{-}O{-}SO_2(OH)$$

peroxodisulphuric acid

The ammonium salt of peroxodisulphuric acid, $(NH_4)_2S_2O_8$, is prepared by the electrolysis of a solution of ammonium sulphate in sulphuric acid. It is used as a bleaching agent. All the persulphates are extremely powerful oxidizing agents, although the reactions are very slow unless silver salts are present.

$$S_2O_8^{2-} + 2e \rightarrow 2SO_4^{2-} \qquad (E^{\ominus} = +2{\cdot}01 \text{ V})$$

No salts of permonosulphuric acid have ever been obtained pure.

Thiosulphuric acid and the thiosulphates

Acidification of an alkali thiosulphate gives a solution of thiosulphuric acid which rapidly decomposes.

$$H_2S_2O_3 \rightarrow H_2SO_3 + S$$

With an ethereal solution of sulphur trioxide and hydrogen sulphide at –75 °C, a solution of the acid is formed which is stable up to –5 °C. The acid

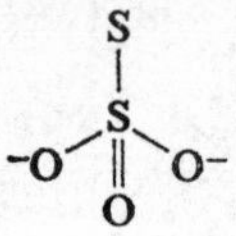

FIGURE 16.20

has also been isolated as an ether complex $H_2S_2O_3.2(C_2H_5)_2O$. X-ray studies of sodium thiosulphate ($Na_2S_2O_3.5H_2O$) reveal a tetrahedral structure for the thiosulphate ion (Figure 16.20).

Alkali metal thiosulphates are readily obtained by boiling sulphite solutions with sulphur.

$$SO_3^{2-} + S \rightarrow S_2O_3^{2-}$$

The sodium salt (hypo) is used:

(*a*) As an antichlor to remove excess chlorine from bleached fabrics.

$$S_2O_3^{2-} + Cl_2 + H_2O \rightarrow SO_4^{2-} + S + 2Cl^- + 2H^+$$

(*b*) As a volumetric standard for the estimation of iodine, when it gives a quantitative yield of sodium tetrathionate.

$$2S_2O_3^{2-} + I_2 \rightarrow 2I^- + S_4O_6^{2-}$$

(*c*) As a fixing agent in photography, when it complexes with unchanged silver halide.

$$Ag^+ + 2S_2O_3^{2-} \rightarrow [Ag(S_2O_3)_2]^{3-}$$

Hyposulphurous acid (dithionous acid)

The sodium salt of hyposulphurous acid is prepared by reducing sodium bisulphite solution with zinc dust and sulphur dioxide, when thin prisms of $Na_2S_2O_4.2H_2O$ can be crystallized.

$$2HSO_3^- + Zn + SO_2 \rightarrow S_2O_4^{2-} + H_2O + Zn^{2+} + SO_3^{2-}$$

The free acid $H_2S_2O_4$ has never been isolated and solutions of the $S_2O_4^{2-}$ ion rapidly disproportionate to bisulphite and thiosulphate.

$$2S_2O_4^{2-} + H_2O \rightarrow S_2O_3^{2-} + 2HSO_3^-$$

The dithionites are powerful reducing agents which bleach indigo and the vat dyes. The dithionite ion has an abnormally long S—S bond (0·239 nm) and the S—O bonds show considerable double bond character (Figure 16.21).

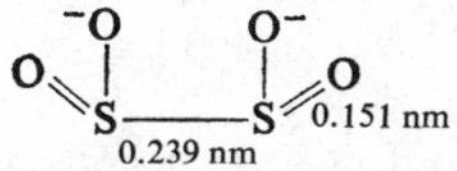

$^-$O O$^-$
O S——S O 0.151 nm
0.239 nm

FIGURE 16.21

The thionic acids

A group of acids exists with the general formula $H_2S_nO_6$:

$n = 2$ $H_2S_2O_6$ dithionic acid
$n = 3$ $H_2S_3O_6$ trithionic acid
$n = 4$ $H_2S_4O_6$ tetrathionic acid
$n = 5$ $H_2S_5O_6$ pentathionic acid
$n = 6$ $H_2S_6O_6$ hexathionic acid

The acids and the salts are difficult to prepare and individual methods are employed for each acid. One general method, however, involves the action of sulphur chlorides on sulphites or thiosulphates in ethereal solution.

$$SCl_2 + 2SO_3^{2-} \rightarrow S_3O_6^{2-} + 2Cl^-$$

$$SCl_2 + 2S_2O_3^{2-} \rightarrow S_5O_6^{2-} + 2Cl^-$$

$$S_2Cl_2 + 2SO_3^{2-} \rightarrow S_4O_6^{2-} + 2Cl^-$$

$$S_2Cl_2 + 2S_2O_3^{2-} \rightarrow S_6O_6^{2-} + 2Cl^-$$

The acids all form crystalline and well-characterized alkali and alkaline earth metal salts which are usually soluble in water, e.g. $K_2S_3O_6$, $K_2S_4O_6$, $Ba_2S_2O_6.2H_2O$. Solutions of the free acids have been made but attempts at concentration cause decomposition.

$$S_nO_6^{2-} \rightarrow SO_4^{2-} + SO_2 + (n-2)S$$

where $n = 2$ to 6.

The structure of the dithionate ($S_2O_6^{2-}$) and the trithionate ($S_3O_6^{2-}$) ions are shown in Figure 16.22.

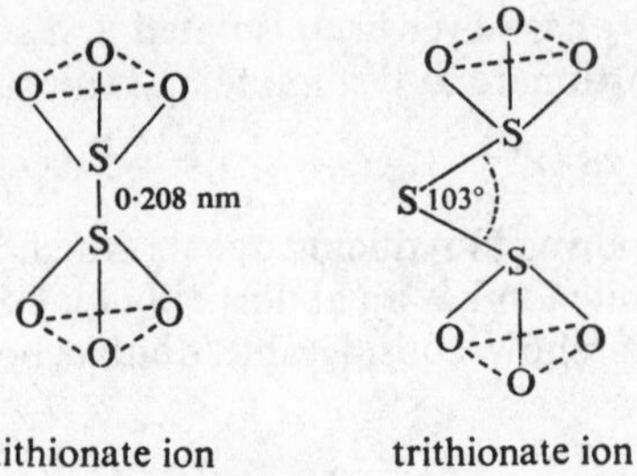

0·208 nm　　103°

dithionate ion　　trithionate ion

FIGURE 16.22

The $S_4O_6^{2-}$, $S_5O_6^{2-}$, and $S_6O_6^{2-}$ ions contain sulphur chains with pyramidal SO_3 groups at each end of the chain.

Peroxo- and poly-acids of selenium and tellurium have not been isolated.

Halides

The octahedral hexafluorides are prepared by direct fluorination of the elements. The sulphur and selenium hexafluorides are inert and sulphur hexafluoride finds some use as a gaseous insulator in high-voltage equipment. Tellurium hexafluoride with considerably more polar bonds is slowly hydrolysed to orthotelluric acid, $Te(OH)_6$.

The tetrafluorides can also be made by controlled fluorination of the elements. They are less stable than the hexafluorides and are hydrolysed by water, depositing the dioxides.

$$SF_4 + 2H_2O \rightarrow SO_2 + 4HF$$

All the tetrafluoride molecules possess a lone pair of electrons and have a trigonal bipyramidal structure (Figure 16.23).

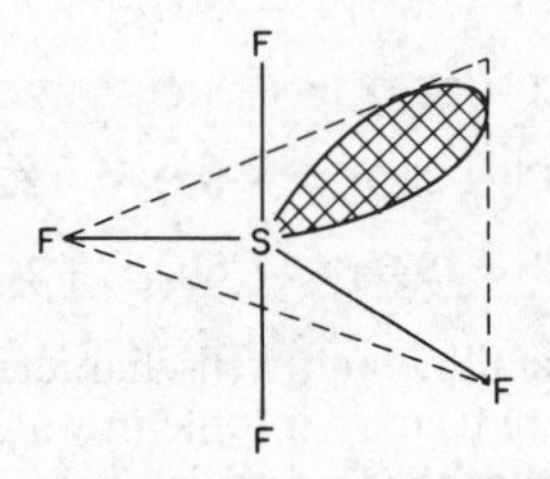

FIGURE 16.23

Sulphur tetrafluoride is used extensively as a fluorinating agent.

Pentafluorides of sulphur and tellurium—$(SF_5)_2$, $(TeF_5)_2$ or S_2F_{10}, Te_2F_{10}—have also been isolated during the fluorination of the elements but no corresponding compound of selenium has yet been isolated.

The principal chlorides of sulphur, and the preparative methods employed to obtain them, are shown in the following reaction sequence:

$$\underset{\text{molten}}{2S} + \underset{\text{dry}}{Cl_2} \rightarrow \underset{\substack{\text{sulphur}\\\text{monochloride}\\\text{(red liquid)}}}{S_2Cl_2} \xrightarrow{\text{excess } Cl_2 \text{ (gas)}} \underset{\substack{\text{sulphur}\\\text{dichloride}\\\text{(red liquid)}}}{SCl_2} \xrightarrow{\text{liquid } Cl_2} \underset{\substack{\text{sulphur}\\\text{tetrachloride}\\\text{(yellow}\\\text{crystals at}\\-31^\circ\text{C)}}}{SCl_4}$$

Sulphur monochloride fumes in air and is hydrolysed by water to sulphur and hydrogen sulphide.

$$Cl\text{—}S\text{—}S\text{—}Cl + 2H_2O \rightarrow 2HCl + H_2S + SO_2$$

$$2H_2S + SO_2 \rightarrow 3S + 2H_2O$$

It is used as a solvent for sulphur in the vulcanization of rubber. The structure of the monochloride is similar to that of hydrogen peroxide (p. 295). The dichloride reverts to the monochloride and chlorine on heating.

Similar compounds are known for selenium and tellurium and the preparative methods are analogous. No tellurium monochloride has been isolated. The compounds are generally more stable than the corresponding sulphur compounds and have a strong tendency to form complexes with concentrated hydrochloric acid, e.g. H_2SeCl_6, H_2TeCl_6, Na_2TeCl_6, K_2TeCl_6.

Some bromides and iodides are also known.

The oxohalides

No oxohalides have been prepared from tellurium and only a few from selenium but sulphur forms a large number of oxohalides, some of which are complex. The two main classes are:

(*a*) The thionyl and selenyl halides typified by thionyl chloride and selenium oxochloride, $SOCl_2$ and $SeOCl_2$.

(*b*) The sulphuryl halides typified by sulphuryl chloride SO_2Cl_2.

The thionyl halides may be prepared by the following reactions:

$$SO_2 + PCl_5 \rightarrow SOCl_2 + POCl_3$$

$$SeO_2 + SeCl_4 \rightarrow 2SeOCl_2 \text{ (in } CCl_4)$$

The oxohalides mentioned are fuming liquids, hydrolysed by water to the dioxides.

$$SeOCl_2 + H_2O \rightarrow SeO_2 + 2HCl$$

Thionyl chloride is used in inorganic chemistry to prepare anhydrous chlorides (p. 330).

In the molecules of $SOCl_2$ and $SeOCl_2$, the sulphur and selenium atoms are sp^3 hybridized, one of the hybrid orbitals containing a lone pair. Double bonding with the oxygen atom is due to overlap between a full *p* orbital on the oxygen atom with an empty *d* orbital on the sulphur (Figure 16.24).

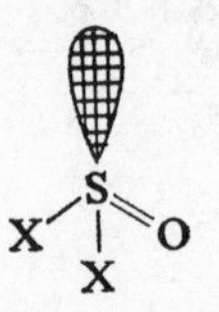

FIGURE 16.24

Sulphuryl chloride SO_2Cl_2 is made by the direct action of sulphur dioxide and chlorine in presence of a charcoal catalyst.

$$SO_2 + Cl_2 \rightleftharpoons SO_2Cl_2$$

It is a fuming liquid, hydrolysed by water to sulphuric acid. The vapour dissociates at 300 °C to sulphur dioxide and chlorine. The molecule has an approximately tetrahedral arrangement.

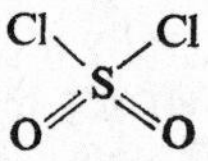

17

Group VIIB

Physical Properties

	Fluorine F	Chlorine Cl	Bromine Br	Iodine I
Atomic mass	18·9984	35·453	79·909	126·9044
Atomic number	9	17	35	53
Isotopes	19	35, 37	79, 81	127
Electronic structure	$1s^2 2s^2 2p^5$	$2p^6 3s^2 3p^5$	$3p^6 3d^{10} 4s^2 4p^5$	$4p^6 4d^{10} 5s^2 5p^5$
Atomic radius (nm)	0·072	0·099	0·114	0·133
Ionic radius (nm) X^-	0·136	0·181	0·195	0·216
Electronegativity	4·0	3·0	2·8	2·5
Ionization potential (kJ mol^{-1})	1686	1266	1146	1016
Density of liquid (g cm^{-3})	1·1	1·5	3·2	4·9 (solid)
Melting point (°C)	−223	−102	−7·3	114
Boiling point (°C)	−187	−34·3	58·8	183
Electrode potential (V) $X(g) + e \rightarrow X^-(aq)$	+2·87	+1·36	+1·065 Br(l)	+0·535 I(s)
Heat of fusion (kJ mol^{-1})	0·26	3·20	5·27	7·87
Heat of dissociation (kJ mol^{-1}) $X_2(g) \rightarrow 2X$	158	242	193	151
Colour and physical state	Pale yellow gas	Greenish yellow gas	Red liquid	Black solid (violet vapour)

General Characteristics

All the atoms have the configuration ns^2np^5 in their outermost orbitals. Atoms can acquire a noble gas configuration with a full outer *p* orbital by:

(*a*) accepting one electron from a donor atom, $X + e \rightarrow X^-$, as in Na^+Cl^-, K^+I^-, etc.;

(*b*) forming a single covalent bond by *p–p* or *p–s* overlap as in the halogen molecule (X—X) and hydrogen halides (H—X).

There are no *d* orbitals available for bonding in the fluorine atom but chlorine, bromine, and iodine all contain empty *d* orbitals and these elements can form compounds containing $d\pi$–$p\pi$ bonding, involving overlap of the *d* orbitals with *p* orbitals of other atoms, particularly oxygen. Examples are chlorine dioxide (ClO_2), chloric acid ($HClO_3$), bromic acid ($HBrO_3$), iodic acid (HIO_3), and perchloric acid ($HClO_4$).

Small radii and high ionization potentials indicate unfavourable conditions for cation formation. Iodine, however, with the largest radius and smallest ionization potential does form a cation in certain compounds, e.g. I_4O_9, iodine iodate ($I^{3+}(IO_3^-)_3$).

The electrode potentials decrease from fluorine to iodine. Fluorine is the strongest electron acceptor and is, therefore, the strongest oxidizing agent.

$$F_2 + 2Cl^- \rightarrow 2F^- + Cl_2$$

$$Cl_2 + 2Br^- \rightarrow 2Cl^- + Br_2, \text{ etc.}$$

The heats of fusion increase from fluorine to iodine but all values are small, indicating weak forces of the van der Waals type between the molecules in the solid state where the halogen molecules are arranged in layers.

The heats of dissociation decrease regularly from chlorine to iodine indicating a decrease in molecular stability. The abnormal value for fluorine is probably due to strong repulsion between non-bonding electrons in the molecule.

Fluorine

Occurrence

Fluorine is a fairly abundant element, occurring in the combined state as fluorspar (CaF_2) in the U.S.A. and as a purple stone, 'Blue John', in England. Cryolite (Na_3AlF_6) is found in Greenland.

Preparation

The preparation of pure fluorine is extremely difficult owing to the great reactivity of the element. Early attempts to isolate fluorine using electrolytic methods failed because electrolysis of aqueous solutions of hydrogen fluoride yielded oxygen, and the anhydrous acid is a non-conductor.

$$2F_2 + 2H_2O \rightarrow 4HF + O_2$$

In 1886, Moissan solved the problem using an electrolyte of anhydrous potassium hydrogen fluoride (KHF_2) in anhydrous hydrogen fluoride. An apparatus suitable for the preparation of elementary fluorine is shown in Figure 17.1. The cell consists of a copper tube which is attacked by fluorine but rapidly forms a protective coating of copper fluoride. The temperature of the cell is maintained at 100 °C when the graphite is not attacked. Hydrogen fluoride is replenished as required.

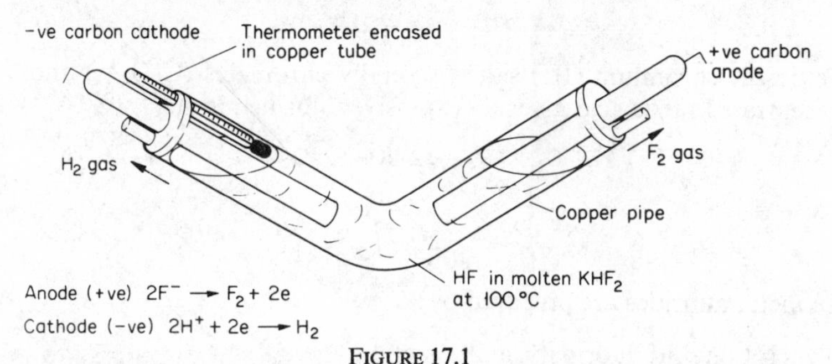

FIGURE 17.1

The fluorine is contaminated with hydrogen fluoride which may be removed by passing the gas over solid sodium fluoride.

$$NaF + HF \rightarrow NaHF_2$$

The rapid development of fluorinated hydrocarbons (CCl_2F_2, $(C_2F_4)_n$, CF_4, etc.) for use as refrigerants, lubricants, and plasticizers has stimulated work on the large-scale production of fluorine. On the industrial scale batteries of cells are employed, the fluorine being compressed into steel cylinders.

Uses of Fluorine and Compounds

Fluorine is used to prepare organic fluorocarbons, useful as plastics, e.g. $(\text{—}CF_2\text{—}CF_2\text{—})_n$ (Teflon), refrigerants CCl_2F_2 (Freon), and insecticides (CCl_3F). Uranium hexafluoride (UF_6) is used in the gaseous diffusion method for separating ^{235}U from ^{238}U. Chlorine trifluoride was used in a liquid incendiary bomb in World War II.

Properties

Fluorine is a pale, greenish yellow gas with a penetrating smell. The gas may be liquefied by cooling in liquid oxygen and solidified by cooling in liquid hydrogen. Fluorine fumes in moist air.

$$2F_2 + 2H_2O \rightarrow 4HF + O_2$$

The element is extremely reactive and readily reacts with non-metals giving volatile fluorides, e.g.

$$2B + 3F_2 \rightarrow 2BF_3$$

The non-metals ignite in most cases when heated in fluorine. It gives fluorides with most metals and will attack platinum and copper at red heat. With hydrogen halides, fluorine acts as an oxidizing agent.

$$2HCl + F_2 \rightarrow 2HF + Cl_2$$

$$HBr + F_2 \rightarrow HF + BrF$$

$$HI + 3F_2 \rightarrow HF + IF_5$$

It oxidizes chromium (III) salts to yellow chromates (CrO_4^{2-}) and with concentrated nitric acid gives an explosive fluorine nitrate.

$$2HNO_3 + \tfrac{1}{2}F_2 \rightarrow FNO_3 + NO_2 + H_2O$$

Fluorides

The metal fluorides are prepared by:

(*a*) Hot or cold fluorination of a metal.

(*b*) The action of hydrofluoric acid on an alkali, followed by evaporation to the point of crystallization.

$$NH_3 + HF \rightarrow NH_4F$$

$$KOH + HF \rightarrow KF + H_2O$$

$$KOH + 2HF \rightarrow KHF_2 + H_2O$$

On heating the hydrogen difluoride, the acid salt decomposes.

$$KHF_2 \rightleftharpoons HF + KF$$

The crystal lattices of the fluorides are almost all ionic, e.g.

Rock-salt lattice	LiF, NaF, KF, AgF
Rutile	MgF_2, ZnF_2, FeF_2, CoF_2, NiF_2
Fluorite	CaF_2, SrF_2, BaF_2, HgF_2, PbF_2

The structures of KHF_2 and NH_4F are modified by hydrogen bonding. Thus K^+ HF_2^- forms a structure of the caesium chloride type, where each potassium ion is surrounded by eight $(HF_2)^-$ neighbours. Ammonium fluoride has a wurtzite lattice with each nitrogen atom forming four tetrahedral N—H···F bonds. Metallic fluorides possess higher melting points and boiling points than the other halides.

Non-metal fluorides are discussed under the appropriate elements.

Hydrogen fluoride

Hydrogen fluoride, HF, is prepared by heating:

(*a*) An acid fluoride, e.g.

$$KHF_2 \rightarrow HF + KF$$

(*b*) A normal fluoride with concentrated sulphuric acid, e.g.

$$CaF_2 + H_2SO_4 \rightarrow CaSO_4 + 2HF$$

The acid is condensed in a freezing mixture or dissolved in water giving a 40 per cent aqueous solution. Purification may be effected by fractionation using a steel apparatus and the acid is now usually stored in polythene bottles.

Pure hydrogen fluoride is a colourless fuming liquid, boiling point 19·4°C, and, like fluorine, is a dangerous compound producing severe burns on the flesh. In the presence of moisture it attacks glass.

$$4HF + SiO_2 \rightarrow 2H_2O + SiF_4$$

The sodium and calcium in the glass form the respective fluorides. Thus glass objects are frosted and etched by hydrogen fluoride.

In pure hydrogen fluoride, the molecules are associated due to hydrogen bonding (p. 49), giving rise to zig-zag chains with an H—F—H bond angle of 140°. The abnormally high boiling point (19·4 °C) compared to the other halogen hydrides (p. 333) is a result of this bonding. The tendency to form hydrogen bonds accounts for the ability of normal fluorides, such as sodium and potassium fluorides, to react with hydrogen fluoride to form hydrogen difluorides:

$$KF + HF \rightleftharpoons K^+ + HF_2^- \text{ (potassium hydrogen difluoride)}$$

The structure of the HF_2^- ion is $(F—H \cdots F)^-$.

Solutions of hydrogen fluoride in water are only slightly ionized and, of all the halogen hydrides, hydrofluoric acid is the weakest acid (p. 333). After the hydrogen bonds and the covalent H—F bonds are broken, the ions hydrate.

$$F \cdots H—F \cdots H— \rightleftharpoons 2H^+ + 2F^- + 2H_2O \rightleftharpoons 2H_3O^+ + 2F^-$$

$$(K - 7 \times 10^{-4} \text{ at } 25\ °C)$$

However, the hydrogen difluoride ion is also formed.

$$F^- + HF \rightarrow F—H \cdots F^- \quad (K = 5{\cdot}5 \text{ at } 25\ °C)$$

Thus, although hydrofluoric acid is a weak acid, $H^+(F—H \cdots F)^-$ is a strong acid and, in a concentrated solution, the proportion of the latter increases. Hence, in this particular case, the degree of ionization increases with increasing concentration.

Oxides of fluorine

Fluorine monoxide (F_2O) is prepared by bubbling fluorine through a 2 per cent solution of sodium hydroxide.

$$2F_2 + 2OH^- \rightarrow 2F^- + F_2O + H_2O$$

It is a pungent-smelling, stable gas with strong oxidizing properties.

$$F_2O + 4I^- + H_2O \rightarrow 2F^- + 2OH^- + 2I_2$$

Difluorine dioxide (F_2O_2) is prepared as an orange solid by the action of an electric discharge on a mixture of fluorine and oxygen at low temperature and pressure. The solid is only stable below −100 °C.

Chlorine, Bromine, and Iodine

Occurrence

The principal sources of these halogens are briefly listed below:

(*a*) Sea water—contains the chlorides, bromides, and iodides of sodium. potassium, magnesium, and calcium. The proportion of sodium chloride is 2·8 per cent.

(*b*) Salt-bed deposits—contain sodium, potassium, magnesium, and calcium chlorides as single and double salts, e.g. carnallite, $KCl.MgCl_2.6H_2O$.

(*c*) Certain deep seaweeds—contain up to 0·5 per cent iodine.

(*d*) Chile saltpetre deposits—contain 0·2 per cent sodium iodate.

Laboratory Preparation

All three halogens may be prepared by heating the appropriate alkali metal halide with dilute sulphuric acid and manganese dioxide.

$$MnO_2 + 4H^+ + 2e \rightarrow Mn^{2+} + 2H_2O$$
$$2X^- \rightarrow X_2 + 2e \qquad (X = Cl, Br, I)$$

The apparatus used in each case is shown in Figure 17.2.

Chlorine is washed with water to remove acid spray and dried with concentrated sulphuric acid.

Chlorine may also be prepared by oxidation of chlorides by strong oxidizing agents.

$$MnO_2 + 2Cl^- + 4H^+ \rightarrow Mn^{2+} + 2H_2O + Cl_2$$

$$2MnO_4^- + 10Cl^- + 16H^+ \rightarrow 2Mn^{2+} + 8H_2O + 5Cl_2$$

Bromine and iodine may be prepared by oxidation of bromides and iodides by chlorine gas.

$$2Br^- + Cl_2 \rightarrow 2Cl^- + Br_2$$

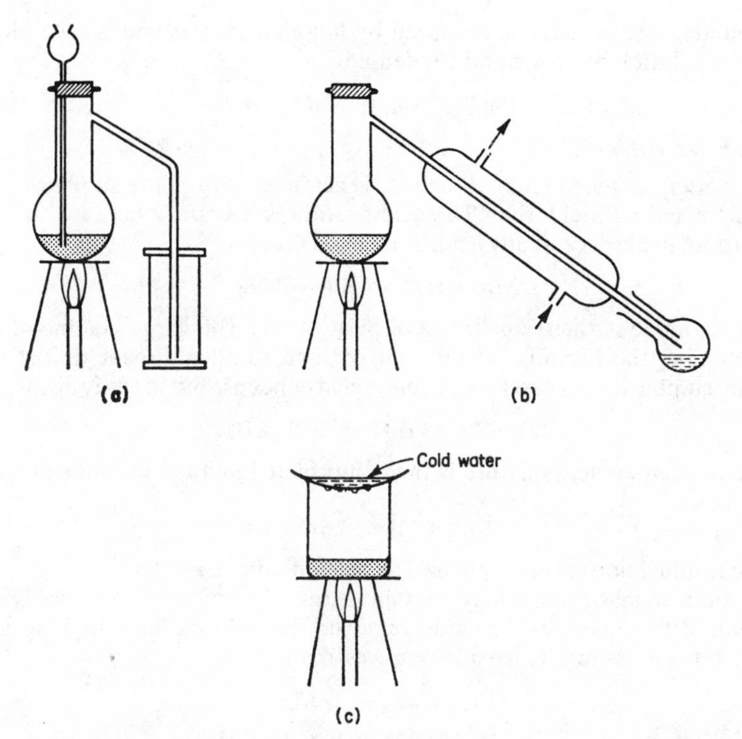

FIGURE 17.2 (*a*) greenish yellow chlorine gas; (*b*) red liquid bromine; (*c*) glittering black scales of solid iodine

Commercial Manufacture of the Halogens

Chlorine

In the electrolysis of (*a*) fused sodium chloride in the Downs cell (p. 166), and (*b*) concentrated solutions of brine in the Castner-Kellner or Billiter cells (p. 168), chlorine is a very important by-product. It is also now produced by a modification of the original Deacon process which consists of direct gas-phase oxidation of hydrogen chloride with air in the presence of a copper catalyst.

$$4HCl + O_2 \rightarrow 2Cl_2 + 2H_2O$$

Bromine

From salt-bed deposits

In the extraction of potassium and magnesium chlorides from carnallite or from Dead Sea water, the hot mother liquor contains 0·2 per cent magnesium bromide. The bromine is displaced by hot chlorine gas and is then blown out of solution by steam and condensed.

$$MgBr_2 + Cl_2 \rightarrow MgCl_2 + Br_2$$

From sea water

Sea water, screened from all debris, is acidified with dilute sulphuric acid and treated with chlorine. The acidification is necessary because, at high dilution, in absence of any acid, hydrolysis occurs.

$$Br_2 + H_2O \rightleftharpoons HBr + HOBr$$

The bromine is then expelled by a blast of air. This large volume of air, containing the bromine, is then mixed with sulphur dioxide and steam, when sulphuric acid and hydrobromic acid condense out in spray absorbers.

$$SO_2 + Br_2 + 2H_2O \rightarrow 2HBr + H_2SO_4$$

The condensed acid mixture is then chlorinated and the bromine expelled by steam.

$$2HBr + Cl_2 \rightarrow 2HCl + Br_2$$

The residual liquors are then used to reacidify the sea water.

Commercial bromine may contain traces of iodine and chlorine. Distillation with potassium bromide removes the chlorine and shaking with aqueous potassium hydroxide removes the iodine.

$$Cl_2 + 2KBr \rightarrow 2KCl + Br_2$$

$$I_2 + 2OH^- \rightarrow I^- + OI^- + H_2O$$

Iodine

From caliche

Caliche ($NaNO_3$) usually contains about 0·2 per cent of sodium iodate and, after crystallizing out the sodium nitrate, sodium bisulphite is added to the mother liquor.

$$IO_3^- + 3HSO_3^- \rightarrow I^- + 3HSO_4^-$$

More iodate is then added and the solution is acidified.

$$5I^- + IO_3^- + 6H^+ \rightarrow 3I_2 + 3H_2O$$

The precipitated iodine is washed, compressed, and purified by sublimation.

From seaweeds

Certain dark red-brown seaweeds are dried in the sun and burned, when the ash (kelp) contains the iodides, carbonates, and sulphates (plus smaller amounts of chlorides and bromides) of potassium, sodium, and magnesium. After extraction with water and filtration, the filtrate is concentrated, when the chlorides, carbonates, and sulphates crystallize out. The mother liquor is either treated with chlorine gas or mixed with manganese dioxide and sulphuric acid and distilled.

Commercial iodine contains traces of iodine monochloride and iodine monobromide (p. 343). These may be removed by distillation with potassium iodide.

$$ICl + I^- \rightarrow I_2 + Cl^-$$

The Properties of Chlorine, Bromine, and Iodine

At normal temperatures and pressures, chlorine is a greenish yellow gas, bromine is a dark red liquid, and iodine a black shiny solid. Vapour densities in the vapour state indicate diatomic molecules, although, at higher temperatures, dissociation takes place.

$$X_2 \rightleftharpoons 2X$$

Chlorine, with a critical temperature of 146 °C, can be liquefied at room temperature by pressure alone and the liquid is easily transported in steel cylinders.

All three halogens react chemically with water producing the hydrohalic and hypohalous acids, e.g.

$$Cl_2 + H_2O \rightleftharpoons \underset{\text{hydrochloric acid}}{HCl} + \underset{\text{hypochlorous acid}}{HOCl}.$$

The extent of the hydrolysis decreases from chlorine to iodine, the equilibrium constants being $4{\cdot}5 \times 10^{-4}$ for chlorine, 5×10^{-9} for bromine, and 3×10^{-13} for iodine at 25 °C. With excess cold dilute alkali, a similar reaction takes place.

$$X_2 + 2OH^- \rightarrow X^- + OX^- + H_2O$$

The hypoiodite ion, however, is unstable even in dilute solution and disproportionates to iodide and iodate.

$$3IO^- \rightarrow 2I^- + IO_3^-$$

At 60 °C in strongly alkaline solution, hypochlorite and hypobromite disproportionate rapidly in a similar manner and quantitative yields of chlorate and bromate are obtained.

When chlorine is passed into ice cold water, crystals of chlorine hydrate separate ($Cl_2 . 7{\cdot}3H_2O$). Bromine water in a freezing mixture deposits red crystals of a bromine hydrate ($Br_2 . 10H_2O$). Violet-red crystals of an iodine hydrate have been reported, formed by subjecting hypoiodite solutions to high pressure. All these hydrates are probably clathrate compounds of the noble gas hydrate type (p. 153).

Organic solvents such as carbon tetrachloride and chloroform extract chlorine, bromine, and iodine from aqueous solutions giving yellow, red, and violet colours respectively. Iodine also gives a brown solution in water, alcohol, and ether due to the formation of 1:1 adducts, e.g.

$$I_2{:}OH_2,\ I_2{:}\underset{}{\overset{\displaystyle H\atop|}{O}}{-}C_2H_5,\ \text{etc.}$$

The depth of colour shown by these adducts varies from brown to red-brown depending upon the extent of overlap of oxygen electron pairs with empty *d* orbitals of iodine. The complexes are referred to as charge transfer complexes. Iodine also dissolves readily in potassium iodide solution and can then be readily titrated with sodium thiosulphate solution (p. 314).

$$I_2 + I^- \rightleftharpoons I_3^- \text{ (potassium triiodide } KI_3)$$

The less stable Br_3^- and Cl_3^- are not known in aqueous solution but solid salts have been isolated from non-aqueous media, e.g. $Cs^+Br_3^-$.

Iodine also gives a deep blue colour with starch which is used as a sensitive test for iodine. The blue colour which disappears on heating is due to absorption of iodine molecules within the helical starch molecules (Figure 17.3). This is another example of a cage or clathrate compound.

The reactivity of the halogens decreases with increasing atomic number, chlorine bringing out the highest oxidation state in the combining element

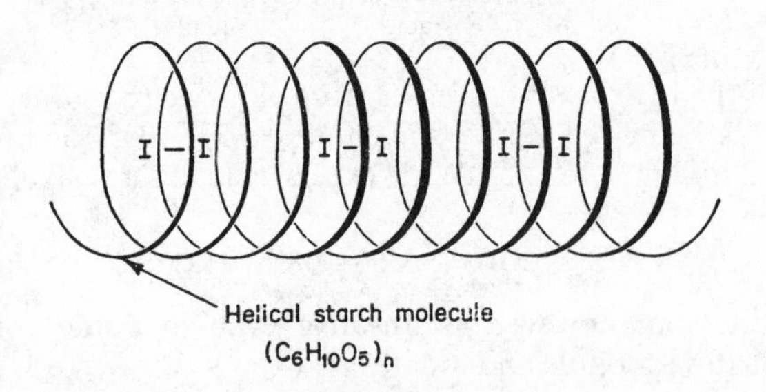

FIGURE 17.3

(p. 330). Thus sodium, white phosphorus, arsenic, powdered antimony, and Dutch metal (Cu–Zn alloy) inflame in chlorine producing sodium chloride, phosphorus pentachloride, arsenic trichloride, antimony pentachloride, and copper and zinc chlorides respectively. Explosive reactions occur in many cases with liquid bromine giving similar products but, although iodine reacts explosively, iodides of lower oxidation state are produced, e.g. antimony triiodide, phosphorus triiodide etc. A cursory examination seems to contradict the statement mentioned above that reactivity decreases with increasing atomic number, but it must be noted that the reactions quoted are for chlorine gas. Liquid chlorine does, in fact, explode violently with the reagents quoted.

The heats of formation of the halogen hydrides show that chlorine has the strongest affinity for hydrogen. It combines explosively with hydrogen in bright sunlight and more slowly in diffuse light, the reaction in each case being photochemical.

$$\underset{\text{quanta}}{Cl_2 + h\nu} \rightarrow Cl^* + Cl^* \quad (\text{* activated atoms})$$

The light energy splits the molecule into *free radicals*, as they are termed, each with a single unpaired electron. These radicals are extremely reactive and the following reactions occur:

$$Cl^* + H_2 \rightarrow HCl + H^*$$

$$H^* + Cl_2 \rightarrow HCl + Cl^*$$

Bromine does not react with hydrogen even in bright sunlight and a platinum catalyst is needed at a temperature of 200°C. Even with a platinum catalyst, the reaction between hydrogen and iodine is reversible and incomplete.

$$H_2 + I_2 \rightleftharpoons 2HI$$

Chlorine also removes covalently bonded hydrogen from certain organic compounds, e.g. warm turpentine.

$$C_{10}H_{16} + 8Cl_2 \rightarrow 10C + 16HCl$$

There are no comparable reactions for bromine and iodine.

$E^{\ominus}$ values for the three halogens show that chlorine is the strongest oxidizing agent and iodine the weakest in aqueous solution. The oxidizing action of chlorine is illustrated by the following equations:

$$H_2S + Cl_2 \rightarrow 2HCl + S$$

$$2Fe^{2+} + Cl_2 \rightarrow 2Fe^{3+} + 2Cl^-$$

$$Cl_2 + 2I^- \rightarrow 2Cl^- + I_2$$

The corresponding reactions of bromine and iodine occur much more slowly and iodine only liberates sulphur from hydrogen sulphide very slowly.

Chlorine and bromine, but not iodine, also oxidize ammonia to nitrogen.

$$2NH_3 + 3Cl_2 \rightarrow N_2 + 6HCl$$

Ammonia reacts with all three halogens when the latter are in excess to give explosive trichlorides, etc.

$$2NH_3 + 6Cl_2 \rightarrow 2NCl_3 + 6HCl$$

In the case of bromine and iodine, double compounds are produced: $NBr_3.6NH_3$ (red oil) and $NI_3.NH_3$ (black solid).

Uses

Large quantities of chlorine are used in the bleaching industry in which the starting material may be bleaching powder (p. 182), sodium hypochlorite, or liquid chlorine. Other uses include chlorinating old tin plate for the recovery of tin, the manufacture of plastics, synthetic rubbers, antiseptics, and insecticides.

Bromine compounds are used in medicine, photography, fire-extinguishers (as methyl bromide), and in the preparation of numerous organic compounds.

Iodine is used in the manufacture of dyes and pharmaceutical chemicals. Tincture of iodine contains iodine dissolved in a solution of potassium iodide diluted with 90 per cent ethanol.

The Halides

The preparative methods used for halides are summarized below:

(*a*) The action of hydrohalic acid on the metal oxide, hydroxide, or carbonate. Evaporation and crystallization of the resulting solution usually produces hydrated crystals, $MgCl_2.6H_2O$, $CaBr_2.6H_2O$, $MgI_2.8H_2O$, etc.

(*b*) Insoluble halides of silver, lead, and mercury(I) are prepared by double decomposition.

$$Pb^{2+} + 2I^- \rightarrow PbI_2\downarrow$$

$$Ag^+ + Cl^- \rightarrow AgCl\downarrow$$

(*c*) The action of dry hydrogen halides or halogens on the heated metal gives the anhydrous halide.

$$2Al + 3X_2 \rightarrow 2AlX_3 \qquad (X = Cl, Br, I)$$

$$Sn + 2HCl \rightarrow SnCl_2 + H_2$$

Chlorination of the oxides mixed with carbon is also often employed, e.g.

$$TiO_2 + C + 2Cl_2 \rightarrow CO_2 + TiCl_4 \qquad \text{(p. 362)}$$

Refluxing the hydrated chloride with thionyl chloride can also produce the anhydrous chloride, e.g.

$$CrCl_3.6H_2O + 6SOCl_2 \rightarrow 6SO_2 + 12HCl + CrCl_3$$

When an element exhibits more than one oxidation state, hydrogen chloride gas gives the lower and chlorine the higher chloride, e.g.

$$Fe + 2HCl \rightarrow Fe^{II}Cl_2 + H_2$$

$$2Fe + 3Cl_2 \rightarrow 2Fe^{III}Cl_3$$

Non-metallic halides are usually prepared by direct action, excess of halogen giving the higher and excess of non-metal giving the lower oxidation state, respectively.

$$2P + 3Cl_2 \rightarrow 2PCl_3$$

$$2P + 5Cl_2 \rightarrow 2PCl_5 \quad \text{(p. 276)}$$

Metallic halide structures

These are summarized in Table 17.1.

TABLE 17.1

Structure	*Examples*
1. Rock salt	Alkali metal halides (except CsCl, CsBr, CsI), AgCl, AgBr
2. Caesium chloride	CsCl, CsBr, CsI, RbCl, RbBr, RbI
3. Fluorite	CaF_2, PbF_2
4. Deformed rutile	$CaCl_2$, $SrCl_2$, $CaBr_2$
5. Layer lattices:	
(*a*) cadmium iodide	$MgBr_2$, MgI_2, CaI_2, ZnI_2, PbI_2, CdI_2, $CdBr_2$, $CoBr_2$, $NiBr_2$
(*b*) cadmium chloride	$CdCl_2$, $FeCl_2$, $CoCl_2$, $NiCl_2$, $MgCl_2$, $MnCl_2$
(*c*) chromic chloride	$CrCl_3$
6. Chain structures	$CuCl_2$, $PdCl_2$, $BeCl_2$
7. Covalent dimers (in vapour)	Al_2Cl_6, Fe_2Cl_6
8. Individual covalent molecules	$HgCl_2$, $SbCl_5$

The majority of structural types indicated in the table have been described in Chapter 5. More detailed descriptions of individual halide structures are to be found in the appropriate section dealing with the element.

The hydrolysis of halides

(*a*) The alkali and alkaline earth metal halides are ionic and give stable hydrated ions in water. Attempts to dehydrate crystalline salts may cause appreciable hydrolysis in certain cases, e.g.

$$Mg(H_2O)_6{}^{2+}Cl_2{}^{-} \xrightarrow{\text{heat}} \underset{\text{oxochloride}}{MgOH^{+}Cl^{-}} + 5H_2O + HCl$$

(*b*) Other halides, such as $SnCl_2$, $SbCl_3$, $BiCl_3$, $FeCl_3$, $AlCl_3$, although predominantly covalent in the pure, dry solid state, produce hydrated cations in water which hydrolyse to varying degrees, functioning as Brönsted acids.

$$[Sn.H_2O]^{2+} + H_2O \rightleftharpoons SnOH^+ + H_3O^+$$

$$4SnOH^+ + 4H_2O \rightleftharpoons Sn_4(OH)_6{}^{2+} + 2H_3O^+$$

(precipitates as basic chloride $Sn_4(OH)_6Cl_2$)

$$\left[\begin{matrix}Sb\\Bi\end{matrix}\right\}H_2O\Big]^{3+} + H_2O \rightleftharpoons \left[\begin{matrix}Sb\\Bi\end{matrix}\right\}OH\Big]^{2+} + H_3O^+$$

$$\left[\begin{matrix}Sb\\Bi\end{matrix}\right\}OH\Big]^{2+} + H_2O \rightleftharpoons \left[\begin{matrix}Sb\\Bi\end{matrix}\right\}O\Big]^{+} + H_3O^+$$

(precipitates as basic oxochlorides, SbOCl, BiOCl)

$$Fe(H_2O)_6{}^{3+} + H_2O \rightleftharpoons Fe(OH)(H_2O)_5{}^{2+} + H_3O^+$$

$$Fe(OH)(H_2O)_5{}^{2+} + H_2O \rightleftharpoons Fe(OH)_2(H_2O)_4{}^{+} + H_3O^+$$

$$Fe(OH)_2(H_2O)_4{}^{+} + H_2O \rightleftharpoons Fe(OH)_3 3H_2O + H_3O^+$$

(precipitates as hydrated iron(III) oxide)

The chlorides of non-metals usually undergo hydrolysis very readily. Initial coordination probably occurs by overlap of full *p* oxygen orbitals with empty *d* orbitals on the non-metallic atoms with subsequent loss of a molecule of hydrogen chloride, e.g.

(i) $Cl_3P \leftarrow OH_2 \longrightarrow Cl_2P{-}OH + HCl$

further hydrolysis ↓

$P(OH)_3 + 3HCl$

(ii) $Cl_3SiCl \leftarrow OH_2 \longrightarrow SiCl_3(OH) + HCl$

further hydrolysis ↓

$SiO_2.2H_2O + 4HCl$

(iii) $PCl_5 \leftarrow OH_2 \longrightarrow PCl_4OH + HCl$

further hydrolysis ↓

$P(OH)_5$

↓ $-H_2O$

H_3PO_4
orthophosphoric acid

Carbon has no available *d* orbitals and initial coordination cannot occur with halides of carbon which are thus resistant to hydrolysis.

(*c*) In nitrogen trichloride (NCl_3), nitrogen and chlorine have similar electronegativities and, although nitrogen has no available *d* orbitals, hydrolysis still occurs. The probable mechanism is shown in Figure 17.4(*a*).

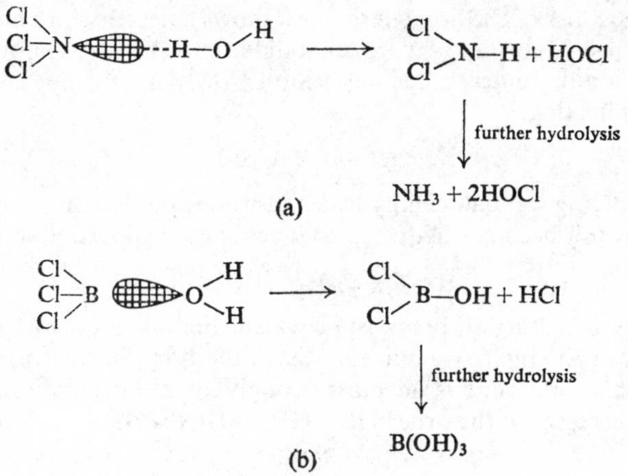

$Cl_3N \cdots H{-}O{-}H \longrightarrow Cl_2N{-}H + HOCl$

further hydrolysis

$NH_3 + 2HOCl$

(a)

$Cl_3B \cdots OH_2 \longrightarrow Cl_2B{-}OH + HCl$

further hydrolysis

$B(OH)_3$

(b)

FIGURE 17.4

(*d*) Boron has no *d* orbitals but in boron trichloride, boron is electron-deficient and the oxygen lone pair overlaps with an empty *p* orbital on the boron (Figure 17.4(*b*)).

The Hydrides

Table 17.2 gives the important physical properties of hydrogen chloride, bromide, and iodide. The figures for hydrogen fluoride are included for comparison.

TABLE 17.2

	H—F	H—Cl	H—Br	H—I
Physical state at 15 °C	Liquid	Gas	Gas	Gas
Melting point (°C)	−83	−114	−87	−51
Boiling point (°C)	19·0	−83	−68	−36
Average bond enthalpies ($kJ\ mol^{-1}$)	562	431	361	299
Bond length (nm)	0·092	0·128	0·142	0·160
Dipole moment (D)	1·90	1·03	0·78	0·38
Percentage ionic bonding	43	17	12	5
Apparent degree of dissociation of 0·1 M solution at 10 °C	0·085	0·92	0·93	0·95

The bond in the halogen hydrides is a σ-bond, formed by overlap between the 1s orbital of hydrogen and the $3p_x$ halogen orbital (p. 35). However, owing to the high electronegativity of fluorine and chlorine, the shared electrons are pulled strongly towards the halogen atom (p. 35). The effect decreases from fluorine to iodine and, in contrast to hydrogen fluoride, none of the other halogen hydrides exhibits hydrogen bonding. An examination of the values of the bond enthalpies shows a decrease in bond strength from hydrogen fluoride to hydrogen iodide. In fact hydrogen bromide and hydrogen iodide undergo decomposition slowly in aqueous solution and rapidly on heating.

$$2HI \xrightarrow{180^\circ} H_2 + I_2$$

On dissolving the halogen hydrides in water, the H—X bond is broken and the proton becomes hydrated producing the hydroxonium ion H_3O^+.

$$H\text{-}\!\mid\!\text{-}X + H_2O \rightleftharpoons H_3O^+ + X^-$$

The energy necessary to break the covalent bond comes from the heat of formation of the hydroxonium ion. Since the bond in hydrogen iodide is weakest, this compound is the most strongly ionized in solution. Thus acid strength decreases in the order HI > HBr > HCl > HF.

Preparation of the halogen hydrides

Hydrogen chloride is prepared by heating a mixture of salt and concentrated sulphuric acid.

$$NaCl + H_2SO_4 \rightarrow NaHSO_4 + HCl$$

$$NaHSO_4 + NaCl \rightarrow Na_2SO_4 + HCl$$

The temperature necessary for the latter reaction is never reached during the laboratory preparation. The gas may be dried by concentrated sulphuric acid and collected by downward delivery. The method cannot be applied to hydrogen bromide and iodide as reduction of the concentrated sulphuric acid occurs. The reduction may be extensive in the case of hydrogen iodide.

$$H_2SO_4 + 2HI \rightarrow SO_2 + 2H_2O + I_2$$

$$SO_2 + 4HI \rightarrow 2H_2O + S + 2I_2$$

$$S + 2HI \rightarrow H_2S + I_2$$

If phosphoric acid is used instead of sulphuric acid, reduction does not occur.

The most convenient method of preparing hydrogen bromide and iodide is by the hydrolysis of phosphorus halides which are usually prepared *in situ*. Thus, for the preparation of hydrogen bromide, bromine is added dropwise to red phosphorus and water, and for hydrogen iodide, water is added to a mixture of red phosphorus and iodine.

$$2P + 3X_2 \rightarrow 2PX_3$$

$$PX_3 + 3H_2O \rightarrow H_3PO_3 + 3HX$$

The apparatus shown in Figure 17.5(*a*) may be used. All three acids are extremely soluble in water and the preparation of aqueous solutions requires the use of a device to prevent 'sucking back' (Figure 17.5(*b*)).

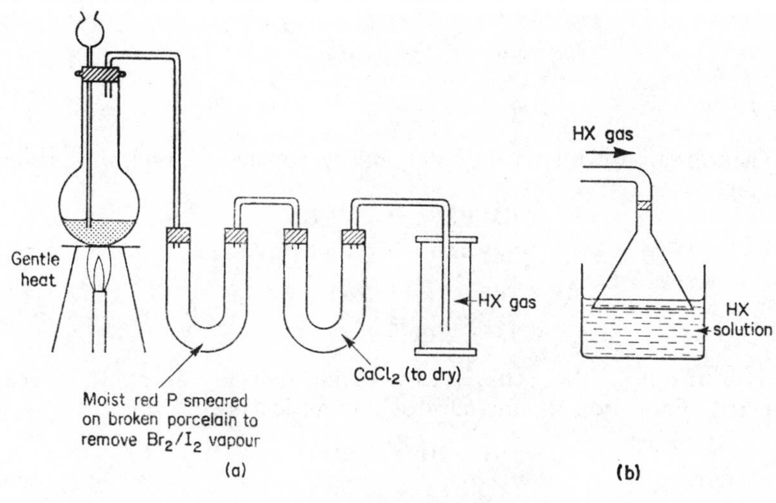

FIGURE 17.5

Manufacture of halogen acids

Hydrogen chloride is made by burning chlorine in an excess of hydrogen in large glass vessels. The gaseous hydrogen chloride is dissolved in cold water to give a 28 per cent solution of hydrochloric acid. Both hydrogen bromide and iodide are made by passing a mixture of halogen vapour and hydrogen over a heated platinum catalyst, but with iodine the reaction is reversible.

Properties of the halogen hydrides

All the halogen hydrides are colourless, pungent smelling gases which fume in air and are extremely soluble in water forming solutions of hydrohalic acids. Hydrates can be frozen out from these solutions at low temperatures, e.g. $HX.3H_2O$, $HX.2H_2O$. Solutions of hydrobromic and hydriodic acids are unstable and liberate free halogen on exposure to the air, turning yellow and brown respectively. Hydriodic acid is most readily oxidized in this respect and is regarded as a reducing agent.

$$4H_3O^+ + 4X^- + O_2 \rightarrow 2X_2 + 6H_2O$$

Each acid forms a constant-boiling mixture with water (Table 17.3).

TABLE 17.3

Acid	HX (%)	Boiling point (°C)
H—Cl	22·24	110
H—Br	48·0	125
H—I	58·0	127

Their reactions with metals, bases, and carbonates are typical of strong monoprotic acids:

$$Zn + 2H^+ \rightarrow Zn^{2+} + H_2$$
$$CuO + 2H^+ \rightarrow Cu^{2+} + H_2O$$
$$OH^- + H^+ \rightarrow H_2O$$
$$CO_3^{2-} + 2H^+ \rightarrow H_2O + CO_2$$

Hydriodic acid also behaves as a strong reducing agent. It reduces iron(III) salts to iron(II), and nitric acid to nitric oxide.

$$2HI + 2H_2O \rightarrow 2H_3O^+ + I_2 + 2e$$
$$2Fe^{3+} + 2e \rightarrow 2Fe^{2+}$$
$$NO_3^- + 4H^+ + 3e \rightarrow 2H_2O + NO$$

The Oxides

The structure and bonding involved in a number of these oxides is not certain. The main oxides of chlorine are shown in Figure 17.6 and the single lines indicate σ-bonds and the double lines σ- and π-bonding. Some promotion of the $3p$ chlorine electron occurs to the $3d$ subshell, with subsequent overlap of these d orbitals with the oxygen p orbitals giving some $d\pi$–$p\pi$ bonding.

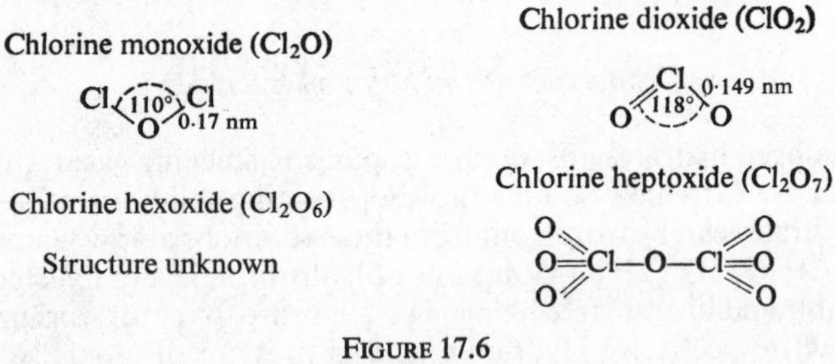

Chlorine monoxide (Cl_2O)

Cl–O–Cl, 110°, 0·17 nm

Chlorine dioxide (ClO_2)

O=Cl=O, 118°, 0·149 nm

Chlorine hexoxide (Cl_2O_6)

Structure unknown

Chlorine heptoxide (Cl_2O_7)

O_3Cl—O—ClO_3

FIGURE 17.6

The structures of the oxides of bromine and iodine (Br_2O, $(BrO_2)_n$, $(BrO_3)_n$, I_2O_4, I_4O_9, I_2O_5) are not definitely known. All the lower oxides are highly explosive compounds, the oxide stability increasing with molecular complexity.

Oxides of chlorine and bromine

The monoxides

Chlorine monoxide (Cl_2O) is a yellow gas and bromine monoxide (Br_2O) is a brown liquid. Each is prepared by passing the halogen over freshly precipitated yellow mercury(II) oxide.

$$2Cl_2 + 2HgO \rightarrow Cl_2O + HgO.HgCl_2$$

Both oxides are explosive, endothermic compounds which decompose violently on warming and dissolve in water or aqueous alkali furnishing the hypohalous acid and hypohalite respectively.

$$Cl_2O + H_2O \rightarrow 2HOCl$$

They are powerful oxidizing agents giving oxides and halides with many metals.

The dioxides

Chlorine dioxide (ClO_2) is an orange gas and bromine dioxide ($(BrO_2)_n$) a yellow solid. The former is prepared by the action of concentrated sulphuric acid on potassium chlorate. A violent crackling results and the explosive dioxide is produced.

$$KClO_3 + H_2SO_4 \rightarrow \underset{\text{chloric acid}}{HClO_3} + KHSO_4$$

$$3HClO_3 \rightarrow \underset{\text{perchloric acid}}{HClO_4} + 2ClO_2 + H_2O$$

Bromine dioxide is prepared by submitting bromine and oxygen to an electrical discharge at very low temperatures. Both compounds are very dangerously explosive. With caustic alkali solution, halites and halates are produced.

$$2ClO_2 + 2OH^- \rightarrow \underset{\text{chlorite}}{ClO_2^-} + \underset{\text{chlorate}}{ClO_3^-} + H_2O$$

The bromite ion, however, rapidly disproportionates to bromide and bromate.

$$3BrO_2^- \rightarrow 2BrO_3^- + Br^-$$

Both oxides are powerful oxidizing agents.

The trioxides

Chlorine hexoxide is a red liquid and bromine trioxide ($(BrO_3)_n$) a white solid. Both are prepared by the action of ozone on the elements at low temperatures. Each compound decomposes readily and dissolves in aqueous alkali, chlorine hexoxide producing chlorate and perchlorate.

$$Cl_2O_6 + 2OH^- \rightarrow ClO_3^- + ClO_4^- + H_2O$$

No perbromate has been isolated from bromine trioxide: bromate, hypobromite, and bromide are produced.

Chlorine heptoxide

Chlorine heptoxide (Cl_2O_7) is a colourless oil, prepared by the action of phosphorus pentoxide on anhydrous perchloric acid at −10 °C.

$$P_4O_{10} + 4HClO_4 \rightarrow 4HPO_3 + 2Cl_2O_7$$

The compound explodes on heating or striking. It is the anhydride of perchloric acid.

$$Cl_2O_7 + H_2O \rightarrow 2HClO_4$$

Oxides of iodine

These are all insoluble solids which decompose without melting. A solvent has not been found for them and hence both their structures and molecular masses are uncertain.

Iodine dioxide

Iodine dioxide, $(I_2O_4)_n$, may be iodyl iodate, $IO^+IO_3^-$. On heating iodic acid with concentrated sulphuric acid, yellow iodyl sulphate is formed ($(IO^+)_2SO_4$), which is decomposed by water precipitating yellow iodine dioxide. On heating, the dioxide decomposes to iodine pentoxide and free iodine.

$$5IO^+IO_3^- \rightarrow 4I_2O_5 + I_2$$

Iodyl sulphate contains polymeric

$$\diagdown \overset{+}{I} \diagup O \diagdown \overset{+}{I} \diagup O \diagdown \overset{+}{I} \diagup$$

chains cross linked by SO_4^{2-} and it appears that I_2O_4 contains similar chains cross linked by IO_3^-.

The oxide I_4O_9 is a pale yellow powder, produced by the action of ozone on iodine. It is a deliquescent compound, forming iodic acid with water. On heating, it decomposes to iodine pentoxide, iodine, and oxygen. The oxide may contain the trivalent cation I^{3+}, i.e. $I^{3+}(IO_3^-)_3$, iodine iodate.

Iodine pentoxide

Iodine pentoxide, I_2O_5, is the most stable halogen oxide and is prepared by the dehydration of iodic acid.

$$2HIO_3 \xrightarrow{200\,^\circ C} I_2O_5 + H_2O$$

The white powder produced decomposes at 300 °C and is the anhydride of iodic acid.

$$2I_2O_5 \rightarrow 2I_2 + 5O_2$$

$$I_2O_5 + H_2O \rightarrow 2HIO_3$$

Iodine pentoxide is a vigorous oxidizing agent and is used in the estimation of traces of carbon monoxide in the air.

$$5CO + I_2O_5 \rightarrow I_2 + 5CO_2$$

The Oxo Acids of the Halogens

The properties of the halogen oxo acids differ to such an extent as to make generalizations difficult. In halous, halic, and perhalic acids, the structures involve double bonding with considerable $d\pi$–$p\pi$ orbital overlap.

Acid strength and stability increase and oxidizing power decreases with increasing oxidation state of the halogen in the oxo acids. There are, however, two exceptions to these trends, i.e. periodic acid (HIO_4) is a stronger oxidizing agent than iodic acid (HIO_3), and perbromic acid ($HBrO_4$) is a stronger oxidizing agent than bromic acid ($HBrO_3$). For oxo acids with the halogen in a fixed oxidation state, the acid strength decreases with decreasing electronegativity of the halogen. Thus for the hypohalous acids the acid strength decreases in the order

$$HOCl > HOBr > HOI$$

The following oxo acids and salts are known (where X = Cl, Br, I):

Hypohalous acids

The acids HOX are only known in solution and the only solid salts isolated are the hypochlorites, e.g. $NaOCl.7H_2O$. The acids are prepared by shaking chlorine water, bromine water, or a fine suspension of iodine in water with precipitated mercury(II) oxide.

$$2X_2 + 2HgO + H_2O \rightarrow HgX_2.HgO + 2HOX$$

The acids show a decrease in acid strength with decreasing electronegativity of the halogen.

$$K(HOCl) = 3 \times 10^{-8} \text{ at } 25\,^\circ C$$

$$K(HOBr) = 2 \times 10^{-9} \text{ at } 25\,^\circ C$$

$$K(HOI) = 4{\cdot}5 \times 10^{-13} \text{ at } 25\,^\circ C$$

Hypoiodous acid is so weak that it is amphoteric in solution.

$$OH^- + I^+ \rightleftharpoons HOI \rightleftharpoons H^+ + OI^-$$

The solubility of these acids and their salts decreases from hypochlorous to hypoiodous acid. The hypohalite ions slowly decompose in solution.

$$3XO^- \rightarrow XO_3^- + 2X^-$$

$$XO_3^- + 5X^- + 6H^+ \rightarrow 3H_2O + 3X_2$$

With hypobromites and hypoiodites, the latter reaction occurs extensively. Hypochlorous acid and the hypochlorites are most important commercially as oxidizing and bleaching agents.

$$OCl^- + 2H^+ + e \rightarrow \tfrac{1}{2}Cl_2 + H_2O \qquad (E^\ominus = +1{\cdot}63\ V)$$

Halous acids

The acid HOXO and salts are only known for chlorine. Chlorous acid (H—O—Cl=O) is known in solution but certain alkali and alkaline earth chlorite salts are known in the solid state, e.g. $NaClO_2.3H_2O$. These contain the non-linear anion $^-$O—Cl=O.

Chlorous acid is prepared in solution by adding chlorine dioxide to a suspension of barium hydroxide in hydrogen peroxide, followed by the acidification of insoluble barium chlorite.

$$2ClO_2 + Ba(OH)_2 + H_2O_2 \rightarrow Ba(ClO_2)_2 + 2H_2O + O_2$$

$$Ba(ClO_2)_2 + H_2SO_4 \rightarrow BaSO_4\downarrow + 2HClO_2$$

A clear solution is obtained which rapidly decomposes.

$$2HClO_2 \rightarrow HClO + HClO_3$$

Salts such as Textone ($NaClO_2.3H_2O$) are used in textiles as bleaching and oxidizing agents.

Halic acids

With halic acids, HXO_3, the chloric and bromic acids are known only in solution but iodic acid is a white crystalline solid. Solid chlorates, bromates, and iodates have all been isolated and the halate ions have a pyramidal configuration (Figure 17.7(*a*)).

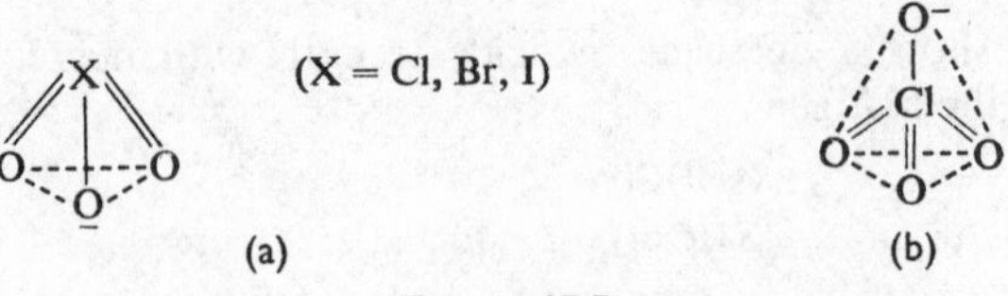

FIGURE 17.7

Chloric and bromic acids are prepared by treating hot concentrated barium hydroxide solution with the appropriate halogen. The resulting barium halate is treated with dilute sulphuric acid.

$$6X_2 + 6Ba(OH)_2 \rightarrow 5BaX_2 + Ba(XO_3)_2 + 6H_2O$$

Iodic acid is a water-soluble deliquescent solid, made by oxidizing iodine with concentrated nitric acid.

$$3I_2 + 10HNO_3 \rightarrow 6HIO_3 + 10NO + 2H_2O$$

Solutions of chloric and bromic acids and solid iodic acid decompose on heating according to the following equations.

$$3HClO_3 \rightarrow HClO_4 + Cl_2 + 2O_2 + H_2O$$

$$4HBrO_3 \rightarrow 2Br_2 + 5O_2 + 2H_2O \quad \text{(No } HBrO_4\text{)}$$

$$2HIO_3 \rightarrow H_2O + I_2O_5$$

$$\qquad \qquad \hookrightarrow I_2 + 2\tfrac{1}{2}O_2$$

Iodic acid is strong acid and solutions dissolve certain metals, e.g.

$$Zn + 2HIO_3 \rightarrow Zn(IO_3)_2 + H_2$$

All the acids are good oxidizing agents.

$$XO_3^- + 6H^+ + 6e \rightarrow X^- + 3H_2O$$

The $E^\ominus$ values for the above reaction where X = Cl, Br, I are +1·48, +1·44, +1·08 V respectively.

The halates

These are prepared by:

(*a*) Dissolving the halogen in hot concentrated alkali.

(*b*) Electrolysis of hot concentrated halide solutions with subsequent mixing of the anode (OH^-) and cathode (Cl^-) products.

Method (*b*) is the commercial method used for the preparation of chlorates. The halates are all soluble except for the silver, lead, and barium salts, which are only sparingly soluble. On heating, they decompose according to the following equation:

$$2KXO_3 \rightarrow 2KX + 3O_2$$

However, the chlorate decomposition occurs in two stages:

$$4KClO_3 \xrightarrow{<370^\circ} 3KClO_4 + KCl$$

$$KClO_4 \xrightarrow{>370^\circ} KCl + 2O_2$$

The bromates and iodates of less electropositive metals tend to give the oxide and free halogen on decomposition.

Perhalic acids

All the perhalic acids (HXO_4) are known and solid perhalates $K^+XO_4^-$ contain the tetrahedral anion XO_4^- (Figure 17.7(*b*)).

Colourless deliquescent crystals of paraperiodic acid (H_5IO_6) and solid salts of the acid, e.g. $Ba_5(IO_6)_2$, have been prepared. These contain the octahedral IO_6^{5-} anion.

Perchloric acid is prepared by distilling potassium perchlorate with concentrated sulphuric acid under reduced pressure.

$$KClO_4 + H_2SO_4 \rightarrow KHSO_4 + HClO_4$$

It is a colourless oily liquid which combines vigorously with water forming hydroxonium perchlorate $H_3O^+ClO_4^-$ (heat of solution $\Delta H = -85$ kJ mol^{-1}). Perchloric acid is one of the strongest acids. The degree of ionization in 0·5 M solution at 25 °C is 0·880, whereas the corresponding values for hydrochloric and hydroidic acid are 0·876 and 0·901 respectively. It is also an oxidizing agent.

$$ClO_4^- + 8H^+ + 7e \rightarrow \tfrac{1}{2}Cl_2 + 4H_2O \qquad (\overline{E^\ominus} = +1{\cdot}36\text{ V})$$

Solid potassium perbromate ($KBrO_4$) has recently been prepared by the oxidation of alkaline potassium bromate with fluorine.

$$BrO_3^- + 2OH^- + F_2 \rightarrow BrO_4^- + H_2O + 2F^-$$

A solution of acidified potassium perbromate passed through a cation exchange resin gave a pure solution of perbromic acid ($HBrO_4$) obtained in concentrations up to 80 per cent by vacuum evaporation at room temperature. The mass spectrum of perbromic acid was obtained by evaporating a solution of potassium perbromate with dilute sulphuric acid from a platinum filament incorporated into a mass spectrometer. The spectrum showed the presence of the parent ion $HBrO_4^+$ and its principal fragments. Spectra from the heated sample persisted for some time indicating perbromic acid to be a relatively stable compound.*

The perhalates are oxidizing agents:

$$XO_4^- + 2H^+ + 2e \rightarrow XO_3^- + H_2O$$

The $E^\ominus$ values for the above reaction where X = Cl, Br, and I, are +1·23, +1·76 and +1·67 V respectively.

Paraperiodic acid is obtained by heating barium iodate. This produces barium paraperiodate, which is then treated with sulphuric acid, filtered, and concentrated; colourless deliquescent crystals of H_5IO_6 are deposited on cooling.

$$3I_2 + 6KOH \rightarrow 5KI + KIO_3 + 3H_2O$$

$$2KIO_3 + BaCl_2 \rightarrow Ba(IO_3)_2 + 2KCl$$

$$5Ba(IO_3)_2 \rightarrow Ba_5(IO_6)_2 + 4I_2 + 9O_2$$

* The hydrate $HBrO_4.2H_2O$ has now been crystallized.

The acid decomposes at 140 °C.

$$2H_5IO_6 \rightarrow 2HIO_3 + 4H_2O + O_2 \quad \text{(plus a little } O_3\text{)}$$

On heating *in vacuo* at 80 °C for a prolonged period, a metaperiodic acid (HIO_4) is produced. Paraperiodic acid contains the octahedral anion IO_6^{5-} and metaperiodic acid the tetrahedral anion IO_4^-. Both acids form salts, e.g. Ag_5IO_6, $Ba_5(IO_6)_2$, KIO_4, $AgIO_4$, etc. Paraperiodic acid and its salts are powerful oxidizing agents and oxidize manganese(II) salts to purple permanganate.

Interhalogen Compounds and Polyhalides

A wide range of compounds of this type is known as shown in Table 17.4.

TABLE 17.4

ClF (gas)	BrF (liquid)	IF_5 (liquid)	IBr (solid)	BrCl (unstable gas)
ClF_3 (gas)	BrF_3 (liquid)	IF_7 (gas)		ICl (solid)
	BrF_5 (liquid)			ICl_3 (solid)

Usually in these compounds the halogen atoms exert a single covalent bond, although fluorine and to a lesser extent chlorine tend to bring out the highest oxidation state of the halogen with which they combine and more than one bond may be involved. All the interhalogen compounds are extremely volatile.

The interhalogen compounds may be prepared by direct action of the elements, excess of one of the reactants giving a compound with one of the halogens in a higher oxidation state. Thus, when fluorine is passed over solid iodine, iodine pentafluoride (IF_5) is produced; excess fluorine on the pentafluoride gives the heptafluoride (IF_7). All the fluorine compounds are extremely reactive chemically and are very powerful fluorinating agents. Bromine trifluoride, produced by disproportionation of bromine monofluoride, is particularly valuable in this respect.

$$3BrF \rightarrow BrF_3 + Br_2$$

Conductivity measurements on bromine trifluoride give a value of 8×10^{-3} ohm^{-1} cm^{-1}, indicating ionization.

$$2BrF_3 \rightarrow BrF_2^+ + BrF_4^-$$

Reactions with alkali and other less electropositive metal fluorides verify this ionization.

$$KF + BrF_3 \rightarrow K^+BrF_4^-$$

$$SnF_4 + 2BrF_3 \rightarrow (BrF_2^+)_2SnF_6^{2-}$$

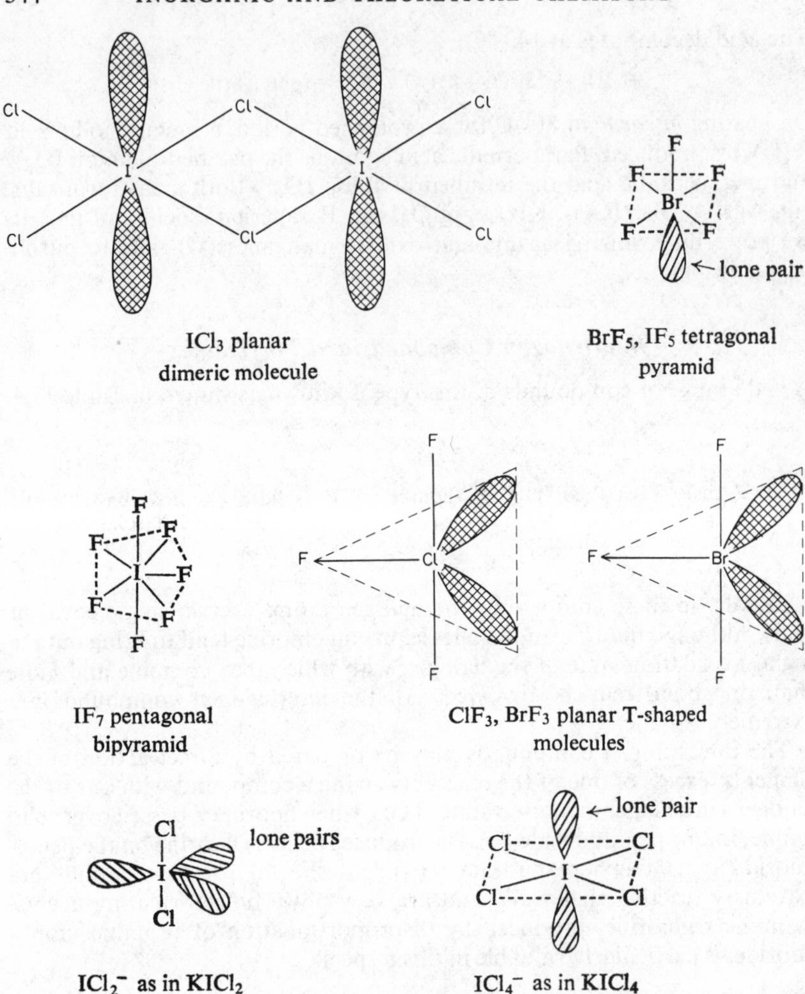

FIGURE 17.8

Among the chlorine compounds, bromine monochloride (BrCl) has a very small heat of formation($\Delta H_f^{\ominus} = +14{\cdot}7$ kJ mol^{-1}) and is only stable when in equilibrium with chlorine or bromine. Iodine monochloride is hydrolysed by water producing hypoiodous acid.

$$ICl + H_2O \rightarrow HCl + HOI$$

The latter rapidly disproportionates to iodate and iodine (p. 340). Iodine trichloride forms deliquescent yellow needles which are also hydrolysed.

$$2ICl_3 + 3H_2O \rightarrow 5HCl + ICl + HIO_3$$

Conductivity values of liquid iodine monochloride and non-aqueous solutions of iodine trichloride again indicate ionization.

$$2ICl \rightleftharpoons I^+ + ICl_2^-$$

$$2ICl_3 \rightleftharpoons ICl_2^+ + ICl_4^-$$

Both compounds add on to alkali metal chlorides giving well defined crystalline salts such as $K^+ICl_4^-$, $K^+ICl_2^-$.

The probable structures of some of the interhalogen compounds and the polyhalides are drawn out in Figure 17.8.

The reader should note that other lone-pair positions are possible with the species ClF_3, ICl_2^-, ICl_4^-, and BrF_5. The structures shown above, however, possess minimum repulsion energy and have been confirmed experimentally.

Cationic Iodine Compounds

Since iodine has the largest radius and hence the smallest ionization energy of the halogens, it is the only halogen to form cationic species of any appreciable stability. The compound, iodyl sulphate, containing the polymerised oxo cation $(IO^+)_n$ has been mentioned previously under the oxides of iodine (p. 338). Solid compounds containing solvated I^+ species have been known for a long time and the equations for the preparation of iodine dipyridyl nitrate and perchlorate are given below.

$$AgNO_3 + I_2 + 2py \xrightarrow{\text{ether}} [py \rightarrow I \leftarrow py]^+NO_3^- + AgI$$
$$AgClO_4 + I_2 + 2py \xrightarrow{\text{ether}} [py \rightarrow I \leftarrow py]^+ClO_4^- + AgI$$

The compounds are hydrolysed by water and liberate iodine quantitatively from acidified potassium iodide.

$$Ipy_2^+ + H_2O \rightarrow HOI + H^+ + 2py$$
$$Ipy_2^+ + I^- \rightarrow I_2 + 2py$$

Solid compounds which contain iodine in oxidation state +3 have also been prepared:

$$\underset{\text{cold conc.}}{I_2 + 12HNO_3} \rightarrow \underset{\text{iodine(III)nitrate}}{2I(NO_3)_3} + 6H_2O + 6NO_2$$

$$I_2 + 6(CH_3CO)_2O + 6H_2O \xrightarrow[\text{conc.}]{HNO_3} \underset{\text{iodine(III)acetate}}{2I(CH_3COO)_3} + 6CH_3COOH$$

$$I_2 + 6HClO_4 + O_3 \rightarrow \underset{\text{iodine(III)perchlorate}}{2I(ClO_4)_3} + 3H_2O$$

Electrolytic studies indicate the presence of I^{3+} cations, stabilized by coordination, in these compounds. They are rapidly hydrolysed by water.

$$5I^{3+} + 9H_2O \rightarrow I_2 + 3IO_3^- + 18H^+$$

18

The Transition Metals

General Characteristics of Elements of the First Transition Series

The physical properties of the transition metals are set out overleaf.

A transition element may be defined as one in which the atoms have an incomplete inner *d* subshell. Moving across the first transition series from scandium to copper, the $3d$ subshell fills to completion at copper. Anomalies in filling the $3d$ subshell occur with chromium and copper and this can be accounted for on the basis that half-filled and completely filled subshells represent particularly stable states. Thus with chromium and copper, one $4s$ electron is drawn into the $3d$ subshell to achieve this state.

Strictly, since the $3d$ shell is full at copper, it should not be classified as a transition element. However, the chemistry of copper is mainly that of the copper(II) ion Cu^{2+} ($3s^2\ 3p^6\ 3d^9$), where the $3d$ subshell is incomplete.

Physical Properties

A slight decrease in atomic radii across the series is coupled with an increase in values of ionization potential. The low values of first ionization potential indicate loosely bound outer electrons which readily enter into metallic bond formation (p. 87). Thus all the elements exhibit the characteristic properties of metals, i.e. bright lustre, good conduction, high density, melting point, and boiling point. The high melting and boiling points of the transition metals are related to their ability to use both $3d$ and $4s$ electrons for metallic bonding. A reduction in the melting and boiling points can be observed as the electrons pair up in the $3d$ subshell and become less available for metallic bonding. The abnormally low values for manganese can be related partly to its complex crystalline structure which involves less efficient packing of the metallic atoms.

As indicated in the tabulation, the majority of the metals can crystallize in more than one form. In the face-centred cubic method of packing, there are more gliding planes than in the hexagonal close-packed or body-centred cubic arrangements. The metals which crystallize with the face-centred cubic

	Scandium Sc	Titanium Ti	Vanadium V	Chromium Cr	Manganese Mn	Iron Fe	Cobalt Co	Nickel Ni	Copper Cu
Atomic mass	44·956	47·90	50·942	51·996	54·9380	55·847	58·9332	58·71	63·54
Atomic number	21	22	23	24	25	26	27	28	29
Most abundant isotope	45 (100%)	48 (73·4%)	51 (99·75%)	52 (83·7%)	55 (100%)	56 (91·6%)	59 (100%)	58 (67·9%)	63 (69%)
Electronic structure	$3d^1 4s^2$	$3d^2 4s^2$	$3d^3 4s^2$	$3d^5 4s^1$	$3d^5 4s^2$	$3d^6 4s^2$	$3d^7 4s^2$	$3d^8 4s^2$	$3d^{10} 4s^1$
Atomic radius (nm)	0·144	0·132	0·122	0·117	0·117	0·116	0·116	0·115	0·117
Ionic radius M^{3+} (nm)	0·081	0·076	0·074	0·069	0·066	0·064	0·063	0·062	—
Electronegativity	1·3	1·6	1·7	1·6	1·5	1·7	1·7	1·8	1·9
Ionization potential (I) (kJ mol^{-1})	638	667	654	659	722	768	763	742	751
Density (g cm^{-3})	3·1	4·5	5·96	7·19	7·21	7·87	8·90	8·90	8·91
Melting point (°C)	1540	1,725	1900	1,890	1,260	1,535	1,490	1,452	1,083
Boiling point (°C)	2730	3,260	3,400	2,475	2,030	2,735	2,950	2,840	2595
Oxidation states	3	2, 3, 4	2, 3, 4, 5	2, 3, 6	1, 2, 3, 4, 6, 7	2, 3, 6	2, 3, 4,	2, 3, 4,	1, 2
Crystal structure of metals*	f.c.c. h.c.p.	h.c.p. b.c.c.	b.c.c.	b.c.c.	3 crystalline forms with complex structures	f.c.c. b.c.c.	f.c.c. h.c.p.	f.c.c. h.c.p.	f.c.c.
Electrode potential (V)									
$M^{2+}(aq) + 2e \rightarrow M(s)$	—	−1·63	−1·18	−0·56	−1·03	−0·44	−0·28	−0·25	+0·34
$M^{3+}(aq) + e \rightarrow M^{2+}(aq)$	—	−0·20	−0·26	−0·41	+1·50	+0·74	+1·82	—	—

* f.c.c. face-centred cubic; h.c.p. hexagonal close packed; b.c.c. body-centred cubic.

system tend to be more malleable and ductile than those with hexagonal and body-centred cubic forms. Thus copper and iron are fairly soft, malleable, and ductile, whereas chromium and vanadium are much harder and more brittle.

The $E^\ominus$ values for the M^{2+}/M system indicate a decreasing tendency to form divalent cations across the series. However, there is an exception with manganese which is the strongest reducing agent and forms the cation Mn^{2+} which has a half-filled $3d$ subshell. The $E^\ominus$ values indicate that all the metals should react with dilute non-oxidizing acids. It should be noted, however, that in practice titanium and vanadium are passive to dilute non-oxidizing acids at room temperature. From an examination of the $E^\ominus$ values for the M^{3+}/M^{2+} system, the Mn^{3+} and Co^{3+} ions are the strongest oxidizing agents in aqueous solution. The ions Ti^{2+}, V^{2+}, and Cr^{2+} will all liberate hydrogen from a dilute acid, e.g.

$$2Cr^{2+} + 2H^{+} \rightarrow 2Cr^{3+} + H_2$$

Interstitial Compounds

The transition metals form interstitial compounds particularly with the non-metals hydrogen, boron, carbon, and nitrogen. The atoms of these non-metals possess small atomic radii (H = 0·030 nm; B = 0·080 nm; C = 0·077 nm; and N = 0·074 nm) and enter into the vacant holes between the packed atoms of the crystalline metal, e.g. TiC, Mn_4N, Fe_8N, TiH_2. Sometimes certain oxides and sulphides of the transition metals are classified as interstitial p. 400). The formulae quoted do not, of course, correspond to any normal oxidation state of the metal and often non-stoichiometric material is obtained with such compositions as $VH_{0 \cdot 56}$ and $TiH_{1 \cdot 7}$, etc.

The interstitial hydrides, borides, carbides, and nitrides are generally prepared by heating the metal with hydrogen, boron, carbon (or a hydrocarbon), and nitrogen (or ammonia), respectively. The compounds are inert chemically except towards strong oxidizing agents. They possess many properties of alloys—high melting points (TiC, 3,410 °C; TiN, 3,200 °C), good electrical conductivity, and extreme hardness. The malleability of these compounds is less than the parent metals because the gliding planes are 'pegged' in position by the non-metallic atoms. The conductivity is also reduced slightly for a similar reason, i.e. the interstitial non-metallic atoms restrict the mobility of the electrons.

Variability in Oxidation States

The various oxidation states known for the transition metals are attributed to the unpaired inner d electrons which require little promotion energy for use as valency electrons. Thus the iron atom ($3d^6 4s^2$) may lose two $4s$ electrons to form the Fe^{2+} ion (oxidation state +2) or, under certain circumstances, a further $3d$ electron may be pulled away to give the Fe^{3+} ion (oxidation state +3). However, *d* electrons may also be used for covalent bonding as in the ferrates (e.g. $K^+_2FeO_4^{2-}$) where the oxidation state of iron is +6 (p. 407). As the electrons continue to pair up in the 3*d* subshell, the high oxidation states disappear with cobalt, nickel and copper.

Magnetism

Many transition-metal cations and compounds possess unpaired electrons in the inner 3*d* subshells and exhibit paramagnetism (p. 76). When the 3*d* subshell is being filled, the added electrons enter the empty orbitals before any pairing takes place. The magnetic moment increases with the number of unpaired electrons and the observed magnetic moment gives a useful indication of the number of unpaired electrons present in the atom, molecule, or ion. The magnetic moments of the divalent cations of the first transition series reach a maximum with the manganese(II) ion (Mn^{2+}) which has five unpaired electrons (Figure 18.1).

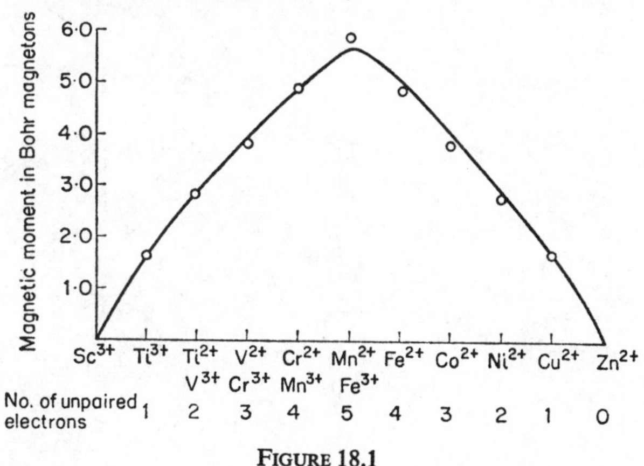

FIGURE 18.1

Ferromagnetism is an extreme form of paramagnetism, due to the aggregation of atoms or ions. Unlike paramagnetism, ferromagnetism remains permanent even in the absence of an applied field. Incomplete *d* (or *f*) subshells are not the only requirement for ferromagnetism as the interionic or interatomic distances and interactions are extremely important. For further details, a more advanced text may be consulted.

Complex Ion Formation

The transition-metal cations are small and highly charged and readily attract polar molecules or ions to form complexes, particularly with cyanide and nitrite ions and with ammonia molecules. Complexes were discussed in Chapter 5, where the bonding of the ligands to the central atom or ion was represented by a coordinate link. There are, however, a number of observations which make this approach unacceptable. Magnetic studies indicate that, in the complex ion hexacyanoferrate(II) ($Fe(CN)_6^{4-}$), there are no unpaired electrons whereas the ferrous ion has four unpaired electrons. Thus, in the complex ion, electron pairing occurs and the bonding must be predominantly covalent. However, with complexes such as $Fe(H_2O)_6^{3+}$ and $FeCl_6^{3-}$, magnetic studies indicate the presence of five unpaired electrons in the complex. The iron(III) ion also has five unpaired electrons. Here, no electron pairing occurs in the complex and the bonding must be predominantly ionic.

The crystal field theory attempts to explain the shapes of the molecules

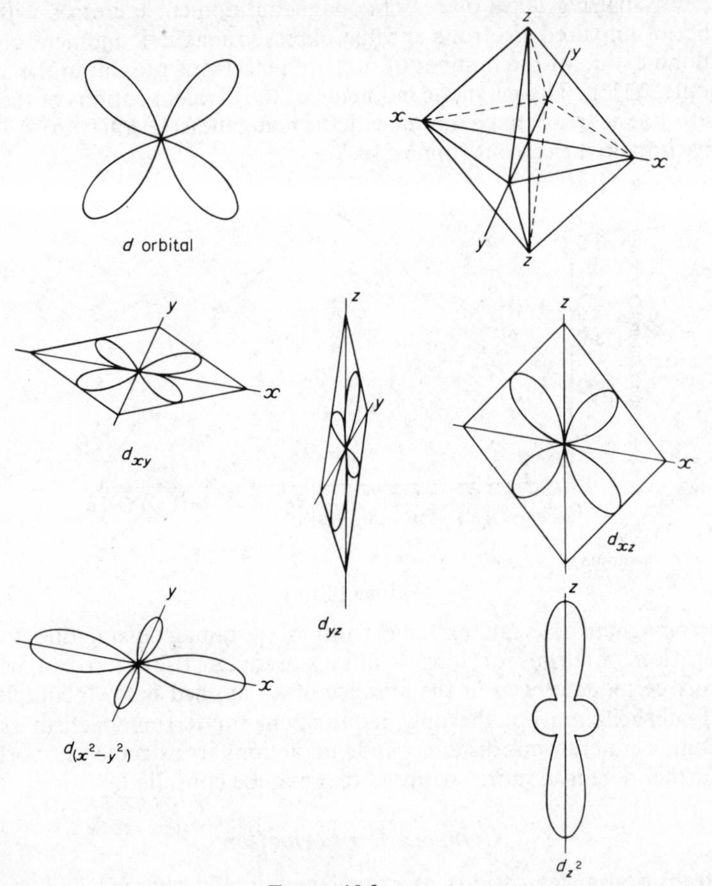

FIGURE 18.2

and the observed magnetic moments of the complexes. The d orbitals occur in sets of five ($l = 2$; $m = 0, \pm1, \pm2$; $s = \pm\frac{1}{2}$). The shape of one such orbital is shown in Figure 18.2. Each d orbital may contain one or two electrons which may be at any point in space within these regions. In the neutral atom, all five d orbitals have the same energy and are said to be degenerate. If six ligands (polar molecules or negative ions) approach a transition-metal cation from the corners of an octahedral structure, the d orbitals no longer have identical energies but are split into two sets. The d_{xz}, d_{xy}, and d_{yz} orbitals (designated by the symbol t_{2g}) lie on the diagonals between the respective axes and the $d_{x^2-y^2}$ and d_{z^2} orbitals (designated e_g) lie on the bond axes as shown in Figure 18.2.

The e_g orbitals directly in line with the approaching ligands become higher in energy than the t_{2g} orbitals.

If a complex is irradiated with visible or ultraviolet light, the electrons in the t_{2g} orbitals can often be excited to such an extent that they jump into the higher energy e_g orbitals. An examination of the resulting absorption spectrum reveals the magnitude of the energy difference (Δ) between the two sets of levels. Thus the absorption maxima for the octahedral complexes $CrCl_6^{3-}$, $Cr(H_2O)_6^{3+}$, $Cr(NH_3)_6^{3+}$ and $Cr(CN)_6^{3-}$ are 13 600, 17 400, 21 600, and 26 300 cm^{-1} respectively. The greater the wave number in cm^{-1}, the larger is the energy difference ($\Delta_{octahedral}$ or Δ_o) between the t_{2g} and e_g levels in the octahedral complex. The energy difference can then be compared for different complexes, giving a relative measure of the ligand field strength. These fall in the following order of decreasing strength: $CN^- > NO_2^- >$ ethylene diamine $> NH_3 > NCS^- > (COO^-)_2 > H_2O > F^- > OH^- > Cl^- > Br^- > I^-$.

The larger the energy difference between the two sets of orbitals, the greater the tendency of the d electrons in the transition metal to fall into the lower energy t_{2g} orbitals. Cyanide and nitrite ions set up strong ligand fields, where the energy difference is large, and water and the chloride ion produce weak fields, where the energy difference is small.

Thus in the $Fe(CN)_6^{4-}$ complex, the six cyanide ligands concentrate the six $3d$ electrons of the iron(II) ions into the t_{2g} orbitals with the result that there are no unpaired electrons in the complex (Figure 18.3(*a*)). The energies of the e_g orbitals lie near to those of the $4s$ and $4p$ orbitals and these are then able to participate in d^2sp^3 hybridization for octahedral bonding with six cyanide ions. The complex is termed an *inner orbital complex* and is said to be either covalent or spin paired.

In the $Fe(CN)_6^{3-}$ complex, one unpaired electron remains in the t_{2g} orbital (Figure 18.3(*b*)), and the resulting octahedral complex possesses one unpaired electron.

In the $Fe(H_2O)_6^{2+}$ complex, the field due to the water molecules is weak and the electrons remain distributed as in the structure of the iron(II) ion and the complex has the same number of unpaired electrons. (Figure 18.3(*c*)). In this case sp^3d^2 hybridization occurs with orbitals from the same quantum shell. The complex is termed an *outer orbital complex* and is said to be ionic or spin free. Figure 18.4(*a*) shows the energy difference between the t_{2g} and e_g orbitals in an octahedral complex, designated Δ_o. It will be noted that

	Ion	Cell diagram	Hybridization	Shape	No. of unpaired electrons
Fe^{2+}	Iron(II)	3d [↑↓][↑][↑][↑][↑] 4s [ ] 4p [ ][ ][ ] 4d [ ][ ]	—	—	4
$Fe(CN)_6^{4-}$ (a)	Hexacyano ferrate(II)	t_{2g} [↑↓][↑↓][↑↓] ←Δ→ 3d e_g [↑↓][↑↓] (dashed) 4s [↑↓] (dashed) 4p [↑↓][↑↓][↑↓] (dashed) 4d [ ][ ]	d^2sp^3	Octahedral	0
Fe^{3+}	Iron(III)	3d [↑][↑][↑][↑][↑] 4s [ ] 4p [ ][ ][ ] 4d [ ][ ]	—	—	5
$Fe(CN)_6^{3-}$ (b)	Hexacyano ferrate(III)	t_{2g} [↑↓][↑↓][↑] ←Δ→ 3d e_g [↑↓][↑↓] (dashed) 4s [↑↓] (dashed) 4p [↑↓][↑↓][↑↓] (dashed) 4d [ ][ ]	d^2sp^3	Octahedral	1
Fe^{2+}	Iron(II)	3d [↑↓][↑][↑][↑][↑] 4s [ ] 4p [ ][ ][ ] 4d [ ][ ]	—	—	4
$Fe(H_2O)_6^{2+}$ (c)	Hexaaquo iron(II)	t_{2g} [↑↓][↑][↑] ←Δ→ 3d e_g [↑][↑] 4s [↑↓] (dashed) 4p [↑↓][↑↓][↑↓] (dashed) 4d [↑↓][↑↓] (dashed)	sp^3d^2	Octahedral	4

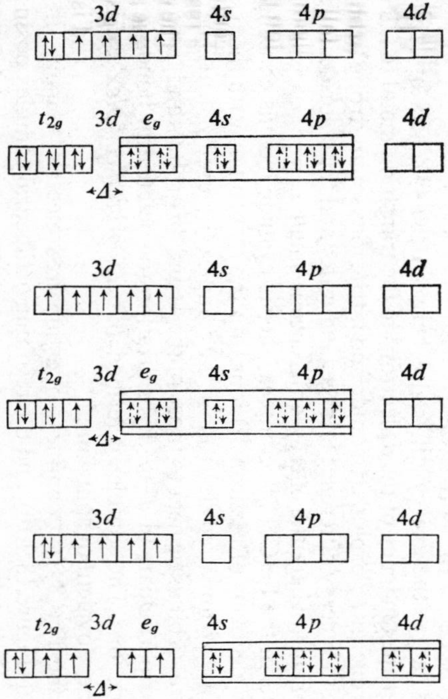

FIGURE 18.3

the two e_g orbitals are $\frac{3}{5}\Delta_o$ above and the three t_{2g} orbitals $\frac{2}{5}\Delta_o$ below the energy of the unsplit 3*d* orbitals.

In tetrahedral complexes, the t_{2g} orbitals point more towards the ligands and thus have higher energy than the e_g orbitals in this case. The electric field set up by four ligands is much smaller than in an octahedral arrange-

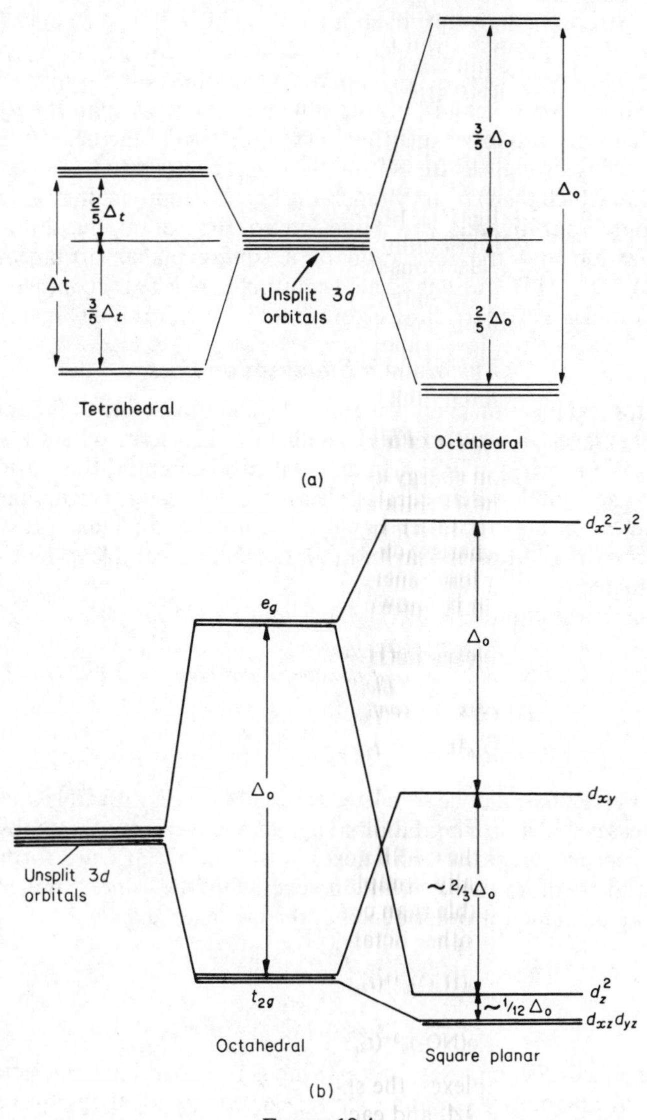

FIGURE 18.4

ment and the resulting energy difference Δ_t is $\frac{4}{9}$ of the value for an octahedral complex, i.e. $\Delta_t = \frac{4}{9} \times \Delta_o$ (Figure 18.4(*a*)). In a strong ligand field, however, electrons should move from t_{2g} orbitals to vacant e_g orbitals but no spin paired tetrahedral complexes have been found, presumably because Δ_t is too small. Thus $Co(NCS)_4^{2-}$, $CoCl_4^{2-}$, $Cu(CN)_4^-$ are all spin free tetrahedral complexes.

If six strong ligands approach a cation with full t_{2g} orbitals, from the corners of an octahedron, electrons cannot be forced from e_g into t_{2g} orbitals. However, the $d_{x^2-y^2}$ orbital, with four lobes pointing in the direction of the approaching ligands, becomes higher in energy than the d_{z^2} orbital which has only two lobes pointing in the direction of the ligands. Since the ligand field is strong, electrons from the $d_{x^2-y^2}$ orbital can be forced into the d_{z^2} orbital which then strongly repels the ligands approaching along its axis. Thus only four ligands can bond on to the cation, resulting in dsp^2 hybridization and the formation of a square planar arrangement (*see* $Ni(CN)_4^{2-}$, p. 413). The ligand field splitting here is fairly complex with the energy of the d_{z^2} orbital falling below that of the d_{xy} orbital (Figure 18.4(*b*)).

Crystal Field Stabilization Energy

The extra stabilization energy acquired by a complex due to the splitting of the energy levels of the *d* orbitals by a ligand field is termed the Crystal Field Stabilization Energy (CFSE). In an octahedral complex, there are three t_{2g} and two e_g orbitals where each e_g electron gains energy by an amount $\frac{3}{5}\Delta_o$ and each t_{2g} electron loses energy by an amount $\frac{2}{5}\Delta_o$. Thus, providing the *d* electron distribution is known in an octahedral complex, the CFSE can be evaluated.

Consider the complexes $Fe(H_2O)_6^{3+}$ and $Fe(CN)_6^{3-}$.

Complex	*Electronic configuration*	*Nature*	*CFSE*
$Fe(H_2O)_6^{3+}$	$t_{2g}^3\, e_g^2$	Spin free	0
$Fe(CN)_6^{3-}$	$t_{2g}^5\, e_g^0$	Spin paired	$\frac{10}{5}\Delta_o$

Hence $Fe(H_2O)_6^{3+}$ can be stabilized by combination with CN^- ligands with an increase of $\frac{10}{5}\Delta_o$ in the CFSE and ligand replacement of water by cyanide readily occurs. Generally complexes with high CFSEs are formed more easily and are more stable than complexes with low values. Similar calculations can be made for other octahedral complexes, e.g.

$$Co(H_2O)_6^{3+}(t_{2g}^4\, e_g^2), \quad CFSE = \frac{2}{5}\Delta_o$$

$$Co(NO_2)_6^{3-}(t_{2g}^6\, e_g^0), \quad CFSE = \frac{12}{5}\Delta_o$$

In tetrahedral complexes, the splitting is reversed. Each e_g electron loses energy by an amount $\frac{3}{5}\Delta_t$ and each t_{2g} electron gains energy by an amount $\frac{2}{5}\Delta_t$. Since $\Delta_t = \frac{4}{9} \times \Delta_o$, the CFSE is much smaller and no spin paired complexes are known. Thus tetrahedral complexes are far less stable than octahedral ones, e.g.

$$CoCl_4^{2-}(t_{2g}^3 e_g^4), \qquad CFSE = \frac{6}{5}\Delta_t = \frac{6}{5} \times \frac{4}{9} \times \Delta_o = \frac{8}{15}\Delta_o$$

Square planar complexes are formed with d^8 and d^9 configurations and, although the d orbital splitting is complex and evaluation of CFSE more difficult, there is a net gain in the CFSE assisting the formation of a square planar complex. It must be noted that, although CFSEs are important, they are not the only factors involved in complex stabilization and considerations such as metal–ligand bond energies, solvation energies and entropy changes all have to be taken into account to obtain a complete understanding of a very complex situation.

Jahn–Teller Effect

In an octahedral copper(II) complex ($t_{2g}^6 e_g^3$) with two electrons in the d_{z^2} and one electron in the $d_{x^2-y^2}$ orbitals, the negative or polar ligands are more screened from the attraction of Cu^{2+} along the z axis than along the x and y axes. A distortion occurs resulting in the elongation of opposite bonds along the z axis and an accompanying shortening of the four planar bonds along the x and y axes. This distortion from perfect symmetry occurs particularly with $t_{2g}^3 e_g^1$, $t_{2g}^6 e_g^1$, and $t_{2g}^6 e_g^3$ configurations, and is termed the Jahn–Teller effect.

Examples of this distortion occur in $Mn^{III}F_6^{3-}$, $Co^{II}(NO_2)_6^{4-}$, and $Cu^{II}(NH_3)_4(H_2O)_2^{2+}$. In fact a number of square planar complexes could be regarded as resulting from Jahn–Teller distortion when the ligands along the z axis have been completely removed.

Reaction Rates in Complexes

Complex ions which undergo rapid ligand exchange are said to be labile whereas those resistant to ligand exchange are termed inert. Spin-paired (inner orbital) complexes with vacant d orbitals are usually labile, the vacant d orbitals permitting the addition of extra ligands, e.g. $V(CN)_6^{3-}$. If all the d orbitals are occupied singly or doubly, the spin-paired complex is usually inert, e.g. $Cr(CN)_6^{3-}$, $Fe(CN)_6^{4-}$. Spin-free (outer orbital) complexes are usually labile but become more inert with an increase in metal–ligand bond strength, e.g. $Ni(en)_3^{2+}$ is more inert than $Ni(NH_3)_6^{2+}$ due to the stronger nickel–nitrogen covalent bonds in the ethylenediamine complex. All tetrahedral complexes are labile but square planar complexes may be labile or inert.

Mechanisms for Ligand Interchange

Octahedral complexes with vacant d orbitals provide a site for lone pairs of an extra ligand to form a seven coordinated activated complex. This is followed by elimination of one of the six original ligands to form a new octahedral complex. The rate of reaction depends upon the activation energy for the formation of the seven coordinate intermediate. Thus, if the activation energy is low because of the presence of empty *d* orbitals, the reaction is rapid. The reaction is a substitution reaction (S) which is nucleophilic (N) and bimolecular (2) and is termed S_N2.

$$\underset{}{ML_6} + \underset{\text{nucleo-phile}}{N} \rightarrow \underset{\text{activated complex}}{ML_6N^*} \rightarrow ML_5N + L$$

The rate-determining step is the formation of the activated complex, ML_6N^*, and the reaction rate $v = k[ML_6][N]$. Thus, the number of molecules taking part in the rate-determining step is two and the reaction is bimolecular.

An alternative mechanism could involve the departure of one of the ligands leaving a square pyramidal activated complex.

$$ML_6 \rightarrow \underset{\text{activated complex}}{ML_5^*} + L$$

$$ML_5^* + \underset{\text{nucleophile}}{N} \rightarrow ML_5N$$

Here the reaction is S_N1, i.e. a substitution reaction which is nucleophilic and unimolecular. It is unimolecular since the rate of reaction

$$v = k[ML_6]$$

Unfortunately, the majority of ligand exchange reactions seldom confirm any one of these mechanisms. The reactions of coordination complexes are nearly always complicated by other factors such as solvent interaction and acid–base catalysis.

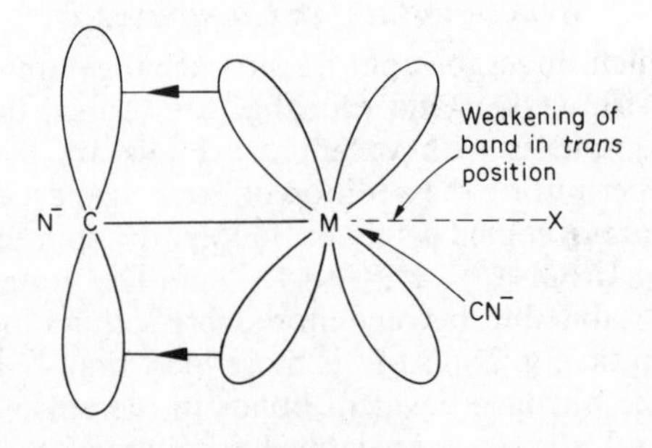

FIGURE 18.5

The Trans Effect

In substitution reactions involving square planar complexes, the lability of a substituted position is found to be influenced by the ligand in the trans position (trans effect). Some ligands listed in decreasing order of their trans directing influence are:

$$CN^- > CO > I^- > NO_2^- > Br^- > Cl^- > NH_3 > OH^- > H_2O$$

The trans effect has considerable application in the preparation of specific square planar complexes:

(*a*)

$$PtCl_4^{2-} + NH_3 \longrightarrow \left[\begin{matrix} Cl & & NH_3 \\ & Pt & \\ Cl & & Cl \end{matrix}\right]^- + Cl^-$$

$$\left[\begin{matrix} Cl & & NH_3 \\ & Pt & \\ Cl & & Cl \end{matrix}\right]^- + NH_3 \longrightarrow \left[\begin{matrix} Cl & & NH_3 \\ & Pt & \\ Cl & & NH_3 \end{matrix}\right] + Cl^-$$

cis isomer

(*b*)

$$Pt(NH_3)_4^{2+} + Cl^- \longrightarrow \left[\begin{matrix} NH_3 & & Cl \\ & Pt & \\ NH_3 & & NH_3 \end{matrix}\right]^+ + NH_3$$

$$\left[\begin{matrix} NH_3 & & Cl \\ & Pt & \\ NH_3 & & NH_3 \end{matrix}\right]^+ + Cl^- \longrightarrow \left[\begin{matrix} NH_3 & & Cl \\ & Pt & \\ Cl & & NH_3 \end{matrix}\right] + NH_3$$

trans isomer

Thus, since Cl^- has a greater trans directing influence than NH_3, the substitution of ammonia in $PtCl_3.NH_3^-$ produces the cis isomer and the substitution of Cl^- in $Pt(NH_3)_3Cl^+$ produces the trans isomer.

There is no completely unifying theory which explains the trans effect. However, ligands such as CN^-, CO, and NO_2^- possess *p* orbitals which can accept electrons from the metal *d* orbitals. This weakens the bond in the trans position by electron withdrawal and makes the trans position more susceptible to attack by a nucleophilic reagent (Figure 18·5).

Colour

If, when white light falls on a compound, the light is totally reflected, the compound appears white and, if totally absorbed, it appears black. Coloured compounds possess electrons which have the ability to absorb certain wavelengths of white light, emitting the remaining wavelengths which appear as coloured light and producing a characteristic absorption spectrum for the compound. The relationship between the colour, wavelength, and energy absorbed and the colour observed is shown in Table 18.1.

From the values of ionization potential (p. 347), it can readily be seen that even violet light (energy $\approx 5 \times 10^{-17}$ J) cannot completely remove an electron from an atom. Visible light can, however, cause electrons to undergo intra-atomic transitions. Thus the transition-metal salts all furnish coloured hydrated cations in solution of the type $M(H_2O)_6^{n+}$. These hydrated cations possess unpaired inner *d* electrons which absorb small

TABLE 18.1

Energy absorbed $\times 10^{-17}$ J	*Wavelength absorbed* in (nm)	*Colour absorbed*	*Colour observed*
5·0	400–435	Violet	Yellow-green
↓	435–480	Blue	Yellow
	480–490	Green-blue	Orange
	490–500	Blue-green	Red
	500–560	Green	Purple
	560–580	Yellow-green	Violet
	580–595	Yellow	Blue
	595–605	Orange	Green-blue
2·8	605–750	Red	Blue-green

quantities of energy in the form of certain wavelengths of light. In these cations the *d* orbitals are split into two sets, the t_{2g} and the e_g orbitals (p. 351).

Spectroscopic laws require that the azimuthal quantum number *l* must change by ±1 before an electronic transition occurs. Nevertheless, although t_{2g} and e_g orbitals have the same *l* value and transition between them is described as 'Laporte forbidden', it does occur to a small extent because of distortion of pure *d* orbitals by the ligands. The intensity of the absorption bands is only low, giving rise to weak colours.

The colour of the ion depends mainly on the energy difference (Δ) between the t_{2g} and e_g orbitals. In the copper(I) (Cu^+) and scandium (Sc^{3+}) ions there are no unpaired electrons, no visible light is absorbed, and the ions are colourless. The titanium(III) ion has one unpaired electron and weak absorption occurs causing a $t_{2g} \rightarrow e_g$ transition, giving rise to a pinkish colour. The ions $Mn(H_2O)_6^{2+}$ and $Fe(H_2O)_6^{3+}$ have the d^5 configuration and excitation of an electron from t_{2g} to e_g will cause a change in the number of unpaired electrons. This transition is said to be spin forbidden as well as Laporte forbidden, and the intensity of the absorption band is extremely low giving rise to very pale colours.

Considering the aquo-complexes of the first transition series, the number of *d* electrons alters: (*a*) in passing across the series; (*b*) with a change in oxidation state of the metal. Both these factors give rise to a change in Δ and hence a change in colour.

Ion	*Colour*	*No. of unpaired* d *electrons*	*Oxidation state*
$V(H_2O)_6^{3+}$	Green	2	+3
$Cr(H_2O)_6^{3+}$	Violet	3	+3
$Cr(H_2O)_6^{2+}$	Blue	4	+2
$Mn(H_2O)_6^{2+}$	Pale pink	5	+2
$Fe(H_2O)_6^{3+}$	Pale violet	5	+3
$Fe(H_2O)_6^{2+}$	Green	4	+2
$Co(H_2O)_6^{2+}$	Pink	3	+2
$Ni(H_2O)_6^{2+}$	Green	2	+2
$Cu(H_2O)_4^{2+}$	Blue	1	+2

A change in the ligand and consequently in the field strength will also bring about a change in Δ. Thus $Cu(H_2O)_4^{2+}$, which has a weak ligand field, is pale blue but $Cu(NH_3)_4^{2+}$ with a strong ligand field has a much higher Δ value and the colour is violet. The change in ligand may also involve a change in stereochemistry.

Thus, the addition of concentrated hydrochloric acid to a solution of a cobalt(II) salt causes the pink octahedral hydrated cation to change to a blue tetrahedral chloro anion.

$$\underset{\text{(octahedral)}}{\underset{\text{pink}}{Co(H_2O)_6^{2+}}} + 4Cl^- \rightarrow \underset{\text{(tetrahedral)}}{\underset{\text{blue}}{CoCl_4^{2-}}} + 6H_2O$$

Changes in colour also occur with geometrical isomers (p. 106), e.g. the $Co(NH_3)_4Cl_2^+$ cation exists as a violet *cis* and a green *trans* isomer. The observed colours are due to different electronic splitting within the *d* orbitals, giving rise to different values of Δ.

Apart from the examples mentioned above, many metal oxides, sulphides, halides, oxoanions such as permanganate (MnO_4^-) and chromate (CrO_4^{2-}), and certain complexes (e.g. $Fe(NCS)^{2+}$), are highly coloured. The observed colours cannot be explained on the basis of *d* orbital transitions, e.g. in permanganates and chromates, manganese and chromium have no unpaired *d* electrons.

In these cases, the colour is due to absorption of light energy causing electronic transitions between atoms. In the iron(III) thiocyanate complex, ($Fe(NCS)^{2+}$), light energy is absorbed, causing transitions from the reducing agent—thiocyanate—to the oxidizing agent—the iron(III) ion—producing the red colour. In aluminium thiocyanate, however, charge transfer from thiocyanate to aluminium is not possible and the resulting compound is colourless.

Intense colour is also observed in compounds containing mixed oxidation states, e.g. in Prussian blue ($KFe^{II}Fe^{III}(CN)_6$), where electronic transitions can occur between Fe(II) and Fe(III), giving the deep blue colour.

Where absorption of light energy causes electronic transitions between atoms, it is called a *charge-transfer transition* and the pattern of the wavelengths absorbed is called the *charge-transfer spectrum*.

Some General Features Concerning Transition Metal Chemistry

The Oxides

The lowest oxidation state oxide of a transition metal is basic and the highest usually acidic. In vanadium(II) oxide ($V^{2+}O^{2-}$) which has a rock salt structure, the V^{2+} cation does not cause any appreciable polarization of the electrons in the O^{2-} anion. However, in V_2O_5, where vanadium has a formal oxidation state of +5, a hypothetical ionic model would involve a very small highly charged cation V^{5+}, exerting a strong attraction for electrons in O^{2-}, causing a large drift of electronic charge towards V^{5+}, and imparting covalency and acidic character to the oxide. These features are typical of many transition metal oxides; e.g. $Cr^{II}O$, $Mn^{II}O$ are basic, $Cr^{VI}O_3$, $Mn_2{}^{VII}O_7$ are acidic.

The Halides

A transition metal exhibits higher oxidation states in its fluorides than in its iodides. Similar arguments to those discussed for oxides can also be applied to fluorides (and to some extent to chlorides) even though the F^- ion is less polarizable than the O^{2-} ion. Thus low oxidation state fluorides (and chlorides) are involatile ionic solids, whereas higher oxidation state fluorides (chlorides) are volatile covalent liquids or solids, e.g.:

$VF_2(s)$ b.p. 2227 °C	$VF_5(s)$ sublimes at 111 °C
$VCl_2(s)$ b.p. 1000 °C	$VCl_4(l)$ b.p. 154 °C

In contrast, the relatively large I^- ion is easily polarized, particularly by transition metal ions in high oxidation states. Thus a hypothetical ionic model of vanadium(V) iodide $V^{5+}(I^-)_5$ would involve such an intense polarization of the I^- ion that electron loss would take place.

$$\overset{\overset{2e}{\longleftarrow}}{V^{5+}(I^-)_5} \rightarrow V^{III}I_3 + I_2$$

Oxo Anions

A transition metal often exhibits its highest oxidation state in an oxo anion. Oxo anions such as $V^{V}O_4{}^{3-}$, $Cr^{VI}O_4{}^{2-}$, $Mn^{VII}O_4{}^{-}$, $Fe^{VI}O_4{}^{2-}$ are quite common for transition metals in high oxidation states. Charge is withdrawn from the oxygen atom reducing the formal oxidation state of the transition metal. This is assisted by $d\pi$–$p\pi$ bonds involving overlap of oxygen *p* orbitals with transition metal *d* orbitals, thus imparting double bond character to the metal–oxygen bond. The charge withdrawal is not excessive and solid compounds containing these oxo anions are readily obtained, e.g. Na_3VO_4, K_2CrO_4, $KMnO_4$, $BaFeO_4$. However, the oxidizing power of these species increases and the stability to air water and heat decreases across the first transition metal series.

Second and Third Transition Series of the Fifth and Sixth Periods

A complete survey of all the elements in the second and third transition series is not attempted and only the essentials of a few of the more important elements are described in this text. The general characteristics outlined for the metals of the first transition series also apply to the metals of the second and third series, but some distinctive features may be noted.

Electronic Structure and Atomic Radius

The second and third transition series involve filling the 4*d* and 5*d* subshells, respectively (p. 58). At the beginning of the third transition series, however, between lanthanum and hafnium, there are the fourteen lanthanide elements in which the inner 4*f* subshell fills from cerium to lutetium. The screening effect of the deep 4*f* electrons is much less effective than that of *d* electrons and hence the nucleus binds the outer electrons more tightly. Thus, as the lanthanides increase in atomic number, there is a considerable contraction in atomic radius. This effect is called the *lanthanide contraction* (p. 440). The increase in size expected for elements of the third transition series, due to the increase in atomic number, is almost entirely offset by this contraction. Thus the atomic radii of elements in the same group of the second and third transition series are very similar, e.g.

Element	*Atomic radius (nm)*	*Element*	*Atomic radius (nm)*
Ti	0·132	Cr	0·117
Zr	0·145	Mo	0·129
Hf	0·144	W	0·130

Oxidation States

Higher oxidation states of the second and third transition-series elements are generally much more stable than the corresponding states for the first transition series. Thus the permanganates (MnO_4^-), dichromates ($Cr_2O_7^{2-}$), and ferrates (FeO_4^{2-}) are strongly oxidizing, whereas the molybdates (MoO_4^{2-}) and tungstates (WO_4^{2-}) are stable and not easily reduced. Often higher oxidation state oxo anions of the second and third series elements have a strong tendency to condense, forming complex polynuclear species (p. 383).

Cation Chemistry, Oxo Salts, and Complexes

The lighter elements of the first transition series have at least one stable lower oxidation state, which furnishes stable hydrated cations and oxo salts. This is rarely found for the metals of the second and third series. Also the large number of complexes which exist for the first transition-series metals in lower oxidation states, particularly with ammonia molecules and cyanide ions, is almost entirely absent with the heavier metals.

Titanium, Zirconium, and Hafnium

There is a strong tendency to form the oxidation state +4 in this group, although the M^{4+} cation does not exist in aqueous solution and the most stable cationic species is the oxo cation $M^{IV}O^{2+}$. Zirconium and hafnium, with almost the same atomic radius, are very similar in their chemical properties.

Occurrence

Titanium occurs as rutile (TiO_2) and ilmenite ($FeTiO_3$). Zirconium and hafnium usually occur together in baddeleyite (ZrO_2 plus 2 per cent Hf) and zircon ($ZrSiO_4$ plus 7 per cent Hf).

Extraction

Titanium

This is extracted by the Kroll process. The ore is concentrated magnetically, mixed with coke, and chlorinated at 1000 °C.

$$TiO_2 + C + 2Cl_2 \rightarrow TiCl_4 + CO_2$$

The titanium tetrachloride is condensed and then reduced with molten magnesium in an atmosphere of argon in a steel reactor.

$$TiCl_4 + 2Mg \rightarrow Ti + 2MgCl_2$$

The molten magnesium chloride is tapped off and the reactor is allowed to cool. The titanium metal is melted into two-tonne ingots in a water-cooled copper crucible by an arc struck between the crucible and a consumable titanium electrode.

Zirconium and hafnium

These metals are extracted from zircon by the following sequence of reactions:

$$\underset{\text{zircon}}{ZrSiO_4} \xrightarrow{\text{fused NaOH}} \underset{\text{sodium silicate and zirconate}}{Na_2SiO_3,\ Na_2ZrO_3} \xrightarrow{\text{boiling } H_2O} ZrO_2.xH_2O + SiO_2.H_2O\downarrow$$

$$\xrightarrow{HCl} \underset{\text{zirconyl chloride}}{ZrOCl_2} + SiO_2.H_2O\downarrow \xrightarrow{NH_3} ZrO_2.xH_2O\ (\text{pure})$$

$$ZrO_2.xH_2O\ (\text{pure}) \xrightarrow{Cl_2 + C} ZrCl_4 \xrightarrow{\text{Mg metal}} Zr/Hf$$

Hafnium reacts in an identical manner and is separated from zirconium by ion-exchange techniques or by fractional crystallization of double fluorides.

Uses

Titanium is used in light-alloy temperature-resistant steels for use in missiles and rocketry. Titanium and its alloys also have a marked resistance to corrosion and chemical action and are used in chemical-plant reaction vessels and marine equipment. Zirconium is used in the production of bullet-proof alloy steels and to remove traces of oxygen from thermionic valves.

Properties

The metals are hard and are good conductors. A summary of their principal reactions is outlined in Figure 18.6 where M = Ti, Zr, Hf.

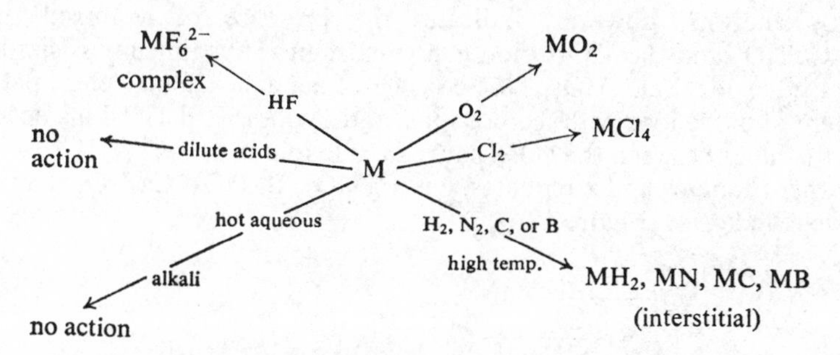

FIGURE 18.6. Principal reactions of titanium, zirconium, and hafnium

Titanium also dissolves in concentrated sulphuric acid producing titanium(III) sulphate (titanous sulphate).

$$2Ti + 6H_2SO_4 \rightarrow Ti_2(SO_4)_3 + 6H_2O + 3SO_2$$

With concentrated nitric acid, however, a hydrated oxide $TiO_2.xH_2O$ is produced (cf. Sn).

The Oxides

Titanium and zirconium dioxides occur naturally and are used as white paint pigments in glazing porcelain and tinting teeth, etc. Titanium dioxide is also manufactured from ilmenite ($FeTiO_3$) by treatment with sulphuric acid.

$$FeTiO_3 + 2H_2SO_4 \rightarrow 2H_2O + FeSO_4 + TiOSO_4$$

The addition of scrap iron converts any iron(III) sulphate back to iron(II) sulphate and, after filtering, evaporation deposits crystals of iron(II) sulphate, $FeSO_4.7H_2O$. The solution remaining contains titanyl sulphate which, on boiling with water, deposits hydrated titanium dioxide. The manufacture of zirconium dioxide from zircon was outlined in the extraction of the metal.

Titanium dioxide exists as three crystalline modifications, rutile (p. 95), where each titanium atom is surrounded octahedrally by six oxygen atoms, anatase and brookite, where the octahedra undergo considerable distortion. Zirconium and hafnium dioxides have an unusual seven-coordinate arrangement.

The hydrated oxides (there are no true hydroxides) are readily soluble in acids giving titanyl and zirconyl salts containing the oxo cation MO^{2+}, but the anhydrous oxides are very resistant to solution. Fusion of the hydrated oxides with caustic soda or sodium carbonate solution gives the so-called titanates and zirconates, e.g.

$$TiO_2 + 2OH^- \rightarrow TiO_3^{2-} + H_2O$$

X-ray analysis, however, indicates the presence of a mixed oxide $Na_2O.TiO_2$ and the TiO_3^{2-} ion is non-existent. Thus ilmenite is similarly a mixed oxide $FeO.TiO_2$, best considered as a hexagonal close-packed system of oxide ions with ferrous ions, and hypothetical Ti^{4+} ions occupying the holes between the close-packed oxide ions.

Other titanates and zirconates, e.g. $CaTiO_3$, $BaTiO_3$, $CaZrO_3$, have the pervoskite lattice (Figure 18.7).

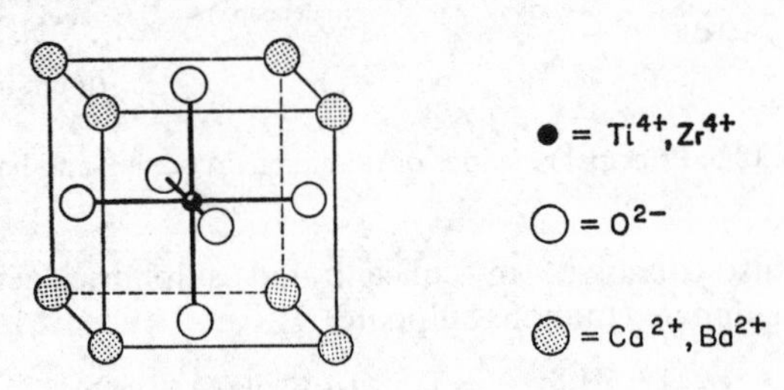

FIGURE 18.7

Since the small M^{4+} ions are free to move in the large octahedral holes, these compounds are polarized reversibly in an electric field.

The Halides

The tetrafluorides are white solids prepared by the action of anhydrous hydrogen fluoride on the tetrachlorides.

$$TiCl_4 + 4HF \rightarrow TiF_4 + 4HCl$$

They are unattacked by water. Excess fluoride ion gives complexes such as TiF_6^{2-}, ZrF_6^{2-} (octahedral), and ZrF_7^{3-} (pentagonal bipyramidal).

The tetrachlorides may all be made from the dioxides and carbon by chlorination.

$$MO_2 + 2Cl_2 + C \rightarrow MCl_4 + CO_2$$

Unlike the fluorides, they are hydrolysed by water.

$$TiCl_4 + 4H_2O \rightarrow TiO_2.2H_2O + 4HCl$$

$$ZrCl_4 + H_2O \rightarrow ZrOCl_2 + 2HCl \quad \text{(Hf is similar)}$$

zirconium oxochloride

Evaporation of a solution of the tetrachloride of zirconium and hafnium in hydrochloric acid gives white needles of $MOCl_2.8H_2O$, which appear to contain $M_4(OH)_8^{8+}$ ions. Excess concentrated hydrochloric acid gives octahedral complexes $TiCl_6^{2-}$, $ZrCl_6^{2-}$, $HfCl_6^{2-}$ and addition of ammonium chloride precipitates salts, e.g. $(NH_4)_2TiCl_6.2H_2O$ (cf. Sn and Pb).

The methods of preparation of lower chlorides of titanium and the colours of these are as follows:

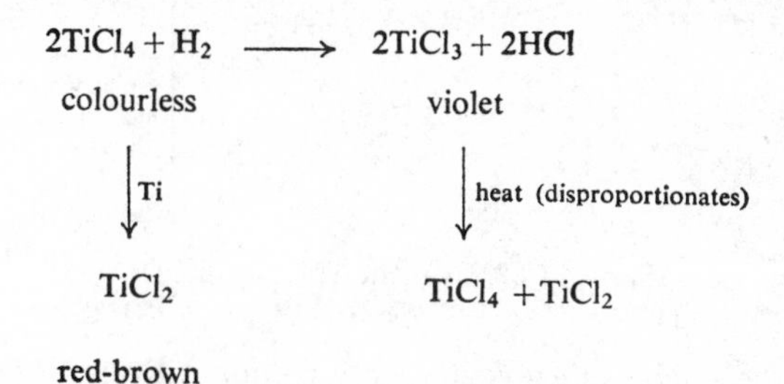

Similar methods are available for the preparation of the lower halides of zirconium and hafnium but these have not been extensively studied.

Titanium(II) chloride inflames on heating in air and with water produces hydrogen.

Oxo Salts

Titanium dioxide dissolves in concentrated sulphuric acid forming titanyl sulphate which can be crystallized from solution as the dihydrate $TiO^{2+}SO_4^{2-}.2H_2O$.

$$TiO_2 + H_2SO_4 \rightarrow TiOSO_4 + H_2O$$

The titanyl ion TiO^{2+} has never been isolated and in the crystalline sulphate there are long —Ti—O—Ti—O— chains:

—O—Ti—O—Ti—O—Ti—O—Ti—O—

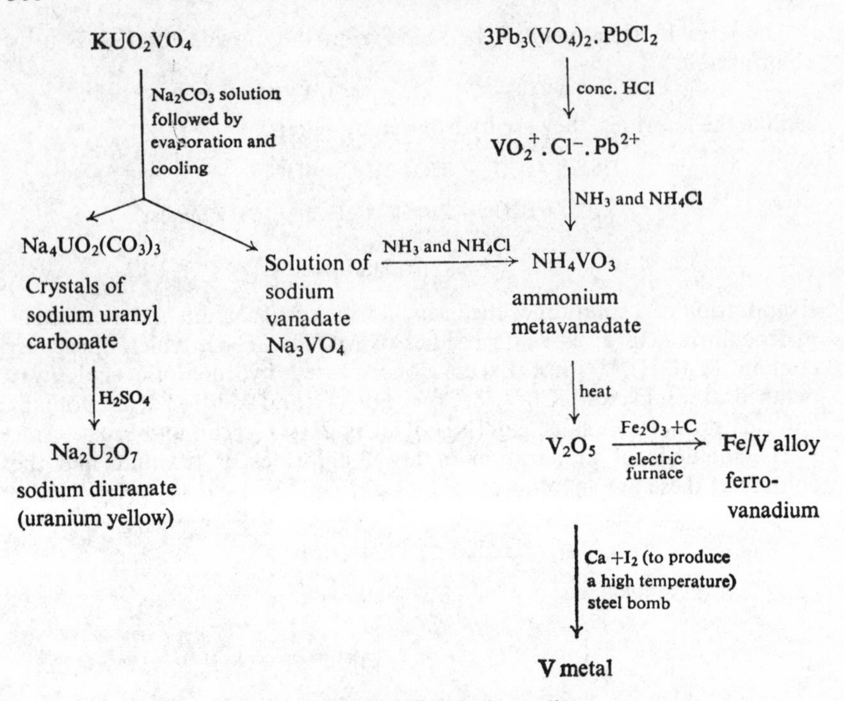

FIGURE 18.8. Extraction of vanadium

Solutions of the above salt give a yellow colour with hydrogen peroxide, said to be due to $H_2TiO_2(SO_4)_2$, peroxo disulphato titanic acid, probably furnishing the yellow TiO_2^{2+} peroxotitanyl ion in aqueous solution (cf. V).

$$TiO^{2+} + O{-}O^{2-} + H_2O \rightarrow Ti\langle{}^{O^{2+}}_{O}| + 2OH^-$$

Titanium(III) sulphate (titanous sulphate) is said to be formed by evaporation of titanium in dilute sulphuric acid. The crystals of $Ti_2(SO_4)_3.8H_2O$ dissolve in water, giving a violet solution containing the cation $Ti(H_2O)_6^{3+}$. The solution is strongly reducing and is only stable for a short time in presence of excess sulphuric acid.

$$Ti^{3+} + H_2O \rightarrow TiO^{2+} + 2H^+ + e \qquad (E^\circ = +0{\cdot}1 \text{ V})$$

No oxo salts of Ti^{II} are known. A solution of $Ti^{IV}(SO_4)_2$ can be made by the following fusion followed by extraction with sulphuric acid:

$$TiO_2 + 2K_2S_2O_7 \rightarrow Ti(SO_4)_2 + 2K_2SO_4$$

Dilution with water causes hydrolysis:

$$Ti^{4+} + H_2O \rightarrow TiO^{2+} + 2H^+$$

Vanadium

Vanadium shows very little resemblance to the Group VB elements. The maximum oxidation state is +5, with lower oxidation states of +4 (most stable), +3, and +2.

Occurrence

Vanadium is widely distributed but there are very few concentrated deposits. The principal ores are vanadinite (lead orthovanadate), $3Pb_3(VO_4)_2.PbCl_2$, and carnotite (potassium uranyl vanadate), $2K(UO_2)VO_4.3H_2O$.

Extraction

A flow diagram, indicating the principal stages in the extraction of vanadium, is shown in Figure 18.8.

Uses

Ferrovanadium is used in the manufacture of alloy steels which are resistant to vibration and shock. Vanadium compounds are used as drying accelerators in paint and in glass manufacture. Vanadium pentoxide is used as a catalyst.

Properties

Pure vanadium metal is fairly soft and ductile but even a small quantity of oxygen renders the metal brittle. The principal reactions of the metal are outlined in Figure 18.9.

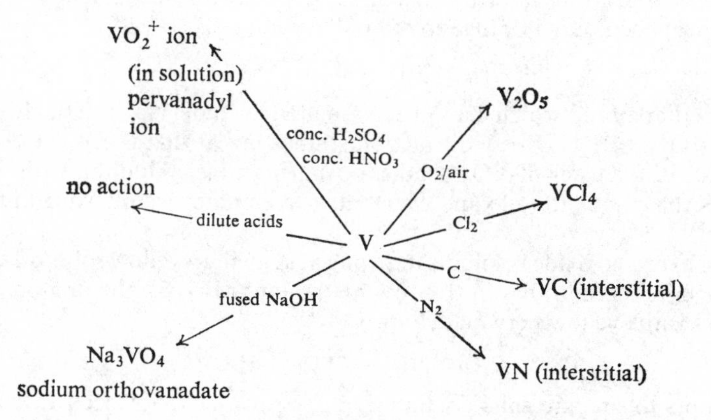

FIGURE 18.9. Reactions of vanadium metal

The Oxides

Vanadium(V) oxide (vanadium pentoxide) is an orange solid prepared by heating ammonium metavanadate.

$$2NH_4VO_3 \rightarrow V_2O_5 + 2NH_3 + H_2O$$

The oxide is acidic and dissolves in alkalis producing orthovanadates, e.g. $Na_3VO_4.12H_2O$.

$$V_2O_5 + 6OH^- \rightarrow 2VO_4^{3-} + 3H_2O$$

The addition of ammonium chloride to such a solution precipitates colourless ammonium metavanadate (NH_4VO_3) which contains the polymeric metavanadate ion consisting of infinite chains of VO_4 tetrahedra sharing corners. The reaction illustrates the strong tendency of vanadium to form condensed oxo anions (cf. phosphorus). The initial process is one of protonation, the extent of which depends upon the pH.

$$VO_4^{3-} + H^+ \rightleftharpoons HVO_4^{2-} \text{ (pH 8–13)}$$
$$HVO_4^{2-} + H^+ \rightleftharpoons H_2VO_4^- \text{ (pH 4–8)}$$

Addition of ammonium ions to an orthovanadate solution lowers the pH to 4–8, producing $H_2VO_4^-$ which then condenses producing the metavanadate.

$$nH_2VO_4^- \rightarrow (VO_3^-)_n + nH_2O$$

With vanadium concentrations $10^{-1} - 10^{-2}$ mol l^{-1} at a pH range 8–13 condensation of the HVO_4^{2-} ions occurs:

$$2HVO_4^{2-} \rightleftharpoons V_2O_7^{4-} + H_2O$$

Solid pyrovanadates have been isolated, e.g. $Cd_2V_2O_7$. These compounds are isostructural with the pyrophosphates and consist of two VO_4 tetrahedra sharing a common edge.

The acidification of orthovanadate solution with acetic acid (pH 6·5) gives deeply coloured orange to red polyvanadates.

$$10H_2VO_4^- + 4H^+ \rightleftharpoons V_{10}O_{28}^{6-} + 12H_2O$$

Solid compounds which have been isolated include $Na_6V_{10}O_{28}.18H_2O$, $Ca_3V_{10}O_{28}.16H_2O$. X-ray diffraction studies reveal that the $V_{10}O_{28}^{6-}$ ion consists of a cluster of VO_6 octahedra sharing edges. Dilution with water causes the colour to fade and eventually precipitates yellow vanadium(V) oxide.

The hydrated oxide is soluble in strong acid giving a yellow solution and it seems fairly certain that in this low pH region (pH < 3), the predominant species is the yellow pervanadyl ion $VO_2^+{}_{(aq)}$.

$$V_2O_5 + 2H^+ \rightarrow 2VO_2^+ + H_2O$$

Attempts to prepare salts containing these ions have not been successful as hydrated vanadium(V) oxide is always precipitated on evaporation.

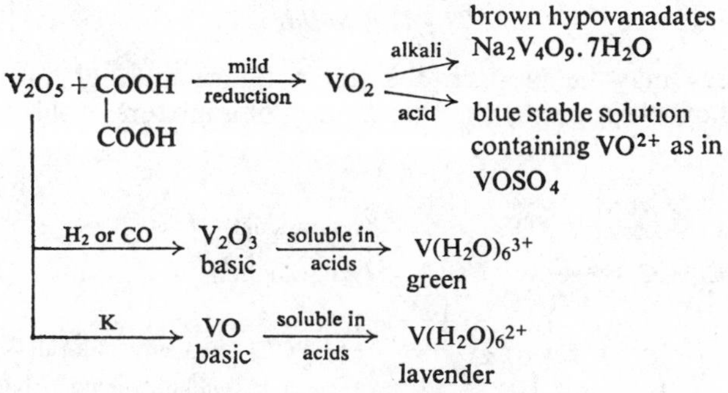

$$V_2O_5 + \begin{array}{c} COOH \\ | \\ COOH \end{array} \xrightarrow[\text{reduction}]{\text{mild}} VO_2 \xrightarrow{\text{alkali}} \text{brown hypovanadates } Na_2V_4O_9.7H_2O$$

$$VO_2 \xrightarrow{\text{acid}} \text{blue stable solution containing } VO^{2+} \text{ as in } VOSO_4$$

$$V_2O_5 \xrightarrow{H_2 \text{ or } CO} \underset{\text{basic}}{V_2O_3} \xrightarrow[\text{acids}]{\text{soluble in}} \underset{\text{green}}{V(H_2O)_6^{3+}}$$

$$V_2O_5 \xrightarrow{K} \underset{\text{basic}}{VO} \xrightarrow[\text{acids}]{\text{soluble in}} \underset{\text{lavender}}{V(H_2O)_6^{2+}}$$

FIGURE 18.10. Vanadium oxides

Other oxides of vanadium include $V^{IV}O_2$ (blue-indigo), $V_2{}^{III}O_3$ (black), and $V^{II}O$ (black). Their preparation and principal reactions are outlined in Figure 18.10. V_2O_5 ($\Delta G_f^{\ominus} = -1440$ kJ mol^{-1}) is the product obtained when all the other oxides are heated strongly in air.

The Halides

Vanadium(V) chloride has not been prepared, the direct chlorination of vanadium yielding the tetrachloride. The methods of preparation of the more important halides are summarized in Figure 18.11.

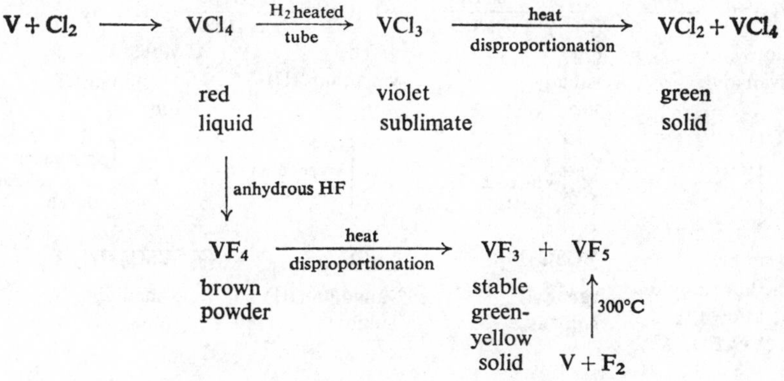

$$V + Cl_2 \longrightarrow \underset{\text{red liquid}}{VCl_4} \xrightarrow[\text{tube}]{H_2 \text{ heated}} \underset{\text{violet sublimate}}{VCl_3} \xrightarrow[\text{disproportionation}]{\text{heat}} \underset{\text{green solid}}{VCl_2} + VCl_4$$

$$VCl_4 \xrightarrow{\text{anhydrous HF}} \underset{\text{brown powder}}{VF_4} \xrightarrow[\text{disproportionation}]{\text{heat}} \underset{\text{stable green-yellow solid}}{VF_3} + VF_5$$

$$V + F_2 \xrightarrow{300°C} VF_5$$

FIGURE 18.11. Preparation of vanadium halides

The lower halides are all strongly reducing and are readily hydrolysed and oxidized. Vanadium(IV) chloride is vigorously hydrolysed to vanadyl chloride.

The Oxohalides

Oxofluorides may be prepared by direct fluorination of vanadium(V) oxide, and oxochlorides by the chlorination of a mixture of this oxide and carbon.

$$V_2O_5 + F_2 \xrightarrow{500\,°C} VOF_3 \quad \text{(yellow solid)}$$

$$V_2O_5 + Cl_2 + C \longrightarrow VOCl_3 \quad \text{(yellow liquid)}$$

$$VOCl_3 \xrightarrow[400\,°C]{Zn} VOCl_2 \quad \text{(vanadyl chloride—green deliquescent crystals)}$$

Oxo Salts

Although salts containing the pervanadyl ion have not been isolated, solutions containing this ion can be reduced to prepare vanadium salts in lower oxidation states (Figure 18.12).

$$VO_2^+ + 2H^+ + e \rightarrow VO^{2+} + H_2O \quad (E^{\ominus} = +1{\cdot}0\ V)$$

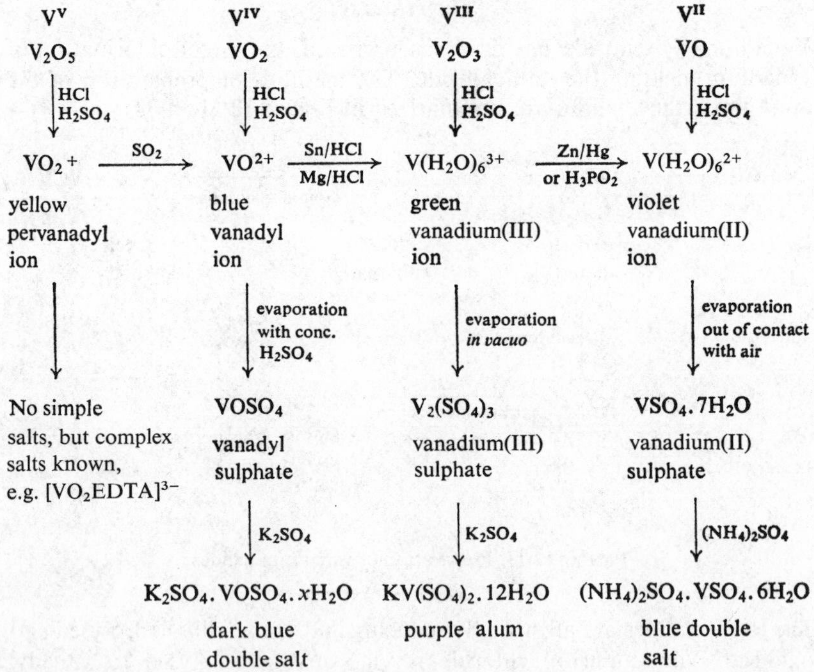

V^V: V_2O_5 —(HCl, H_2SO_4)→ VO_2^+ (yellow pervanadyl ion) → No simple salts, but complex salts known, e.g. $[VO_2EDTA]^{3-}$

V^{IV}: VO_2 —(HCl, H_2SO_4)→ VO^{2+} (blue vanadyl ion) —(evaporation with conc. H_2SO_4)→ $VOSO_4$ (vanadyl sulphate) —(K_2SO_4)→ $K_2SO_4.VOSO_4.xH_2O$ (dark blue double salt)

V^{III}: V_2O_3 —(HCl, H_2SO_4)→ $V(H_2O)_6^{3+}$ (green vanadium(III) ion) —(evaporation *in vacuo*)→ $V_2(SO_4)_3$ (vanadium(III) sulphate) —(K_2SO_4)→ $KV(SO_4)_2.12H_2O$ (purple alum)

V^{II}: VO —(HCl, H_2SO_4)→ $V(H_2O)_6^{2+}$ (violet vanadium(II) ion) —(evaporation out of contact with air)→ $VSO_4.7H_2O$ (vanadium(II) sulphate) —($(NH_4)_2SO_4$)→ $(NH_4)_2SO_4.VSO_4.6H_2O$ (blue double salt)

$$VO_2^+ \xrightarrow{SO_2} VO^{2+} \xrightarrow[Mg/HCl]{Sn/HCl} V(H_2O)_6^{3+} \xrightarrow[\text{or } H_3PO_2]{Zn/Hg} V(H_2O)_6^{2+}$$

FIGURE 18.12. Vanadium oxosalts

Solutions of vanadium(II) and vanadium(III) salts are unstable and reducing.

$$V^{3+} + H_2O \rightarrow VO^{2+} + 2H^+ + e \qquad (E^\ominus = +0{\cdot}361\ V)$$

$$V^{2+} \rightarrow V^{3+} + e \qquad (E^\ominus = -0{\cdot}255\ V)$$

Solutions of vanadium(II) salts liberate hydrogen from water and deposit copper from copper sulphate solution.

$$2V^{2+} + Cu^{2+} \rightarrow 2V^{3+} + Cu\downarrow$$

Complexes

A number of peroxo complexes of vanadium(V) are known. If the yellow solution produced from vanadium(V) oxide and sulphuric acid is treated with hydrogen peroxide, a red coloration, insoluble in ether is produced.

$$VO_2^+ + H_2O_2 \rightarrow VO(O{-}O)^+ + H_2O$$

In weakly alkaline solution, a fairly stable yellow anion $VO_2(O{-}O)_2^{3-}$ is produced and, in strongly alkaline solution at 0 °C, a violet anion $V(O{-}O)_4^{3-}$ is formed. Alkali metal and ammonium salts of these anions have been made, e.g.

$$(NH_4)_2HVO_2(O{-}O)_2.xH_2O, \quad K_3V(O{-}O)_4.xH_2O$$

Among other complexes of note are potassium hexacyanovanadate(III) (potassium vanadicyanide, $K_3V(CN)_6$), prepared by adding potassium cyanide to a solution of vanadium(III) chloride, followed by addition of alcohol. Potassium hexacyanovanadate(II) (potassium vanadocyanide, $K_4V(CN)_6.3H_2O$) may be prepared by the addition of potassium cyanide, followed by alcohol, to a freshly prepared solution of vanadium(II) chloride in an atmosphere of hydrogen.

Chromium

The electronic structure of $3d^5 4s^1$ indicates a maximum oxidation state of +6 which occurs in the chromates (CrO_4^{2-}) and dichromates ($Cr_2O_7^{2-}$) and in chromyl chloride (CrO_2Cl_2). All these compounds are strong oxidizing agents. The only stable species in aqueous solution is that with the chromium in oxidation state +3. Chromium(IV) and (V) compounds have no stable aqueous chemistry and chromium(II) compounds in solution are strongly reducing.

Occurrence

The main ore is iron(II) chromite or chrome iron stone ($Fe^{II}Cr_2^{III}O_4$). This has a spinel structure which may be regarded as cubic close packing of oxide ions with vacant lattice sites occupied by Cr^{3+} and Fe^{2+} ions. This is similar to the structure of the mineral spinel ($Mg^{II}Al_2^{III}O_4$) and other mixed oxides such as Mn_3O_4 ($Mn^{II}Mn_2^{III}O_4$) and Fe_3O_4 ($Fe^{II}Fe_2^{III}O_4$). Rarer minerals of chromium include chrome ochre (Cr_2O_3) and crocoisite ($PbCrO_4$).

Extraction

The alloy ferrochrome (70 per cent chromium, 30 per cent iron), used for making tough corrosion-resistant steels, is manufactured by reducing iron(II) chromite with carbon in an electric furnace.

$$FeCr_2O_4 + 4C \rightarrow \underbrace{Fe + 2Cr}_{\text{ferrochrome}} + 4CO$$

In the manufacture of pure chromium, powdered iron(II) chromite is first converted to potassium chromate by heating in a reverberatory furnace with sodium or potassium carbonate and free access of air.

$$4FeCr_2O_4 + 8Na_2CO_3 + 7O_2 \rightarrow 8Na_2CrO_4 + 2Fe_2O_3 + 8CO_2$$

The residue is extracted with water and, after evaporation, yellow sodium or potassium chromate ($Na_2CrO_4.10H_2O$ or K_2CrO_4) crystallizes out on cooling. Acidification of yellow aqueous chromate solution with sulphuric acid yields orange dichromates.

$$2CrO_4^{2-} + 2H^+ \rightleftharpoons Cr_2O_7^{2-} + H_2O$$

Orange-red $Na_2Cr_2O_7.2H_2O$ or $K_2Cr_2O_7$ crystallize out. The dichromate is then reduced with carbon to the green sesquioxide ($Cr_2^{III}O_3$), which is then further reduced to the metal with aluminium powder.

$$Na_2Cr_2O_7 + 2C \rightarrow Cr_2O_3 + Na_2CO_3 + CO$$

$$Cr_2O_3 + 2Al \rightarrow Al_2O_3 + 2Cr \qquad (\Delta H = -469\text{kJ})$$

Uses

Chromium is used in the manufacture of steels such as stainless steel (Fe 86 per cent, Cr 13 per cent, Ni 1 per cent) and in the production of Nichrome (Ni 60 per cent, Cr 15 per cent, Fe 25 per cent). Chromium-plating is usually carried out by electrodeposition from hot chromic acid solution with a little chromic sulphate and using a lead anode.

Properties

Chromium is a silvery white metal with a body-centred cubic structure. The principal reactions of the metal are indicated in Figure 18.13.

The metal dissolves rapidly in dilute hydrochloric and sulphuric acid

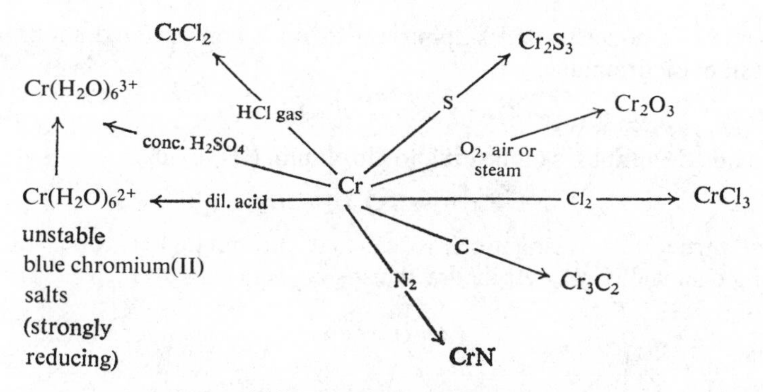

FIGURE 18.13. Reactions of chromium metal

($E^{\ominus}(Cr^{2+}/Cr) = -0{\cdot}91$ V) but the pure metal is insoluble in dilute nitric acid and the concentrated acid renders the metal passive. Concentrated sulphuric acid dissolves the metal rapidly, producing chromium(III) sulphate and sulphur dioxide. Although $Cr(H_2O)_6^{3+}$ is violet, solutions in sulphuric acid are often green due to the formation of sulphate complexes, e.g. $Cr_2^{III}(SO_4)_4^{2-}$. Chromium nitride is interstitial but the carbide has a peculiar structure consisting of long chains of carbon atoms. Non-stoichiometric interstitial hydrides have been prepared by indirect methods.

The Oxides

Black chromium(II) oxide (chromous oxide) is made by the action of dilute nitric acid on chromium amalgam. It is a basic oxide, and readily forms Cr_2O_3 on warming in air.

Green chromium(III) oxide (chromium sesquioxide) possesses an α-alumina structure (p. 196). It may be prepared from the elements or, more conveniently, by heating orange ammonium dichromate.

$$(NH_4)_2Cr_2O_7 \rightarrow Cr_2O_3 + 4H_2O + N_2$$

The oxide so prepared is almost insoluble in acids but fused potassium hydroxide and chlorate convert the oxide to chromate. Chromium(III) oxide is the most thermodynamically stable oxide ($\Delta G_f^{\ominus} = -1047$ kJ mol^{-1}) and is the ultimate product when all the other oxides are heated in air.

The thermal decomposition of chromyl chloride vapour produces a black compound, chromium(IV) oxide.

$$CrO_2Cl_2 \rightarrow CrO_2 + Cl_2$$

Alkaline earth salts, e.g. $Ba_2Cr^{IV}O_4$, have been isolated.

Chromium(VI) oxide (chromium trioxide) is prepared as red needles by the action of concentrated sulphuric acid on a concentrated solution of potassium dichromate.

$$Cr_2O_7^{2-} + 2H^+ \rightarrow 2CrO_3 + H_2O$$

The oxide decomposes on heating to chromium (III) oxide.

$$4CrO_3 \rightarrow 2Cr_2O_3 + 3O_2$$

Recent structural investigations reveal that chromium(VI) oxide consists of long chains of CrO_4 tetrahedra sharing corners (Figure 18.14).

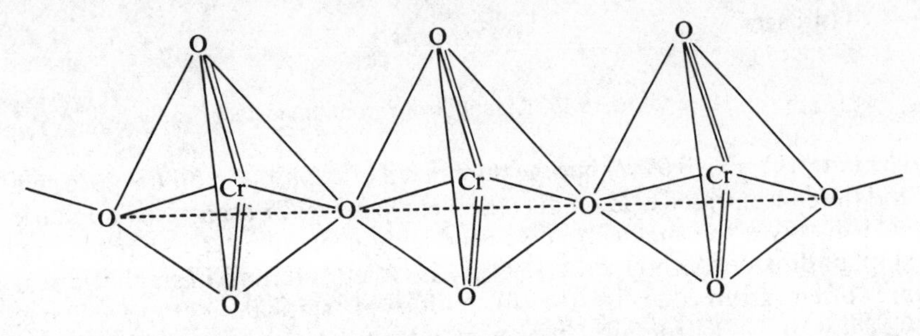

FIGURE 18.14

Chromium(VI) oxide is a strongly acidic oxide, dissolving in water and producing a red acidic solution which probably contains chromate, dichromate, and polychromate ions such as $Cr_3O_{10}^{2-}$, $Cr_4O_{13}^{2-}$.

$$CrO_3 + H_2O \rightleftharpoons CrO_4^{2-} + 2H^+$$

$$2CrO_3 + H_2O \rightleftharpoons Cr_2O_7^{2-} + 2H^+$$

$$3CrO_3 + H_2O \rightleftharpoons Cr_3O_{10}^{2-} + 2H^+$$

The Hydroxides

Yellow chromium(II) hydroxide (chromous hydroxide) is precipitated by the action of caustic soda solution on a solution of a chromium(II) salt. It is readily oxidized to hydrated chromium(III) oxide by atmospheric oxygen.

Pale green chromium(III) hydroxide (chromic hydroxide) is probably not a true hydroxide but a hydrated oxide. It is prepared by the action of alkali on a soluble chromium(III) salt. With excess alkali, the 'hydroxide' gives a deep green solution probably containing a chromite (hexahydroxochromate(III)).

$$Cr(OH)_3 + 3OH^- \rightarrow Cr(OH)_6^{3-}$$

Solid alkali metal chromites are known, e.g. $Na_3Cr(OH)_6$ or $Na_3CrO_3 \cdot 3H_2O$. Fusion of chromium(III) oxide with some divalent metal oxides also yields so-called chromites, e.g. $Ba^{II}Cr_2^{III}O_4$ (barium chromite) and

$Cu^{II}Cr_2^{III}O_4$ (copper chromite). These latter compounds are in fact mixed oxides with the spinel structure characteristic of iron(II) chromite. Chromium(III) hydroxide dissolves readily in alkaline hydrogen peroxide to give yellow chromates.

$$2Cr(OH)_3 + 3(O{-}O)^{2-} \rightarrow 2CrO_4^{2-} + 2OH^- + 2H_2O$$

The Halides

Chromium(II) chloride (chromous chloride) may be prepared dry as white needles by the action of hydrogen chloride gas on chromium metal at 700 °C. Reduction of the green solution of chromium(III) chloride with zinc dust and hydrochloric acid causes the solution to turn blue and crystallization in an atmosphere of hydrogen gives blue needles of $CrCl_2.4H_2O$.

$$2Cr^{3+} + H_2 \rightarrow 2Cr^{2+} + 2H^+$$

Chromium(II) chloride reacts with anhydrous hydrogen fluoride gas to give a white insoluble solid, chromium(II) fluoride.

Chromium(III) chloride (chromic chloride) is a violet solid prepared anhydrous from the elements at 600 °C. It is not very soluble in water but dissolves rapidly in the presence of tin(II) chloride to give a green solution. Three solid crystalline hydrates of chromium(III) chloride have been isolated:

$$[Cr(H_2O)_4Cl_2]^+Cl^-.2H_2O \text{ (deep green)}$$

$$[Cr(H_2O)_5Cl]^{2+}2Cl^-.H_2O \text{ (pale green)}$$

$$[Cr(H_2O)_6]^{3+}3Cl^- \text{ (violet)}$$

In aqueous solution all the chloride is precipitated by silver nitrate on the violet chloride. Two-thirds of the chloride is precipitated from the pale green compound and one-third from the dark green chloride. The above halides are isomeric (p. 106).

Sparingly soluble dark green chromium(III) fluoride is prepared by the action of hydrogen fluoride on anhydrous chromium(III) chloride. Both chromium(II) and chromium(III) fluorides are probably predominantly ionic. Chromium(III) chloride has an unusual layer lattice (Figure 18.15),

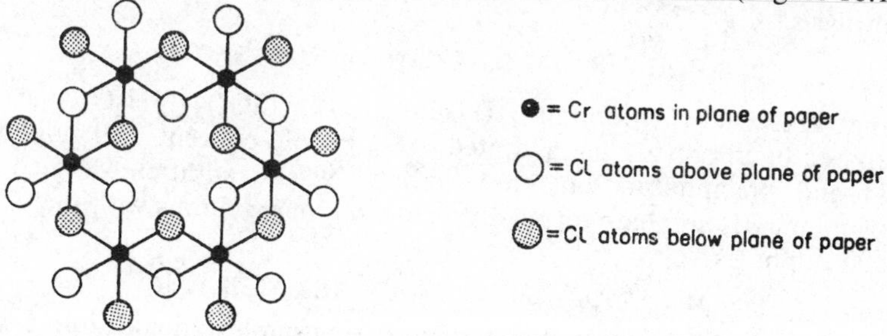

FIGURE 18.15

where the bonding between chromium and chlorine has considerable covalent character.

Essentially the structure consists of cubic close-packed halogen atoms with one-third of the octahedral holes occupied by chromium atoms. In the cadmium chloride lattice one-half of the holes are occupied by the metal atoms (p. 97).

A chromium(IV) chloride has been obtained by the reaction:

$$2CrCl_3 + Cl_2 \xrightarrow{600\,^\circ C} 2CrCl_4$$

A chromium(IV), (V) and (VI) fluoride have been made from the elements.

Oxo Salts of Chromium(II) and (III)

Reduction of chromium(III) chloride solution with zinc dust and hydrochloric acid (p. 375) gives a blue solution containing the hydrated cation $Cr(H_2O)_6^{2+}$. Solutions of chromium(II) salts are, however, strongly reducing and readily pass to the chromium(III) state.

$$Cr^{2+} \rightarrow Cr^{3+} + e \qquad (E^\ominus = -0{\cdot}41\ V)$$

Chromium(II) salts will liberate hydrogen from acidified water.

$$2Cr^{2+} + 2H^+ \rightarrow 2Cr^{3+} + H_2$$

The addition of sodium acetate to blue chromium(II) salt solutions precipitates a red stable acetate which is useful as a starting point for preparing other chromium(II) salts.

$$2Cr^{2+} + 4CH_3COO^- + 2H_2O \rightarrow Cr_2(CH_3COO)_4.2H_2O$$

This, on dissolution in dilute sulphuric acid and crystallizing in an inert atmosphere, gives blue needles of chromium(II) sulphate, $CrSO_4.7H_2O$.

$$Cr_2(CH_3COO)_4 + 2H_2SO_4 \rightarrow 2CrSO_4 + 4CH_3COOH$$

Chromium(III) salts form violet solutions in water which are acidic by hydrolysis.

$$Cr(H_2O)_6^{3+} + H_2O \rightarrow Cr(H_2O)_5OH^{2+} + H_3O^+$$

Violet crystals of chromium(III) sulphate, $Cr_2(SO_4)_3.18H_2O$, may be prepared by dissolving the hydrated oxide in cold concentrated sulphuric acid and precipitating out the sulphate by the addition of alcohol. Chromium(III) sulphate forms deep violet-coloured alums with the alkali metal sulphates:

$$M^ICr^{III}(SO_4)_2.12H_2O \qquad (M = NH_4, Na, K, etc.)$$

Potassium chrome alum is obtained by passing sulphur dioxide into a well-cooled mixture of potassium dichromate and concentrated sulphuric acid, keeping the temperature below 60 °C. Alcohol is added and the solution allowed to crystallize.

$$Cr_2O_7^{2-} + 2H^+ + 3SO_2 \rightarrow 2Cr^{3+} + H_2O + 3SO_4^{2-}$$

Chromium(VI) Compounds

Chromates and dichromates

The sodium and potassium chromates and dichromates are manufactured from chrome iron stone as indicated in the extraction (p. 372).

The yellow chromate ion (CrO_4^{2-}) is tetrahedral and the orange dichromate ion ($Cr_2O_7^{2-}$) consists of two CrO_4 tetrahedra, linked together by a common oxygen atom (Figure 18.16). The Cr—O—Cr angle is 115°.

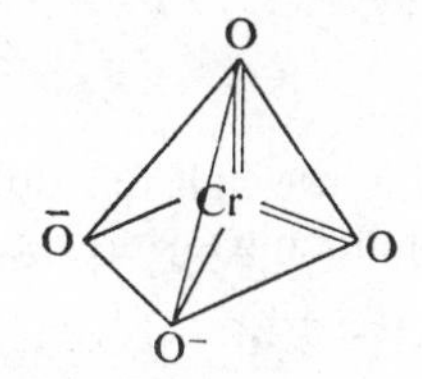

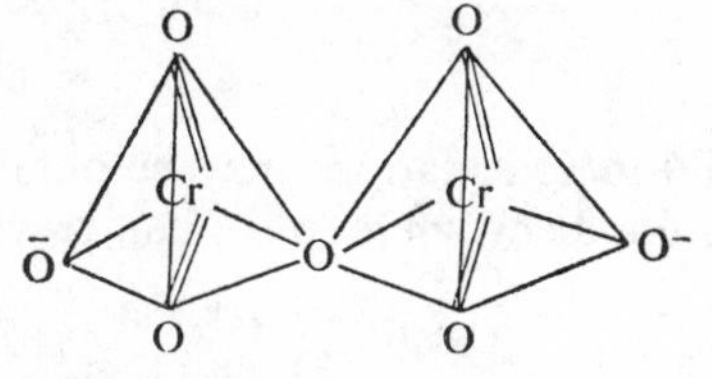

FIGURE 18.16

Double bonding arises from the overlap of oxygen *p* orbitals with chromium *d* orbitals. The bonds are equalized by resonance and only one canonical form is shown in Figure 18.16. The chromates and dichromates are interconvertible in aqueous solution. Addition of acids causes the change from chromate to dichromate, while the addition of alkali causes the change from dichromate to chromate.

$$2CrO_4^{2-} + 2H^+ \rightarrow Cr_2O_7^{2-} + H_2O$$

$$Cr_2O_7^{2-} + 2OH^- \rightarrow 2CrO_4^{2-} + H_2O$$

Among important insoluble chromates are red silver chromate (Ag_2CrO_4 —used as an indicator in silver nitrate titrations), yellow barium and lead chromates ($BaCrO_4$ and $PbCrO_4$—used as pigments and as a test for lead and barium in qualitative analysis).

The only dichromate of importance apart from the sodium and potassium salts is the orange ammonium dichromate, $(NH_4)_2Cr_2O_7$, prepared by adding ammonia to a solution of chromium(VI) oxide in water. In contrast to the alkali metal salts, it decomposes readily on warming.

$$(NH_4)_2Cr_2O_7 \rightarrow Cr_2O_3 + 2N_2 + 4H_2O$$

$$4K_2Cr_2O_7 \xrightarrow{\text{red heat}} 4K_2CrO_4 + 2Cr_2O_3 + 3O_2$$

Sodium and potassium dichromates are vigorous oxidizing agents; the sodium salt has a greater solubility in water and is used extensively as an oxidizing agent in organic chemistry where concentrated solutions are required. Extreme caution must be exercised in handling sodium dichromate which can cause painful sores on the skin. Potassium dichromate is used as a primary standard in volumetric analysis. In acid solution the reaction of dichromate with reducing agents proceeds as follows (p. 135):

$$Cr_2O_7^{2-} + 14H^+ + 6e \rightarrow 2Cr^{3+} + 7H_2O \qquad (E^{\ominus} = +1{\cdot}33\ V)$$

Thus, acidified potassium dichromate will oxidize iodides to iodine, sulphides to sulphur, tin(II) salts to tin(IV), and iron(II) salts to iron(III). The half reactions are quoted below:

$$6I^- \rightarrow 3I_2 + 6e$$

$$3H_2S \rightarrow 6H^+ + 3S + 6e$$

$$3Sn^{2+} \rightarrow 3Sn^{4+} + 6e$$

$$6Fe^{2+} \rightarrow 6Fe^{3+} + 6e$$

The full ionic equations may be obtained by adding the half reaction for potassium dichromate to the half reaction for the reducing agent, e.g.

$$Cr_2O_7^{2-} + 14H^+ + 6Fe^{2+} \rightarrow 2Cr^{3+} + 6Fe^{3+} + 7H_2O$$

Chromyl chloride

Chromyl chloride, CrO_2Cl_2, may be prepared by distilling potassium dichromate with sodium chloride and concentrated sulphuric acid or by dissolving chromium(VI) oxide in cold concentrated hydrochloric acid, when careful addition of concentrated sulphuric acid causes the chromyl chloride to separate as a red oil.

$$Cr_2O_7^{2-} + 4Cl^- + 6H^+ \rightarrow 2CrO_2Cl_2 + 3H_2O$$

$$CrO_3 + 2HCl \rightarrow CrO_2Cl_2 + H_2O$$

Chromyl chloride is a dark red fuming liquid, similar in appearance to liquid bromine. It is a vigorous oxidizing agent, hydrolysed by water.

$$CrO_2Cl_2 + H_2O \rightleftharpoons CrO_3 + 2HCl$$

The addition of saturated potassium chloride solution to chromyl chloride precipitates potassium chlorochromate (Peligot's salt) as orange crystals.

$$CrO_2Cl_2 + KCl + H_2O \rightarrow K^+CrO_3Cl^- + 2HCl$$

The bromide and iodide are also known and may be precipitated in a similar manner. Potassium chlorochromate is hydrolysed by water.

$$2CrO_3Cl^- + H_2O \rightarrow Cr_2O_7^{2-} + 2HCl$$

Chromyl chloride and potassium chlorochromate may be regarded as

Chromic acid—not isolated

Chromyl chloride—distorted tetrahedral structure

Potassium chlorochromate—tetrahedral

FIGURE 18.17

derivatives of the theoretical chromic acid which has not been isolated (Figure 18.17).

Complex peroxochromates

Hydrogen peroxide, added to a dilute acidified solution of potassium dichromate, gives a blue coloration which rapidly changes to the green colour of chromium(III) salt. The blue colour is due to chromium peroxide (CrO_5) and is stable in ether with which it forms a loose compound $CrO_5(C_2H_5—O—C_2H_5)$. A blue explosive solid ($CrO_5.py$) has been isolated when pyridine is used in place of ether.

$$Cr_2O_7^{2-} + 2H^+ + 4H_2O_2 \rightarrow 2CrO_5 + 5H_2O$$

An alkaline solution of a chromate, treated with 30 per cent hydrogen peroxide gives solutions from which red peroxochromate salts can be isolated, e.g. Na_3CrO_8.

$$2CrO_4^{2-} + 2OH^- + 7H_2O_2 \rightarrow 2CrO_8^{3-} + 8H_2O$$

Magnetic studies indicate the following structures for chromium peroxide pyridine and the peroxochromate ion:

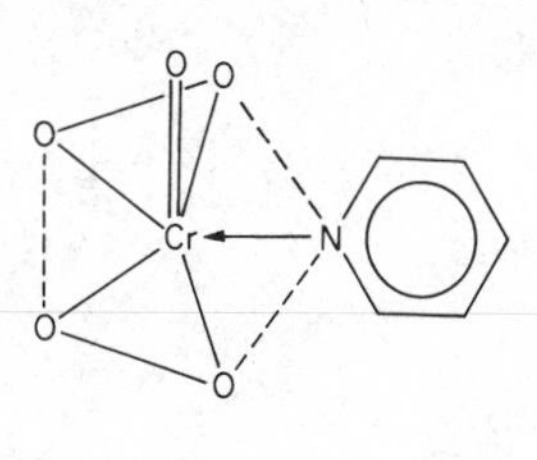

CrO_5 pyridine

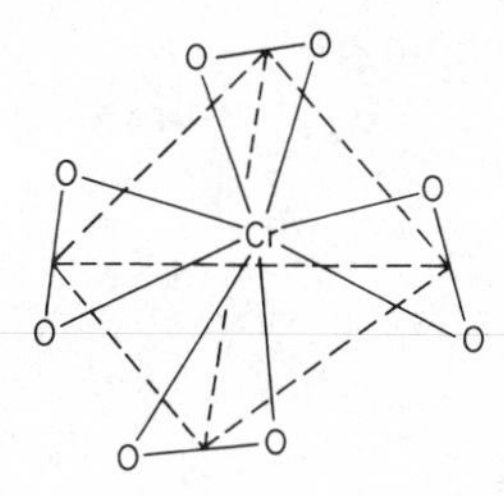

CrO_8^{3-}

4 peroxo groups tetrahedrally disposed round the chromium atom

Complex cyanides

Chromium(II) acetate with excess potassium cyanide solution in an inert atmosphere produces blue $K_4Cr^{II}(CN)_6$. Air and acid cause oxidation to yellow $K_3Cr^{III}(CN)_6$. Magnetic measurements indicate t_{2g}^4 and t_{2g}^3 configurations for chromium in $Cr(CN)_6^{4-}$ and $Cr(CN)_6^{3-}$ respectively.

Molybdenum and Tungsten

These two elements are quite similar chemically and show only a superficial resemblance to chromium. Little is known concerning molybdenum and tungsten in oxidation state +2 and, in contrast to chromium, there are no stable complexes of molybdenum and tungsten in oxidation state +3. In general, the higher oxidation states are more common and more stable.

Occurrence

The principal ores are the sulphides, molybdenite (MoS_2) and tungstenite (WS_2). Certain molybdates and tungstates also occur naturally, e.g. wulfenite ($PbMoO_4$) and wolframite (Fe, Mn)WO_4.

Extraction

Molybdenum

Molybdenite ore is concentrated by oil flotation and roasted in air to give the molybdenum(VI) oxide (MoO_3). Hot concentrated ammonia is added and, on cooling, ammonium molybdate crystallizes out. This, on heating, gives pure molybdenum(VI) oxide which is reduced with hydrogen to the metal.

$$MoS_2 \xrightarrow{O_2/air} \underset{\text{impure}}{MoO_3} \xrightarrow{NH_3} (NH_4)_2MoO_4 \xrightarrow{heat} \underset{\text{pure}}{MoO_3} \xrightarrow{H_2} Mo$$

Tungsten

Wolframite is concentrated magnetically and then fused with sodium carbonate under oxidizing conditions and leached with water, when the tungsten goes into solution as sodium tungstate (Na_2WO_4). Acidification of this solution precipitates pure tungsten(VI) oxide (WO_3) which is heated and reduced to the metal with hydrogen.

$$FeWO_4 \xrightarrow[\text{leach } H_2O]{\text{fuse } Na_2CO_3} Na_2WO_4 \xrightarrow{HCl} WO_3 \xrightarrow{H_2} W$$

Properties

Molybdenum and tungsten are lustrous silvery white metals with high melting points (Mo 3410 °C; W 3380 °C). The principal reactions of the metals are indicated in Figure 18.18.

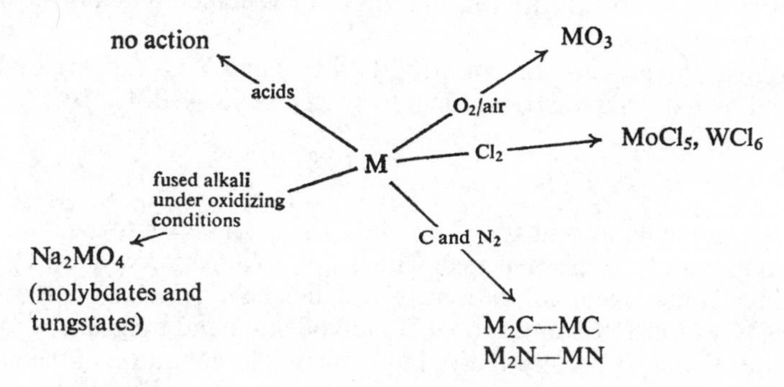

FIGURE 18.18

The metals are not readily attacked by acids but dissolve in a mixture of concentrated nitric and hydrofluoric acids. Fused alkali and sodium peroxide attacks the metals rapidly giving simple molybdates and tungstates. When heated in air or oxygen at red heat, MoO_3 and WO_3 are produced. Chlorination of the metals at elevated temperatures yields molybdenum(V) chloride and tungsten(VI) chloride. The metals combine with carbon, boron, silicon, and nitrogen to give hard refractory and chemically inert interstitial compounds, e.g. M_2C—MC, and M_2N—MN). The carbides and nitrides are incorporated in furnace linings and cutting tools, etc.

Uses

Apart from the uses mentioned above, both metals are used in the production of alloy steels, giving hardness and strength. Tungsten–chromium alloy is added to steel to make cutting tools, which remain hard at red heat. Tungsten is used in lamp filaments.

Oxides

The important oxides of molybdenum and tungsten are the trioxides, MoO_3 which is white when cold and yellow hot and WO_3 which is white. They are prepared by the action of heat on the metal or sulphide in air or oxygen. The trioxides are unattacked by most acids although molybdenum(VI) oxide dissolves in concentrated sulphuric acid forming molybdenyl sulphate ($MoO_2^{2+}SO_4^{2-}$), but the nature of this compound is uncertain.

Both oxides are, however, readily soluble in alkali giving molybdate and tungstate solutions.

$$MoO_3 + 2OH^- \rightarrow MoO_4^{2-} + H_2O$$

Tungsten(VI) oxide and molybdenum(VI) oxide both possess continuous structures of MO_6 octahedra sharing edges. The octahedra in molybdenum (VI) oxide are distorted.

Coloured dioxides are known, MoO_2 (violet) and WO_2 (green), and are prepared by reducing the trioxides in hydrogen below 470 °C.

Molybdenum and tungsten blue

Acidified molybdates and tungstates or suspensions of molybdenum(VI) and tungsten(VI) oxide, reduced with sulphur dioxide or a tin(II) salt give blue pigments and colorations. The solids have variable composition from $M^{V}O_{2.5}$ to $M^{VI}O_3$ and probably molybdenum and tungsten in oxidation states V and VI occupy vacant lattice sites. The condition of the oxides is probably colloidal.

Tungsten bronzes

These may be prepared by heating the trioxide with an alkali metal or an alkali metal tungstate with zinc or hydrogen. The compounds produced are chemically inert and bronze-like in appearance. They are good electrical conductors. They are non-stoichiometric with a range of compositions from $Na_{0.3}WO_3$ to $Na_{0.9}WO_3$. Structurally, they are regarded as defective perovskite lattices (p. 364) with face-centred cubic packing of oxide and sodium ions, with tungsten(VI) in the octahedral holes. Numbers of the sites normally occupied by sodium ions are vacant.

The Sulphides

Numbers of sulphides are known, the most important being $W^{IV}S_2$ (grey) and $Mo^{IV}S_2$ (black), both prepared from the elements. Trisulphides $W^{VI}S_3$ (dark brown) and $Mo^{VI}S_3$ (black) can be prepared by dissolving the trioxides in alkaline sulphide solution and then acidifying the resulting thiomolybdate and tungstate complexes.

$$WO_3 \xrightarrow[Na_2S_x]{NaOH} Na_2WS_4 \xrightarrow{H^+} WS_3$$

Simple Molybdates and Tungstates

Alkali metal tungstates and molybdates (M_2WO_4, M_2MoO_4) are prepared by dissolving the trioxides in the appropriate alkali.

$$MoO_3 + 2OH^- \rightarrow MoO_4^{2-} + H_2O$$

These salts contain discrete tetrahedral MoO_4^{2-} and WO_4^{2-} anions. The ammonium and alkali metal salts are stable (cf. CrO_4^{2-}) and soluble in water. The acidification of boiling molybdate and tungstate solutions causes the precipitation of yellow molybdic and tungstic acids.

$$MoO_4^{2-} + 2H^+ \rightarrow \underset{\text{molybdic acid}}{H_2MoO_4\downarrow}$$

Proton magnetic resonance studies indicate that the hydrogen is present as a water molecule and these compounds are in reality hydrated oxides, better written as $Mo(W)O_3 . H_2O$.

Iso- and Heteropolyacids

Molybdenum and tungsten form two very complex series of acids and anions:

(*a*) Isopolyacids and anions containing molybdenum or tungsten and oxygen and hydrogen.

(*b*) Heteropolyacids and anions which, apart from molybdenum, tungsten, oxygen, and hydrogen, contain atoms of other elements such as phosphorus, arsenic, silicon, etc.

Isopolyacids and anions

If strongly basic solutions of MoO_4^{2-} ions are slowly acidified, the simple tetrahedral anions polymerize forming polymolybdate ions. The extent of polymerization increases as the pH is lowered and complex species such as $Mo_7O_{24}^{6-}$ and $Mo_8O_{26}^{4-}$ are obtained:

$$\text{pH} = 6 \qquad 7MoO_4^{2-} + 8H^+ \rightarrow \underset{\text{hepta-molybdate or para-molybdate}}{Mo_7O_{24}^{6-}} + 4H_2O$$

$$\text{pH} = 2 \qquad Mo_7O_{24}^{6-} + MoO_4^{2-} + 4H^+ \rightarrow \underset{\text{octa-molybdate ion}}{Mo_8O_{26}^{4-}} + 2H_2O$$

It appears that, at some stage during the polymerization, the tetrahedral unit (MoO_4^{2-}) changes to an octahedral configuration (MoO_6). Dimolybdates can be prepared by oxide fusions.

$$2MoO_3 + Na_2O \rightarrow Na_2Mo_2O_7$$

These compounds, however, do not contain discrete $Mo_2O_7^{2-}$ anions and their structure consists of complicated chains of MoO_4 tetrahedra and MoO_6 octahedra.

The simple tungstate ion also undergoes polymerization as the pH is lowered, furnishing ionic species such as $W_{12}O_{46}^{20-}$ (paratungstate) and $W_{12}O_{40}^{8-}$ (metatungstate).

$$\text{pH} = 6 \quad 12WO_4^{2-} + 4H^+ \rightarrow W_{12}O_{46}^{20-} + 2H_2O$$

$$\text{pH} = 2 \quad 12WO_4^{2-} + 16H^+ \rightarrow W_{12}O_{40}^{8-} + 8H_2O$$

Structure of the isopolyacids

X-ray diffraction shows the unit of the structure to be an MoO_6 (or WO_6) octahedron. The complex anions are built up from these octahedra by sharing corners and edges but never faces. Thus the common structural unit for heptamolybdate and octamolybdate ions is as in Figure 18.19. This

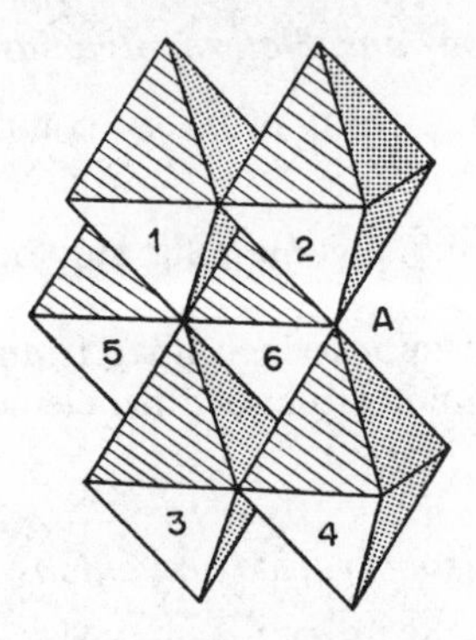

FIGURE 18.19

consists of six octahedra numbered 1 to 6, where octahedra 5 and 6 are behind 1, 2, 3, and 4.

In the heptamolybdate ion $Mo_7O_{24}^{6-}$ the seventh octahedron goes in position *A* behind 1, 2, 3, 4, and shares edges with 6, 2 and 4. In the octamolybdate ion $Mo_8O_{26}^{4-}$, the two additional octahedra sharing a common edge go into positions in front of octahedra 1, 2, 3, 4, one in the space between these and the other sharing edges with 2 and 4.

Heteropolyacids

These are produced when alkaline solutions of molybdate and tungstate ions are mixed with phosphates, silicates or other anions and then acidified. One typical example is $[X^{n+}M_{12}O_{40}]^{(8-n)-}$, where M = Mo or W; X = the hetero atom, P, Si, etc.; and n = the maximum oxidation state of the hetero atom. When X is phosphorus, $n = 5$ and for a hetero atom: molybdenum ratio of 1:12, the formula becomes $[P.Mo_{12}O_{40}]^{3-}$. This is the phosphomolybdate ion found in ammonium phosphomolybdate, $(NH_4)_3PMo_{12}O_{40}$. Commercial ammonium molybdate is in fact ammonium hepta- or paramolybdate, $(NH_4)_6Mo_7O_{24}.4H_2O$. A large number of elements can function as hetero atoms and many different hetero atom: molybdenum (tungsten) ratios occur, namely, 1:10, 1:6, 1:9, etc. In all heteropolyacids the hetero atom sits at the centre of complex clusters of MoO_6 octahedra, but for further details, a more advanced text must be consulted.

The Halides

Direct fluorination of molybdenum and tungsten produces colourless liquids, the hexafluorides. Direct chlorination of the elements produces molybdenum(V) chloride and tungsten(VI) chloride. Molybdenum(VI) chloride has recently been prepared as a black, moisture-sensitive powder by refluxing molybdic acid with thionyl chloride. Some $MoOCl_4$ and $MoCl_5$ are produced in this reaction. Controlled reduction of these chlorides gives rise to lower halides:

$$W + 3Cl_2 \rightarrow \underset{\text{green}}{WCl_6} \xrightarrow{H_2} \underset{\text{black}}{WCl_5} \rightarrow \underset{\text{brown}}{WCl_4} \rightarrow \underset{\text{grey}}{WCl_2}$$

$$2Mo + 5Cl_2 \rightarrow \underset{\text{green}}{MoCl_5} \xrightarrow{H_2} \underset{\text{red}}{MoCl_3} \xrightarrow[\text{heat}]{CO_2} \underset{\text{yellow}}{MoCl_2} + \underset{\text{brown}}{MoCl_4}$$

The bromides and iodides may be prepared in a similar manner.

The Oxohalides

A number of oxohalides are known and are prepared by reactions such as:

$$2M + O_2 + 4F_2 \rightarrow 2MOF_4 \qquad (M = \text{Mo, W})$$

$$MO_2 + Cl_2 \rightarrow \underset{\text{(yellow solid)}}{MO_2Cl_2}$$

The molybdenum compounds tend to be less stable than those of tungsten.

Complexes

The complexes that are known are mainly anions. Electrolytic reduction of $Mo^{VI}O_3$ in concentrated hydrochloric acid gives the octahedral complex anion $Mo^{III}Cl_6^{3-}$. A red potassium salt, K_3MoCl_6, has been isolated. Reduction of tungstates in concentrated hydrochloric acid also gives rise to complex anions, e.g. $W_2^{III}Cl_9^{3-}$, giving a yellow salt $K_3W_2Cl_9$. The structure of $W_2Cl_9^{3-}$ is interesting in that it consists of two octahedra joined by one face with the nine remaining corners occupied by chlorine atoms. Solutions of molybdenum and tungsten in oxidation state +3, with excess potassium cyanide solution and oxygen, give dodecahedral cyanides of molybdenum and tungsten, $K_4Mo(CN)_8$ (yellow) and $K_4W(CN)_8$ (yellow), in which the oxidation state of the metals is +4. These are remarkably stable complexes, oxidized to yellow octacyanomolybdate(V) and octacyanotungstate(V) ($K_3Mo(CN)_8$, $K_3W(CN)_8$) by means of potassium permanganate.

Manganese

The electronic structure $3d^5 4s^2$ indicates a maximum oxidation state of +7 which occurs in the permanganates (MnO_4^-) and manganese heptoxide. These compounds are powerful oxidizing agents, being readily reduced to the manganese(II) state (manganous). The +2 oxidation state, with a half-filled *d* subshell, is the most stable lower oxidation state. Other less stable oxidation states include +3, +4, and +6. Apart from the volatile, acidic, and explosive heptoxides (Cl_2O_7 and Mn_2O_7) and the isomorphous perchlorates ($KClO_4$) and permanganates ($KMnO_4$), manganese shows little resemblance to the halogens (Group VIIB).

Occurrence

The chief ore is pyrolusite, MnO_2. Other ores of less importance include braunite, Mn_2O_3, and manganese blende, MnS.

Extraction

Direct reduction of the dioxide is too vigorous and pyrolusite is heated strongly to the red trimanganic tetroxide which is then reduced with carbon in an electric furnace.

$$3MnO_2 \rightarrow Mn_3O_4 + O_2$$

$$Mn_3O_4 + 4C \rightarrow 3Mn + 4CO$$

A purer product can be obtained using the aluminothermic method.

$$3Mn_3O_4 + 8Al \rightarrow 4Al_2O_3 + 9Mn \qquad (\Delta H = -2520 \text{ kJ})$$

Electrolytic manganese of high purity is prepared by the electrolysis of manganese (II) sulphate solutions.

Uses

Alloys of iron and manganese, obtained in the blast furnace (p. 392), are known as ferromanganese. Ferromanganese alloys are used for making steel. The open-hearth process uses an alloy of 80 per cent Mn and $<0{\cdot}3$ per cent C and the Bessemer process an alloy of 25 per cent Mn and 0·4 per cent C (spiegeleisen). Manganese steel is extremely hard and is used for the jaws of rock crushers, railway points, and heavy machinery. A Cu–Mn–Ni alloy (Manganin) shows little change in resistance with change in temperature and is used in certain electrical measuring instruments.

Properties

Manganese is a hard white metal, stable in dry air and slowly attacked by water.

$$Mn + 2H_2O \rightarrow Mn(OH)_2 + H_2$$

The reaction can be speeded up by the addition of ammonium chloride which dissolves the insoluble manganese(II) hydroxide. The principal reactions of the metal are indicated in Figure 18.20.

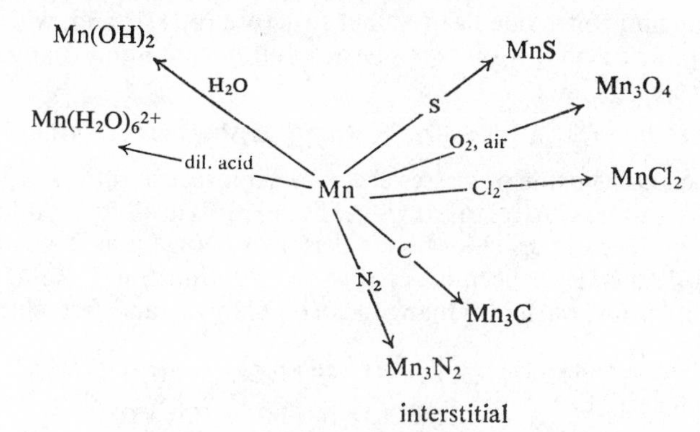

FIGURE 18.20. Reactions of manganese metal

The metal is slowly attacked by dilute mineral acids giving pink solutions of manganese(II) salts which probably contain the pink hydrated cation $Mn(H_2O)_6^{2+}$. Manganese combines on heating with many non-metals such as chlorine, sulphur, oxygen, carbon, and nitrogen.

The Oxides

The common naturally occurring oxide is manganese(IV) oxide (manganese dioxide, $Mn^{IV}O_2$). All the lower oxides may be prepared from manganese(IV) oxide.

$$\underset{\text{manganese dioxide}}{MnO_2} + H_2 \rightarrow \underset{\text{manganese(II) oxide}}{MnO} + O_2 \xrightarrow{300\,°C} \underset{\text{manganese sesquioxide}}{Mn_2O_3}$$

$$MnO_2 \xrightarrow{1{,}000°C} \underset{\text{trimanganic tetroxide}}{Mn_3O_4[Mn^{II}(Mn^{III}O_2)_2]}$$

The standard free energies of formation for these oxides in kJ mol^{-1} are: $MnO = -363$, $Mn_2O_3 = -893$, $MnO_2 = -466$, $Mn_3O_4 = -1280$. Thus Mn_3O_4 is the final product obtained when any other oxide is heated in air.

Green manganese(II) oxide (manganese oxide) possesses a rock-salt structure and is basic, dissolving in acids to give pink solutions of manganese(II) salts.

Brown manganese(III) oxide (manganese sesquioxide) exists in two forms (*see* Fe_2O_3, p. 400). With concentrated sulphuric acid it yields a green manganese(III) salt.

$$Mn_2O_3 + 3H_2SO_4 \rightarrow Mn_2(SO_4)_3 + 3H_2O$$

Red trimanganic tetroxide has a spinel structure (p. 371) and, with concentrated sulphuric acid, dissolves to give a solution containing manganese(II) and (III) salts.

$$Mn^{II}(Mn^{III}O_2)_2 + 4H_2SO_4 \rightarrow Mn^{II}SO_4 + Mn_2^{III}(SO_4)_3 + 4H_2O$$

Manganese(IV) oxide possesses a non-stoichiometric rutile lattice whose composition approximates to $MnO_{1\cdot95}$. The oxide is insoluble in dilute acids. Cold concentrated hydrochloric acid dissolves manganese(IV) oxide but a tetrachloride has never been isolated as the solution contains the complex ion $MnCl_6^{2-}$. On warming, manganese(II) chloride and free chlorine are produced.

$$MnO_2 + 6Cl^- + 4H^+ \rightarrow MnCl_6^{2-} + 2H_2O$$

$$MnCl_6^{2-} \rightarrow Mn^{2+} + 4Cl^- + Cl_2$$

Concentrated sulphuric acid dissolves manganese(IV) oxide producing violet solutions containing $Mn_2^{III}(SO_4)_3$. Manganese(IV) sulphate $Mn^{IV}(SO_4)_2$ can be prepared by the action of sulphuric acid and potassium permanganate on manganese(II) sulphate. The black crystals of manganese(IV) sulphate are rapidly hydrolysed by water to manganese(IV) oxide.

With the alkaline-earth metal oxides, manganese(IV) oxide gives double oxides of unknown structure, sometimes called manganites, e.g. $CaO.MnO_2$ ($CaMnO_3$). Manganese(IV) oxide is used to decolorize glass and as a paint drier.

The addition of powdered potassium permanganate to cold concentrated sulphuric acid gives a dark green solution. Dilution with ice-cold water causes brown drops of manganese(VII) oxide to separate.

$$2MnO_4^- + 2H^+ \rightarrow H_2O + Mn_2O_7$$

The oxide is liable to explode violently.

$$2Mn_2O_7 \rightarrow 4MnO_2 + 3O_2$$

Manganese(VII) oxide is a covalent acidic oxide which, with water, yields a purple solution of permanganic acid (p. 391).

$$Mn_2O_7 + H_2O \rightarrow 2HMnO_4$$

Manganese(VII) oxide is covalent with the structure, as shown in Figure 18.21.

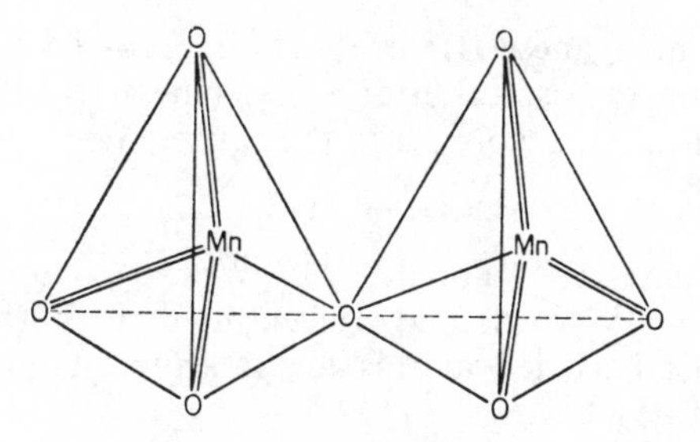

FIGURE 18.21

The Hydroxides

The only true hydroxide is manganese(II) hydroxide, $Mn^{II}(OH)_2$, precipitated as a white powder by the addition of alkali to a solution of a manganese(II) salt.

$$Mn^{2+} + 2OH^- \rightarrow Mn(OH)_2$$

The precipitate, however, readily undergoes atmospheric oxidation to brown hydrated manganese(III) oxide.

$$4Mn(OH)_2 + O_2 \rightarrow 2Mn_2O_3.H_2O + 2H_2O$$

The Halides

Red anhydrous manganese(II) chloride ($Mn^{II}Cl_2$) may be prepared by treating manganese with chlorine at elevated temperatures. It is, however, more commonly prepared by the action of hot concentrated hydrochloric acid on manganese(IV) oxide (p. 387). Pink crystals of $MnCl_2.4H_2O$ separate which have an octahedral cis configuration (p. 105).

A manganese(II) fluoride is also known but, unlike the chloride, it is rapidly hydrolysed by water.

No manganese(III) bromides or iodides are known and the chloride is only stable below −35 °C. A red fluoride ($MnF_3.3H_2O$) has been made by the action of hydrofluoric acid on manganese(III) oxide. This fluoride complexes with alkali metal fluorides (e.g. K_2MnF_5).

There are no manganese(IV) halides.

The Oxo Salts

Pink manganese(II) salts are fairly stable and well characterized: $MnSO_4.4H_2O$ and deliquescent $Mn(NO_3)_2.6H_2O$ may be made by heating manganese(II) oxide with the appropriate acid. The sulphate forms crystalline double salts with ammonium and the alkali metal sulphates which are isomorphous with Mohr's salt, e.g. $(NH_4)_2SO_4.MnSO_4.6H_2O$.

Buff-coloured manganese(II) carbonate is prepared by adding sodium carbonate solution to a solution of a manganese(II) salt. The carbonate decomposes on heating according to the equation:

$$MnCO_3 \rightarrow MnO + CO_2$$

Green manganese(III) sulphate, $Mn_2^{III}(SO_4)_3$, can be prepared by heating manganese(IV) oxide and concentrated sulphuric acid at 140 °C. The crystals must be dried at this temperature. It forms violet alums, $NH_4Mn(SO_4)_2.12H_2O$.

All manganese(III) salts disproportionate in aqueous solution forming manganese(II) salts and depositing manganese(IV) oxide.

$$2Mn^{3+} + 2H_2O \rightarrow Mn^{2+} + MnO_2\downarrow + 4H^+$$

Complex Cyanides

The addition of excess potassium cyanide to a manganese(III) solution produces deep blue potassium hexacyanomanganate(II) ($K_4Mn(CN)_6$). Evaporation of an aqueous solution of $K_4Mn(CN)_6$ in air gives red crystals of potassium hexacyanomanganate(III) ($K_3Mn(CN)_6$). Magnetic measurements indicate a t_{2g}^5 configuration for $Mn^{II}(CN)_6^{4-}$ and a t_{2g}^4 configuration for $Mn^{III}(CN)_6^{3-}$.

Manganese(VI) and (VII) Compounds

Manganates and permanganates

Manganese in oxidation state +6 is known only in the green manganates, e.g. K_2MnO_4, which, in solution, furnish the deep green manganate ion MnO_4^{2-}. The parent acid H_2MnO_4 is not known. The sodium and potassium salts can be prepared by fusing manganese(IV) oxide with sodium or potassium hydroxide and a little potassium chlorate, with free access of air.

$$2MnO_2 + 4OH^- + O_2 \rightarrow 2MnO_4^{2-} + 2H_2O$$

The cooled dark-green mass obtained is dissolved in a little water and evaporation in vacuum gives green crystals of K_2MnO_4 or $Na_2MnO_4.10H_2O$.

The manganates may also be obtained by boiling purple permanganate solutions with strong alkali.

$$\underset{\text{purple}}{4MnO_4^-} + 4OH^- = \underset{\text{green}}{4MnO_4^{2-}} + 2H_2O + O_2$$

The green manganate solutions are stable in excess alkali but, on dilution, acidification, or passage of chlorine, purple permanganate is produced.

$$3MnO_4^{2-} + 2H_2O \rightarrow 2MnO_4^- + MnO_2 + 4OH^-$$

$$3MnO_4^{2-} + 4H^+ \rightarrow 2MnO_4^- + MnO_2 + 2H_2O$$

$$2MnO_4^{2-} + Cl_2 \rightarrow 2MnO_4^- + 2Cl^-$$

On evaporation of the solutions, sodium and potassium permanganates form deep purple crystals ($NaMnO_4.3H_2O$, $KMnO_4$).

A solution of purple permanganic acid can be made by oxidizing manganese(II) salts in nitric acid and sodium bismuthate. This is used as a qualitative test for manganese.

$$2Mn^{2+} + 5BiO_3^- + 16H^+ \rightarrow 2HMnO_4 + 7H_2O + 5Bi^{3+}$$

A pure solution of permanganic acid can be made by treating barium permanganate with dilute sulphuric acid and filtering off the precipitated barium sulphate. Concentration of the solution of permanganic acid at −73 °C under high vacuo produces a distillate of pure $HMnO_4.2H_2O$ and a residue of $HMnO_4$ containing a little manganese dioxide. Both $HMnO_4.2H_2O$ and $HMnO_4$ are strong oxidizing agents but the latter reacts explosively.

Solid alkali-metal permanganates decompose at 240 °C.

$$2KMnO_4 \rightarrow K_2MnO_4 + MnO_2 + O_2$$

The permanganates are powerful oxidizing agents in either alkaline or acid media. In alkaline solution, the permanganate is quickly reduced to the green manganate which then deposits brown manganese(IV) oxide. The overall half reaction is:

$$MnO_4^- + 2H_2O + 3e \rightarrow MnO_2\downarrow + 4OH^- \qquad (E^\ominus = +1{\cdot}23\ V)$$

Alkaline permanganate oxidizes iodides to iodates.

$$I^- + 3H_2O \rightarrow IO_3^- + 6H^+ + 6e$$

In acid solution, sulphuric acid is used, since hydrochloric acid is oxidized to chlorine. The half reaction is:

$$MnO_4^- + 8H^+ + 5e \rightarrow Mn^{2+} + 4H_2O \qquad (E^\ominus = +1{\cdot}52\ V)$$

Acidified potassium permanganate solution oxidizes oxalates to carbon dioxide, iron(II) salts to iron(III), nitrites to nitrates, and liberates iodine from potassium iodide.

$$5\begin{array}{l}COO^- \\ | \\ COO^-\end{array} \rightarrow 10CO_2 + 10e$$

$$5Fe^{2+} \rightarrow 5Fe^{3+} + 5e$$

$$5NO_2^- + 5H_2O \rightarrow 5NO_3^- + 10H^+ + 10e$$

$$10I^- \rightarrow 5I_2 + 10e$$

The full ionic equation is obtained by adding the half reaction for potassium permanganate to the half reaction for the reducing agent, balancing where necessary.

The manganate and permanganate ions are tetrahedral. The green manganate ion is paramagnetic with one unpaired electron but the purple permanganate ion is diamagnetic with no unpaired electrons:

$$\begin{array}{c}{}^{-}O \quad O \\ \diagdown Mn \diagup\!\!\diagup \\ {}^{-}O \diagup \quad \diagdown\!\!\diagdown O\end{array} \qquad \begin{array}{c}O \quad O \\ \diagdown\!\!\diagdown Mn \diagup\!\!\diagup \\ {}^{-}O \diagup \quad \diagdown\!\!\diagdown O\end{array}$$

tetrahedral green manganate ion | tetrahedral purple permanganate ion

Double bonding arises from the overlap of oxygen *p* orbitals with manganese *d* orbitals. The bonds are equalized by resonance and only one canonical form is shown. Addition of sulphite to strongly alkaline manganate(VI) gives a blue solution which deposits $Na_3Mn^VO_4$ on cooling.

Iron

The iron atom, with electronic structure $3d^64s^2$, shows no oxidation state corresponding to the total number of electrons in the valency shells. The maximum oxidation state is +6 and is only found in the ferrates (e.g. K_2FeO_4). These compounds are stronger oxidizing agents than the permanganates and chromates but are much less stable. The most important oxidation states are +2 and +3, found in iron(II) and iron(III) compounds. Since both the iron(II) and iron(III) ions have small radii, there is a strong tendency towards complex formation.

Occurrence

The metallurgical processes employed for the production of iron and steel depend to a large extent upon the phosphorus content of the iron ores and the ores are conveniently classified into:

(*a*) Low-phosphorus ores—red haematite (Fe_2O_3), magnetite (Fe_3O_4).
(*b*) High-phosphorus ores—brown haematite, siderite ($FeCO_3$).

Iron pyrites (FeS_2) is not worked for iron because of the high sulphur content; its industrial importance is as a sulphur ore.

Extraction

The oxide ores are reduced directly in a blast furnace whereas the carbonate ores are first roasted in a kiln and then reduced. A charge of the ore, coke, and limestone is heated by a blast of hot air, introduced by means of narrow pipes (tuyères) at the base of the blast furnace (Figure 18.22). The approximate temperatures and the probable reactions which occur at the various zones in the furnace are shown in the diagram.

The molten iron runs off at the base of the furnace and is known as pig iron. The silicate slag is tapped off, mixed with tar, and used in road making.

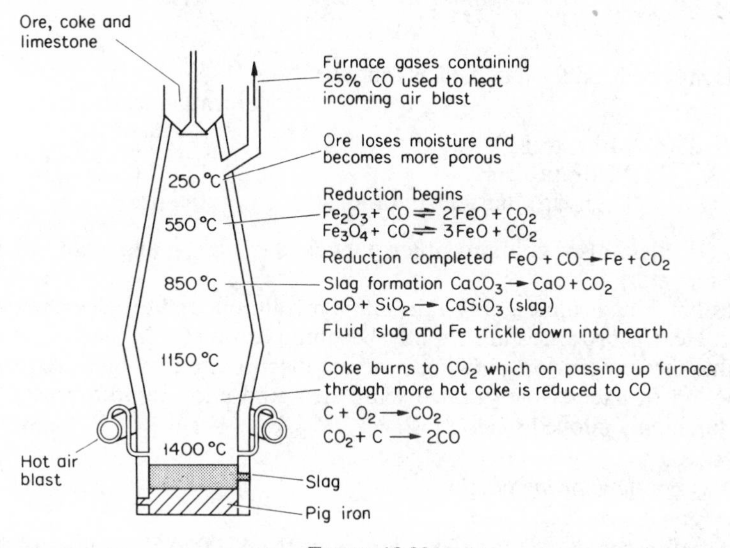

FIGURE 18.22

Pig iron, or *cast iron*, is very brittle and contains up to 5 per cent impurities, namely carbon (3–4 per cent), phosphorus, sulphur, silicon, and manganese. Some of the carbon may be in the form of iron carbide, or cementite (Fe_3C). The melting point is 1200 °C and, when molten, can be cast into a variety of shapes in moulds; it is used for fire grates, gutter piping, railings, machine beddings, etc.

Wrought iron

Wrought iron, or *malleable iron*, the purest form of commercial iron, is made by oxidizing the impurities in cast iron in a reverberatory furnace lined with haematite. The haematite oxidizes the carbon to carbon monoxide.

$$Fe_2O_3 + 3C \rightarrow 2Fe + 3CO$$

Limestone is added as a flux and sulphur, silicon, and phosphorus are oxidized and pass into the slag. The metal is removed and freed from the slag by passing through rollers. Wrought iron possesses a fibrous structure due to the thin films of slag trapped between the layers of pure iron. Wrought iron contains approximately 0·5 per cent of impurities of which about half is carbon. The melting point is 1400 °C but the metal can be welded at 1000 °C. It is tough, malleable, and ductile and can be used for making chains, bolts, frameworks, etc. For structural purposes it has been largely replaced by mild steel.

Steel

Steels may be broadly classified as follows:

	percentage carbon
(*a*) Mild steel	0·1–0·3
(*b*) Medium steel	0·3–0·6
(*c*) High-carbon steel and tool steel	0·6–1·3

Steels may also contain other metals such as manganese, nickel, tungsten, etc.

Steel making is undergoing a dramatic revolution, and the Bessemer and Open Hearth processes are rapidly becoming obsolete. Modern furnaces involve the oxygen top blowing process. Electric-arc and high-frequency induction furnaces which utilize scrap steel and alloy scrap are used for making highly alloyed steels.

The oxygen top blowing process

Modern furnaces have the capacity to refine from 200 to 300 tonnes of steel with a reaction time of below 45 minutes. Liquid iron from the blast furnace is charged into the converter, scrap steel is added and oxygen is blown in down a water-cooled copper lance on to the top of the liquid metal (Figure 18.23(*a*)). A violent reaction ensues in which carbon, silicon, and phosphorus are oxidized and, with addition of lime, calcium silicates and phosphates slag off. When steel of the desired composition is obtained, the oxygen is turned off and the molten steel is poured into ladles for casting into ingots. The refining part of the cycle (at the end of the oxygen blow) which reduces the carbon content down to the level required, takes only from three to five minutes. Thus one of the major difficulties is the very rapid analysis that must be carried out to ensure that, when the oxygen blow has finished, the steel has the correct carbon content. One solution to this analytical problem has been the development of a secondary lance parallel to the oxygen lance which sucks up the liquid steel, atomizes it, and passes the vaporized particles to a vacuum spectrograph.

The electric-arc process

A charge of scrap steel and turnings is fed into the furnace and is melted by electric arcs struck between adjustable carbon electrodes. Again acidic and basic linings are employed for scrap differing in phosphorus content.

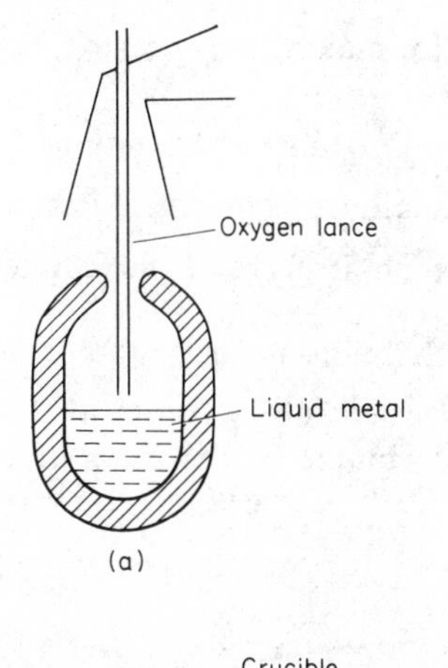

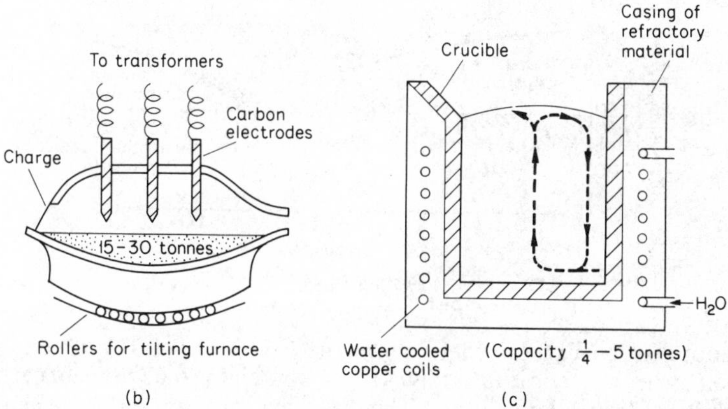

FIGURE 18.23. (*a*) The oxygen top blowing process; (*b*) electric-arc process; (*c*) high-frequency induction process

This method is used for making high-quality alloy steels such as stainless steel, heat-resistant steel, and high-speed cutting steel. A diagram of the furnace is shown in Figure 18.23(*b*).

The high-frequency induction process

A charge of alloy scrap of known composition, together with iron, is fed into the furnace, a diagram of which is shown in Figure 18.23(*c*). Alternating current at 500–2000 Hz per second passes through insulated water-cooled copper coils and the resulting magnetic field sets up eddy currents which generate heat in the metal. The circulation of the melt is due to these currents and the stirring effect is so pronounced that the surface of the melt becomes shaped like an inverted saucer.

The induction furnace is capable of producing high-quality alloy steels, containing tungsten, vanadium, chromium, manganese, molybdenum, cobalt, and nickel for making ball-bearing, magnet, die and tool steels, etc.

The Allotropy of Iron

There are two main allotropic forms of iron:

(*a*) α-ferrite, which is ferromagnetic and has a body-centred cubic structure, and

(*b*) γ-ferrite which is non-magnetic and has a face-centred cubic structure.

α-Ferrite is converted into γ-ferrite at 910 °C with an accompanying contraction in the volume of the mass of the metal. α-Ferrite loses its magnetism at 760 °C (the Curie point) and becomes paramagnetic. α-Ferrite

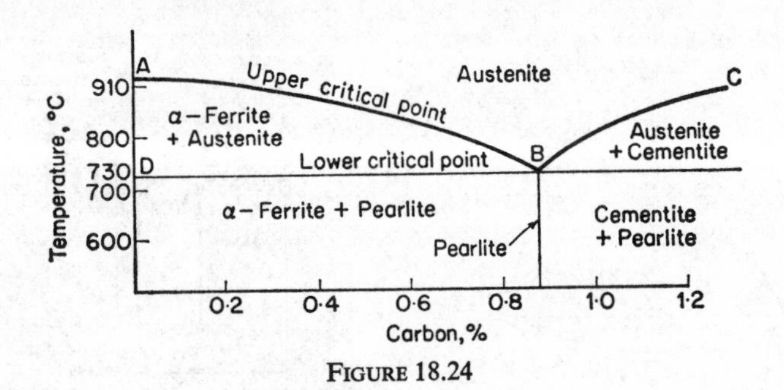

FIGURE 18.24

does not dissolve carbon whereas γ-ferrite (stable above 910 °C) dissolves carbon as *cementite* or iron carbide (Fe_3C, containing 6·67 per cent carbon), which it holds in solid solution. The dissolved carbon in γ-ferrite causes a depression of the transition temperature between α- and γ-forms but, when the carbon content exceeds 0·85 per cent, the temperature rises again. From Figure 18.24 it can be seen that, as the carbon content increases from 0 to 0.85 per cent, the transition temperature of the α- to γ-forms falls from 910 °C to 730 °C. When the carbon content exceeds 0·85 per cent, however, the temperature of the change begins to increase again.

If a horizontal line is drawn through the minimum transition temperature at *B*, meeting the temperature axis at *D*, an equilibrium diagram is produced, indicating phases over a particular range of temperature and carbon content:

(*a*) Above the curves *AB*, *BC*, cementite is held in γ-iron as a solid solution known as *austenite*.

(*b*) At *B* (where the carbon content = 0·85 per cent), where the change from γ- to α-iron takes place, all the cementite is thrown out of solid solution, giving interleaved plates of α-ferrite and cementite called *pearlite*.

(*c*) Where the carbon content is less than 0·85 per cent, α-ferrite appears until the carbon content of the remaining γ-iron reaches 0·85 per cent when the change to pearlite occurs and, where the carbon content is greater than 0·85 per cent, cementite (Fe_3C) appears until the carbon content of the remaining γ-iron is reduced to 0·85 per cent, when again the change to pearlite occurs.

The physical properties of ferrite, pearlite, and cementite

(*a*) *Ferrite*—this is chemically pure iron, soft and ductile but tough (resistant to shock loads) and will impart these properties to steel if present in large amounts.

(*b*) *Pearlite*—this is very hard, owing to the close interleaving of ferrite and cementite which gives no slip and little ductility. As the percentage of pearlite increases, hardness increases but toughness and ductility decrease.

(*c*) *Cementite* (Fe_3C)—this is typical of a metal carbide (p. 211) and is very hard and brittle.

By altering the rate at which the temperature falls from the upper curve *AB*, steels of different structures and properties can be obtained with the same carbon content. This is because, with increased rates of cooling, the equilibrium is frozen and is not a true equilibrium. For example, rapid cooling in cold water gives a hard and brittle steel known as martensite (p. 92).

Alloys

Alloying elements do not behave in the same way as carbon. They may:

(*a*) be held in solid solution in the steel and toughen the ferrite phase, e.g. silicon and nickel. Silicon steel is used in magnets, transformers, and springs: nickel steel is used in the drives of shafts and gears;

(*b*) slow up the γ- to α-change and avoid cracking of steel during fast cooling processes. They form carbides associated with cementite and pearlite, e.g. nickel, chromium, molybdenum, tungsten, and vanadium. Molybdenum and tungsten alloys are used in high-speed cutting tools. Vanadium and chromium steels are strongly strain-resistant and used in axles, etc.

(*c*) stabilize γ-iron so that austenite can exist at room temperature, e.g. manganese. Such steel can stand intense wear and is used in railway crossings and points, dredger buckets, armour plate, etc.

A chromium (18 per cent), nickel (8 per cent) alloy is corrosion-resistant and is used for the manufacture of stainless steel.

Properties of Iron

The principal reactions of the metal are indicated in Figure 18.25. The metal dissolves readily in dilute hydrochloric acid and sulphuric acid giving

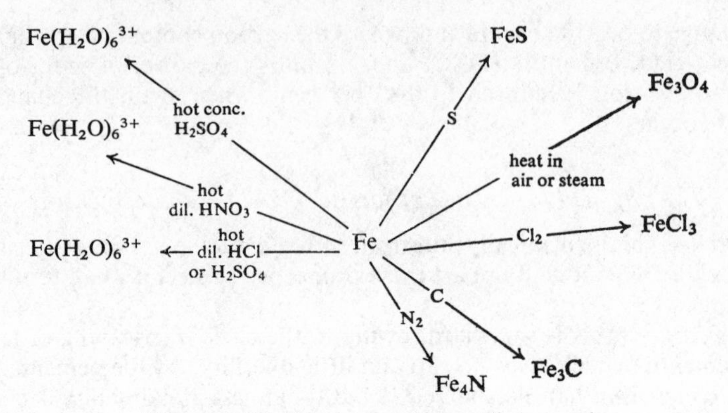

FIGURE 18.25. Principal reactions of iron metal

iron(II) salts but, with hot dilute nitric and concentrated sulphuric acid, the iron(II) salt is oxidized to iron(III) with the production of oxides of nitrogen and sulphur respectively. Concentrated nitric acid renders the metal *passive*, i.e. it will not dissolve in dilute acids or displace copper from copper sulphate. The passivity is due to a coherent film of γ-Fe_2O_3 (p. 400), which covers the metal surface. The film can be destroyed by mechanical scraping or by heating the passive metal in a current of hydrogen to reduce the oxide.

An oxide film forms a coherent and protective layer on a metal surface if the structure and density of the oxide film is similar to that of the parent metal. Otherwise expansion or contraction of the surface layer with respect to the metal occurs continually exposing fresh metallic surfaces, e.g. rust, $Fe_2O_3 . xH_2O$.

Iron carbide, unlike the nitride, is not interstitial (p. 211).

The rusting of iron

Rust consists of hydrated iron(III) oxide, $Fe_2O_3 . H_2O$ or FeO(OH) (p. 401). For rusting to occur, there must be a thin film of water on the surface of the metal and oxygen or air present. Even so, chemically pure, completely homogeneous iron will not rust. The impurities or strained portions cause the formation of numerous small electrolytic cells with anodes of pure iron and cathodes of impure or strained portions. A schematic diagram showing one such cell is given in Figure 18.26.

Iron dissolves at the anode, producing iron(II) ions, and the electrons travel along the surface of the metal to the cathode, forming hydroxyl ions with oxygen and water. The hydroxyl ions react with the iron(II) ions, forming iron(II) hydroxide, which then undergoes atmospheric oxidation to give hydrated iron(III) oxide or rust. If no oxygen is present, the reaction

$2H^+ + 2e \rightarrow H_2$ occurs at the cathode and the evolved hydrogen clings to the metal surface polarizing the cell and preventing rusting. Carbon dioxide (in the air) assists rusting by making the solution acidic and facilitating the formation of iron(II) ions. Sodium chloride also increases the rate

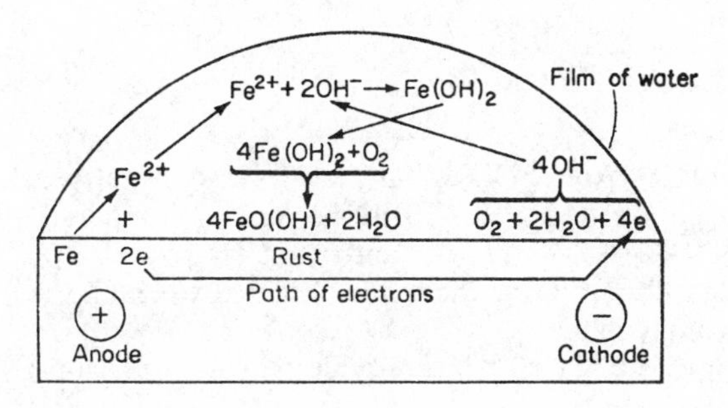

FIGURE 18.26

of rusting by making the water more conducting, whereas sodium hydroxide retards rusting by removing hydrogen ions.

Protective processes to limit rusting include:

(*a*) Application of paint.

(*b*) Application of a coating of sacrificial metal such as zinc (galvanizing) or tin (tinning).

(*c*) Electroplating with nickel or chromium (p. 372).

(*d*) Dipping the iron into a phosphate bath (orthophosphoric acid with zinc and manganese phosphates) which gives a coherent protective phosphate film and may be followed by the application of paint or enamel.

(*e*) Cathodic protection. For example, underground piping is connected at intervals to a strip of more active metal, such as magnesium, zinc, or aluminium. The magnesium corrodes to form magnesium ions whereas hydrogen ions are discharged at the iron cathode, thus preventing the formation of rust. The application of zinc in galvanizing is also an example of cathodic protection.

Compounds of Iron

The oxides

There are three oxides of iron, black iron(II) oxide ($Fe^{II}O$) or ferrous oxide, red iron(III) oxide ($Fe_2^{III}O_3$) or ferric oxide, and black iron(II) diiron(III) oxide ($Fe^{II}Fe_2^{III}O_4$). The reactions used to prepare these oxides are indicated in the following scheme:

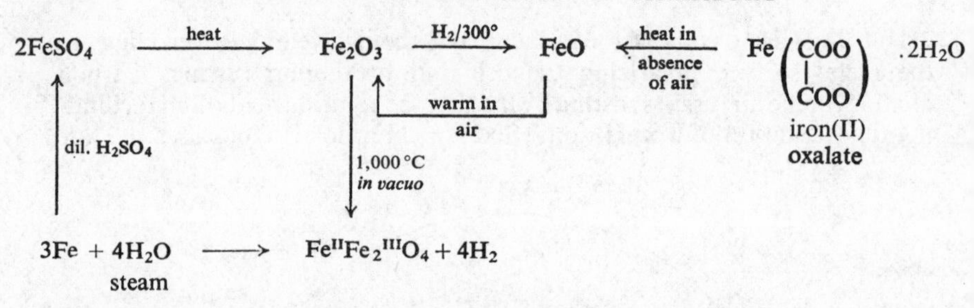

Iron(II) diiron(III) oxide $Fe^{II}Fe_2^{III}O_4$, has a large, negative, standard free energy of formation ($\Delta G_f^\ominus = -10142$ kJ mol^{-1}) compared with iron(II) oxide ($\Delta G_f^\ominus = -244{\cdot}3$ kJ mol^{-1}) and iron(III) oxide ($\Delta G_f^\ominus = -741$ kJ mol^{-1}). Hence $Fe^{II}Fe_2^{III}O_4$ is the oxide obtained when all the other oxides are heated strongly in air.

Iron(II) oxide, $Fe^{II}O$, is a basic oxide which gives iron(II) salts with dilute acids.

Iron(III) oxide, $Fe_2^{III}O_3$, exists in two forms. The α-form is prepared as shown above and the γ-form by heating a iron(III) hydroxide gel or sol below 300°C. The α-form possesses the α-alumina structure (p. 196) and is not very soluble in acids. The γ-form, with the structure indicated below dissolves in strong acids producing iron(III) salts.

Iron(II) diiron(III) oxide, $Fe^{II}Fe_2^{III}O_4$, with a spinel structure, dissolves in strong acids to give a mixture of iron(II) and iron(III) salts.

$$FeFe_2O_4 + 8HCl \rightarrow FeCl_2 + 2FeCl_3 + 4H_2O$$

Iron(II) oxide, γ-iron(III) oxide, and iron(II) diiron(III) oxide all have a tendency towards non-stoichiometry and are related structurally. Consider a cubic close-packed system of 32 oxide ions. If these accommodate 32 iron(II) ions between the oxide ions, we have the ideal structure FeO. If iron(II) ions are replaced by 4 iron(III) ions to maintain electrical neutrality, we now have non-stoichiometry ($Fe_{30/32}O$ or $Fe_{0{\cdot}94}O$), the compound being iron-deficient. Continuing this process, replacement of 24 iron(II) by 16 iron(III) ions gives Fe_3O_4. Finally, if all the iron(II) ions are replaced by $21\frac{1}{3}$ iron(III) ions in the system, we have γ-Fe_2O_3. The stepwise replacement of iron(II) ions by iron(III) is shown in Table 18.2.

TABLE 18.2

Fe^{2+}	Fe^{3+}	O^{2-}	
32	0	32	Ideal FeO
26	4	32	Non-stoichiometric $Fe_{0{\cdot}94}O$
8	16	32	$Fe^{II}Fe_2^{III}O_4$
0	$21\frac{1}{3}$*	32	Fe_2O_3

* In an infinite close-packed system of ions this would be a whole number.

The hydroxides

Iron(II) hydroxide ($Fe(OH)_2$) or ferrous hydroxide is a white solid precipitated from an aqueous solution of a iron(II) salt by sodium or potassium hydroxide in the absence of air. The white iron(II) hydroxide rapidly oxidizes in the air to iron(III) oxohydroxide or rust.

$$4Fe(OH)_2 + O_2 \rightarrow 4FeO(OH) + 2H_2O$$

Iron(II) hydroxide is a true hydroxide with the cadmium iodide lattice. It is slightly amphoteric, dissolving in acids to give iron(II) salts and in concentrated alkalis producing the tetrahydroxoferrate(II) ion $Fe(OH)_4{}^{2-}$.

The compound corresponding to the formula $Fe(OH)_3$ does not exist and the red-brown precipitate formed by the action of alkali on a iron(III) salt is probably a mixture of hydrated iron(III) oxide and an oxohydroxide, (FeO(OH)). The oxohydroxide exists in two crystalline forms, α-FeO(OH) (goethite) and γ-FeO(OH) (lepidocrocite). γ-FeO(OH) is formed during the atmospheric oxidation of iron(II) hydroxide or during rusting. α-FeO(OH) is precipitated by the action of alkaline hypochlorite on a iron(II) salt. On heating, α-FeO(OH) yields α-Fe_2O_3 and γ-FeO(OH) yields γ-Fe_2O_3.

The sulphides

Black iron(II) sulphide $Fe^{II}S$ (ferrous sulphide) is prepared from the elements and is always contaminated with iron. It dissolves readily in mineral acids yielding hydrogen sulphide. Iron(II) sulphide possesses the nickel arsenide structure but is non-stoichiometric and the formula varies from $Fe_{0\cdot 86}S$ to $Fe_{0\cdot 89}S$ and is never actually FeS.

Naturally occurring golden-yellow pyrites FeS_2 has a distorted rock-salt structure. Iron(II) ions take the place of sodium ions and $(S—S)^{2-}$ ions have their centres at the chloride ion positions.

The halides

Iron(II) halides (ferrous halides) can be prepared anhydrous by passing the vapour of the appropriate halogen acid over heated iron, e.g.

$$Fe + 2HCl \rightarrow FeCl_2 + H_2$$

Solutions of iron in dilute hydrochloric acid produce green crystals of $FeCl_2.4H_2O$ which have an octahedral trans configuration (p. 105).

Iron(III) halides (ferric halides) are made by direct halogenation of the heated metal, e.g.

$$2Fe + 3Cl_2 \rightarrow 2FeCl_3$$

The anhydrous chloride is black, but crystalline iron(III) chloride, $FeCl_3.6H_2O$, is yellow and contains octahedral trans $Fe(H_2O)_4Cl_2{}^+$ and

$Cl(H_2O)_2^-$ ions. Iron(III) iodide does not exist. In the vapour state, the chloride exists as dimeric molecules, Fe_2Cl_6 (cf. Al_2Cl_6), but in the solid state it has a similar structure to chromium(III) chloride.

Oxo salts

Iron(II) compounds (ferrous salts)

The nitrate, $Fe(NO_3)_2.6H_2O$, and the sulphate, $FeSO_4.7H_2O$, are prepared as light green crystals by dissolving the metal oxide or carbonate in the appropriate acid and crystallizing. Both compounds decompose on heating.

$$4Fe(NO_3)_2 \rightarrow 2Fe_2O_3 + 8NO_2 + O_2$$

$$2FeSO_4 \rightarrow Fe_2O_3 + SO_2 + SO_3$$

Iron(II) sulphate forms double salts with ammonium and alkali metal sulphates, e.g. $M^I{}_2SO_4.FeSO_4.6H_2O$ (M = K, Rb, Cs, and NH_4). The ammonium salt, Mohr's salt, is used in volumetric analysis. Aqueous solutions of iron(II) salts furnish the hydrated octahedral cation $Fe(H_2O)_6{}^{2+}$, which is pale green and readily oxidized.

Thus even molecular oxygen can convert iron(II) to iron(III) salts.

$$4Fe^{2+} + O_2 + 2H_2O \longrightarrow \underbrace{4Fe^{3+} + 4OH^-}_{\downarrow}$$

$$FeO(OH)\downarrow \text{ (brown)}$$

The solution eventually deposits iron(III) oxohydroxide. The hydrated iron(II) ion forms a dark brown complex with nitric oxide $Fe(H_2O)_5NO^{2+}$. The complex contains NO^+ with iron in oxidation state +1.

Iron(II) carbonate, which occurs naturally as siderite, has the same structure as calcite. On addition of sodium carbonate to a solution of a iron(II) salt, a white precipitate of iron(II) carbonate is produced which rapidly turns green and then brown.

$$4FeCO_3 + 2H_2O + O_2 \rightarrow 4FeO.OH + 4CO_2$$

Iron(III) compounds (ferric salts)

Iron(III) sulphate, $Fe_2(SO_4)_3.9H_2O$, may be prepared by oxidizing iron(II) sulphate with concentrated sulphuric acid; iron(III) nitrate, $Fe(NO_3)_3.9H_2O$, is prepared by dissolving iron in hot dilute nitric acid. Iron(III) sulphate decomposes on strong heating.

$$Fe_2(SO_4)_3 \rightarrow Fe_2O_3 + 3SO_3$$

It forms violet alums with potassium and ammonium salts, e.g.

$M^IFe^{III}(SO_4)_2.12H_2O$. Aqueous solutions of iron(III) salts contain the violet hydrated cation $Fe(H_2O)_6{}^{3+}$ which functions as a Lowry and Brönsted acid and, unless the solution is made strongly acid, hydrolysis occurs rapidly, the solution turning yellow, then brown and finally precipitating iron(III) oxohydroxide.

$$Fe(H_2O)_6{}^{3+} + H_2O \rightleftharpoons Fe(OH)(H_2O)_5{}^{2+} + H_3O^+$$
$$Fe(OH)(H_2O)_5{}^{2+} + H_2O \rightleftharpoons Fe(OH)_2(H_2O)_4{}^{+} + H_3O^+$$
$$Fe(OH)_2(H_2O)_4{}^{+} + H_2O \rightleftharpoons Fe(OH)_3.(H_2O)_3 + H_3O^+$$
$$\downarrow \text{precipitates}$$
$$FeO(OH)$$

The reactions outlined above are somewhat oversimplified since the monomeric ionic species can dimerize by the formation of hydroxo bridges.

$$2Fe(H_2O)_5OH^{2+} \rightleftharpoons \left[(H_2O)_4Fe\begin{smallmatrix}H\\O\\ \\O\\H\end{smallmatrix}Fe(H_2O)_4\right]^{4+} + 2H_2O$$

The above reaction is termed olation and the dimers produced can then split off protons.

$$\left[(H_2O)_4Fe\begin{smallmatrix}H\\O\\ \\O\\H\end{smallmatrix}Fe(H_2O)_4\right]^{4+} + H_2O$$
$$\rightleftharpoons$$
$$\left[(H_2O)_4Fe\begin{smallmatrix}O\\ \\O\\H\end{smallmatrix}Fe(H_2O)_4\right]^{3+} + H_3O^+$$

The species formed contains oxo bridges and the reaction is termed oxolation. Oxolation proceeds very slowly, finally producing macromolecules containing very strong metal–oxygen bridges. Olation and oxolation processes explain the increasing chemical stability which accompanies the 'ageing' of many hydroxide precipitates. Similar olation and oxolation processes probably occur during the hydrolysis of the chromium(III) and aluminium(III) hydrated cations.

Reducing agents such as hydrogen sulphide, sulphur dioxide, and tin(II) chloride reduce iron(III) salts to iron(II) (p. 141).

The complexes of iron

The most important complexes of iron(II) and iron(III) are the hexacyanoferrates, i.e. yellow potassium hexacyanoferrate(II), $K_4Fe(CN)_6.3H_2O$, and red potassium hexacyanoferrate(III), $K_3Fe(CN)_6$. Potassium hexacyanoferrate(II) (potassium ferrocyanide) is manufactured from the 'spent oxide' from the gas works. The bog iron ore in the gas-works purifiers removes cyanogen (C_2N_2) from coal gas and forms Prussian blue (*see* below). This is then treated with lime and potassium carbonate when potassium hexacyanoferrate(II) is formed. In the laboratory preparation, iron(II) sulphate solution is treated with potassium cyanide solution.

$$Fe^{2+} + 6(CN)^- \rightarrow Fe(CN)_6^{4-}$$

Potassium hexacyanoferrate(II) yields hydrogen cyanide with dilute acids but with concentrated sulphuric acid, carbon monoxide is produced.

$$Fe(CN)_6^{4-} + 6H^+ \rightarrow Fe^{2+} + 6HCN$$

$$Fe(CN)_6^{4-} + 12H^+ + 6H_2O \rightarrow Fe^{2+} + 6NH_4^+ + 6CO$$

Oxidizing agents such as chlorine convert the hexacyanoferrate(II) anion into the hexacyanoferrate(III) anion.

$$2Fe(CN)_6^{4-} + Cl_2 \rightarrow 2Fe(CN)_6^{3-} + 2Cl^-$$

On evaporation, deep red crystals of the hexacyanoferrate(III) are formed which give a yellow aqueous solution. The hexacyanoferrate(III) anion is a mild oxidizing agent.

$$2Fe(CN)_6^{3-} + 2I^- \rightarrow 2Fe(CN)_6^{4-} + I_2$$

The hexacyanoferrates are used in the production of blue pigments such as Prussian blue and to detect the presence of iron(II) and iron(III) ions in qualitative analysis. A summary of the reactions of iron(II) and iron(III) ions with the hexacyanoferrates is shown in Table 18.3.

TABLE 18.3

	Fe^{2+}	Fe^{3+}
$K_4Fe^{II}(CN)_6$	White ppt. of $K_2Fe^{II}Fe^{II}(CN)_6$; rapid atmospheric oxidation to Prussian blue	$KFe^{III}Fe^{II}(CN)_6$ Prussian blue
$K_3Fe^{III}(CN)_6$	$KFe^{II}Fe^{III}(CN)_6$ Turnbull's blue	Dark brown solution; probably non-ionized $Fe^{III}Fe^{III}(CN)_6$

Prussian blue and Turnbull's blue appear to be identical chemically and are simply written as $KFeFe(CN)_6$. X-ray analysis work on Prussian blue indicates a unit cube or framework (Figure 18.27), in which half the cube centres are occupied by potassium ions.

In the white precipitate of $K_2Fe^{II}Fe^{II}(CN)_6$ (Everitt's salt), all the ions are iron(II) and all the cube centres are occupied. Oxidation of Prussian blue with nitric acid or chlorine gives another green pigment, Berlin green, $Fe^{III}Fe^{III}(CN)_6$, where all the corners are occupied by iron(III) ions and

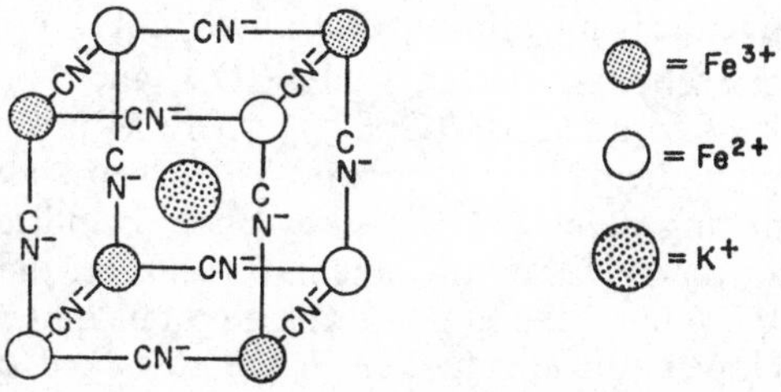

FIGURE 18.27

there are no potassium ions at the cube centres. The deep red coloration, obtained by adding potassium thiocyanate to a solution of a iron(III) salt, is used as a sensitive test for the presence of iron(III) ions. The red colour probably contains all the complexes from $FeNCS^{2+}$ to $Fe(NCS)_6{}^{3-}$.

$$Fe^{3+} + NCS^- \rightarrow FeNCS^{2+}$$
$$FeNCS^{2+} + NCS^- \rightarrow Fe(NCS)_2{}^+, \text{ etc.}$$

Sodium nitroprusside, $Na^+{}_2[Fe(CN)_5NO]^{2-}.2H_2O$, is prepared by boiling potassium hexacyanoferrate(II) with concentrated nitric acid, diluting and making alkaline with sodium hydroxide. Sodium nitroprusside forms red rhombic crystals and, in freshly prepared solution, gives an intense violet coloration with alkali sulphides.

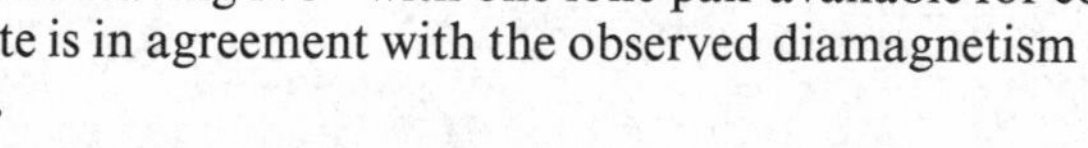

$$[Fe(CN)_5NO]^{2-} + S^{2-} \longrightarrow \left[Fe(CN)_5N^{+}\begin{matrix} \nearrow O \\ \searrow S^- \end{matrix} \right]^{4-}$$

This is used as a sensitive test for the sulphide ion in alkaline solution. The structure of the nitroprusside ion is in doubt but it is probable that the odd electron in NO (p. 252) enters the shell of the iron(III) ion giving a iron(II) ion and leaving NO^+ with one lone pair available for coordination. This postulate is in agreement with the observed diamagnetism of the nitroprusside ion.

'*Sandwich*' *organo-metallic compounds*

A remarkable series of organo-metallic compounds with 'sandwich' structures has now been synthesized. Limitations of space preclude any detailed discussion of these. One general method of preparation is to react the metallic carbonyl at 300 °C with cyclopentadiene:

```
      CH2
     /   \
   CH     CH
   \\     //
   CH --- CH
```

This molecule is acidic readily forming the anion ($C_5H_5^-$). Thus chromium, iron, and nickel carbonyls ($Cr(CO)_6$, $Fe(CO)_5$, and $Ni(CO)_4$) react to give compounds of formula $M^{2+}(C_5H_5^-)_2$ where M is Cr, Fe, or Ni.

Iron dicyclopentadienyl (or ferrocene) is a very stable diamagnetic orange solid, melting point 173 °C. X-ray analysis indicates a sandwich structure in which the C—C bond distance is 0·14 nm and the Fe—C distance is 0·204 nm. In the vapour, the rings are able to rotate quite freely but in the crystals the cyclopentadiene rings are in the staggered configuration of Figure 18.28. Bonding is thought to be by overlap of empty d_{xz} and d_{yz} orbitals in the iron atom with the delocalized π orbitals of the cyclopentadiene ring system. Ferrocene is insoluble in water but oxidizing agents such as dilute nitric acid convert ferrocene into a blue water-soluble unstable cation $Fe(C_5H_5)_2^+$. Purely ionic cyclopentadienyl derivatives have been made with sodium and magnesium, e.g.

$$C_5H_6 + Na \xrightarrow[\text{(THF)}]{\text{tetrohydrofuran}} Na^+ [C_5H_5^-] + \tfrac{1}{2}H_2$$

$$\underset{\text{magnesium ethyl bromide}}{MgC_2H_5Br} + 2C_5H_6 \longrightarrow Mg^{2+}[C_5H_5^-]_2 + HBr + C_2H_6$$

The reaction of sodium cyclopentadienide with a metal halide in THF gives good yields (~ 70 per cent) of cyclopentadienyl compounds, e.g.

$$2C_5H_5^-Na^+ + Fe^{2+}Cl_2^- \xrightarrow{\text{THF}} \underset{\text{ferrocene}}{Fe^{2+}(C_5H_5^-)_2} + 2Na^+Cl^-$$

Since sodium and magnesium ions have no *d* orbitals available for π-bonding, the ionic nature of the cyclopentadienyl compounds and their rapid hydrolysis by water can be appreciated.

$$C_5H_5^-Na^+ + H_2O \rightarrow C_5H_6 + Na^+OH^-$$

A large number of sandwich compounds have now been made with other ringed systems, e.g. benzene C_6H_6. Thus, if chromium(III) chloride, aluminium(III) chloride and aluminium powder are heated to 150 °C with benzene, a metal sandwich compound is obtained in which chromium is bonded to the benzene molecules.

$$2C_6H_6 + CrCl_3 + Al \xrightarrow[\text{cat}]{AlCl_3} (C_6H_6)Cr(C_6H_6) + AlCl_3$$

The dark brown crystals are air sensitive and dissolve in hydrochloric acid, producing the cation $Cr(C_6H_6)_2^+$.

Iron(VI) compounds—the ferrates

The ferrates are either prepared by fusing finely divided iron with alkali, followed by extraction with water, or by oxidizing iron(III) hydroxide in

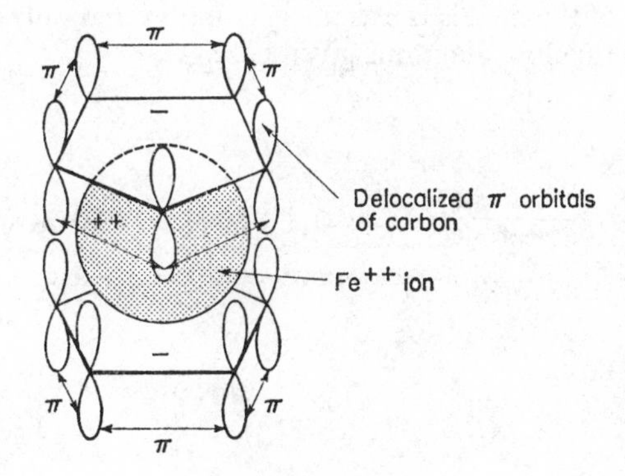

FIGURE 18.28

concentrated alkali with chlorine or bromine. The sodium and potassium salts are dark red, giving violet solutions. The addition of barium chloride to these solutions precipitates a stable red barium salt, $BaFeO_4.H_2O$. The ferrates are stronger oxidizing agents than the permanganates but are only stable in alkali. Dilution or acidification causes rapid decomposition.

$$2FeO_4^{2-} + 10H^+ \rightarrow 2Fe^{3+} + 5H_2O + 1\tfrac{1}{2}O_2$$

Cobalt and Nickel

The electronic structures of the outermost orbitals of cobalt and nickel are $3d^7 4s^2$ and $3d^8 4s^2$, respectively. Oxidation states of +2 and +3 are shown by both metals but the divalent state is the more stable. The higher oxidation states (+6, +7), known for chromium and manganese become very unstable with iron and disappear entirely with cobalt and nickel. In its +3 oxidation state, cobalt forms an extensive series of octahedral complexes but nickel in oxidation state +3 is only found in an oxide of doubtful composition and a few complexes, e.g. $Ni[(C_2H_5)_3P]_2Br_3$.

Occurrence

The main ores are cobalt glance, a complex sulphide of cobalt, arsenic, copper, iron, and nickel (Co(As, Cu, Fe, Ni), S), and pentlandite, a sulphide of nickel and iron (NiS.2FeS). Other ores include smaltite ($CoAs_2$) and garnierite, a hydrated silicate of magnesium and nickel.

Extraction

The extraction of the metals is extremely complex and only the main stages are outlined in the flow diagrams given here.

Cobalt

Co(As, Fe, Cu, Ni)S (cobalt glance) $\xrightarrow[\text{smelt with } CaCO_3 \text{ and } SiO_2]{\text{roast and}}$ $As_2O_3\uparrow + SO_2\uparrow + FeSiO_3$ (slag) + speiss containing Ni, Cu, Fe, Co arsenides

Speiss $\xrightarrow{\text{roast with NaCl and extract with } H_2O}$ solution of chlorides ($CuCl_2$, $FeCl_2$, $NiCl_2$, $CoCl_2$)

$Cu\downarrow \xleftarrow{Fe} CuCl_2$

$FeCO_3\downarrow \xleftarrow{CaCO_3} FeCl_2$

$NiCO_3Ni(OH)_2\downarrow \xleftarrow{Na_2CO_3} NiCl_2$

$Co_3O_4 \xleftarrow{\text{heat}} Co_2O_3H_2O \xleftarrow[OCl^-]{\text{bleaching powder}} CoCl_2$

$Co_3O_4 \xrightarrow{Al}$ Co metal

Electrolytic refining of double sulphates $NiSO_4.(NH_4)_2SO_4.6H_2O$ and $CoSO_4.(NH_4)_2SO_4.6H_2O$ is also employed as a method of obtaining pure metals.

Nickel

$$\underset{\text{pentlandite}}{(Ni, Fe)S} \xrightarrow[\text{smelt with } CaCO_3,\ SiO_2 \text{ and coke}]{\text{roast and}} SO_2\uparrow + FeSiO_3 + \text{matte of Ni, Fe sulphides}$$

$$\text{matte of Ni, Fe sulphides} \xrightarrow{\text{heat in blast-furnace with a basic lining to slag-off Fe as } FeSiO_3} NiO$$

$$NiO \xrightarrow{\text{water gas}} \text{Ni (impure, finely divided)}$$

$$\underset{\text{98·8 per cent}}{Ni} \xleftarrow[\text{over Ni pellets}]{160°C} Ni(CO)_4 \xleftarrow[60°C]{CO} \text{Ni (impure, finely divided)}$$

Properties of Cobalt and Nickel

Both metals are hard, white, and ferromagnetic; they have a hexagonal close-packed structure. The principal reactions of the metals are shown in Figure 18.29.

Uses

Finely divided nickel finds extensive use as a catalyst, e.g. in the conversion of vegetable oils into solid fats during the manufacture of margarine. Important alloys include Monel metal (Cu–Fe–Ni) for turbines and propellers, nickel silver (Cu–Zn–Ni), Nichrome (Cr–Fe–Ni), and 'silver' coinage (Cu–Zn–Ni). Cobalt, alloyed with chromium, tungsten, vanadium, and iron gives alloys suitable for high-speed cutting tools and magnets.

The Oxides

Green cobalt(II) oxide and nickel(II) oxide possess rock-salt structures, and may be prepared by heating the nitrates.

$$2Ni(NO_3)_2 \rightarrow 2NiO + 4NO_2 + O_2$$

They are basic oxides which dissolve slowly in dilute acids giving solutions of green nickel or pink cobalt salts. Cobalt(II) oxide ($\Delta G_f^\ominus = -213$ kJ mol^{-1}) on heating to red heat forms black tricobaltic tetroxide ($\Delta G_f^\ominus = -758$ kJ mol^{-1}) which has a spinel structure $Co^{II}Co_2^{III}O_4$.

The Hydroxides

Addition of concentrated sodium hydroxide solution to solutions of cobalt(II) and nickel(II) salts precipitates coloured hydroxides ($Co(OH)_2$ is

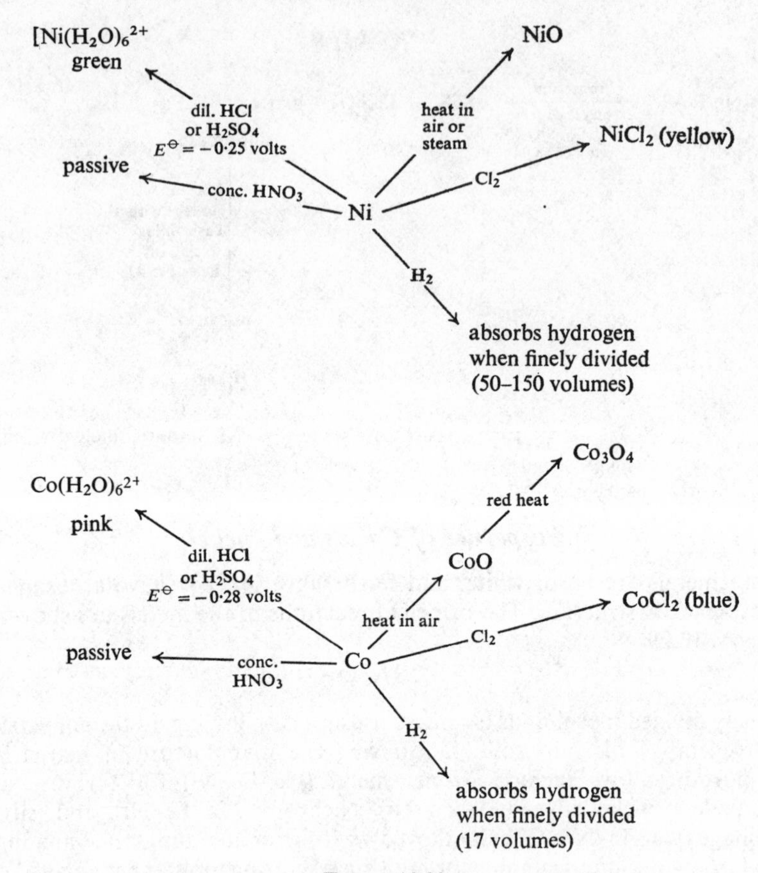

FIGURE 18.29

violet and $Ni(OH)_2$ is green). These are true hydroxides with the cadmium iodide layer lattice. Cobalt(II) hydroxide is soluble in hot concentrated alkali giving violet solutions containing the tetrahydroxo and hexahydroxo-cobaltate(II) ions $Co(OH)_4^{2-}$ and $Co(OH)_6^{4-}$. Both hydroxides dissolve in excess ammonia producing octahedral hexammine complexes.

$$Ni(OH)_2 + 6NH_3 \rightarrow Ni(NH_3)_6^{2+}(OH)^-_2$$

blue solution containing
hexammine nickel(II) hydroxide

$$Co(OH)_2 + 6NH_3 \rightarrow Co(NH_3)_6^{2+}(OH)^-_2$$

brown solution containing
hexammine cobalt (II) hydroxide

The brown cobalt(II) complex, however, rapidly undergoes atmospheric oxidation to the pink hexammine cobalt(III) complex.

$$Co(NH_3)_6^{2+} \rightarrow Co(NH_3)_6^{3+} + e$$

Literally thousands of cobaltammine complexes of this kind have been prepared.

Black cobalt(III) oxohydroxide (CoO(OH)) may be prepared by oxidation of a cobalt(II) solution with alkaline hypochlorite. The solid decomposes at 300 °C, losing water and oxygen.

$$12CoO(OH) \rightarrow 4Co_3O_4 + O_2 + 6H_2O$$

The Sulphides

The black sulphides are precipitated in alkaline or neutral solution by hydrogen sulphide. Whereas cobalt(II) sulphide is soluble in dilute hydrochloric acid, nickel(II) sulphide only dissolves in the concentrated acid.

The Halides

The action of hydrogen fluoride on anhydrous cobalt(II) chloride gives red cobalt(II) fluoride which, with excess fluorine, produces brown cobalt(III) fluoride. The latter compound, which can also be made by treating the metal with fluorine, is a good fluorinating agent and, like fluorine, liberates oxygen from water. Yellow green nickel(II) fluoride can be prepared by the action of hydrogen fluoride on the anhydrous chloride or by direct fluorination of the metal. Nickel(III) fluoride is unknown.

The other anhydrous nickel(II) and cobalt(II) halides can be prepared by direct halogenation of the metal. Hydrated halides $MCl_2.6H_2O$ (pink for cobalt and green for nickel) are best prepared by dissolving the oxides in dilute hydrochloric acid and crystallizing. Cobalt(II) chloride has the cadmium iodide and nickel(II) chloride the cadmium chloride layer lattice. Complex double halides are known, e.g. Cs_3CoCl_5 and Cs_2NiCl_4. The former contains the ions Cs^+, Cl^- and $CoCl_4^{2-}$ and the latter Cs^+ and $NiCl_4^{2-}$. Fluorination of these complex halides yields Cs_2CoF_6 and Cs_2NiF_6 which appear to contain cobalt(IV) and nickel(IV) respectively

Oxo Salts

Crystalline hydrated sulphates and nitrates are stable and well characterized. The nickel(II) oxo salts are green and the cobalt(II) oxo salts pink, e.g. $NiSO_4.7H_2O$, $CoSO_4.7H_2O$, $Ni(NO_3)_2.6H_2O$, and $Co(NO_3)_2.6H_2O$. In solution the nickel oxo salts furnish the green hydrated cation $Ni(H_2O)_6^{2+}$ and cobalt oxo salts the pink hydrated cation $Co(H_2O)_6^{2+}$. Double salts are known for both sulphates, $(NH_4)_2SO_4.MSO_4.6H_2O$. The addition of concentrated hydrochloric acid or sodium chloride to the pink cobalt(II) ion gives a blue colour due to the formation of a tetrahedral blue chloro-complex.

$$Co(H_2O)_6^{2+} + 4Cl^- \rightarrow CoCl_4^{2-} + 6H_2O$$

Insoluble carbonates (green $NiCO_3$ and pink $CoCO_3$) are precipitated by the addition of sodium bicarbonate to a solution of the metal salt. The addition of a normal carbonate precipitates the basic salts (p. 217).

Complexes

Cobalt

The cobaltammines and ethylenediamine complexes have already been mentioned (pp. 103 and 106). When excess potassium cyanide solution, followed by alcohol, is added to a solution of a cobalt(II) salt, an amethyst powder of potassium pentacyanocobaltate(II) is precipitated. The violet solid is believed to be $K_6Co_2(CN)_{10}$ containing a Co–Co metal bond, but the solution from which this solid precipitates certainly contains $Co(CN)_5^{3-}$. The cell diagrams for Co^{2+} and the complex are shown in Figure 18.30(*a*). The strong cyanide ligand forces an electron from the high-energy e_g orbitals into the lower energy t_{2g} orbitals. This leaves one vacant *d* orbital and allows for dsp^3 hybridization, giving a trigonal bipyramidal configuration. The complex is inner orbital and paramagnetic. With weaker ligands such as NCS^- and Cl^-, blue tetrahedral complexes ($Co(NCS)_4^{2-}$, $CoCl_4^{2-}$) are formed. The ligands are not strong enough to cause spin pairing and the resulting complexes are outer orbital and spin-free.

Potassium or sodium hexanitrocobaltate(III) (potassium and sodium cobaltinitrites), $K_3(Na_3)Co(NO_2)_6$, may be prepared by adding potassium nitrite or sodium nitrite and acetic acid to a solution of a cobalt(II) salt. The reactions probably proceed in the following stages:

$$Co^{2+} + 6NO_2^- \rightarrow Co^{II}(NO_2)_6^{4-}$$

(hexanitrocobaltate (II))

$$Co^{II}(NO_2)_6^{4-} + NO_2^- + 2H^+ \rightarrow Co^{III}(NO_2)_6^{3-} + NO + H_2O$$

(hexanitro-
cobaltate
(III))

Hexanitrocobaltate(III) compounds are used as yellow pigments. Sodium hexanitrocobaltate(III) is used to detect the presence of potassium in qualitative analysis. Yellow potassium hexacyanocobaltate(III) is prepared by acidification of the cobalt(II) pentacyano-complex. The majority of cobalt(III) complexes are octahedral. Ligand fields set up by NH_3, CN^-, NO_2^- cause unpaired electrons in the e_g orbitals to pair up in the lower energy t_{2g} orbitals and allow for octahedral d^2sp^3 hybridization. The resulting complexes are diamagnetic, inner orbital, and spin paired.

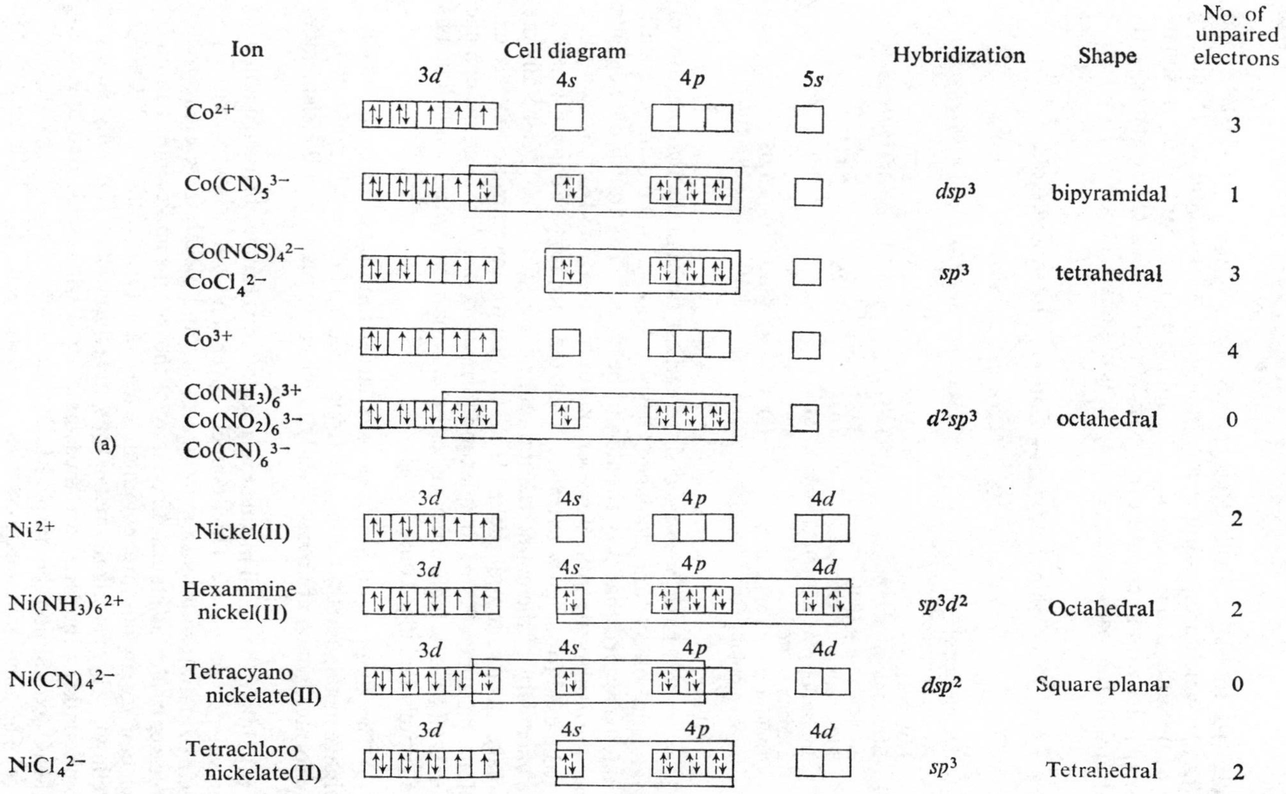

	Ion	Cell diagram 3d	4s	4p	5s	Hybridization	Shape	No. of unpaired electrons
(a)	Co^{2+}	[↑↓][↑↓][↑][↑][↑]	[ ]	[ ][ ][ ]	[ ]			3
	$Co(CN)_5{}^{3-}$	[↑↓][↑↓][↑↓][↑][↑↓]	[↑↓]	[↑↓][↑↓][↑↓]	[ ]	dsp^3	bipyramidal	1
	$Co(NCS)_4{}^{2-}$ $CoCl_4{}^{2-}$	[↑↓][↑↓][↑][↑][↑]	[↑↓]	[↑↓][↑↓][↑↓]	[ ]	sp^3	tetrahedral	3
	Co^{3+}	[↑↓][↑][↑][↑][↑]	[ ]	[ ][ ][ ]	[ ]			4
	$Co(NH_3)_6{}^{3+}$ $Co(NO_2)_6{}^{3-}$ $Co(CN)_6{}^{3-}$	[↑↓][↑↓][↑↓][↑↓][↑↓]	[↑↓]	[↑↓][↑↓][↑↓]	[ ]	d^2sp^3	octahedral	0

(b)	Ion		3d	4s	4p	4d	Hybridization	Shape	No. of unpaired electrons
	Ni^{2+}	Nickel(II)	[↑↓][↑↓][↑↓][↑][↑]	[ ]	[ ][ ][ ]	[ ][ ]			2
	$Ni(NH_3)_6{}^{2+}$	Hexammine nickel(II)	[↑↓][↑↓][↑↓][↑][↑]	[↑↓]	[↑↓][↑↓][↑↓]	[↑↓][↑↓]	sp^3d^2	Octahedral	2
	$Ni(CN)_4{}^{2-}$	Tetracyano nickelate(II)	[↑↓][↑↓][↑↓][↑↓][↑↓]	[↑↓]	[↑↓][↑↓][ ]	[ ][ ]	dsp^2	Square planar	0
	$NiCl_4{}^{2-}$	Tetrachloro nickelate(II)	[↑↓][↑↓][↑↓][↑][↑]	[↑↓]	[↑↓][↑↓][↑↓]	[ ][ ]	sp^3	Tetrahedral	2

FIGURE 18.30. Cell diagrams for (*a*) cobalt; (*b*) nickel

Nickel

Nickel(II) forms octahedral complexes, for example $Ni(NH_3)_6^{2+}$, and $Ni(en)_3^{2+}$ which are paramagnetic, indicating outer orbital sp^3d^2 binding (p. 413). With strong CN^- ligands, spin pairing occurs giving a square planar inner-orbital complex (p. 353). With the weak Cl^- ligand, a tetrahedral spin-free paramagnetic complex is obtained (Figure 18.30(*b*)).

The Coinage Metals—Copper, Silver, and Gold

	Copper	Silver	Gold
Atomic mass	63·546	107·870	196·967
Atomic number	29	47	79
Isotopes	63, 65	107, 109	197
Electronic structure	$3d^{10}4s^1$	$4d^{10}5s^1$	$4f^{14}5d^{10}6s^1$
Atomic radius (nm)	0·117	0·134	0·134
Ionic radius(nm)M^+	0·096	0·126	0·137
Ionization potential (I) (kJ mol^{-1})	751	738	897
Density (gcm^{-3})	8·92	10·50	19·3
Melting point (°C)	1,083	960·5	1,062
Boiling point (°C)	2595	2210	2970

These metals have been used since earliest times in the manufacture of ornamental objects and coinage. They are known as the coinage metals. It may be argued that they should not be classified as transition elements since the *d* subshell is full. However, one or two electrons from the *d* subshell may be used for valency purposes. Thus the chemistry of copper is mainly the chemistry of the copper(II) ion (cupric ion), $Cu^{2+}(3d^9)$. Generally the coinage metals show all the tendencies noted for the transition metals, e.g. colour, paramagnetism, complex ion formation, etc. Although the coinage metals all have a single outer *s* electron, comparisons with the alkali metals are entirely lacking. Thus:

(*a*) The alkali metal atoms possess larger radii than the corresponding coinage metals and the outer *s* electrons are much more easily removed from the former, which are consequently much more reactive in the elementary state.

(*b*) Unlike the alkali metal atoms, *d* electrons can be abstracted and used for valency purposes. Thus copper, silver, and gold show variability in their oxidation states although, like cobalt and nickel, these are restricted to lower oxidation states. Also many of the cations and complex anions of the coinage metals are coloured, e.g. $Cu(H_2O)_4^{2+}$ (blue) and $CuCl_4^{2-}$ (yellow), whereas alkali metal ions are colourless. The only coloured compounds of the latter are those associated with a coloured anion, e.g. $KMnO_4$ containing the purple MnO_4^- ion.

(*c*) The alkali metal cations with their large radii have little tendency to attract polar molecules or ions and do not form complexes. The smaller coinage-metal cations readily form many stable complexes, $CuCl_4^{2-}$, $Cu(NH_3)_4^{2+}$, $Ag(NH_3)_2^+$, $Au(CN)_2^-$, etc.

Trends in the properties of the coinage metals are not very regular; for example, the stable oxidation state of copper is +2, of silver +1, and of gold +3. Among the few regular trends are:

(*a*) The resistance of the elements to chemical attack increases from copper to gold.

(*b*) The thermal stability of the oxides and the salts decreases from copper to gold.

Copper

Occurrence

The chief ore is copper pyrites, $CuFeS_2$. Carbonate and oxide ores are of much less importance and include malachite, $CuCO_3.Cu(OH)_2$, and cuprite, Cu_2O. Some copper occurs native in Canada.

Extraction

The oxide and carbonate ores are roasted with coke and a flux.

$$CuCO_3.Cu(OH)_2 \rightarrow 2CuO + CO_2 + H_2O$$

$$CuO + C \rightarrow Cu + CO$$

The sulphide ores, however, are usually very low grade (<10 per cent of sulphide) and are associated with iron(II) sulphide, gangue, and smaller quantities of arsenic, antimony, bismuth, selenium, tellurium, silver, gold, and platinum. The sulphide ore is first concentrated by oil flotation and then roasted in a current of air in a reverberatory furnace below the fusion point, when arsenic and sulphur are driven off as volatile oxides. The temperature is then allowed to rise above the fusion point and limestone or silica is added. Iron(II) oxide slags off as iron silicate and the copper matte remaining consists of a mixture of copper(I) sulphide and iron(II) sulphide.

$$FeO + SiO_2 \rightarrow \underset{\text{slag}}{FeSiO_3}$$

The matte is transferred to a silica-lined converter through which hot compressed air is blown. The remaining iron slags off as silicate and the copper(I) sulphide is reduced to copper in two stages.

$$2Cu_2S + 3O_2 \rightarrow 2Cu_2O + 2SO_2$$

$$2Cu_2O + Cu_2S \rightarrow 6Cu + SO_2$$

The molten copper is run off into moulds and, on cooling, sulphur dioxide, nitrogen, and oxygen escape from the metal giving the surface a blistered appearance (blister copper). Refining is carried out electrolytically using anodes of impure copper and pure copper-strip cathodes. The electrolyte is acidified copper sulphate solution and the net result of the electrolysis is the transfer of pure copper from the anode to the cathode.

$$\text{At the anode} \quad Cu \rightarrow Cu^{2+} + 2e$$

$$\text{At the cathode} \quad Cu^{2+} + 2e \rightarrow Cu$$

Impurities in the blister copper deposit as an anode slime, which contains antimony, bismuth, selenium, tellurium, silver, gold, and platinum, recovery of which may pay for the cost of refining.

Chrysocolla ($CuSiO_3.2H_2O$), often bluish green in colour, is found extensively associated with quartz but until recently was not considered as a starting material for the extraction of copper. The TORCO process (Treatment Of Refractory Copper Ores) is based upon the reaction of chrysocolla with sodium chloride, carbon, and steam at 800 °C.

$$CuSiO_3.2H_2O \rightarrow CuO + SiO_2 + 2H_2O$$

$$4CuO \rightarrow 2Cu_2O + O_2$$

$$2NaCl + H_2O + SiO_2 \rightarrow 2HCl + Na_2SiO_3$$

$$Cu_2O + 2HCl \rightarrow 2CuCl(g) + H_2O$$

$$2CuCl + H_2 \rightarrow 2Cu + 2HCl$$

The hydrogen for the last stage is produced by the reaction of steam with carbon. The hydrogen chloride goes through the same cycle repeatedly. Recent improvements in furnace design have enabled the smelting, converting, and slag formation to be carried out in separate zones of a single furnace.

Uses

Copper is used for electrical wiring, roofing material, and in electrotyping. The metal forms many alloys, e.g. brass (Cu–Zn), bronze (Cu–Sn), phosphor-bronze (Cu–Sn–P), copper coinage (Cu 98 per cent plus Sn–Zn), silver coinage (Cu 75 per cent plus Ni).

Properties

Copper is a salmon-pink coloured metal with a cubic close-packed structure. It is ductile, malleable, and a good conductor of heat and electricity. On weathering, a green protective coating of basic sulphate is formed, $CuSO_4.3Cu(OH)_2$. The main chemical reactions of copper are summarized in Figure 18.31.

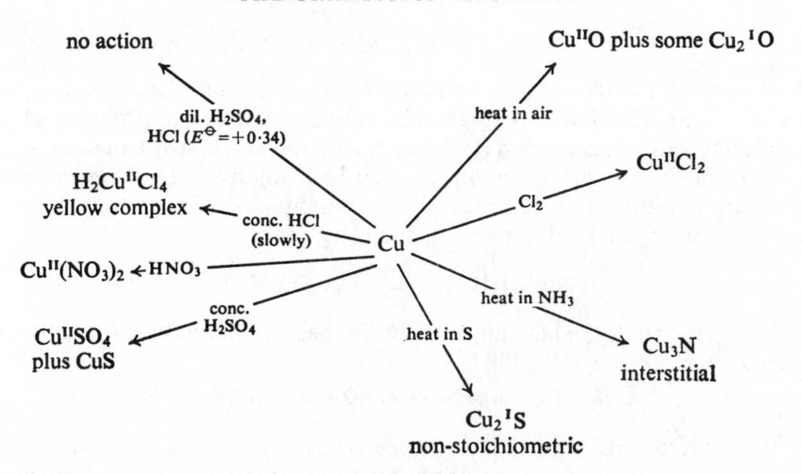

FIGURE 18.31. Principal reactions of copper metal

The main oxidation states of copper are +1 (cuprous state) and +2 (cupric state). Two points which are fundamental to the chemistry of copper compounds are:

(*a*) The covalent compounds of copper(I) are much more stable than the corresponding copper(II) compounds. Thus covalent copper(II) chloride ($CuCl_2$) and bromide ($CuBr_2$) decompose on heating to the corresponding copper(I) halide.

$$2CuCl_2 \xrightarrow{\text{red heat}} 2CuCl + Cl_2$$

$$2CuBr_2 \xrightarrow{\text{gentle heat}} 2CuBr + Br_2$$

(*b*) The copper(I) ion does not exist in aqueous solution and rapidly disproportionates into the copper(II) ion and copper.

$$2Cu^+ \rightleftharpoons Cu^{2+} + Cu \qquad (E^\ominus = +0{\cdot}37\ V)$$

$$K = [Cu^{2+}]/[Cu^+]^2 = 1{\cdot}2 \times 10^6$$

The above equilibrium is well over to the right hand side for all non-polarizable ligands, e.g. H_2O, ClO_4^-, en(ethylenediamine), NH_3,

$$2Cuen^+ \rightleftharpoons Cu(en)_2{}^{2+} + Cu$$

although with ammonia the copper(I) state is favoured slightly ($K = 10^{-2}$)

$$Cu(NH_3)_4{}^{2+} + Cu \rightleftharpoons 2Cu(NH_3)_2{}^+$$

However, the equilibrium is well over to the left hand side for polarizable (reducing) ligands, e.g. CN^-, I^-, tu(thiourea), and the copper(I) state is favoured.

$$2Cu^{2+} + 4I^- \rightleftharpoons 2CuI + I_2$$

The oxides

Red copper(I) oxide may be prepared by boiling copper(I) chloride with caustic soda solution (no hydroxide exists), or by the reduction of a copper(II) salt. Thus copper(II) sulphate solution added to alkaline sodium potassium tartrate gives a complex deep blue solution of copper tartrate (Fehling's solution). This unstable solution can be reduced to copper(I) oxide by an organic reducing agent such as glucose.

$$2Cu^{2+} + 2OH^- + 2e \rightarrow Cu_2O + H_2O$$

Black copper(II) oxide can be made by heating the basic carbonate or nitrate.

$$2CuCO_3.Cu(OH)_2 \rightarrow 4CuO + 2CO_2 + 2H_2O$$

Both oxides are basic but the copper(I) salts produced from copper(I) oxide and acid rapidly disproportionate to copper(II) salts and copper.

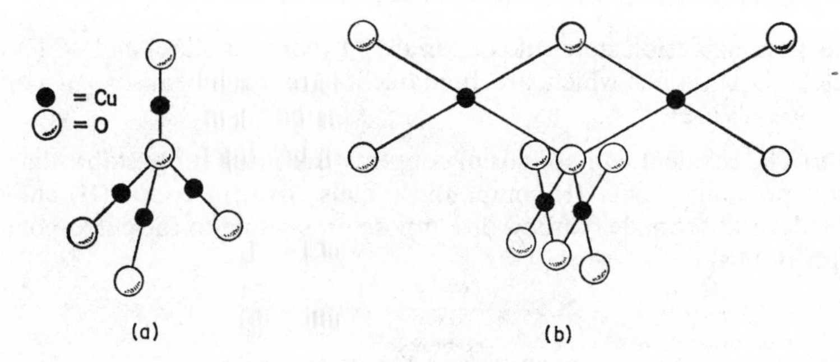

FIGURE 18.32

Both oxides have macromolecular covalent structures. Copper(I) oxide consists of two independent interpenetrating cubic frameworks of atoms. In these frameworks, each metal atom has two close oxygen neighbours and each oxygen atom is surrounded tetrahedrally by four copper atoms (Figure 18.32(*a*)). In copper(II) oxide there are four coplanar bonds round each copper atom and four tetrahedral bonds round each oxygen atom, giving 4-coordination (Figure 18.32(*b*)). The oxides are readily reduced to the metal by hydrogen or carbon monoxide.

Hydroxides

No copper(I) hydroxide has been isolated. The action of alkali on a solution of a copper(II) salt gives a blue gelatinous precipitate of copper(II) hydroxide.

$$Cu^{2+} + 2OH^- \rightarrow Cu(OH)_2$$

The hydroxide readily decomposes on heating to black copper(II) oxide. The hydroxide has slightly amphoteric properties, dissolving in acids to give copper(II) salts and in very strong alkali to give cuprates.

$$Cu(OH)_2 + 2OH^- \rightarrow Cu(OH)_4^{2-}$$

square planar
tetrahydroxocuprate (II)
ion

The hydroxide is also readily soluble in ammonia giving a deep blue solution of the tetrammine copper(II) ion (cuprammonium ion), $Cu(NH_3)_4^{2+}$.

$$Cu(OH)_2 + 4NH_3 \rightarrow Cu(NH_3)_4^{2+} + 2OH^-$$

square planar
tetrammine
copper (II)

If excess ammonia is added to copper(II) sulphate solution and a layer of alcohol poured on to the top of the deep blue complex solution, deep blue crystals of the complex sulphate, tetrammine copper(II) sulphate, $Cu(NH_3)_4SO_4 . H_2O$, separate.

The halides

Copper(II) fluoride is a white solid, prepared by the action of fluorine on copper or anhydrous hydrogen fluoride on copper(II) oxide. The compound is water-soluble and loses fluorine at 950 °C,

$$2CuF_2 \rightarrow 2CuF + F_2$$

The residue, copper(I) fluoride, is a red solid, stable in air but rapidly hydrolysed by water.

Copper(I) chloride may be prepared by boiling copper(II) chloride with copper and concentrated hydrochloric acid. The resulting solution probably contains complexes such as the $CuCl_2^-$ ion and, on pouring into a large volume of distilled water, a white precipitate of copper(I) chloride is thrown down.

$$Cu + CuCl_2 \rightarrow 2CuCl$$

The chloride turns green on exposure to the air and dissolves in ammonia giving a colourless solution containing the linear complex ion $[NH_3—Cu—NH_3]^+$, the solution containing this complex ion gradually turns blue on exposure to the air

$$Cu(NH_3)_2^+ + 2NH_3 \rightarrow Cu(NH_3)_4^{2+} + e$$

Ammoniacal copper(I) chloride dissolves carbon monoxide, and the dimeric complex shown below separates as colourless crystals.

$$\begin{array}{c} CO \searrow \quad \nearrow Cl \searrow \quad \swarrow CO \\ \quad Cu \quad\quad Cu \quad \\ H_2O \nearrow \quad \nwarrow Cl \nearrow \quad \nwarrow H_2O \end{array}$$

With acetylene, ammoniacal copper(I) chloride gives a red, insoluble polymeric acetylide of empirical formula $Cu—C{\equiv}C—Cu.H_2O$.

The bromide is similar to the chloride but the iodide is best prepared by adding potassium iodide solution to a solution of a copper(II) salt (p. 417). This reaction is used in the volumetric estimation of copper, the liberated iodine being titrated with sodium thiosulphate solution.

The anhydrous copper(II) halides may be made from the elements and crystalline hydrates from the oxide and the appropriate halogen acid. Copper(II) chloride, for example, crystallizes from solution as a green dihydrate $CuCl_2.2H_2O$ with a planar configuration.

```
        Cl
        |
H2O—Cu—H2O
        |
        Cl
```

The anhydrous salt is brown and possesses a chain structure.

```
  Cl    Cl    Cl    Cl
>Cu<  >Cu<  >Cu<  >Cu<
  Cl    Cl    Cl    Cl
```

In concentrated hydrochloric acid, the anhydrous salt gives a dark brown solution in which most of the chains are probably still intact. On dilution with water, the chains break forming yellow complex ions such as $CuCl_4^{2-}$ and $CuCl_3^-$. Further dilution causes the formation of blue hydrated copper(II) ions $Cu(H_2O)_4^{2+}$, producing a green colour which ultimately turns blue.

$$\underbrace{\underset{\text{yellow}}{CuCl_4^{2-}} + 4H_2O \rightleftharpoons \underset{\text{blue}}{Cu(H_2O)_4^{2+}}}_{\text{green}} + 4Cl^-$$

The stability of the halides to heat decreases with increasing atomic size of the halogen. Thus copper(II) iodide has never been isolated.

The cyanides

The addition of potassium cyanide solution to a solution of a copper(II) salt precipitates the white copper(I) cyanide, CuCN.

$$2Cu^{2+} + 4CN^- \rightarrow 2CuCN\downarrow + (CN)_2$$

Copper(I) cyanide is soluble in excess potassium cyanide producing a solution containing the complex tetrahedral tetracyanocuprate(I) anion.

$$CuCN + 3CN^- \rightarrow Cu^I(CN)_4^{3-}$$

Oxo salts

Copper(I) oxo salts have no aqueous chemistry, as disproportionation takes place to copper(II) salts and metallic copper (p. 417,. Oxo salts such as copper(I) sulphate, Cu_2SO_4, are known but are prepared in non-aqueous media, e.g.

$$\underset{\text{dimethyl sulphate}}{Cu_2O + (CH_3)_2SO_4} \rightarrow \underset{\text{dimethyl ether}}{Cu_2SO_4 + (CH_3)_2O}$$

The Cu^+ ion can be stabilized with a variety of coordinating agents, e.g. thiourea in $[Cu(tu)_3]_2SO_4.2H_2O$, where tu = thiourea. Coordination occurs through the sulphur atoms, and the solid consists of chains of $Cutu_4$ tetrahedra sharing one corner, with sulphate ions and water molecules between the chains. The complex is a colourless salt, soluble and stable in aqueous solutions which are very slightly acidic.

The most common oxo salt of copper(II) is blue vitriol, $CuSO_4.5H_2O$, prepared industrially by blowing a current of air through scrap copper and dilute sulphuric acid.

$$2Cu + 2H_2SO_4 + O_2 \rightarrow 2CuSO_4 + 2H_2O$$

The crude copper(II) sulphate solution obtained contains iron(II) sulphate as an impurity and, on crystallization, mixed crystals of $Cu(Fe)SO_4.5H_2O$ would form. Dilute nitric acid is added to oxidize the iron(II) to iron(III) sulphate which remains in the mother liquor after crystallization. Modern X-ray crystallographic work has shown that in the pentahydrate four water molecules are coordinated to the central copper cation at the centre of a square, but the fifth water molecule is held by hydrogen bonds between a sulphate ion and a coordinated water molecule. On heating, the pentahydrate decomposes in the following stages:

$$CuSO_4.5H_2O \xrightarrow[\text{desiccator}]{30\,^\circ C} CuSO_4.3H_2O \xrightarrow{100\,^\circ C} CuSO_4.H_2O \xrightarrow{300\,^\circ C} CuSO_4 \xrightarrow{350\,^\circ C} CuO + SO_3$$

The fifth hydrogen-bonded water molecule is difficult to remove because it is deeply embedded in the crystal lattice.

Double salts of copper(II) sulphate are also known, e.g. green copper ammonium sulphate $CuSO_4.(NH_4)_2SO_4.6H_2O$. Copper(II) sulphate is used in copper plating, electrotyping, and as a germicide and fungicide (Bordeaux mixture is $CuSO_4$ and $Ca(OH)_2$).

Blue copper nitrate, $Cu(NO_3)_2.3H_2O$, can be prepared by normal methods but the salt is extremely deliquescent. Blue anhydrous copper(II) nitrate can be prepared by the dissolution of copper in N_2O_4 and ethyl acetate. The double salt $Cu(NO_3)_2.N_2O_4$ can be crystallized out from this solution and the N_2O_4 sublimed off at 85 °C, leaving anhydrous copper(II) nitrate. Solid anhydrous copper(II) nitrate has a complex structure consisting of Cu^{2+} ions linked together by planar nitrate ions.

Solutions of all copper(II) salts furnish the blue hydrated cation

$Cu(H_2O)_6^{2+}$. It must be noted, however, that two of the water molecules are further away from the cation than the others. Since the addition of aqueous ammonia or chloride ion will displace four of these water molecules, forming, for example, $Cu(NH_3)_4(H_2O)_2^{2+}$, $CuCl_4.2H_2O^{2-}$, it is usually quite acceptable to represent the hydrated copper(II) cation as $Cu(H_2O)_4^{2+}$. This cation undergoes hydrolysis, functioning as a Lowry and Brönsted acid.

$$Cu(H_2O)_4^{2+} + H_2O \rightleftharpoons [Cu(H_2O)_3OH]^+ + H_3O^+$$

A normal copper(II) carbonate seems to be unknown, as both sodium carbonate and sodium bicarbonate precipitate a green basic salt, $2CuCO_3.Cu(OH)_2$, from copper(II) solutions. Copper(II) ions, with potassium hexacyanoferrate(II), give a brown precipitate of $Cu^{II}Cu^{II}Fe^{II}(CN)_6$, used as a sensitive test for the presence of copper.

Silver and Gold

Occurrence

Silver and gold occur native, often alloyed with each other and with copper and platinum metals. An important ore of silver is argentite, Ag_2S, but gold is only found combined in a few minerals such as calaverite, $AuTe_2$.

Extraction

(*a*) Both metals can be extracted by the cyanide process, which involves leaching the crude metal or ore with sodium cyanide solution through which air is blown. The silver and gold dissolve as complex cyanides and the pure metals are precipitated by the addition of zinc,

$$4M + 8CN^- + O_2 + 2H_2O = 4M(CN)_2^- + 4OH^-$$

$$2M(CN)_2^- + Zn \rightarrow 2M\downarrow + Zn(CN)_4^{2-} \qquad (M = Ag \text{ or } Au)$$

Both metals are refined electrolytically, using an electrolyte of silver nitrate for silver and chloroauric acid, $HAuCl_4$, for gold, with cathodes of pure, and anodes of impure, metal.

(*b*) The Parkes process for silver (desilverization of lead). Crude lead, manufactured from galena, contains small amounts of silver. The silver is concentrated by the Parkes process, which is based on the almost mutual insolubility of molten zinc and lead (Figure 18.33).

In a molten mixture of these metals, the silver concentrates in the upper layer so that:

$$\frac{\text{concentration of silver in upper layer}}{\text{concentration of silver in lower layer}} = \frac{300}{1}$$

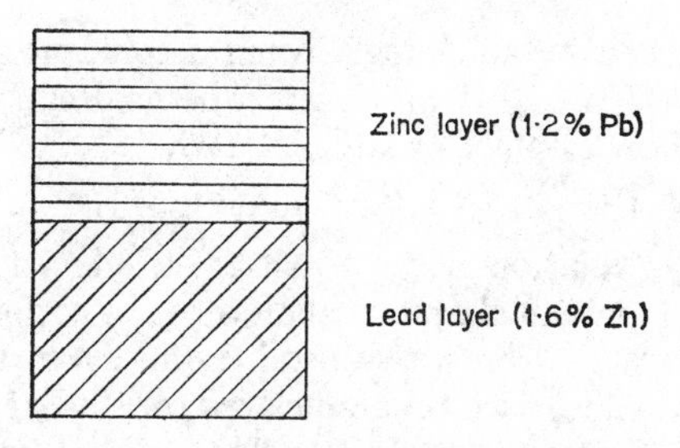

FIGURE 18.33

The silver–zinc–lead alloy floats to the surface, solidifies on cooling, and is skimmed off. About 1 per cent of zinc is added for each extraction and the process repeated until the lead remaining contains only 0·0005 per cent silver. The zinc is then distilled off in retorts and the silver–lead alloy is refined by cupellation, a process which consists of heating the alloy to a high temperature in a shallow crucible in a current of air. The lead is oxidized to lead monoxide, which is blown away leaving behind the silver metal. Any gold or copper present dissolves in the zinc and the silver produced will be alloyed with these metals. On electrolytic refining, however, the copper passes into the electrolyte and the gold forms a valuable anode slime.

Uses

Silver and gold are extensively used in coinage and jewellery. Gold metal, however, is too soft to be used alone and is alloyed with other metals; white gold, for example, is an alloy of gold, palladium, nickel, and zinc. The purity of these alloys is expressed in carats. An 18-carat gold alloy contains $\frac{18}{24}$ of gold by weight. A 24-carat gold consists of the pure metal. Silver salts are also used in photography and in silver plating in which the electrolyte used is potassium dicyanoargentate(I), which gives an even coherent film of silver.

Properties

Silver is a white and gold a yellow metal. Both possess face-centred cubic lattices and are very malleable and ductile. Silver and gold are unreactive metals and remain unaffected by heating in air or oxygen. Both, however, combine with chlorine on heating, producing silver(I) chloride and gold(III) chloride, respectively. Silver, but not gold, also combines with sulphur on heating to give silver(I) sulphide. Neither metal dissolves in dilute mineral acids. Silver, but not gold, is attacked by concentrated sulphuric acid and 50 per cent nitric acid.

$$3Ag + 4H^+ + NO_3^- \rightarrow 3Ag^+ + 2H_2O + NO$$

$$2Ag + 4H^+ + SO_4^{2-} \rightarrow 2Ag^+ + 2H_2O + SO_2$$

Gold is only attacked by aqua regia when a solution of hydrogen tetrachloroaurate(III) ($HAuCl_4$) is produced. The acid can be obtained from the solution as yellow crystals, $HAuCl_4.4H_2O$.

$$Au + 4H^+ + NO_3^- + 4Cl^- \rightarrow AuCl_4^- + NO + 2H_2O$$

The important oxidation state of silver is +1, and compounds are considerably more ionic in character than those of copper. Compounds of gold in oxidation state +1 are very unstable and the stable oxidation state of gold is +3. Oxo salts are well characterized for silver(I) but are unknown for gold(I) and uncommon for gold(III). The gold(III) ion is unknown and the oxo salts isolated are the acid salts, e.g. $AuOHSO_4$, $HAu(NO_3)_4$.

Oxidation state +1

Oxides

Black silver(I) oxide is precipitated by the action of caustic alkali solution on a solution of a silver salt. The oxide completely dissociates at 300 °C.

$$2Ag_2O \rightarrow 4Ag + O_2$$

The oxide possesses a copper(I) oxide lattice. The existence of a gold(I) oxide seems doubtful, as caustic soda solution on a solution of a gold(I) compound produces a mixture of metallic gold and gold(III) oxide (auric oxide Au_2O_3).

Sulphides

Black silver(I) sulphide may be precipitated by saturating a solution of a silver(I) salt with hydrogen sulphide. Brown gold(I) sulphide is obtained by saturating potassium bicyanoaurate(I), $KAu(CN)_2$, with hydrogen sulphide and adding concentrated hydrochloric acid. Both sulphides have extremely small solubilities.

Halides

All the halides of silver(I) are known. The fluoride is soluble whereas the chloride, bromide, and iodide are insoluble in water. Silver(I) fluoride (yellow) may be prepared by the action of hydrofluoric acid on silver(I) oxide. The white chloride and the yellow bromide and iodide are most conveniently made by double decomposition.

$$Ag^+ + X^- \rightarrow AgX$$

The fluoride, chloride, and bromide have a rock-salt structure but the iodide is trimorphic and can crystallize with wurtzite, blende, and cubic lattices. The chloride and bromide are soluble in ammonia (the chloride very readily) to give a solution containing the linear complex ion $Ag(NH_3)_2^+$. All the halides dissolve in thiosulphate and cyanide solutions giving bithiosulphato- and bicyano-complexes of silver(I).

$$\left[\begin{array}{c} O{=}S(O)(S)\text{—}Ag\text{—}(O)(S)S{=}O \end{array}\right]^{3-}$$

bithiosulphatoargentate(I) ion

$$[N{\equiv}C\text{—}Ag\text{—}C{\equiv}N]^-$$

bicyanoargentate(I) ion

No fluoride of gold(I) has been isolated and the other monohalides are best prepared by heating the corresponding gold(III) halide, e.g.

$$AuCl_3 \rightarrow \underset{\text{yellow}}{AuCl} + Cl_2$$

In water, the gold(I) halides rapidly disproportionate to gold and the gold(III) halide.

$$3AuCl \rightarrow AuCl_3 + Au$$

Cyanides

A white precipitate of silver cyanide is produced by adding potassium cyanide solution to a solution of a silver(I) salt. The solid has a chain structure.

$$\text{—}Ag\text{—}C{\equiv}N\text{—}Ag\text{—}C{\equiv}N\text{—}$$

With excess potassium cyanide the chains break and the solid dissolves giving a clear solution containing the linear bicyanoargentate(I) ion $Ag(CN)_2^-$.

The addition of aqueous cyanide solution to a gold(III) solution gives a yellow precipitate of gold(I) cyanide which possesses a similar chain structure and gives a similar complex ion $Au(CN)_2^-$ with excess potassium cyanide.

Oxo Salts

Gold(I) forms no oxo salts. Soluble silver nitrate and silver sulphate may be prepared by the action of the appropriate concentrated acid on the metal. In solution, silver nitrate furnishes the colourless silver(I) ion Ag^+, solutions of which are rather susceptible to decomposition by light. Both the nitrate and sulphate decompose on heating, the former readily.

$$AgNO_3 \xrightarrow{450\,°C} AgNO_2 + O_2$$

$$AgNO_2 \rightarrow Ag + NO_2$$

$$Ag_2SO_4 \xrightarrow{920\,°C} 2Ag + SO_2 + O_2$$

Oxidation state +2

Some silver(II) (argentic) compounds are known but they are relatively unimportant. A black unstable silver(II) oxide formed at the anode during the electrolysis of silver nitrate is diamagnetic and is probably $Ag^{I}Ag^{II}O_2$. Paramagnetic silver(II) fluoride can be prepared from the elements at 300 °C. It is a powerful oxidizing and fluorinating agent.

Oxidation state +3

This is confined to gold(III) (auric) compounds.

The oxides

The addition of alkali to a solution containing tetrachloroaurate(III) anions ($AuCl_4^-$) produces a red-brown precipitate of gold(III) hydroxide which, on dehydration, yields brown gold(III) oxide, Au_2O_3. Both the oxide and the hydroxide are soluble in alkali, giving solutions containing the aurate anion $Au(OH)_4^-$ which is square planar. Yellow needles of a monohydrate ($KAu(OH)_4 . H_2O$) have been isolated.

The halides

Red gold(III) chloride and black gold(III) bromide can be formed from the elements at 200 °C. These halides are dimeric in the vapour state with planar bridged configurations.

$$\begin{matrix} Cl \diagdown & & \swarrow Cl \diagdown & & \diagup Cl \\ & Au & & Au & \\ Cl \diagup & & \diagdown Cl \nearrow & & \diagdown Cl \end{matrix}$$

The halides are soluble in water undergoing some hydrolysis. They decompose on heating to the gold(I) halide and free halogen.

$$AuCl_3 \rightarrow AuCl + Cl_2$$

With excess halogen acid, solutions containing the complex tetrahaloaurate(III) anions $AuCl_4^-$ and $AuBr_4^-$ are formed, from which the free acids and salts such as $HAuCl_4 . 4H_2O$, $HAuBr_4$, and $KAuCl_4$ have been isolated. The addition of tin(II) chloride to a solution of gold(III) chloride gives a deep-purple gold sol (purple of Cassius), used in ruby glass and as a sensitive test for gold.

19

The Post-transition Elements

Physical Properties

	Zinc Zn	Cadmium Cd	Mercury Hg
Atomic mass	65·37	112·40	200·59
Atomic number	30	48	80
Isotopes	64, 66, 68	114, 112, 111, 110, 113, 106	202, 200, 199, 201, 198
Electronic structure	$3d^{10}4s^2$	$4d^{10}5s^2$	$4f^{14}5d^{10}6s^2$
Atomic radius (nm)	0·125	0·141	0·144
Ionic radius (nm) M^{2+}	0·074	0·097	0·110
Ionization potential (kJ mol^{-1}) (I)	914	872	1016
(II)	1736	1636	1816
(III)	3834	3626	3306
Electronegativity	1·6	1·6	1·6
Electrode potential (V) $M_{(aq)}^{2+} + 2e \rightarrow M(s)$	−0·76	−0·40	+0·80
Density (g cm^{-3})	7·1	8·6	13·6
Melting point (°C)	419	321	−39
Boiling point (°C)	910	778	357

General Characteristics

All the atoms have the configuration $(n-1)d^{10}ns^2$ in their outermost orbitals. The *d* subshell is full and the high values of the third ionization potentials indicate the difficulty of removing electrons from this subshell. Thus zinc, cadmium, and mercury cannot utilize *d* electrons for valency purposes and the common oxidation state is +2. Mercury is anomalous in showing an oxidation state of +1 (mercurous state) but compounds of mercury(I) are unique in involving a metal–metal bond (p. 433). It is thus obvious that zinc, cadmium, and mercury are not classified as transition elements.

The non-availability of *d* orbitals for metallic bonding, together with the very distorted metallic structures of zinc, cadmium, and mercury, account to some extent for their low melting points and boiling points compared with the transition metals. The very low values for mercury indicate incomplete delocalization of the two 6*s* electrons which are available for metallic bonding. The first and second ionization potentials for mercury, which are higher than the values for zinc and cadmium, support this view.

Although the metals do form the divalent cations M^{2+} in certain salts, an examination of $E^{\ominus}$ values indicates a decreasing tendency to form divalent cations with increasing relative atomic mass. Thus, although the nitrates, sulphates, and carbonates are predominantly ionic, there is considerable covalent character in the halides, oxides, and sulphides, which are poor electrolytes in aqueous solution.

The ions, when formed, are small and doubly charged, giving rise to a strong field at the periphery of the ion which readily attracts polar molecules and ions to form complexes. Coordination numbers of four and two are common and, since the *d* electrons are not available for bonding, tetrahedral and linear complexes are produced involving sp^3 and sp hybridization respectively, e.g.

$$Zn(NH_3)_4^{2+};\ Zn(CN)_4^{2-};\ CdI_4^{2-}\ \text{(tetrahedral)}$$

$$Hg(NH_3)_2^{2+}\ \text{(linear)}$$

$E^{\ominus}$ values indicate that zinc and cadmium dissolve in dilute non-oxidizing acids but mercury is unattacked.

Zinc and cadmium show many common features in their chemistry and these elements will be discussed together. Mercury, however, is an element with many anomalies and is dealt with separately.

Zinc and Cadmium

Occurrence

The main ore of zinc is the sulphide, zinc blende ZnS. It is also found as the carbonate, calamine $ZnCO_3$. Although cadmium minerals are known they are extremely rare, e.g. CdS, greenockite. The most important source of cadmium is zinc blende which may contain cadmium sulphide in concentrations up to 5 per cent.

Extraction

The retort method

The ore is concentrated by oil flotation and then roasted on a sintering machine to give an oxide sinter. The sulphur dioxide produced is purified and used to manufacture sulphuric acid.

$$ZnCO_3 \rightarrow ZnO + CO_2$$

$$2ZnS + 3O_2 \rightarrow 2ZnO + 2SO_2$$

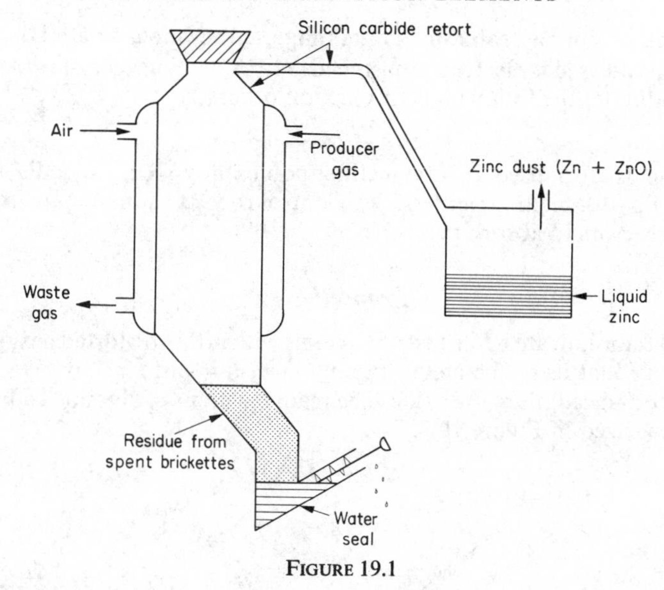

FIGURE 19.1

The oxide is then made into brickettes with coke and clay and heated by producer gas in vertical retorts at 1400 °C. Zinc, boiling point 910 °C, distils off and is collected as shown in Figure 19.1.

$$ZnO + C \rightarrow Zn + CO$$

To obtain the pure metal, fractional distillation is now employed using silicon carbide columns. The least volatile, lead, impurity collects in the first column, the most volatile, cadmium, in the last.

The electrolytic process

Here the blende is roasted in air below 700 °C to a mixture of the oxide and sulphate.

$$ZnS + 2O_2 \rightarrow ZnSO_4$$

This is leached with dilute sulphuric acid, when insoluble lead sulphate is precipitated. Milk of lime is added to precipitate iron and aluminium as their hydroxides. Zinc dust is then stirred in to precipitate cadmium. The solution is filtered, acidified, and electrolysed using a lead anode and an aluminium cathode.

Uses

Zinc is used to protect iron against rusting (galvanizing), in the outer casings of dry batteries, and in certain alloys, e.g. brass (Cu–Zn). Electrodeposition from a solution containing tetracyanocadmate(II) ions, $Cd(CN)_4^{2-}$, is used in electroplating metals as a protection against corrosion. At the cathode, the following half reaction occurs:

$$Cd(CN)_4^{2-} + 2e \rightarrow Cd + 4CN^-$$

Cadmium is also used in low-melting-point alloys, e.g. Wood's metal (Cd–Sn–Pb–Bi), melting point 65 °C. Control rods of cadmium are used in nuclear reactors to absorb neutrons.

Properties

Zinc and cadmium are white lustrous metals with rather distorted hexagonal close-packed lattices. The metals rapidly tarnish in air, zinc giving a basic carbonate and cadmium the oxide. The main reactions of zinc and cadmium are summarized in Figure 19.2.

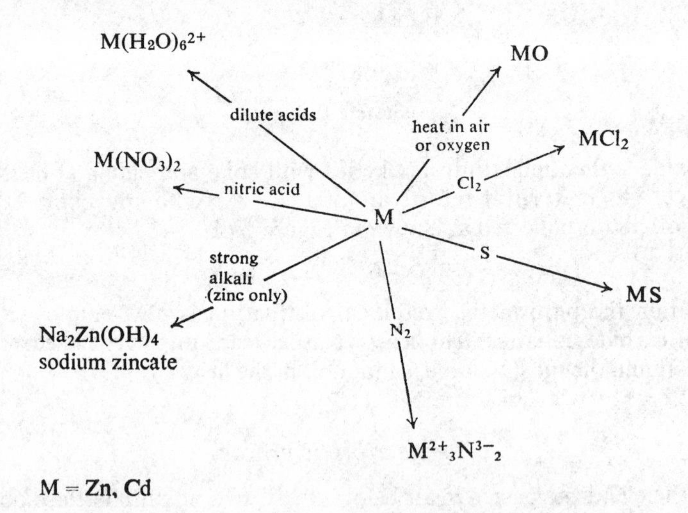

FIGURE 19.2. Reactions of zinc and cadmium metals

The Oxides

These may be prepared by heating the carbonates or nitrates. Zinc oxide has a wurtzite lattice and is amphoteric, yielding zincate solutions, $Zn(OH)_4^{2-}$, with alkalis. Cadmium oxide has a rock-salt lattice and is basic.

The Hydroxides

White insoluble hydroxides are precipitated by double decomposition.

$$Zn^{2+} + 2OH^- \rightarrow Zn(OH)_2$$

Zinc hydroxide is amphoteric.

$$Zn(OH)_4^{2-} + 2H^+ \underset{}{\overset{2H_2O}{\rightleftharpoons}} Zn(OH)_2 \rightleftharpoons Zn^{2+} + 2OH^-$$

Cadmium hydroxide does not dissolve in alkali. Both hydroxides dissolve in ammonia to give tetrahedral complexes $[M(NH_3)_4]^{2+}(OH^-)_2$.

The Sulphides

White zinc sulphide and yellow cadmium sulphide can be prepared from the elements or by passing hydrogen sulphide through a metal-salt solution. Zinc sulphide is soluble in dilute acids and thus the solution must be neutral or alkaline for precipitation to occur. Zinc sulphide and barium sulphate as used as a basis of the white paint pigment, lithopone.

The Halides

All the halides are known for zinc and cadmium. They are prepared by dissolving the oxide or carbonate in the appropriate halogen acid. The halides commonly encountered are the chlorides, $ZnCl_2.H_2O$ (deliquescent) and $CdCl_2.4H_2O$. Cadmium chloride possess a layer structure (p. 97) but zinc chloride is trimorphic, two forms having a three-dimensional structure of $ZnCl_4$ tetrahedra and one form the mercury(II) iodide structure (p. 436). Both chlorides are poor electrolytes in aqueous solution, indicating considerable covalent character. In concentrated chloride solutions, complex ions appear to be formed.

$$CdCl_2 + 2Cl^- \rightleftharpoons CdCl_4^{2-}$$

On heating the hydrates, some hydrolysis occurs to hydroxochlorides.

$$ZnCl_2 + H_2O \rightleftharpoons Zn(OH)Cl + HCl$$

Zinc chloride combines with ammonium chloride to give the complex salt $NH_4^+ZnCl_3^-$. Zinc chloride dissolves cellulose and is used in the production of hardboard.

The Oxo Salts

The following soluble nitrates and sulphates are known and are prepared by normal methods: $ZnSO_4.7H_2O$, $Zn(NO_3)_2.6H_2O$, $CdSO_4$ (variable water), and $Cd(NO_3)_2.4H_2O$. The zinc ion is hydrated with six water molecules, $Zn(H_2O)_6^{2+}$, and solutions undergo hydrolysis.

$$Zn(H_2O)_6^{2+} + H_2O \rightleftharpoons Zn(OH)(H_2O)_5^+ + H_3O^+$$

Both zinc and cadmium form double salts with potassium and ammonium sulphates of the type $M_2SO_4.YSO_4.6H_2O$, where $M = K$ or NH_4 and $Y = Zn$ or Cd.

The carbonates are insoluble and are prepared by adding sodium bicarbonate solution to a solution of a zinc or cadmium salts.

$$Zn^{2+} + 2HCO_3^- \rightarrow ZnCO_3\downarrow + H_2O + CO_2$$

Normal carbonates precipitate a basic salt.

Mercury

Occurrence

Mercury is sometimes found native in rocks and as a sliver in gold amalgam. The most important commercial ore is red cinnabar, HgS, found in the U.S.A., U.S.S.R., and Spain.

Extraction

The sulphide ore is roasted in air above 500 °C or mixed with quicklime and heated in absence of air.

$$HgS + O_2 \rightarrow Hg + SO_2$$

$$4HgS + 4CaO \rightarrow 4Hg + CaSO_4 + 3CaS$$

Mercury vapour distils over from the furnace and is condensed in water-cooled earthenware pipes. Commercial mercury contains a scum of oxidized metals (zinc, cadmium, lead, and copper) which cause the mercury to tail. The scum is removed by squeezing through a chamois leather or passing through a funnel holding a filter paper containing a small pin hole at the apex. Dissolved metals are removed by allowing the mercury to flow in a stream of fine droplets through a solution of nitric acid in a burette. Some of the mercury dissolves, forming mercurous nitrate (mercury(I) nitrate, $Hg_2(NO_3)_2$), and the metallic impurities displace mercury from the nitrate and go into solution.

$$Zn + Hg_2^{2+} \rightarrow Zn^{2+} + 2Hg\downarrow$$

Further purification is carried out by distillation under reduced pressure.

Uses

Mercury dissolves many metals (excluding iron) to form amalgams. Sodium amalgam (Na/Hg) is used as a reducing agent.

$$2Na/Hg + 2H_2O \rightarrow 2NaOH + 2Hg + H_2$$

A plastic amalgam of tin is used for coating mirrors and a mercury–silver–tin alloy for filling teeth. Other uses of mercury include the production of mercurial drugs, detonators (e.g. mercury fulminate $Hg(OCN)_2$), its application in barometers and vacuum pumps, and the use of the vapour in fluorescent lamps.

Properties

Mercury is a liquid metal of high density giving a monatomic vapour. Some of the main chemical reactions of the element are summarized in Figure 19.3.

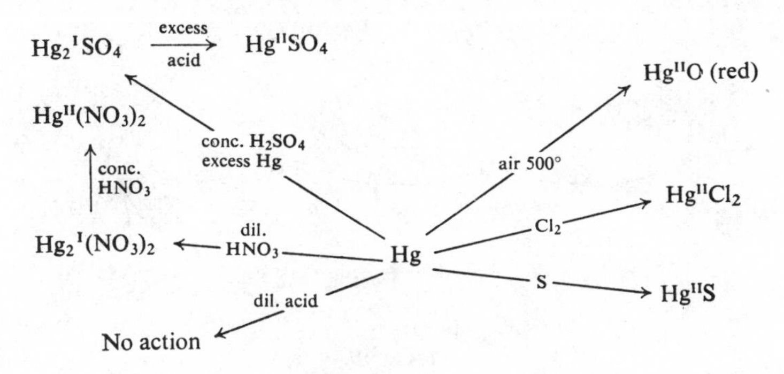

FIGURE 19.3. Reactions of mercury metal

Mercury is anomalous in forming the unique mercury(I) ion $Hg_2{}^{2+}$ or $(Hg—Hg)^{2+}$, the only true ionic species containing a metal–metal bond. In this ion mercury exhibits an oxidation state of +1. Evidence which confirms the existence of this ion is summarized below:

(*a*) Magnetic studies show that the ion is diamagnetic, indicating the absence of unpaired electrons. The simple Hg^+ ion ($5d^{10}6s^1$) would contain an unpaired electron and the ion would be paramagnetic. The ion $(Hg—Hg)^{2+}$ would possess no unpaired electrons and be diamagnetic.

(*b*) If excess mercury is added to mercury(II) nitrate solution, an equilibrium mixture of mercury(I) and (II) salts is obtained. If the mercury(I) ion was Hg^+, the following equilibrium would exist.

$$Hg + Hg^{2+} \rightleftharpoons 2Hg^+ \quad \text{where} \quad K_1 = [Hg^+]^2/[Hg^{2+}]$$

since [Hg] is constant.

If, however, the mercury(I) species were $Hg_2{}^{2+}$ the following equilibrium would be valid:

$$Hg + Hg^{2+} \rightleftharpoons Hg_2{}^{2+} \quad \text{where} \quad K_2 = [Hg_2{}^{2+}]/[Hg^{2+}]$$

On analysis of various equilibrium mixtures, it was found that K_1 varied but K_2 remained constant. Thus the correct formula for the mercury(I) ion is $Hg_2{}^{2+}$.

(*c*) X-ray analysis of covalent mercury (I) halides, Hg_2X_2, also shows the presence of Hg—Hg bonds. In these compounds, one 6*s* electron in each mercury atom is promoted to a 6*p* orbital. *sp*-Hybridization occurs and one of the hybrid orbitals from each mercury atom overlap forming a σ-bond. In the covalent halides, e.g. Hg_2Cl_2, the other hybrid orbital overlaps with the singly occupied $3p_x$ orbital of the halogen, giving rise to a linear molecule (Figure 19.4).

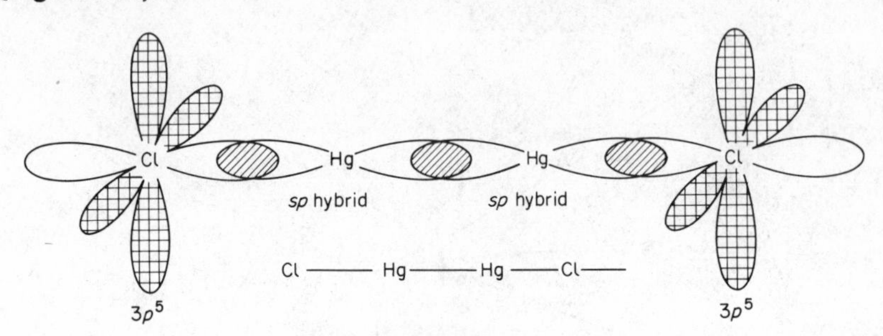

FIGURE 19.4

The Oxides

No mercury(I) oxide or hydroxide exists. The black precipitate obtained by adding hydroxide ions to a solution of a mercury(I) salt is an intimate mixture of mercury and mercury(II) oxide.

$$Hg_2^{2+} + 2OH^- \rightarrow Hg + HgO + H_2O$$

Mercury(II) oxide exists in a yellow and a red form but both are built up from infinite planar zig-zag chains, the colour difference being due to a different particle size. The red form is prepared by heating the metal in air or by heating the nitrate. The yellow form is precipitated by adding alkali to a solution of mercury(II) chloride.

$$HgCl_2 + 2OH^- \rightarrow HgO\downarrow + 2Cl^- + H_2O$$

On heating the yellow form, the colour changes to red and finally black: at about 500 °C decomposition to the elements occurs.

Dilute ammonia on mercury(II) oxide gives a yellow explosive powder called Millon's base, $Hg_2NOH.2H_2O$. This has a three-dimensional structure consisting of Hg_2N^+ cations and OH^- anions with water molecules occupying the vacant holes in the lattice and being held there by ionic interaction and hydrogen bonding.

The Sulphides

Mercury(I) sulphide has not been isolated. Black mercury(II) sulphide may be precipitated by passing hydrogen sulphide through an acidified solution of a mercury(II) halide. Black amorphous mercury(II) sulphide, on heating, gives the red pigment vermilion, which is built up from helical chains. Although insoluble in all acids, apart from aqua regia, it dissolves in alkaline sulphide solutions producing soluble thiomercurate ions.

$$HgS + S^{2-} \rightarrow HgS_2^{2-}$$

The Halides

All the mercury(I) halides are known. The fluoride is anomalous in being rapidly hydrolysed by water to mercury and hydrofluoric acid. The other halides are insoluble. The most common halide is white mercury(I) chloride or calomel, (Hg_2Cl_2), prepared by heating a mixture of mercury(II) chloride and mercury.

$$HgCl_2 + Hg \rightarrow Hg_2Cl_2$$

The calomel sublimes and is extracted with water to remove sparingly soluble mercury(II) chloride, which also sublimes. A wet method involves the double decomposition of mercury(I) nitrate solution and hydrochloric acid or sodium chloride.

$$Hg_2^{2+} + 2Cl^- \rightarrow Hg_2Cl_2$$

Vapour density measurements on perfectly dry calomel correspond to the formula Hg_2Cl_2 but, if precautions to dry the calomel are not observed, the vapour will amalgamate a gold leaf and give yellow mercury(II) oxide with caustic soda solution. This is due to the thermal dissociation to mercury and mercury(II) chloride.

$$Hg_2{}^{I}Cl_2 \rightarrow Hg + Hg^{II}Cl_2$$

The solid is blackened by aqueous ammonia, forming a mixture of black, finely divided mercury and white mercury aminochloride.

$$Hg_2Cl_2 + 2NH_3 \rightarrow Hg + HgNH_2Cl + NH_4Cl$$

Mercury(I) chloride is a linear covalent halide (p. 434).

The mercury(II) halides are also all covalent apart from mercury(II) fluoride which is ionic, with a fluorite lattice. Mercuric chloride ($HgCl_2$) is known as corrosive sublimate and may be prepared by heating the metal in chlorine gas or heating mercury(II) sulphate with common salt and manganese dioxide, when mercury(II) chloride sublimes off.

$$HgSO_4 + 2NaCl \rightarrow HgCl_2 + Na_2SO_4$$

The manganese dioxide prevents the formation of mercury(I) chloride. Mercury(II) chloride is a covalent compound with a molecular lattice (p. 97). It is sparingly soluble in water and slightly ionized, the solubility being increased by the addition of concentrated hydrochloric acid, when a soluble tetrachloromercurate(II) complex ion is formed.

$$HgCl_2 + 2Cl^- \rightarrow HgCl_4^{2-}$$

With aqueous ammonia, an infusible white precipitate of mercury amino-chloride is produced.

$$HgCl_2 + 2NH_3 \rightarrow HgNH_2Cl + NH_4Cl$$

This possesses a chain structure ($Hg—\overset{+}{N}H_2—Hg—\overset{+}{N}H_2—$) with the chloride ions between the chains. Gaseous ammonia or ammonium chloride on mercury(II) chloride, however, produces a fusible white precipitate, $Hg(NH_3)_2Cl_2$, the crystal of which is built up of $Hg(NH_3)_2^{2+}$ ions and chloride ions.

Mercury(II) chloride solution is reduced by many reducing agents, e.g. formaldehyde, tin(II) chloride, sulphur dioxide, etc., precipitating white mercury(I) chloride which, with excess of reducing agent, turns black owing to the formation of metallic mercury.

$$2HgCl_2 + SnCl_2 \rightarrow \underset{\text{white}}{Hg_2Cl_2\downarrow} + SnCl_4$$

$$Hg_2Cl_2 + SnCl_2 \rightarrow \underset{\text{black}}{2Hg} + SnCl_4$$

Mercury(II) iodide is dimorphic, with a yellow form stable above 126 °C and a scarlet form stable below 126 °C. The addition of potassium

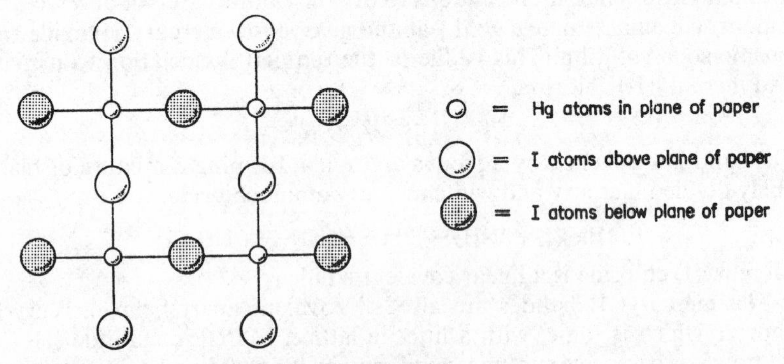

FIGURE 19.5

iodide to a mercury(II) chloride solution gives a precipitate of mercury(II) iodide which, although yellow initially, very rapidly turns scarlet. The iodide possesses a layer lattice (Figure 19.5).

In yellow mercury(II) iodide the four iodine atoms round the mercury atom are at 0·346 nm but in red mercury(II) iodide the Hg—I distance is 0·278 nm. Although sparingly soluble in water, the iodide dissolves readily in potassium iodide solution giving the tetrahedral tetraiodomercurate(II) anion.

$$HgI_2 + 2I^- \rightarrow HgI_4^{2-}$$

The complex dissolves in potassium hydroxide solution to give Nessler's solution, which gives a brown coloration with ammonia, owing to the formation of the iodide of Millon's base, $Hg_2NI.2H_2O$. The structural framework of Millon's base is unaltered with the OH^- ions replaced by I^- ions.

The Cyanides

Mercury(II) cyanide $Hg(CN)_2$ may be prepared by dissolving mercury(II) oxide in aqueous hydrocyanic acid and crystallizing. Mercury(II) cyanide is a covalent linear molecule with a molecular lattice. On heating, it evolves cyanogen and with excess cyanide, the tetrahedral complex, tetracyanomercurate(II) $Hg(CN)_4^{2-}$, is produced. Mercury fulminate $Hg(ONC)_2$ is an explosive compound used in detonators and percussion caps. It is prepared by the action of excess nitric acid on mercury in the presence of alcohol.

Oxo Salts

Mercury(I) nitrate, $Hg_2(NO_3)_2.2H_2O$, and mercury(I) sulphate, Hg_2SO_4, are prepared from the metal and the appropriate acid. The sulphate is hydrolysed by water and forms a basic salt. The nitrate is the only common soluble mercury(I) salt and contains the linear hydrated cation $[H_2O—Hg—Hg—OH_2]^{2+}$. With water, some hydrolysis occurs, eventually producing a basic salt.

$$[Hg_2(H_2O)_2]^{2+} + H_2O \rightleftharpoons H_3O^+ + [Hg_2OH.H_2O]^+$$

On heating the solid above 500 °C, it decomposes to the metal.

$$Hg_2(NO_3)_2 \longrightarrow 2HgO + 2NO_2$$

$$2HgO \xrightarrow{500\ °C} 2Hg + O_2$$

Mercury(II) oxo salts such as the nitrate and sulphate can be made by the action of heat and concentrated acid on the metal. Both salts are, however, very rapidly hydrolysed to give basic salts.

20

The Lanthanides

In the atoms of the fourteen elements following lanthanum, the inner 4*f* subshell is filled (p. 59). These elements are very similar chemically and are usually placed at the foot of the periodic table.

Occurrence

The term *rare earth*, originally applied to these elements, is now no longer significant since fairly large deposits have been found in the U.S.A., U.S.S.R., and Scandinavia. The most important ore is a complex phosphate, monazite, found as a dark heavy sand. There are also many silicates of which cerite, $H_3(Ca, Fe)Ce_3Si_3O_{13}$, found in Sweden, is important. Both the above ores contain several lanthanides.

Extraction

The ore is ground and treated with concentrated sulphuric acid. The soluble lanthanide sulphates are extracted with excess water and then precipitated as insoluble oxalates with oxalic acid.

Separation

This originally involved the long and tedious process of fractional crystallization from double salts, either sulphates or nitrates, e.g. $M(NO_3)_3 . 2NH_4NO_3$. This method has been rendered obsolete by the ion-exchange separation method. The lanthanide cations are adsorbed on a cationic exchange resin and the ions distribute themselves down a column of resin according to the ease of adsorption.

$$M^{3+} + 3RSO_3^-H^+ \rightarrow M^{3+}(RSO_3^-)_3 + 3H^+ \qquad M = \text{lanthanide}$$

$$M(H_2O)_n{}^{3+} + 3RSO_3^-H^+ \rightarrow M(H_2O)_n{}^{3+}(RSO_3^-)_3 + 3H^+$$

The theory of the separation is as follows. The smaller the radius of the lanthanide ion, the greater is the tendency to hydration and the larger the radius of the resulting hydrated ion. Thus there is less electrostatic attraction between the hydrated cations and the anionic portion of the exchange resin and the lanthanides with the smallest radii are less strongly adsorbed on the resin than those with larger radii. Elution of the column with a suitable complexing agent, such as citric acid and ammonium citrate, enhances the separation. The water of hydration is now replaced by the complexing agent and, since the lanthanides with the smallest radii possess the strongest tendency to form complexes, these are the first to be eluted from the column. The pure metals may be made either by the electrolysis of the anhydrous trichloride using carbon electrodes or by reduction of the heated trichlorides or trifluorides using calcium metal in an argon atmosphere.

Recently solvent extraction methods have been applied. A nitric acid solution of the lanthanides flows counter to a solution of tributyl phosphate in kerosene. The elements are extracted into the latter according to their extraction coefficients. Effective separation is obtained after a series of such extractions.

$$La_{(aq)}^{3+} + 3NO_{3(aq)}^{-} + 3TBP_{(org)} \rightarrow La(NO_3)_3(TBP)_{3(org)} + aq$$

The extent of extraction into the organic layer increases with the decrease in ionic radii. More effective separation has been obtained using 2-ethylhexyl phosphoric acid as a complexing agent.

Properties

Table 20.1 shows the electronic structures, ionic radii, and colours of the lanthanide ions. It will be noted that ions with *n* electrons in the 4*f* subshell exhibit the same light absorption characteristics as those with 14-*n* electrons.

TABLE 20.1

Element	*Name*	*Atomic number*	*Electronic structure of atoms* 4*f*	5*d*	6*s*	*Electronic structure of ions* M^{3+}	*Ionic radius* M^{3+} (*nm*)	*Colour of ions*
La	Lanthanum	57	0	1	2	$4f^0$	0·115	Colourless
Ce	Cerium	58	2	0	2	$4f^1$	0·111	Colourless
Pr	Praseodymium	59	3	0	2	$4f^2$	0·109	Green
Nd	Neodymium	60	4	0	2	$4f^3$	0·108	Red
Pm	Promethium	61	5	0	2	$4f^4$	0·106	Pink
Sm	Samarium	62	6	0	2	$4f^5$	0·104	Yellow
Eu	Europium	63	7	0	2	$4f^6$	—	Pink
Gd	Gadolinium	64	7	1	2	$4f^7$	0·102	Colourless
Tb	Terbium	65	9	0	2	$4f^8$	0·100	Pink
Dy	Dysprosium	66	10	0	2	$4f^9$	0·099	Yellow
Ho	Holmium	67	11	0	2	$4f^{10}$	0·097	Orange
Er	Erbium	68	12	0	2	$4f^{11}$	0·096	Red
Tm	Thulium	69	13	0	2	$4f^{12}$	0·095	Green
Yb	Ytterbium	70	14	0	2	$4f^{13}$	0·113	Colourless
Lu	Lutetium	71	14	1	2	$4f^{14}$	0·093	Colourless

All the lanthanide atoms have complete 5*s* and 5*p* subshells but certain irregularities are noted in the filling of the 4*f* subshells. After europium, which possesses a half-full 4*f* subshell, the extra electron goes into the 5*d* subshell and gadolinium has the electronic structure $4f^7 5d^1 6s^2$. The 4*f* subshell is complete at ytterbium and again, with the next element, lutetium, the extra electron goes into the 5*d* subshell. Similar anomalies were noted for chromium and copper in the first transition series (p. 346).

The principal oxidation state of the lanthanides is +3 and their chemistry is mainly that of the M^{3+} cations. Variability in oxidation states is not so marked as with the elements of the first transition series, although oxidation states of +4 are known for Ce, Pr, and Tb, and oxidation states of +2 for Eu, Sm, and Yb.

Since the 4*f* subshell is incomplete, the lanthanide ions are nearly all coloured and paramagnetic. However, the lanthanum ion, with an empty 4*f* subshell and lutetium, with a complete 4*f* subshell, have no unpaired electrons and are diamagnetic. Unlike the transition metals, the lanthanides form few complexes with unidentate ligands (NH_3, CN^-, etc.). Citrate, oxalate, and EDTA chelate complexes are far more common. Two main factors are responsible for this:

(*a*) The rather larger cationic radii compared with the radii of the transition metal cations.

(*b*) The inability of the inner 4*f* orbitals to form strong hybrid orbitals.

From Table 20.1 it will have been noted that a steady, but not perfectly regular decrease in atomic radius occurs with increasing atomic number. A similar but not so pronounced decrease was observed with elements of the first transition series. Here, however, the screening effect of the inner 4*f* electrons is very weak indeed and the increase in atomic number causes the reduction in atomic radii. The effect is called the *lanthanide contraction*.

The lanthanides are silvery white metals with melting points in the region of 1000 °C and densities which range from 6·5 g cm^{-3} at lanthanum to 9·8 g cm^{-3} at lutetium. All the metals burn in air, giving ionic sesquioxides M_2O_3, although cerium gives cerium(IV) oxide CeO_2. They combine directly on heating with hydrogen, carbon, ammonia, sulphur, and chlorine giving hydrides MH_3, carbides MC_2, nitrides MN, sulphides M_2S_3, and chlorides MCl_3. The hydrides, sulphides, and chlorides are ionic but the carbides and nitrides are interstitial. $E^{\ominus}$ values range from −2·5 V at lanthanum to −2·25 V at lutetium. Thus the elements attack water in the cold, liberating hydrogen.

Oxides and Hydroxides

The oxides, M_2O_3, are prepared by heating the nitrates or carbonates. They are soluble in water, producing solutions from which definite hydroxides,

$M(OH)_3$, can be isolated. As expected, hydroxide strength decreases with decreasing atomic radii from lanthanum to lutetium (p. 439).

The Halides

All the halides are known, the fluorides being somewhat unusual because of their insolubility in water. The chlorides usually crystallize with six molecules of water. These hydrates readily form oxochlorides of formula MOCl on heating. The anhydrous chlorides are best prepared by heating the oxides with ammonium chloride.

The Oxo Salts

Insoluble carbonates, $M_2(CO_3)_3$, are prepared by passing carbon dioxide into an aqueous solution of the hydroxide. Nitrates and sulphates, $M_2(SO_4)_3$ and $M(NO_3)_3$, are well characterized. The nitrates are often deliquescent and crystallize with six molecules of water. Both the nitrates and sulphates form double salts with alkali metal salts, e.g. $M_2(SO_4)_3 . 3Na_2SO_4 . 12H_2O$. Solutions of the oxo salts furnish hydrated cations, $M(H_2O)_x{}^{3+}$, which tend to undergo slight hydrolysis in aqueous solution. Salts of the cations with smallest radii hydrolyse most extensively.

$$M(H_2O)_6{}^{3+} + H_2O \rightleftharpoons M(H_2O)_5OH^{2+} + H_3O^+$$

When cerium(IV) oxide is warmed with concentrated sulphuric acid, yellow crystalline cerium(IV) sulphate ($Ce(SO_4)_2$) is obtained. Solutions of cerium(IV) sulphate in dilute sulphuric acid are powerful oxidizing agents.

$$Ce^{4+} + e \rightarrow Ce^{3+} \qquad (E^{\ominus} = +1{\cdot}44 \text{ V})$$

21

The Actinides

A second series of elements, the *actinides*, in which the $5f$ subshell fills, follows actinium ($6d^1 7s^2$) in Group III. The series consists of fourteen elements from thorium to lawrencium and these are again included at the foot of the periodic table. The atomic numbers, symbols, names, and electronic structures of the actinides are shown in Table 21.1.

TABLE 21.1

Atomic number	*Symbol*	*Name*	*Electronic structure*	*Ionic radii (M^{3+}) (nm)*
90	Th	Thorium	$6d^2\ 5f^0$	0·114
91	Pa	Protoactinium	$6d^1\ 5f^2$	0·112
92	U	Uranium	$6d^1\ 5f^3$	0·111
93	Np	Neptunium	$6d^1\ 5f^4$	0·109
94	Pu	Plutonium	$6d^0\ 5f^6$	0·107
95	Am	Americium	$6d^0\ 5f^7$	0·106
96	Cm	Curium	$6d^1\ 5f^7$	0·099
97	Bk	Berkelium	$6d^1\ 5f^8$	0·098
98	Cf	Californium	$6d^0\ 5f^{10}$	—
99	Es	Einsteinium	$6d^0\ 5f^{11}$	—
100	Fm	Fermium	$6d^0\ 5f^{12}$	—
101	Md	Mendelevium	$6d^0\ 5f^{13}$	—
102	No	Nobelium	$6d^0\ 5f^{14}$	—
103	Lw	Lawrencium	$6d^1\ 5f^{14}$	—

The $5f$ subshell fills rather irregularly from thorium to lawrencium and, since electrons in the $5f$ and $6d$ orbitals have similar energies, the electronic structures given are not conclusively proven.

The common oxidation state is +3 and the stability of actinide compounds with this oxidation state increases with increasing atomic number. Solutions containing the trivalent cation M^{3+} have been prepared for a number of actinides usually by reducing the oxycations $M^{VI}O_2^{2+}$ with zinc amalgam, sulphur dioxide, etc. These solutions are often coloured (U^{3+} red, Np^{3+}

purple, and Pu^{3+} blue) and exhibit paramagnetism. The radii of the M^{3+} cations (where known) show an *actinide contraction* analogous to the lanthanide contraction (p. 440).

Oxidation states of +4, +5, +6 are also known but the stability of these states falls off rapidly with increasing atomic number and are unknown for the actinides following berkelium. The common aqueous species of oxidation state +6 are the coloured oxo cations $M^{VI}O_2^{2+}$, e.g.

UO_2^{2+} (uranyl ion)—yellow
NpO_2^{2+} (neptunyl ion)—pink
PuO_2^{2+} (plutonyl ion)—yellow
AmO_2^{2+} (americyl ion)—brown.

All these cations are linear and the actinide uses 5*f* and 7*s* orbitals in linear *sf* hybridization.

As mentioned above, the energy difference between the 5*f*, 6*d*, and even the 7*s* orbitals is very small and transitions can easily occur. Thus bonding may involve any one of these orbitals. In UCl_4 and $NpCl_4$, sf^3 hybridization occurs to give tetrahedral molecules and, in UCl_6 and PuF_6, d^2sf^3 hybridization occurs giving octahedral arrangements. The actinides show a far greater tendency to form complex ions than the lanthanides, e.g. $UO_2F_6^{4-}$, $UO_2(CO_3)_3^{4-}$, $PuEDTA^-$.

The first three members of the series occur naturally in uranium and thorium ores. Although minute traces of neptunium and plutonium occur in these minerals, these and other actinides are usually obtained in small amounts by nuclear transmutation reactions (p. 12). Since the elements are very similar chemically, separation is difficult and, as with the lanthanides, the ion exchange technique is invaluable. Aqueous solutions containing the tripositive actinide ions (M^{3+}) are absorbed on cation exchange resins

$$M^{3+} + 3RSO_3^-H^+ \rightarrow M^{3+}(RSO_3^-)_3 + 3H^+$$

The actinides are eluted from the column with citrate, lactate, and ethylenediaminetetraacetate ions. Actinides with the smallest ionic radii possess the strongest tendency to form these complexes and are the first to be eluted from the column. Solvent extraction methods are also employed. Nitric acid solutions containing actinide MO_2^{2+} and M^{4+} cations form complexes with organic compounds such as tributyl phosphate, methylisobutyl ketone, and glycol ethers. These complexes are immiscible with water, and hence the actinide is extracted from the aqueous solution. The extraction coefficients are generally larger, the smaller and more highly charged the cation. Thus the large-scale separation of uranium, plutonium, and other fissionable products has been effected by this method. Of the actinides mentioned, only the chemistry of uranium will be considered further.

Uranium

The electronic structure of uranium, $7s^2 6d^1 5f^3$, indicates a maximum oxidation state of +6 but lower oxidation states of +5, +4, +3, and +2 are known. The most important compounds are the uranyl salts which contain the uranyl ion UO_2^{2+} with uranium in oxidation state +6. Diuranates such as $Na_2U_2O_7$ exist and are analogous to the pyrosulphates and the dichromates. Natural uranium contains three isotopes $^{238}_{92}U$ (99·274 per cent), $^{235}_{92}U$ (0·72 per cent), and $^{234}_{92}U$ (0·006 per cent). Only $^{235}_{92}U$ undergoes fission with neutrons and the separation of the isotope from natural uranium presents a formidable problem. Among the techniques employed are:

(*a*) Gaseous diffusion of uranium(VI) fluoride, UF_6, when the lighter isotopes diffuse more rapidly than the heavier isotopes.

(*b*) Electromagnetic separation based upon the mass spectrometer principle.

When bombarded with neutrons, $^{235}_{92}U$ splits into barium and krypton fragments with an average emission of $2\frac{1}{2}$ neutrons for each fission. Thus a loss in mass occurs, causing the release of a large quantity of energy.

$$^{235}_{92}U + ^{1}_{0}n \rightarrow ^{140}_{56}Ba + ^{92}_{36}Kr + 2\tfrac{1}{2}\,^{1}_{0}n \qquad (\Delta H = -1{\cdot}9 \times 10^{10}\ \text{kJ mol}^{-1})$$

The energetic neutrons emitted from each fission can cause other uranium nuclei to split and give rise to a chain reaction. If the mass of the ^{235}U exceeds a few kilogrammes (the critical mass) a nuclear explosion results. The neutrons can be slowed down (moderated) in a reactor or pile by graphite or heavy water which have little tendency to become radioactive. A self-sustaining reaction then takes place using natural uranium.

$$^{238}_{92}U + ^{1}_{0}n \longrightarrow ^{239}_{92}U \longrightarrow ^{239}_{93}Np + \beta^-$$

$$^{239}_{93}Np \longrightarrow ^{239}_{94}Pu + \beta^-$$

The final product, $^{239}_{94}Pu$, accumulates in the pile, is fissionable, and can be separated and used for nuclear bombs.

Occurrence

Uranium occurs naturally as pitchblende, U_3O_8, which contains iron, bismuth, lead, and radium salts, and in clevite (also U_3O_8) which contains thorium and the rare earths. Other ores include carnotite, $K_2(UO_2)_2(VO_4)_2 \cdot 3H_2O$; the ash of Swedish bituminous coal (kolm) can contain up to 3 per cent of U_3O_8.

Extraction

The extraction of uranium from pitchblende is extremely complex and only the main stages will be outlined. The ore is fused with a mixture of sodium carbonate and nitrate and extracted with boiling water. On heating the extract with dilute sulphuric acid, lead, bismuth, and radium precipitate as the sulphates. On boiling the filtrate with excess ammonium carbonate solution, the uranium forms ammonium uranyl carbonate (p. 447). This yields pure U_3O_8 on heating. The metal is then obtained by the following sequence of reactions:

$$U_3O_8 \xrightarrow[\text{C or H}_2]{\text{heat}} UO_2 \xrightarrow{\text{HF}} UF_4 \xrightarrow{\text{Mg}} U$$

Properties

Uranium is a silvery white metal which is malleable and ductile. In air, at ordinary temperatures, the surface goes yellow (UO_3) and finally black (U_3O_8). The metal is resistant to attack by alkali but dissolves rapidly in dilute hydrochloric and nitric acids giving uranium(IV) salts. It combines, on heating, with a number of non-metals, the compounds often being non-stoichiometric (Figure 21.1). The carbides and the sulphides find use as refractories.

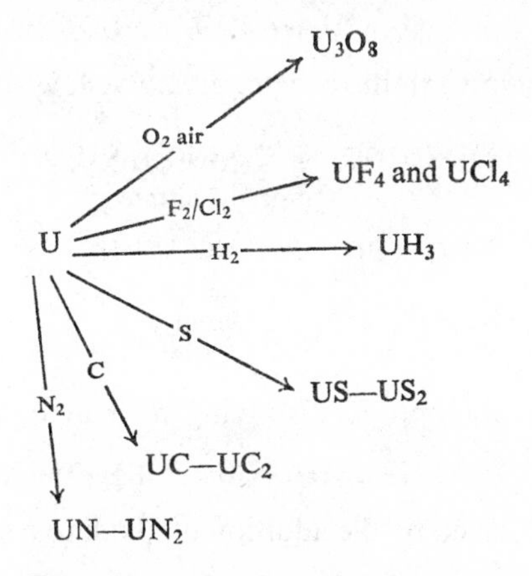

FIGURE 21.1. Reactions of uranium metal

The Hydrides

Uranium hydride is prepared from the elements at 250–300 °C and is a grey powder which is pyrophoric in air above 300 °C, often being non-stoichiometric with a deficiency of hydrogen.

$$2U + 3H_2 \rightarrow 2UH_3$$

The hydride is more reactive than the metal and is a suitable material for the preparation of uranium compounds.

$$2UH_3 + 4H_2O \rightarrow 2UO_2 + 7H_2$$

$$3HCl + UH_3 \rightarrow UCl_3 + 3H_2$$

$$2UH_3 + 4X_2 \rightarrow 2UX_4 + 3H_2 \qquad (X = \text{F or Cl})$$

The Oxides

These form a very complex system, often showing large deviations from stoichiometry. The stoichiometric oxides include U_3O_8 (black), $U^{VI}O_3$ (orange), and $U^{IV}O_2$ (brown-black). Both U_3O_8 ($U^{IV}(U^{VI}O_4)_2$) and UO_3 are dimorphic, possessing rather complex structures. The compound UO_2 crystallizes with a rutile lattice. The addition of ammonia to a uranyl salt precipitates yellow ammonium diuranate, which, on heating yields UO_3.

$$(NH_4)_2U_2O_7 \rightarrow 2UO_3 + H_2O + 2NH_3$$

Strong heating to 700 °C converts the trioxide to U_3O_8 which, on reduction with hydrogen, yields the dioxide.

$$3UO_3 \rightarrow U_3O_8 + \tfrac{1}{2}O_2$$

$$U_3O_8 + 2H_2 \rightarrow 3UO_2 + 2H_2O$$

All the oxides dissolve in nitric acid, giving yellow solutions containing uranyl salts.

$$UO_3 + 2HNO_3 \rightarrow \underset{\text{uranyl nitrate}}{UO_2(NO_3)_2} + H_2O$$

$$UO_3 + 2H^+ \rightarrow \underset{\text{yellow uranyl cation}}{UO_2^{2+}} + H_2O$$

The trioxide UO_3 is amphoteric, dissolving in alkali to produce uranates.

$$UO_3 + 2OH^- \rightarrow UO_4^{2-} + H_2O$$

Uranates are also formed by the addition of alkali to a solution of a uranyl salt.

$$UO_2^{2+} + 4OH^- \rightarrow UO_4^{2-} + 2H_2O$$

Simple uranates such as Na_2UO_4 have never been isolated because the UO_4^{2-} ion rapidly polymerizes and a yellow precipitate of the diuranate is produced.

$$2UO_4^{2-} + H_2O \rightarrow U_2O_7^{2-} + 2OH^-$$

Thus the action of caustic soda or ammonia on a uranyl salt precipitates yellow sodium diuranate ($Na_2U_2O_7.6H_2O$, uranium yellow) and $(NH_4)_2U_2O_7$.

The Halides

A summary of some of the preparative methods employed for the preparation of the fluoride and chlorides is shown below.

$$U + 2X_2 \rightarrow UX_4 \xrightarrow{\text{excess } X_2} UX_5 \xrightarrow{\text{excess } X_2} UF_6$$

$$UO_2 + 4HX \rightarrow UX_4 \xrightarrow{H_2} UX_3 \qquad (X = F \text{ or } Cl)$$

The pentahalides tend to disproportionate.

$$2UX_5 \rightarrow UX_4 + UX_6$$

Oxohalides such as uranyl fluoride, UO_2F_2, and uranyl chloride, UO_2Cl_2, are also known.

$$UF_6 + 2H_2O \rightarrow UO_2F_2 + 4HF$$

$$UCl_6 + 2H_2O \rightarrow UO_2Cl_2 + 4HCl$$

Oxo Salts

The main cationic species is the yellow oxo anion (uranyl ion), UO_2^{2+}. Solutions of uranyl salts are acidic due to the hydrolysis of this ion and the solution appears to contain polynuclear species.

$$UO_2^{2+} + 2H_2O \rightleftharpoons [UO_2(OH)]^+ + H_3O^+$$

$$[UO_2(OH)]^+ + 2H_2O \rightleftharpoons UO_2(OH)_2 + H_3O^+$$

$$UO_2(OH)_2 + 2H_2O \rightleftharpoons [UO_2(OH)_3]^- + H_3O^+$$

$$2[UO_2(OH)_3]^- \rightarrow (UO_2)_2(OH)_6^{2-}$$

This explains the precipitation of $Na_2U_2O_7$ when solutions of uranyl salts are made alkaline.

$$2UO_2(OH)_3^- \rightarrow U_2O_7^{2-}.3H_2O$$

The addition of sodium carbonate and ammonium carbonate to uranyl salts precipitates white uranyl carbonate, $(UO_2)CO_3$, but the precipitate is soluble in excess carbonate giving clear solutions containing sodium or ammonium uranyl carbonate, $Na_4UO_2(CO_3)_3$ or $(NH_4)_4UO_2(CO_3)_3$.

The trioxide UO_3 dissolves in the appropriate acid, giving uranyl salts which can be crystallized, e.g. $UO_2(NO_3)_2.6H_2O$ (uranyl nitrate), $UO_2SO_4.H_2O$ (uranyl sulphate), and $UO_2(CH_3COO)_2.2H_2O$ (uranyl acetate). The latter salt with zinc acetate gives zinc uranyl acetate, used as a delicate test for sodium in which a yellow crystalline precipitate of sodium zinc uranyl acetate is produced: $NaZn(UO_2)_3(CH_3COO)_9.9H_2O$. A uranium(IV) salt (uranic salt) can be made by dissolving UO_2 in concentrated sulphuric acid.

$$UO_2 + 2H_2SO_4 \rightarrow U(SO_4)_2 + 2H_2O$$

Solutions of the salt are acidic owing to hydrolysis.

$$U^{4+} + H_2O \rightleftharpoons U(OH)^{3+} + H^+$$

Complexes

The action of hydrogen peroxide on a solution of a uranyl salt precipitates a pale yellow peroxide, $UO_4.2H_2O$, with the probable structure

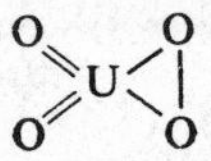

O   O
  ⟍U⟨ |
O⟋   O

When potassium hexacyanoferrate(II) solution is added to a uranyl salt, a brown uranyl hexacyanoferrate(II) is obtained: $(UO_2)_2Fe(CN)_6$. With caustic soda solution, this turns yellow owing to the formation of $Na_2U_2O_7$ (cf. Cu).

REVISION QUESTIONS

Chapters 1–8

1. Use the elements of the first three periods to illustrate the relationship between atomic structure and the position of an element in the periodic table. Show, with the aid of diagrams, the type of valency which unites the atoms in the following compounds: (*a*) CCl_4; (*b*) NH_4Cl; (*c*) NaCl.

2. Define the term atomic number. Write down the electronic configuration of the atoms of carbon, fluorine, sodium, and calcium and indicate their position in the periodic table. Illustrate, by means of examples, the difference between electrovalency and covalency.

3. Explain the terms (*a*) relative atomic mass, (*b*) atomic number, (*c*) isotopy. Illustrate each answer with one example. Give a concise account of the modern method for the determination of relative atomic masses.

4. Classify as electrovalent or covalent and give the electronic structures for: (*a*) HCl; (*b*) CaS; (*c*) $MgCl_2$; (*d*) NH_3; (*e*) $CHCl_3$.

5. Give the electron distribution in terms of *s*, *p*, *d*, and *f* subshells for each of the following and classify each as to whether it is a representative element, a transition element, an inner transition element, or a noble gas:

	V	Ca	B	Gd	Xe
atomic number	23	20	5	64	54

6. Decide, giving reasons, which atom in each of the following pairs has the higher first ionization potential: strontium and rubidium; fluorine and iodine; neon and sodium; chlorine and argon; copper and zinc; magnesium and sodium; boron and carbon; potassium and calcium.

7. Explain and illustrate, with examples, the basic concepts of the electron-pair repulsion theory. Write Lewis structures for each of the following ions and predict their configurations: NO_3^-, PO_4^{3-}, ClO_2^-, SO_3^{2-}, IO_6^{5-}, SO_4^{2-}, ClO_3^-.

8. What experimental evidence shows that: (*a*) all atoms contain electrons and protons; (*b*) electrons are small negatively charged particles?

9. Describe (*a*) the types of rays emitted by radioactive materials and (*b*) the Rutherford nuclear theory of the atom.

10. Give the nuclear structure and the electronic configuration of the following isotopes: $^{20}_{10}Ne$, $^{22}_{10}Ne$, $^{63}_{29}Cu$, $^{65}_{29}Cu$, $^{12}_{6}C$, $^{13}_{6}C$, $^{14}_{6}C$.

11. Explain the terms electrovalency, covalency, and coordinate valency, giving examples. What is meant by the term electronegativity? How is the nature of the chemical bond related to the difference in electronegativity between atoms involved in the bond?

12. Using the electronegativity scale on p. 64, state, with reasons, which of the following compounds would be covalent and which ionic: $MgCl_2$, ICl, Na_2S, CS_2, Na_2O, CaO.

13. Explain the terms oxidation and reduction in terms of (*a*) electron transfer and (*b*) change in oxidation number. Give the oxidation numbers of each element in the following: K_2MnO_4; MnO_4^-; $Cr_2O_7^{2-}$; Na_2SO_4; VO^{2+}; $Fe(CN)_6^{4-}$; $K_3Fe(CN)_6$; Na_2O_2; PbO_2; H_3PO_4.

14. Balance the following redox equations by (*a*) the oxidation number and (*b*) the ion electron method:

1. $Zn + H^+ + NO_3^- \rightarrow Zn^{2+} + N_2O + H_2O$
2. $MnO_4^- + NO_2^- + H^+ \rightarrow Mn^{2+} + NO_3^- + H_2O$
3. $Cr_2O_7^{2-} + H^+ + Fe^{2+} \rightarrow Cr^{3+} + Fe^{3+} + H_2O$
4. $H_2O_2 + MnO_4^- + H^+ \rightarrow Mn^{2+} + H_2O + O_2$
5. $S_2O_3^{2-} + I_2 \rightarrow S_4O_6^{2-} + I^-$
6. $NO_3^- + Zn + OH^- + H_2O \rightarrow NH_3 + Zn(OH)_4^{2-}$
7. $Fe(CN)_6^{4-} + Cl_2 \rightarrow Fe(CN)_6^{3-} + Cl^-$

15. Draw a labelled diagram to show the electronic transitions responsible for the Lyman, Balmer, Paschen, and Brackett series of spectral lines. Indicate on your diagram the approximate regions of the electromagnetic spectrum where these series of lines are to be found. Calculate (*a*) the wavelength of the first line in the Balmer series (*b*) the wavelength of the line at the series limit for the Lyman series of spectral lines, hence evaluate the ionization energy (kJ mol^{-1}) of the hydrogen atom. ($R = 1{\cdot}097 \times 10^7$ m^{-1}, $c = 300 \times 10^6$ m s^{-1}, $N_A = 602 \times 10^{21}$ mol^{-1}, $h = 633 \times 10^{-36}$ J s.)

16. What is meant by the term isoelectronic ion? Explain why there is a decrease in radius in the following series of isoelectronic ions: Mg^{2+}, Al^{3+}, Si^{4+} and P^{3-}, S^{2-}, Cl^-.

17. Account for the position of potassium and argon, and cobalt and nickel in the periodic table. Explain the meaning of the following terms with regard to the periodic table: short period, long period, group, transition element.

18. Predict and explain which of the following is:

(*a*) the strongest alkali, (i) LiOH or NaOH; (ii) $Al(OH)_3$ or $Si(OH)_4$.
(*b*) the strongest acid, (i) HOCl or HOBr; (ii) HCl or HI.
(*c*) the strongest oxidizing agent, (i) Cl_2 or S; (ii) F_2 or Br_2.
(*d*) the strongest reducing agent, (i) HBr or HI; (ii) Li or C.
(*e*) more metallic, (i) I_2 or Cl_2; (ii) Al or Si.

19. Explain why the water molecule is polar and account for the abnormally high boiling point, melting point, and heat of vaporization of water compared with the other hydrides of Group VI. Why is ice less dense than liquid water?

20. Compare and contrast the nature of the covalent and the metallic bond. Outline the basic ideas behind the band theory of metals.

21. Prepare a Born–Haber cycle diagram for the reaction: $K(s)+\frac{1}{2}Cl_2(g) \rightarrow KCl(s)$. The lattice energy of potassium chloride is -701 kJ mol^{-1}. Using available data (p. 34), calculate the heat of formation of the salt.

22. The lattice energy of magnesium oxide is calculated to be -3889 kJ mol^{-1}. Calculate the electron affinity of oxygen in kJ mol^{-1}. Explain why the electron affinity of the oxide ion is positive.

23. Calculate the radius ratios for the following ionic solids: NaBr, KI, CsBr, CdF_2, CoF_2, SnO_2. Deduce the coordination number of the cation and the anion and state the type of structure which these compounds exhibit.

24. Explain carefully, with the aid of diagrams, the essential differences between the zinc blende and wurtzite structures. List other common substances which crystallize with these structures.

25. Describe the concept of hybridization. Sketch the molecular orbital configuration of (*a*) SiH_4, (*b*) BCl_3, (*c*) ClO_4^-, (*d*) CS_2, (*e*) HBr. Indicate the σ- and π-bonds where necessary.

26. Complete the following transmutation equations:

$$^{27}_{13}Al + ^{4}_{2}He \rightarrow \ldots + ^{1}_{0}n$$
$$^{10}_{5}B + ^{1}_{0}n \rightarrow ^{3}_{1}H + \ldots$$
$$^{14}_{7}N + ^{1}_{1}H \rightarrow \ldots + \gamma$$

$$^{238}_{92}U + ^{1}_{0}n \rightarrow ^{239}_{92}U + \cdots$$

$$^{31}_{15}P + \cdots \rightarrow ^{32}_{15}P + ^{1}_{1}H$$

$$^{23}_{11}Na + ^{1}_{0}n \rightarrow ^{24}_{11}Na + \cdots$$

$$\hookrightarrow ^{24}_{12}Mg + \cdots$$

27. Calculate the binding energy per nucleon for $^{9}_{4}Be$, given that the isotopic mass of beryllium is 9·012, the mass of the proton is 1·0073, and that of the neutron is 1·0087.

28. The atomic mass of the proton is 1·0073, that of helium is 4·0026 and of tritium 3·01605. Calculate how many kilojoules of energy will be released in the reaction:

$$^{3}_{1}H + ^{1}_{1}H \rightarrow ^{4}_{2}He$$

29. Given the following bond energies and the electronegativity of hydrogen as 2·1, calculate the electronegativities of the halogens.

H—H	436 kJ mol^{-1}	H—Cl	431 kJ mol^{-1}
Cl—Cl	242	H—Br	366
Br—Br	193	H—I	299
I—I	151		

30. Explain carefully, with examples, the meaning of the following terms: (*a*) Lewis acid; (*b*) Brönsted base; (*c*) amphiprotic; (*d*) disproportionation; (*e*) oxidation number.

31. Discuss (*a*) the governing principles and (*b*) the general methods available for the extraction of a metal from one of its naturally occurring compounds. Outline the methods that are available for the refining of metals.

32. Explain what is meant by the terms (*a*) effective atomic number; (*b*) ligand; (*c*) coordination number; (*d*) chelating group. Give (*a*) orbital diagrams, (*b*) the type of hybridization, and (*c*) the geometrical structure for each of the following: (i) $Fe(CN)_6^{4-}$; (ii) $Ni(CN)_4^{2-}$; (iii) $Zn(CN)_4^{2-}$.

Chapters 9–17

1. Draw out and describe the structure of each of the following halides: (*a*) $NaCl$; (*b*) $CsCl$; (*c*) $CdCl_2$; (*d*) $CrCl_3$; (*e*) $CuCl_2$; (*f*) $AlCl_3$.

2. Describe the preparation of (*a*) xenon difluoride; (*b*) xenon tetrafluoride; (*c*) xenon hexafluoride. What are the structures of these compounds? Describe the preparation and structure of the compound formed when quinol reacts with xenon.

3. Describe the preparation and the structures of: (*a*) red phosphorus; (*b*) P_4O_6; (*c*) P_4O_{10}; (*d*) P_4S_3; (*e*) PCl_5; (*f*) HPO_3.

4. Discuss the preparation and the structure of the following oxides: (*a*) Na_2O_2; (*b*) KO_2; (*c*) BaO_2; (*d*) PbO_2; (*e*) Pb_3O_4; (*f*) CrO_3; (*g*) Cu_2O; (*h*) CuO; (*i*) ZnO.

5. Give a brief account of the production of ammonia by the Haber process indicating the underlying physico-chemical principles. How is ammonia converted into nitric acid in industry? How and under what conditions does nitric acid react with: (*a*) copper; (*b*) iodine; (*c*) iron(II) chloride solution; (*d*) tin?

6. How may carbon monoxide be prepared? How does carbon monoxide react with: (*a*) nickel; (*b*) copper(I) chloride; (*c*) sodium hydroxide; (*d*) hydrogen?

7. Compare the methods employed for the preparation of the halogen hydrides. Compare and contrast the properties of these hydrides with regard to (*a*) thermal stability; (*b*) acid strength; (*c*) reducing action.

8. Describe the principal properties and structures of four of the following: (*a*) graphite; (*b*) nitrogen(V) oxide; (*c*) phosphorus(V) chloride; (*d*) boric acid; (*e*) red lead; (*f*) sodium metaphosphate.

9. Write a comparative account of the chemistry of sulphur, selenium, and tellurium with regard to their hydrides, oxides, halides, and oxo acids.

10. Lithium and beryllium show similarities in chemical properties to their respective diagonal neighbours magnesium and aluminium. In each case account for and illustrate the chemical similarity between these pairs of elements.

(Ionic radii: $Li^+ = 0{\cdot}060$ nm; $Mg^{2+} = 0{\cdot}065$ nm; $Be^{2+} = 0{\cdot}031$ nm; $Al^{3+} = 0{\cdot}050$ nm.)

11. Describe the production of white phosphorus. How would you convert (*a*) white phosphorus into red phosphorus; (*b*) red phosphorus into white phosphorus; (*c*) white phosphorus into black phosphorus? Discuss the structure of these allotropic forms. Starting with white phosphorus, how would you obtain a solution of orthophosphoric acid?

12. Starting with elementary arsenic, how would you prepare a solution of (*a*) sodium arsenite; (*b*) sodium arsenate? Explain the reactions which take place when (*c*) hydrogen sulphide is passed into acidified sodium arsenite solution; (*d*) the precipitate from (*c*) is dissolved in sodium hydroxide solution; (*e*) the solution from (*d*) is acidified with dilute hydrochloric acid.

13. Write a comparative account of the chemistry of phosphorus, arsenic, antimony, and bismuth with regard to the elements, the hydrides, the oxides, and the halides.

14. Compare and contrast the properties of the oxides and chlorides of the Group IV elements, carbon, silicon, germanium, tin, and lead.

15. Write notes on the following topics: (*a*) structure of water and ice; (*b*) clathrate compounds; (*c*) metallic halide structures; (*d*) metallic carbides.

16. Indicate carefully the stages in the following preparations: (*a*) fluorine from calcium fluoride; (*b*) iodine pentoxide from iodine; (*c*) chlorine heptoxide from potassium chlorate; (*d*) a solution of chloric acid from barium hydroxide.

17. Explain carefully: (*a*) the nature and properties of solutions of the alkali metals in liquid ammonia; (*b*) evidence for the existence of the nitronium ion; (*c*) the reduction products obtained by the action of nitric acid on metals.

18. Describe the structure and principal properties of the allotropes of carbon, relating the properties mentioned to the structure. What is the nature of amorphous carbon? How does graphite react with strong oxidizing agents?

19. Compare and contrast the properties of ammonia and phosphine with reference to (*a*) thermal stability; (*b*) action of dilute halogen acids; (*c*) oxidation.

20. From metallic lead, how would you prepare specimens of (*a*) red lead; (*b*) lead(IV) oxide; (*c*) lead(IV) chloride; (*d*) white lead; (*e*) lead sulphate?

21. Give reasons for the following properties of the alkali metals:

(*a*) They are soft metals with low melting points and are excellent conductors.

(*b*) In solution the ions are colourless and monovalent with little tendency to form complex ions.

(*c*) The salts are generally all soluble in water.

(*d*) The hydroxides are all soluble in water, alkali strength increasing with increase in atomic number down the group.

(*e*) The metals are strong reducing agents.

22. Explain carefully why: (*a*) the metals of Group IIA are harder and have higher melting points than those in Group IA; (*b*) the hydroxides of Group IIA metals are weaker alkalis than those of Group IA; (*c*) the carbonates of Group IIA metals are less stable to heat than those of Group IA.

23. Write equations for the action of heat on the following ammonium salts: NH_4Cl; NH_4NO_2; NH_4NO_3; $(NH_4)_2SO_4$; $(NH_4)_2Cr_2O_7$. How would you detect the presence of the ammonium ion?

24. Compare and contrast the reactions of the alkaline earth metals, magnesium, calcium, strontium, and barium with: (*a*) oxygen; (*b*) carbon dioxide; (*c*) hydrogen; (*d*) nitrogen; (*e*) sulphur; (*f*) water.

25. Write down the formulae of the following and state how they may be prepared: (*a*) quicklime; (*b*) slaked lime; (*c*) milk of lime; (*d*) soda lime.

26. Starting with aluminium metal, how would you prepare: (*a*) potash alum crystals; (*b*) ammonium alum crystals; (*c*) anhydrous aluminium chloride; (*d*) aluminium nitride; (*e*) aluminium oxide; (*f*) a solution of sodium aluminate?

27. What is the action of (*a*) HCl; (*b*) HNO_3; (*c*) NaOH on metallic tin? Starting with metallic tin, how would you prepare crystals of tin(II) chloride and anhydrous tin(IV) chloride?

28. Write the electronic structures for carbon monoxide, phosgene, carbon disulphide, hydrogen cyanide, and calcium carbide.

29. What is meant by the term basic carbonate? Account for the formation of basic carbonates of zinc, copper, lead, and magnesium. How are normal carbonates prepared for these metals?

30. Write short notes on: (*a*) structure of glass; (*b*) the natural silicates; (*c*) the silicones.

31. For the alkaline earth metals in Group IIA explain the trends which occur in (*a*) solubilities of the hydroxides; (*b*) the melting points of the chlorides; (*c*) the melting points of the fluoride, chloride, bromide, and iodide of calcium. Explain carefully all the theoretical principles involved in these trends.

32. Starting with concentrated nitric acid and any other chemicals you require, how would you prepare: (*a*) nitrogen(I) oxide; (*b*) nitrogen(II) oxide; (*c*) nitrogen(III) oxide; (*d*) nitrogen(IV) oxide; (*e*) nitrogen(V) oxide?

33. Explain and account for the hydrolysis of the following chlorides: (*a*) BCl_3; (*b*) PCl_3; (*c*) $AlCl_3$; (*d*) $SnCl_2$; (*e*) $BiCl_3$; (*f*) $SiCl_4$.

34. Outline one general method that can be employed for the preparation of chlorine, bromine, and iodine. Compare and contrast the properties of chlorine, bromine, and iodine with regard to their action on (*a*) water (*b*) alkalis; (*c*) organic solvents; (*d*) hydrogen.

35. Describe briefly the preparation of aqueous solutions of hydrochloric, hydrobromic, and hydriodic acids. Explain why hydrobromic and hydriodic acids cannot be prepared by the action of concentrated sulphuric acid on the appropriate alkali metal halide. Compare and contrast the properties of the three acids with regard to: (*a*) acid strength; (*b*) thermal stability; (*c*) reducing action.

Chapters 18–21

1. Explain the meaning of the term transition element. Give an account of the principal physical and chemical properties of the elements in the first transition series.

2. Given a supply of potassium dichromate and any other reagent you require, how would you prepare: (*a*) a solution of potassium chromate; (*b*) crystals of chromium(VI) oxide; (*c*) crystals of chrome alum; (*d*) crystals of chromium(II) acetate?

3. Outline the extraction of mercury from the sulphide ore. How is the metal purified? Starting from the metal, how would you prepare specimens of: (*a*) mercury(I) chloride; (*b*) mercury(II) chloride? Discuss the evidence for the existence of the mercury(I) ion.

4. Describe how copper is extracted on the large scale from its sulphide ore. Starting from metallic copper, how would you prepare specimens of: (*a*) crystalline copper sulphate; (*b*) copper(I) chloride; (*c*) copper(I) oxide; (*d*) tetrammine copper(II) sulphate; (*e*) copper ammonium sulphate?

5. How many unpaired electrons would you expect for the complex ions $Fe(CN)_6^{3-}$ and $Fe(H_2O)_6^{3+}$? Explain your answer in terms of the crystal field theory. Draw diagrams for the isomers which could exist for $Co(en)_2NO_2Cl^+$ and $Cr(en)_3^{3+}$ (where en = ethylenediamine).

6. Outline a method for the extraction of uranium from pitchblende. Give an account of the fundamental chemistry of uranium.

7. Carefully explain each stage in the following preparations: (*a*) potassium hexacyanoferrate(II) from metallic iron; (*b*) barium ferrate from metallic iron; (*c*) manganese heptoxide from manganese dioxide; (*d*) Berlin green from iron(III) chloride; (*e*) chromium(II) acetate from chromium(III) chloride.

8. Describe the extraction of vanadium metal from vanadinite. Starting with vanadium pentoxide, how would you prepare: (*a*) vanadyl sulphate; (*b*) vanadium trichloride; (*c*) vanadium(II) sulphate; (*d*) ammonium metavanadate; (*e*) vanadium oxotrichloride?

9. Write notes on the following topics: (*a*) interstitial compounds; (*b*) paramagnetism and transition metal ions; (*c*) colour of the transition metal cations.

10. State the colour and write Lewis structures for the following ions: (*a*) permanganate ion; (*b*) manganate ion; (*c*) chromate ion; (*d*) dichromate ion. State the conditions and write ionic equations for the conversion of (*a*) manganate to permanganate; (*b*) permanganate to manganate; (*c*) chromate to dichromate; (*d*) dichromate to chromate.

11. An orange solid A dissolves in sodium hydroxide solution from which a white solid B is precipitated on addition of ammonium chloride. B dissolves in dilute sulphuric acid to give a yellow solution which on the passage of sulphur dioxide turns blue. Prolonged action of zinc and dilute sulphuric acid converts the blue solution to green and finally violet. Identify A and B and explain carefully all the reactions involved. Describe what happens when A is treated with (*a*) oxalic acid and (*b*) potassium.

12. Compare and contrast the chemistry of chromium, molybdenum, and tungsten with regard to (*a*) the chlorides; (*b*) the oxides; (*c*) oxo acids.

13. Outline the production of (*a*) titanium metal from rutile; (*b*) titanium dioxide from ilmenite. How and under what conditions does titanium react with (*a*) sulphuric acid; (*b*) nitric acid; (*c*) hydrofluoric acid?

14. What happens when a solution of potassium hexacyanoferrate(II) and a solution of potassium hexacyanoferrate(III) is added separately to a solution of (*a*) an iron(II) salt and (*b*) an iron(III) salt? Discuss and interrelate the structures of the compounds produced.

15. Compare and contrast the chemistry of cobalt and nickel with regard to their (*a*) oxides; (*b*) halides; (*c*) complexes.

16. Compare and contrast the preparation, properties, and structures of the following pairs of oxides: (*a*) chromium trioxide and sulphur trioxide; (*b*) vanadium pentoxide and phosphorus pentoxide; (*c*) manganese heptoxide and chlorine heptoxide; (*d*) titanium dioxide and silica.

17. Compare and contrast the chemistry of chromium and iron in the +2 and +3 oxidation states with regard to their (*a*) oxides; (*b*) halides; (*c*) oxo salts.

18. Describe the cyanide process for the extraction of silver and gold. Give a comparative survey of the chemistry of silver and gold.

19. Write an essay on (*a*) the iso- and heteropolyacids of molybdenum and tungsten or (*b*) the extraction and separation of the lanthanides or (*c*) the electronic structure and general chemistry of the actinides.

Some More Advanced Questions

1. (*a*) Given the following thermochemical quantities in kJ mol^{-1}, calculate the lattice energy of copper(I) chloride. $\Delta H_f = -135$, $\Delta H_{at} = +339$, $\Delta H_{diss} = +242$, $I_p = +751$, $E_A = -364$.

(*b*) Given Madelung's constant $M = 1{\cdot}638$, $n = 9$, $r_0 = 0{\cdot}235$ nm, $\epsilon_0 = 8{\cdot}854 \times 10^{-12}$ kg^{-1} m^{-3} s^4 A^2, $N_A = 602 \times 10^{21}$ mol^{-1} and $e = 160 \times 10^{-21}$ C, calculate the lattice energy of copper(I) chloride. Comment on the values you have obtained.

2. Give diagrammatic representations of molecular bonding and anti-bonding orbitals formed from *s* and *p* valency orbitals in a homonuclear diatomic molecule. Show the order of molecular orbital energy levels for homonuclear diatomic molecules. Discuss the electron structure and bonding properties of the following ions: (*a*) O_2^-; (*b*) CN^-; (*c*) O_2^{2-}; (*d*) NO^+.

3. Discuss the significance of the term charge-radius ratio. Place the following chlorides in the order of increasing ionic character of metal–chlorine bonds: $AlCl_3$, $BaCl_2$, $BeCl_2$, $CaCl_2$, CsCl, LiCl, NaCl. Comment briefly on the structures of these chlorides. (Ionic radii (nm) $Al^{3+} = 0{\cdot}050$, $Ba^{2+} = 0{\cdot}135$, $Be^{2+} = 0{\cdot}031$, $Ca^{2+} = 0{\cdot}099$, $Cs^+ = 0{\cdot}169$, $Li^+ = 0{\cdot}060$, $Na^+ = 0{\cdot}095$).

4. Give systematic names and discuss the types of isomerism possible for the following complexes:

(*a*) $(Co(NH_3)_4(H_2O)Cl)Br_2$;
(*b*) $(Co(NH_3)_5NO_2)SO_4$.

5. (*a*) The standard redox potentials ($E^\ominus$) involving Mn^{3+} ions in acid solution are:

$$Mn^{2+} = Mn^{3+} + e \qquad E^\ominus = +1{\cdot}51 \text{ V}$$

$$Mn^{3+} + 2H_2O = MnO_2 + 4H^+ + e \qquad E^\ominus = +0{\cdot}95 \text{ V}$$

Calculate the standard redox potential and the equilibrium constant of the following reaction at 25 °C assuming that all the reactants are at unit activity:

$$2Mn^{3+} + 2H_2O = Mn^{2+} + MnO_2 + 4H^+$$

(*b*) Some standard redox potentials are as follows: $Cr_2O_7^{2-}/Cr^{3+}$, $E^\ominus = +1{\cdot}33$ V; $\frac{1}{2}Cl_2/Cl^-$, $E^\ominus = +1{\cdot}36$ V; HNO_2/NO_3^-, $E^\ominus = +0{\cdot}94$ V; $\frac{1}{2}Hg_2^{2+}/Hg^{2+}$, $E^\ominus = +0{\cdot}79$ V. State which oxidations are possible with acidified potassium dichromate. Give ionic equations for the reaction chosen.

6. (*a*) Calculate the equilibrium constant of the following reaction at 25 °C, assuming that all the reactants are at unit activity.

$$Cr_2O_7^{2-} + 14H^+ + 6I^- \rightleftharpoons 2Cr^{3+} + 7H_2O + 3I_2$$

The standard redox potentials are: $Cr_2O_7^{2-}/Cr^{3+}$, $E^\ominus = +1{\cdot}33$ V; $\frac{1}{2}I_2/I^-$, $E^\ominus = +0{\cdot}54$ V.

(*b*) Calculate the standard free energy change and standard electrode potential for the following reaction $3Fe^{2+} \rightleftharpoons 2Fe^{3+} + Fe$. The standard redox potentials are: Fe^{2+}/Fe, $E^\ominus = -0{\cdot}44$ V; Fe^{3+}/Fe^{2+}, $E^\ominus = +0{\cdot}76$ V. Comment on your results.

7. Give a concise classification of the typical hydrides formed by members of the various groups of the periodic table. Starting from deuterium oxide, how would you prepare: (*a*) ND_3; (*b*) CD_4; (*c*) C_2D_2; (*d*) D_2SO_4; (*e*) $Ca(OD)_2$?

8. Describe the behaviour of the alkali metals with the following reagents: (*a*) oxygen; (*b*) hydrogen; (*c*) gaseous ammonia; (*d*) nitrogen; (*e*) cyclopentadiene in tetrahydrofuran. What is the action of water on the products obtained in these reactions?

9. Compare and contrast the chemistry of beryllium with the other Group IIA elements, emphasizing the structures and properties of: (*a*) the elements; (*b*) the oxides; (*c*) the halides; (*d*) the complexes.

10. Describe the bonding in the following chlorides: (*a*) boron(III) chloride; (*b*) silicon(IV) chloride; (*c*) nitrogen(III) chloride; (*d*) aluminium(III) chloride. Indicate the mechanisms involved in the hydrolysis of these chlorides.

11. Describe concisely the structure of: (*a*) calcium metaborate: (*b*) borax; (*c*) boric acid. Discuss the behaviour of boric acid in water. Explain how the acid strength of a solution of boric acid is changed by the addition of certain polyhydroxo compounds.

12. (*a*) *A* is a reddish-violet solid, insoluble in water, organic solvents and sodium hydroxide solution. However, it dissolves freely when warmed with concentrated nitric acid, forming an acid *B*. When heated in either oxygen or chlorine, *A* burns forming dense white fumes which, in each case, dissolve in water to form acid solutions.

(*b*) *X* forms an unstable blue solution in water which rapidly decomposes even in the cold. The solution decolorizes bromine water and an acid solution of potassium permanganate, and oxidizes a solution of tin(II) chloride in dilute hydrochloric acid. Identify *A*, *B* and *X*, and explain all the reactions involved.

13. Write notes on two of the following topics: (*a*) allotropy of sulphur, selenium, and tellurium; (*b*) the Born–Haber Cycle; (*c*) the structures of the natural silicate minerals.

14. Explain the structure and function of (*a*) the zeolite minerals, and (*b*) the synthetic organic ion exchange resins.

15. Describe how the following compounds can be prepared from sulphuric acid: (*a*) chlorosulphonic acid; (*b*) sulphamic acid; (*c*) sulphamide; (*d*) peroxomonosulphuric acid; (*e*) peroxodisulphuric acid.

16. Write notes on two of the following topics: (*a*) the allotropy of sulphur, selenium, and tellurium; (*b*) the preparation, properties, and structures of the thionic acids; (*c*) the oxo acids of selenium and tellurium.

17. Describe with the aid of simple sketches (where appropriate) the structures of the following: (*a*) chlorine hydrate; (*b*) the iodine starch complex; (*c*) magnesium oxide; (*d*) tin(IV) oxide; (*e*) calcium fluoride.

18. Tabulate the oxo acids of chlorine indicating (*a*) the oxidation state of chlorine; (*b*) the name and structure of the oxo acid; (*c*) the trends in stability, acid strength, and oxidizing power. Starting with elementary iodine and using any other reagent you may require, how would you prepare: (*d*) paraperiodic acid; (*e*) mesodiperiodic acid; (*f*) metaperiodic acid?

19. Write a concise account of *two* of the following topics. (*a*) cationic compounds of iodine; (*b*) interhalogen compounds and polyhalides; (*c*) oxo acids of the halogens.

20. Discuss the structure of five of the following compounds in the solid state: (*a*) ice; (*b*) boric acid; (*c*) sulphuric acid; (*d*) potassium hydrogen difluoride; (*e*) paraperiodic acid; (*f*) phosphorus pentachloride.

21. Write an essay on either (*a*) the structure and bonding of the metallic carbonyls of the first transition series or (*b*) the lamellar compounds of graphite.

22. Give examples, and suggest reasons for the following features of transition metal chemistry:

(*a*) The lowest oxide of a transition metal is basic, the highest is usually acidic.

(*b*) A transition metal usually exhibits higher oxidation states in its fluorides than its iodides.

(*c*) A transition metal usually exhibits its highest oxidation state in an oxo anion.

23. Draw out and explain the energy level diagrams to show the way in which the degeneracy of a set of 3*d* orbitals is removed by an (*a*) octahedral, (*b*) tetrahedral, (*c*) square planar crystal field. Indicate the occupancy of orbitals in the following complex ions: (*d*) $Mn(CN)_6^{4-}$; (*e*) $MnCl_6^{2-}$; (*f*) $Ni(CN)_4^{2-}$; (*g*) $NiCl_4^{2-}$.

24. How would you prepare: (*a*) crystals of titanyl sulphate; (*b*) crystals of titanium(III) sulphate; (*c*) a solution of titanium(IV) sulphate. Describe the changes which occur when solutions of (*b*) and (*c*) are diluted with water and boiled. What is the action of hydrogen peroxide on a solution of (*a*)?

25. Starting from ammonium metavanadate how would you prepare an aqueous solution of sodium orthovanadate? Explain carefully with reasons what happens when (*a*) ammonium chloride, (*b*) acetic acid, are added separately to a solution of sodium orthovanadate. Comment briefly on the nature of the solid compounds that have been isolated from these reactions.

26. Describe carefully how the following compounds are prepared from chromium(VI) oxide: (*a*) potassium dichromate; (*b*) potassium chlorochromate; (*c*) chromyl fluoride; (*d*) pyridine perchromate; (*e*) chromium(III) oxide.

27. Describe carefully the changes which take place when strongly basic solutions of molybdates and tungstates are slowly acidified. Discuss briefly the structures of the solid compounds of molybdenum which have been isolated from these solutions.

28. Outline the method of preparation and discuss the structures of the following compounds: (*a*) molybdenum blue; (*b*) tungsten bronze; (*c*) molybdic acid; (*d*) ammonium paramolybdate; (*e*) molybdenum(IV) oxide.

29. Discuss the bonding and structure of (*a*) the manganate ion; (*b*) the permanganate ion. Discuss the behaviour of the manganate and permangate ions in basic and acidic solutions.

30. Starting from metallic iron, explain carefully how you would prepare a specimen of Prussian blue. How would you convert Prussian Blue to the green pigment Berlin green? Describe briefly the structures of these two coloured pigments.

31. Discuss the chemical and structural inter-relationships between (*a*) the oxides of iron; (*b*) Prussian blue, Berlin green and potassium iron(II) hexacyanoferrate(II) (Everitt's salt).

32. Explain carefully the reactions which occur and the changes in colour which are observed when:

(*a*) cobalt(II) nitrate solution is treated with concentrated hydrochloric acid;
(*b*) iodine is dissolved in alcohol;
(*c*) iron(III) sulphate solution is treated with potassium thiocyanate;
(*d*) copper(II) sulphate solution is treated with an excess of aqueous ammonia;
(*e*) iron(II) sulphate solution is treated with potassium hexacyanoferrate(III).

Indicate briefly the factors responsible for the colour changes.

33. Suggest explanations for the following observations:

(*a*) The $NiCl_4^{2-}$ ion is paramagnetic (μ(calc) = 2·8 Böhr magnetons), but the $Ni(CN)_4^{2-}$ ion is diamagnetic.
(*b*) The pH of an aqueous solution of sodium chloride is 7, but that of iron(III) chloride is less than 7.
(*c*) The ICl_2^- ion is linear, the ICl_4^- ion is square planar, and the ClF_3 molecule is T-shaped.

34. Discuss the preparation and structure of the following organometallic compounds: (*a*) monocyclopentadienyl sodium; (*b*) dicyclopentadienyl magnesium; (*c*) dicyclopentadienyl iron; (*d*) dibenzene chromium.

35. A black compound *A* dissolves readily in hydrochloric acid to give a green solution *B*. When solution *B* is heated with copper turnings, a colourless solution *C* is obtained which on dilution with water, deposits a white solid *D*. *D* dissolves in aqueous ammonia to give a colourless solution *E* which, on exposure to the air rapidly turns blue. The blue solution *F* is decolorized when aqueous potassium cyanide is added and solution *G* is obtained. Treatment of *G* with zinc dust gives a red-brown precipitate *H* which reacts with hot dilute nitric acid to give a blue solution *I*. Strong evaporation of solution *I* to dryness produces the original compound *A*. Identify the lettered compounds and solutions and explain carefully all the reactions involved.

36. (*a*) Outline the chemistry of the TORCO process for the extraction of copper from copper silicate ores.

(*b*) A metal *A* dissolved in concentrated nitric acid with the evolution of brown fumes. The resulting solution gave a yellow precipitate *B* with potassium iodide solution. *B* was insoluble in ammonia but was soluble in sodium thiosulphate and potassium cyanide solutions. The latter solution in potassium cyanide gave a precipitate of *A* when zinc dust was added. Identify *A* and *B*, and explain all the reactions involved.

37. A white solid *X* is sparingly soluble in water. An aqueous solution of *X* gives white precipitates with (*a*) aqueous ammonia (*b*) silver nitrate solution and (*c*) dilute tin(II) chloride solution. The precipitate obtained in (*b*) dissolves in aqueous ammonia but the precipitate in (*c*) turns black with this reagent. A concentrated solution of potassium iodide slowly added to an aqueous solution of *X* gives a yellow precipitate which rapidly turns scarlet and then redissolves. The addition of excess potassium hydroxide and ammonia to this solution gives a brown coloration. Identify *X* and explain carefully the reactions involved.

38. Outline two methods used for the separation of the lanthanides. Give a concise account of the general characteristics of the lanthanide elements.

39. Describe two methods used for the separation of the actinide elements. Give a concise account of the general characteristics of the actinide elements.

40. Outline the production of uranium from pitchblende. Starting with uranium(VI) oxide how would you prepare: (*a*) uranium(IV) oxide; (*b*) uranyl nitrate; (*c*) ammonium diuranate?

41. Carefully explain the meaning of the following, giving one illustrative example in each case: (*a*) clathrate compound; (*b*) interstitial compound; (*c*) chelate complex; (*d*) spinel; (*e*) sandwich compound.

42. A white solid *A* (unstable in air) dissolves in excess ammonia to give a colourless solution *B* which produces a red precipitate *C* with ethyne gas. Solution *B* slowly oxidizes in air to give a deep-blue solution *D* which changes to a pale-blue solution *E* on the addition of EDTA. When *A* is boiled with sodium hydroxide solution a red solid *F* is obtained. Dissolution of *F* in dilute sulphuric acid produces a red precipitate *G* and a blue solution *H*. Treatment of *H* with aqueous thiourea and cooling produces white crystals *I* which are soluble and stable in aqueous solutions which are slightly acidic. Identify the lettered compounds and explain carefully all the reactions involved.

43. A metal *M* dissolved slowly in hot 20% potassium hydroxide to produce a clear solution *A*. Addition of oxalic acid to solution *A* produced a white gelatinous precipitate *B* which disappeared on the addition of further oxalic acid. On cooling and addition of ethanol, white crystals *C* were formed. Solid *C*, when heated with concentrated sulphuric acid, diluted, and made neutral with ammonia, deposited a white precipitate *D* with 8-hydroxyquinoline. A diluted ammoniacal solution of solid *C* gave no precipitate with this reagent. Identify the lettered compounds and explain all the reactions involved. Describe briefly the structure of the solid compounds *C* and *D*.

44. Tabulate the high oxidation state oxoanions of the elements of the first transition series including the name and formula of the oxoanion together with the oxidation state of the transition metal. Discuss the behaviour of these oxoanions in basic and acidic solutions.

BIBLIOGRAPHY

Some suggestions for further reading:

Textbooks

Bell, C. F. and Lott, K. A. *Modern Approach to Inorganic Chemistry* (London: Butterworth, 1963).

Cotton, F. A. and Wilkinson, G. *Advanced Inorganic Chemistry* (London: Interscience, 2nd Edn, 1966).

Day, F. H. *The Chemical Elements in Nature* (New York: Harrap, 1963).

Durrant, J. and Durrant, B. *Introduction to Advanced Inorganic Chemistry* (London: Longmans, 1960).

Glasstone, S. and Lewis, D. *Elements of Physical Chemistry* (London: Macmillan, 1961).

Heslop, R. B. and Robinson, P. L. *Inorganic Chemistry* (Amsterdam, London: Longmans, 2nd Edn, 1970).

Keenan, C. W. and Wood, J. H. *General College Chemistry* (New York: Harrap, 1963).

Kleinberg, J., Argersinger, W. J., and Griswold, E. *Inorganic Chemistry* (Boston: Heath, 1963).

Nebergall, W. H., Schmidt, F., and Holtzclaw, T. *General College Chemistry* (Boston: Heath, 1963).

Partington, J. R. *General and Inorganic Chemistry* (London: Macmillan, 1961).

Sanderson, R. T. *Inorganic Chemistry* (Rheinhold, 1968).

Wells, A. F. *Structural Inorganic Chemistry* (London: Oxford University Press, 1962).

Specialized Monographs, Reports, etc.

Annual Reports—Chemical Society, *A*, 1969–70.

Campbell, J. A. *Why do Chemical Reactions Occur?* (New Jersey, U.S.A.: Prentice Hall, 1965).

Gillespie, R. J. and Nyholm, R. S. *Quarterly Reviews*, **11**, 339, 1957.

Hargreaves, G. *Elements of Chemical Thermodynamics*. (London: Butterworth, 1963).

Chemistry Today—A Guide for Teachers (Geneva: Organization for Economic Cooperation and Development, 1963).

Copper and its Uses, Mining, and Extraction (London: C.D.A. Publications, 1959).

Manufacture and Uses of Phosphorus and some of its Compounds (Albright and Wilson, 1954).
Silicones, Properties and Products (I.C.I. Ltd., 1963).
A Simple Guide to the Structure and Properties of Steel (London: British Iron and Steel Federation, 1959).
Stark, J. G. and Wallace, H. G. *Chemistry Data Book* (J. Murray, 1971).

Royal Institute of Chemistry Lectures, Monographs, and Reports

Ives, D. H. G. *Principles of Extraction of Metals*, No. 3, 1960.
Sharpe, A. G. *Principles of Oxidation and Reduction*, No. 2, 1960.
Chemical Background to Aluminium Industry. No. 3. 1955.
Chemistry and Metallurgy of Titanium Production. No. 1. 1958.
Coal and Coal Chemicals. No. 4. 1957.
Coal as a Raw Material. No. 3. 1956.
Hydrazine and Its Derivatives. No. 5. 1957.
Hydrogen Peroxide. No. 2. 1954.
Production of Sulphuric Acid from Calcium Sulphate. No. 3. 1952.
Zone Refining. No. 3. 1957.

TABLE OF ELEMENTS

Element	*Symbol*	*Atomic number*	*Relative atomic mass*
Actinium	Ac	89	227
Aluminium	Al	13	26·9815
Americium	Am	95	243
Antimony	Sb	51	121·75
Argon	Ar	18	39·948
Arsenic	As	33	74·9216
Astatine	At	85	210
Barium	Ba	56	137·34
Berkelium	Bk	97	247
Beryllium	Be	4	9·0122
Bismuth	Bi	83	208·980
Boron	B	5	10·811
Bromine	Br	35	79·904
Cadmium	Cd	48	112·40
Caesium	Cs	55	132·905
Calcium	Ca	20	40·08
Californium	Cf	98	251
Carbon	C	6	12·01115
Cerium	Ce	58	140·12
Chlorine	Cl	17	35·453
Chromium	Cr	24	51·996
Cobalt	Co	27	58·9332
Copper	Cu	29	63·546
Curium	Cm	96	247
Dysprosium	Dy	66	162·50
Einsteinium	Es	99	254
Erbium	Er	68	167·26
Europium	Eu	63	151·96
Fermium	Fm	100	253
Fluorine	F	9	18·9984
Francium	Fr	87	223
Gadolinium	Gd	64	157·25
Gallium	Ga	31	69·72
Germanium	Ge	32	72·59
Gold	Au	79	196·967
Hafnium	Hf	72	178·49
Helium	He	2	4·0026

Element	*Symbol*	*Atomic number*	*Relative atomic mass*
Holmium	Ho	67	164·930
Hydrogen	H	1	1·0080
Indium	In	49	114·82
Iodine	I	53	126·9044
Iridium	Ir	77	192·2
Iron	Fe	26	55·847
Krypton	Kr	36	83·80
Kurchatovium	Ku	104	(260)
Lanthanum	La	57	138·91
Lawrencium	Lw	103	257
Lead	Pb	82	207·19
Lithium	Li	3	6·941
Lutetium	Lu	71	174·97
Magnesium	Mg	12	24·305
Manganese	Mn	25	54·9380
Mendelevium	Md	101	256
Mercury	Hg	80	200·59
Molybdenum	Mo	42	95·94
Neodymium	Nd	60	144·24
Neon	Ne	10	20·179
Neptunium	Np	93	237
Nickel	Ni	28	58·71
Niobium	Nb	41	92·906
Nitrogen	N	7	14·0067
Nobelium	No	102	254
Osmium	Os	76	190·2
Oxygen	O	8	15·9994
Palladium	Pd	46	106·4
Phosphorus	P	15	30·9738
Platinum	Pt	78	195·09
Plutonium	Pu	94	242
Polonium	Po	84	210
Potassium	K	19	39·102
Praseodymium	Pr	59	140·907
Promethium	Pm	61	147
Protoactinium	Pa	91	231
Radium	Ra	88	226
Radon	Rn	86	222
Rhenium	Re	75	186·20
Rhodium	Rh	45	102·905
Rubidium	Rb	37	85·47
Ruthenium	Ru	44	101·07
Samarium	Sm	62	150·35

Element	*Symbol*	*Atomic number*	*Relative atomic mass*
Scandium	Sc	21	44·956
Selenium	Se	34	78·96
Silicon	Si	14	28·086
Silver	Ag	47	107·870
Sodium	Na	11	22·9898
Strontium	Sr	38	87·62
Sulphur	S	16	32·064
Tantalum	Ta	73	180·948
Technetium	Tc	43	99
Tellurium	Te	52	127·60
Terbium	Tb	65	158·924
Thallium	Tl	81	204·37
Thorium	Th	90	232·038
Thulium	Tm	69	168·934
Tin	Sn	50	118·69
Titanium	Ti	22	47·90
Tungsten	W	74	183·85
Uranium	U	92	238·03
Vanadium	V	23	50·942
Xenon	Xe	54	131·30
Ytterbium	Yb	70	173·04
Yttrium	Y	39	88·905
Zinc	Zn	30	65·37
Zirconium	Zr	40	91·22

The relative atomic masses are quoted to different numbers of decimal places because of the variation in experimental accuracy for different elements: e.g. $Br = 79{\cdot}904 \pm 0{\cdot}002$; $Ag = 107{\cdot}870 \pm 0{\cdot}003$.

Index